Water Resources
of the World

WATER INFORMATION CENTER, Inc.

PERIODICALS

Water Newsletter
Research and Development News
Ground Water Newsletter

BOOKS

Geraghty, Miller, van der Leeden, and Troise — *Water Atlas of the United States*
Todd — *The Water Encyclopedia*
van der Leeden — *Ground Water — A Selected Bibliography*
Giefer and Todd — *Water Publications of State Agencies*
Soil Conservation Service — *Drainage of Agricultural Land*
Gray — *Handbook on the Principles of Hydrology*
National Water Commission — *Water Policies for the Future*
Officials of NOAA — *Climates of the States*
The Water Research Association — *Groundwater Pollution in Europe*
Litton, Tetlow, Sorensen and Beatty — *Water and Landscape*
Meta Systems, Inc. — *Systems Analysis in Water Resources Planning*
Giefer — *Sources of Information in Water Resources*
van der Leeden — *Water Resources of the World*

Water Resources of the World

Selected Statistics

Compiled and Edited by

Frits van der Leeden

Consulting Hydrogeologist

Geraghty & Miller, Inc.
Port Washington, New York

Published by WATER INFORMATION CENTER, INC.
Port Washington, New York

Acknowledgments

The author wishes to thank the many persons and organizations that contributed statistics on water resources and gave permission to reproduce data.

The guidance and assistance in preparing the book received from James J. Geraghty, President of Geraghty & Miller, Inc., and Fred L. Troise, Vice President of Water Information Center, Inc., are gratefully acknowledged. Jon A Pollack of Yale University assisted with research, selection and proofing of tabular data. Miklos Pinther, Chief Cartographer of the American Geographical Society, prepared the maps of the continents. Irene Berkenfield typed and composed the statistical material.

WATER RESOURCES OF THE WORLD

Library of Congress Catalog Card Number: 75-20952

ISBN: 0-912394-14-5

Printed in the United States of America

Preface

Solving critical water problems will be one of the principal tasks of planners, scientists and engineers in the years ahead. The tremendous growth of the world's population accompanied by industrial and agricultural development is creating heavy demands on the water resource everywhere on earth. Projections of water demand and water supplies by the World Health Organization show that the situation in most developing countries is critical and that the rate of increase in community water supplies is not sufficient to keep pace with the growing population. It is estimated that by 1980, 55 percent of the urban population, or 390 million people, will not be served with drinking water. By the year 2000, the world's total annual water demand, now 2,000 cubic kilometers, will triple to 6,000 cubic kilometers. Because the world's fresh-water supplies are limited and unevenly distributed over the surface of the earth, the search for, and development and distribution of new water sources will become increasingly important and crucial to man's continuing existence on earth.

Information on water availability and water use is of key importance for sound water planning and water management. It is with this in mind that these important water statistics have been selected for handy reference to hydrologists, engineers, planners, developers, managers, and other interested persons. In compiling this book, every effort has been made to include, as far as possible, uniform data on streamflow and runoff, water demand and water use for public supply, irrigation, industry and power generation. In addition, miscellaneous factual data of interest are also included. The tables represented reflect the nature and type of statistics available. On a worldwide basis, information on streamflow proved to be fairly complete, but data on the use of water resources were found to be quite inadequate. Nevertheless, a significant start has been made in putting together essential worldwide water statistics. Undoubtedly, further editions will show an improved and broadened worldwide coverage as more of this type of material is published and cooperative efforts among nations increase.

A detailed index has been provided to guide the reader. Statistical data are frequently found on more than one table and some overlapping material and occasional discrepancies in specific data will be noticed. In each case, however, the origin and date of the material are clearly identified on the table and in the bibliography that follows each continental section.

The editor believes that the book will prove to be useful for water resources studies and he would like to regard it as his personal contribution to the goals and success of the International Hydrological Program.

Westbury, New York   *Frits van der Leeden*
May 1975

Contents

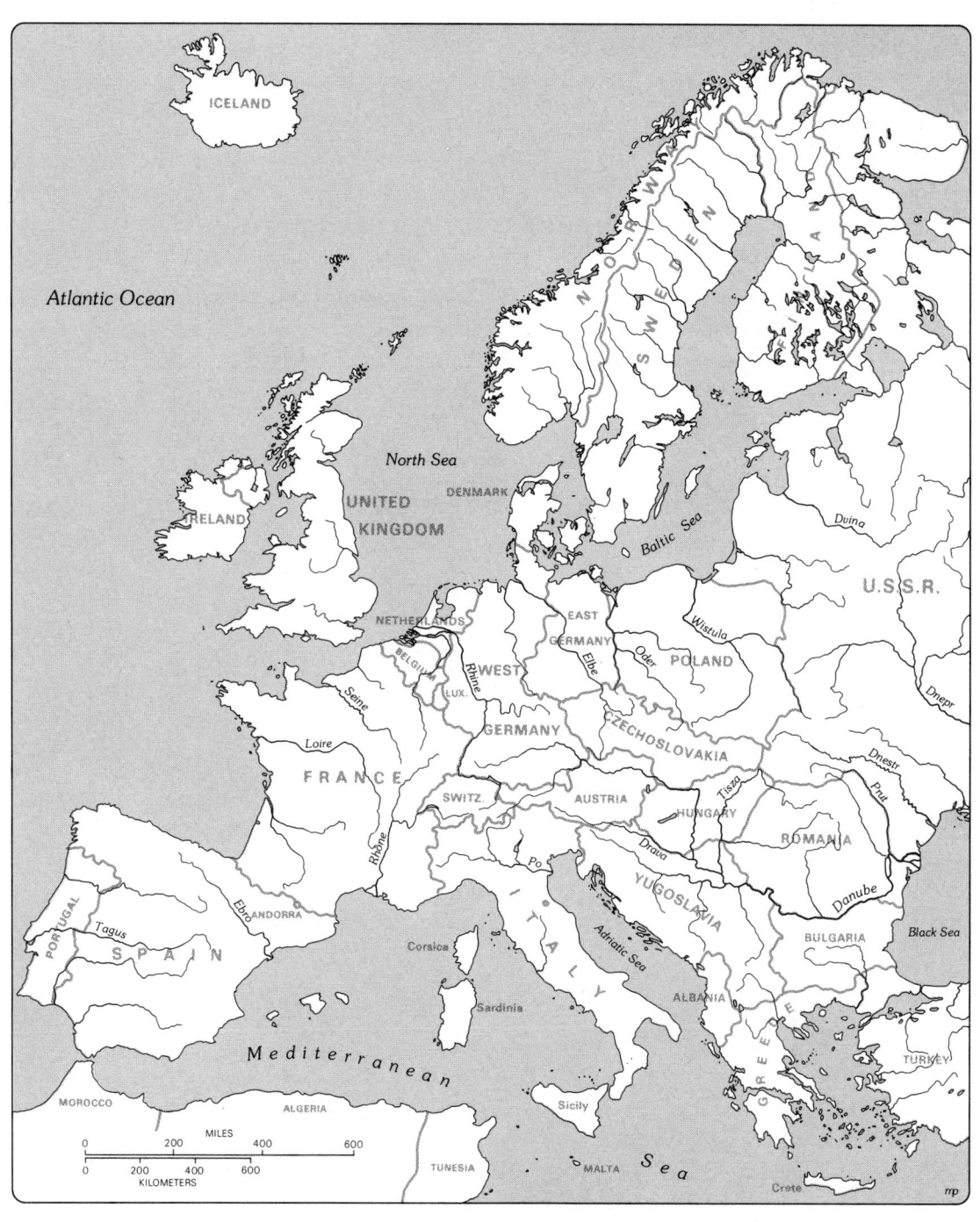

Atlantic Ocean

ICELAND

North Sea

IRELAND

UNITED
KINGDOM

DENMARK

Baltic Sea

Dvina

U.S.S.R.

NETHERLANDS

BELGIUM

LUX.

EAST
GERMANY

Elbe

Oder

Wistula

POLAND

Dnepr

WEST
GERMANY

Seine

Rhine

CZECHOSLOVAKIA

Dnestr

Loire

FRANCE

SWITZ.

AUSTRIA

HUNGARY

Tisza

Drava

ROMANIA

Prut

Rhône

Po

YUGOSLAVIA

Danube

PORTUGAL

Tagus

Ebro

ANDORRA

S P A I N

Corsica

ITALY

Adriatic Sea

ALBANIA

BULGARIA

Black Sea

Sardinia

GREECE

TURKEY

M e d i t e r r a n e a n

MOROCCO

ALGERIA

Sicily

MALTA

S e a

Crete

TUNESIA

MILES
0   200   400   600

0   200   400   600
KILOMETERS

rrp

TABLE 1-1. EUROPE—ANNUAL PRECIPITATION AND RUNOFF

(Source: Economic Commission for Europe, 1967)

Country	Population	Area	Density of population	Approximate average precipitation		Approximate average runoff		Average flow of rivers	
---	---	---	---	---	---	---	---	received from countries situated upstream	leaving the country
	millions	1,000 km²	persons/km²	mm/year	km³/year	mm/year	km³/year	km³/year	km³/year
Albania	1.6	29	56	1,200	34	350	10	3	13
Austria (1966)	7.1	84	84	1,191	100	661	55	35	90
Belgium (1966)	9.5	30.5	311	850	25.9	360	11	5	16
Bulgaria (1964)	8.2	111	74	672	74.5	162	18	179	197
Cyprus (1965)	0.6	9.2	65	503	4.5	37	0.3	–	–
Czechoslovakia	13.8	128	108	717	91	220	28	62	90
Denmark (1966)	4.6	43	106	660	28	260	11	–	11
Finland (1966)	4.5	337	13	550	185	300	100	4	104
France (1966)	49.2	551	89	750	415	300	168	39	207
East Germany (1966)	17.0	108	158	600	65	150	16.2	10	26.2
West Germany (1966)	54.2	248	219	803	200	307	77	85	162
Greece	8.4	131	64	650	86	150	20	23	43
Hungary (1966)	10.0	93	108	640	60	64	6	114	120
Iceland (1966)	0.2	103	2	1,400	140	1,750	170	–	170
Ireland (1966)	2.9	70	41	1,194	84	710	50	2	50
Italy	49.6	301	164	1,000	301	615	185	4	187
Luxembourg (1966)	0.3	3	131	850	2	300	1	–	5
Malta (1965)	0.3	0.2	1,301	508	0.1	10	–	–	–
Netherlands (1965)	12.2	34	360	750	30	250	10	80	90
Norway (1966)	3.6	324	11	1,450	470	1,250	405	8	413
Poland (1966)	31.5	312	101	583	182	158	49	6	55
Portugal	8.5	89.3	95	896	80	224	20	17	37
Romania (1965)	19.0	238	80	700	167	190	39	155	194
Spain	30.1	496	60	600	300	153	76	–	76
Sweden (1966)	7.8	450	17	700	315	400	180	3	183
Switzerland (1965)	5.9	41.2	144	1,500	61.5	1,000	42	8	50
Turkey in Europe (1966)	2.5	24	104	620	14.9	140	3.3	5.8	9.1
Yugoslavia	18.5	256	72	975	248	430	110	117	227
United Kingdom (1966)	53.3	241	221	1,064	256	508	122	–	122
USSR in Europe (1965)	165.7	5,326	31.1	570	3,160	201	1,117	204	1,321
Totals (or averages)	602.6	10,211.4	av. 59	av. 983	10,674	av. 389	3,104.8	–	–
Turkey (entire) (1966)	31.4	781	40	670	518	215	167	–	–
USSR (entire) (1965)	231.9	22,402	10.3	502	11,200	195	4,353	–	–

TABLE 1-2. EUROPE—ANNUAL WATER WITHDRAWAL

(Source: Economic Commission for Europe, 1967)

Country	Water withdrawal			
	Municipal	Industrial	Agricultural	Total
			million m^3/year	
Albania	60	120	20	200
Austria (1966)	500	1,000	350	1,850
Belgium (1966)	284.9	277.6	133.4	695.9
Bulgaria (1964)	566	1,120	3,800	5,486
Cyprus (1965)	26	—	422	448
Czechoslovakia	650	3,700	250	4,600
Denmark (1966)	295	190	105	590
Finland (1966)	200	2,100	50	2,350
France (1966)	3,500	13,000	10,000	26,500
East Germany (1966)	700	4,900	400	6,000
West Germany (1966)	2,600	9,050	1,260	12,910
Greece	200	300	2,500	3,000
Hungary (1966)	480	1,790	1,660	3,930
Iceland (1966)	40	—	10	50
Ireland (1966)	85	80	70	235
Italy	1,500	3,500	10,000	15,000
Luxembourg (1966)	17	150	8	175
Malta (1965)	2.5	0.2	0.1	2.8
Netherlands (1965)	430	840	700	1,970
Norway (1966)	200	1,200	30	1,430
Poland (1966)	1,116	5,348	1,303	7,767
Portugal	150	600	1,600	2,350
Romania (1965)	642	3,764	1,746	6,152
Spain	1,200	1,000	3,500	5,700
Sweden (1966)	850	3,600	50	4,500
Switzerland (1965)	985	1,100	360	2,445
Turkey in Europe (1966)	75	60	125	260
Yugoslavia	450	1,500	550	2,500
United Kingdom (1966)	3,300	7,350	70	10,720
USSR in Europe (1965)	4,402	20,033	26,540	84,584
Totals (or averages)	25,506.4	87,692.8	67,612.5	214,400.7
Turkey (entire) (1966)	—	—	—	—
USSR (entire) (1965)	9,855	54,275	120,470	240,100

TABLE 1-3. EUROPE—WATER WITHDRAWN FOR PUBLIC SUPPLIES IN SELECTED COUNTRIES, 1969
(Source: Verband der deutschen Gas-und Wasserwerke e.V., 1971)

Country	Year	Ground Water * Million m3	%	Spring Water Million m3	%	Surface Water Million m3	%	Total Water Withdrawn Million m3	%
Belgium	1968	313.5	74	—	—	109.5	26	423.0	100
	1969	310.3	73	—	—	114.7	27	425.0	100
West Germany	1968	2,850.0	75	653.0	17	317.0	8	3,820.0 6	100
	1969	2,973.0	75	678.0	17	339.0	8	3,990.0 6	100
United Kingdom [1]	1968	2,723.0 7	34	—	—	5,214.0	66	7,937.0	100
	1969	2,713.0 7	34	—	—	5,195.0	66	7,908.0	100
Italy [2]	1968	1,470.0	55	911.2	34	296.4	11	2,677.6	100
	1969	1,550.8	56	933.1	33	315.3	11	2,799.2	100
Luxembourg [3]	1968	—	—	11.1	100	—	—	11.1	100
	1969	—	—	12.4	100	—	—	12.4	100
Netherlands	1968	620.1 8	81	—	—	149.3	19	769.4	100
	1969	660.1 8	80	—	—	160.0	20	820.1	100
Austria	1968	165.1	45	194.6	54	3.8	1	363.5	100
	1969	175.8	47	195.5	52	4.6	1	375.9	100
Sweden	1968	370.0	45	—	—	459.0	55	829.0	100
	1969	367.0	42	—	—	508.0	58	875.0	100
Switzerland	1968	277.0	44	192.0	31	159.0	25	628.0	100
	1969	284.0	44	185.0	28	179.0	28	648.0	100
Spain	1968	423.9	27	31.3	2	1,111.7	71	1,566.9	100
	1969	444.3	27	32.5	2	1,167.1	71	1,643.9	100

For explanation of footnotes see Table 1-4.

TABLE 1-4. EUROPE—WATER USED FOR PUBLIC SUPPLIES IN SELECTED COUNTRIES, 1969

(Source: Verband der deutschen Gas-und Wasserwerke e.V., 1971)

Country	Year	Population		Consumption (Million m3)				Daily Per Capita Consumption (liters)	
		Total Population 1,000	Served 1,000	Household including small business	Other	Industrial included in "Other"	Total	Household water use **	Total water use **
Belgium	1968	9,632	8,885	247.7	133.3	—	381.0	76.2	117.2
	1969	9,692	8,885	250.7	135.0	—	385.7	77.3	118.9
West Germany	1968	60,184	57,200	2,235.0	1,265.0	995.0	3,500.0	106.8	167.2
	1969	60,848	58,100	2,345.0	1,315.0	1,035.0	3,660.0	110.6	172.6
United Kingdom [1]	1968	48,593	48,320	2,933.0	1,793.0 [4]	—	4,726.0 [4]	165.8	267.2
	1969	48,827	48,717	2,976.0	1,849.0 [4]	—	4,825.0 [4]	167.4	271.3
Italy [2]	1968	53,940	23,835	1,968.3	404.2	285.4	2,372.5	225.6	271.9
	1969	54,250	24,403	2,073.4	425.6	299.4	2,499.0	232.8	280.6
Luxembourg [3]	1968	336	130	6.1	3.3	3.3	9.4	128.2	197.6
	1969	338	132	6.9	3.6	3.6	10.5	143.2	217.9
Netherlands	1968	12,798	12,670	404.1	294.0	—	698.1	87.4	150.5
	1969	12,958	12,828	425.0	322.1	—	747.1	90.8	159.6
Austria	1968	7,362	3,352	147.8	162.7	—	348.2	120.5	283.8
	1969	7,384	3,415	163.4	117.5	—	362.9	131.1	291.1
Sweden	1968	7,942	5,739	494.0	253.0	168.0	747.0	235.2	355.6
	1969	8,004	5,770	457.0	260.0	167.0	717.0	217.0	340.4
Switzerland	1968	6,147 [5]	3,551	347.0	170.0	126.0	517.0	267.0	397.8
	1969	6,224 [5]	3,604	359.0	178.0	132.0	537.0	272.9	408.2
Spain	1968	32,788	26,786	1,255.1	311.4	—	1,566.9	128.0	159.8
	1969	33,100	27,035	1,329.0	314.9	—	1,643.9	134.7	166.6

* Artificial recharge included.
** Based on population served.
1 England and Wales only.
2 Information applies only to 50% of population.
3 Information applies only to area served with drinking water by the "Syndicat des Eaux du Sud".

4 Industrial water use excluded.
5 Population covered by SVGW Statistical Survey only.
6 Purchased water excluded.
7 Includes spring water.
8 Includes infiltrated surface water.

TABLE 1-5. EUROPE—HYDROELECTRIC AND THERMAL POWER GENERATING CAPACITY AND PRODUCTION, 1968

(Source: U.S. Federal Power Commission, 1971)

Country	Installed capacity (MW.)[1]			Energy production (GWh.)[2]			Population (1,000)	Kwh. per capita
	Hydro	Thermal	Total	Hydro	Thermal	Total		
Albania	116	113	229	460	190	650	2,019	322
Austria	4,821	2,235	7,056	18,003	7,154	25,157	7,349	3,423
Belgium	65	6,665	6,730	242	24,781	25,023	9,619	2,601
Bulgaria	771	2,691	3,462	1,292	13,018	14,310	8,370	1,710
Czechoslovakia	1,540	8,531	10,071	3,115	35,223	38,338	14,362	2,669
Denmark	9	3,911	3,920	21	12,076	12,097	4,870	2,484
Finland	2,093	2,542	4,635	10,384	6,980	17,364	4,689	3,703
France	14,512	19,621	34,133	50,342	67,583	117,925	49,914	2,362
Germany (East)	667	11,006	11,673	1,185	58,934	60,119	17,084	3,519
Germany (West)[3]	4,741	42,313	47,054	16,515	173,186	189,701	60,165	3,153
Greece	703	1,212	1,915	1,352	5,987	7,339	8,803	834
Hungary	21	2,339	2,360	86	11,756	11,842	10,256	1,155
Iceland	127	66	193	684	37	721	201	3,587
Ireland	219	1,126	1,345	763	3,897	4,660	2,910	1,601
Italy	14,765	15,499	30,264	43,262	56,987	100,249	52,750	1,900
Luxembourg	925	228	1,153	707	1,255	1,962	336	5,839
Netherlands	0	9,296	9,296	0	31,847	31,847	12,725	2,503
Norway	11,981	141	12,122	59,609	92	59,701	3,819	15,633
Poland	486	11,105	11,591	1,046	50,398	51,444	32,305	1,592
Portugal	1,555	605	2,160	5,165	938	6,103	9,465	645
Romania	831	4,780	5,611	1,550	24,956	26,506	19,721	1,344
Spain	8,560	5,575	14,135	24,040	19,833	43,873	32,621	1,345
Sweden	10,423	3,308	13,731	48,284	7,063	55,347	7,918	6,990
Switzerland	8,940	560	9,500	29,402	1,150	30,552	6,147	4,970
United Kingdom	2,164	57,464	59,628	3,869	205,306	209,175	55,283	3,784
USSR[4]	27,035	115,469	142,504	103,000	507,891	610,891	237,798	2,569
Yugoslavia	2,832	2,044	4,876	11,650	8,434	20,084	20,154	997
Islands[5]	18	237	255	56	713	769	1,249	616
Total	120,920	330,682	451,602	436,084	1,337,665	1,773,749	692,902	2,560

[1] MW. — Megawatts—Thousand Kilowatts.

[2] GWh.— Gigawatt-hours—Million kilowatt-hours.

[3] Includes West Berlin.

[4] Includes all of USSR.

[5] Includes Cape Verde, Cyprus, Faroe Islands, Gibraltar, Malta.

TABLE 1-6. ALBANIA—DISCHARGE OF SELECTED RIVERS
(Source: UNESCO, 1971)

River and station	Basin area km²	Jan.	Feb.	Mar.	Apr.	May	Jun.	Jul.	Aug.	Sep.	Oct.	Nov.	Dec.	Max flow m³/s	Date	Period of record or year
Drin River, Vau Deje	12,368	472	462	506	541	559	330	135	72.4	70.7	173	397	564	–	–	1960-65
		576	772	372	482	459	295	134	88	90	212	498	556	2,180	Feb. 12	1966
		286	238	290	477	444	200	171	91	101	145	120	284	850	Oct. 5	1967
		445	452	294	310	270	254	91	13	240	199	479	639	2,330	Dec. 18	1968
Vijose River, Dorze	5,200	366	378	282	279	192	111	66.8	43.3	38.9	56.3	199	360	–	–	1960-65
		595	285	240	183	147	104	50	28	29	49	389	494	2,730	Jan. 13	1966
		316	130	124	172	182	100	69	44	47	46	37	188	1,340	Dec.28	1967
		311	255	183	151	118	96	42	39	35	33	52	165	1,070	Feb. 18	1968

TABLE 1-7. AUSTRIA—DISCHARGE OF SELECTED RIVERS
(Source: Hydrographische Zentralburo, Vienna, 1972)

River	Gaging station	Basin area km²	Mean annual flow, m³/s Period 1951-1960	Year 1967
Rhine	Lustenau	6,110	230	278
Bregenzer Ach	Kennelbach	826	47.1	50.8
Inn	Innsbruck	5,794	167	205
Salzach	Oberndorf	6,111	241	269
Inn	Scharding	25,664	757	853
Danube	Linz	79,490	1,477	1,717
			2,468 *	
Traun	Wels	3,499	130	139
Enns	Liezen	2,116	63.3	79.0
Steyr	Pergern	898	36.4	36.2
Enns	Enns	6,071	199	232
Ybbs	Opponitz	507	20.4	23.0
Kamp	Stiefern	1,493	11.2	11.8
Danube	Wien-Nussdorf	101,700	1,916	2,227
			1,920 *	
Raab	Feldbach	689	5.81	5.43
Mur	Leoben	4,392	80.1	90.5
Mur	Landscha	8,340	139	149
Isel	Lienz	1,199	38.3	43.0
Drau	Villach	5,266	158	166
Gail	Notsch	936	35.7	35.8
Drau	Annabrucke	7,566	230	240

* Period 1901-1950

TABLE 1-8. AUSTRIA—DISCHARGE OF THE INN RIVER
(Source: UNESCO, 1971)

River and station	Basin area km²	Mean monthly discharge, m³/s												Max flow m³/s	Date	Min flow m³/s	Period of record or year
		Jan.	Feb.	Mar.	Apr.	May	Jun.	Jul.	Aug.	Sep.	Oct.	Nov.	Dec.				
Inn River, Scharding	25,665	376	400	556	734	1,030	1,330	1,380	1,070	743	569	465	403	—	—	—	1951-60
		410	346	591	908	1,627	2,372	1,592	1,084	928	500	346	444	3,720	Jun. 12	241	1965
		376	552	518	765	1,255	1,479	1,678	1,685	953	565	527	563	3,650	Jul. 24	276	1966
		513	545	725	937	1,415	1,679	1,422	999	767	491	371	357	2,551	Jun. 9	248	1967
		452	416	474	904	911	1,081	1,177	1,149	782	838	429	337	2,510	Aug. 8	262	1968

TABLE 1-9. AUSTRIA—MUNICIPAL WATER SUPPLY SYSTEMS
(Source: Die Wiener Wasserversorgung, Verlag fur Jugend und Volk, Vienna, 1967)

[Data as of 1964]

City	Population	Total consumption m³	Source of Water			Per capita consumption l/day	Population served
			Spring water %	Ground water %	Surface water %		
Bregenz	22,904	2,500,592	—	100	—	192	22,904
Eisenstadt	70,000	2,811,693	15.8	84.2	—	109.2	70,000
Graz	248,752	18,482,683	—	100	—	234	220,000
Innsbruck	109,500	15,049,950	99	1	—	347	105,000
Klagenfurt	70,500	5,246,618	7.9	92.1	—	195	65,000
Linz	215,385	18,558,287	76.3	23.7	—	ca. 240	203,149
Salzburg	115,018	10,166,086	—	100	—	225	107,500
Vienna	1,638,700	171,281,150	73.3	25.5	1.2	255	1,538,194
Vienna (1965)	1,637,000	176,396,910	75.3	22.8	1.9	262	1,554,029

TABLE 1-10. BELGIUM–DISCHARGE OF SELECTED RIVERS, 1969
(Source: Hydrologisch Jaarboek van Belgie, 1971)

River and station	Basin area km^2	Jan.	Feb.	Mar.	Apr.	May	Jun.	Jul.	Aug.	Sep.	Oct.	Nov.	Dec.	Maximum daily flow m^3/s	Date
Schelde (Escaut) River Basin															
Leie River, St. Eloois-Vijve	3,190	47.5	58.1	45.9	25.8	35.8	32.9	22.2	19.3	14.6	10.9	26.0	46.0	104	Jan. 16, 1968
Escaut River, Kain	5,091	17.1	30.5	29.9	24.7	20.9	30.8	25.6	20.8	17.2	14.2	18.3	22.6	170	Jan. 1, 1967
Maas (Meuse) River Basin															
Ourthe River, Angleur	3,626	74.6	84.6	77.5	86.3	39.4	26.7	19.1	72.0	33.1	14.1	39.5	53.4	635	Dec. 11, 1967
Vesdre River, Chaudfontaine	680	15.2	17.5	15.9	16.2	6.31	5.12	3.66	19.9	6.35	2.56	8.18	8.76	–	–
Ambleve River, Martinrive	1,044	21.6	22.4	23.9	26.2	11.8	7.74	5.63	25.9	12.1	3.39	12.4	15.1	–	–
Ourthe River, Hamoir	1,597	30.5	36.4	30.3	37.0	18.7	11.3	8.26	18.9	10.6	4.34	18.8	27.1	–	–
Meuse River, Ampsin-Neuville	16,400	242	285	294	268	189	157	103	73.5	63.1	41.6	122	188	2,000	Jan. 3, 1967
Mehaigne River, Moha	345	2.65	4.11	2.55	1.89	1.63	2.09	1.71	2.08	1.29	.790	1.81	3.07	–	–
Eau d'Heure River, Jamioulx	324	4.81	6.26	3.86	4.18	2.57	1.82	1.28	1.52	1.14	.789	2.18	3.58	–	–
Lesse River, Gendron-Pont	1,314	22.7	28.9	21.1	26.9	15.6	9.43	7.54	9.42	4.99	2.35	13.5	19.8	–	–
Viroin River, Treignes	554	11.4	12.1	7.87	12.7	6.61	6.19	3.41	2.97	1.73	.929	6.99	8.31	–	–
Semois River, Membre	1,235	31.2	26.8	34.3	36.0	17.9	15.2	9.19	5.86	5.12	3.74	21.8	24.1	–	–

TABLE 1-11. BELGIUM—WATER DEMAND FOR PUBLIC SUPPLIES, 1965-80

(Source: Commissariat royal au probleme de l'eau, 1967)

[By province]

| | Total population 1965 | Population served 1965 | Per capita consumption l/d 1965 | Estimated population served in 1980 | Estimated per capita consumption l/d 1980 | Comparison of water demand | | | |
| | | | | | | 1965 | | 1980 | |
						m³/d	m³/yr	m³/d	m³/yr
Brabant	2,108,296	1,996,287	106	2,390,000	170	212,060	77,400,000	407,000	148,300,000
Antwerp (Anvers)	1,494,062	1,244,616	135	1,710,000	264	168,450	61,400,000	452,000	165,000,000
Limbourg	624,446	550,687	46.7	813,000	98	25,730	9,400,000	80,000	29,200,000
Liege	1,017,582	998,454	103	1,084,000	192	103,900	38,900,000	207,000	75,600,000
Hainaut	1,333,432	1,286,105	89.8	1,460,000	131	115,300	42,100,000	191,000	69,800,000
Namur	378,106	371,059	90.7	398,000	142	33,600	12,200,000	57,000	20,800,000
Flanders East	1,294,695	993,283	80.2	1,350,000	125	79,450	28,950,000	169,000	62,800,000
Flanders West	1,029,165	815,129	63.5	1,064,000	77	51,800	18,900,000	82,000	29,900,000
Luxembourg	219,450	210,716	86.3	223,000	126	17,500	6,400,000	28,000	10,600,000
Total Nation	9,499,234	8,466,336	95.4	10,492,000	157	807,790	295,650,000	1,673,000	612,000,000

TABLE 1-12. BELGIUM—INDUSTRIAL WATER DEMAND AND GROUNDWATER USE

(Source: Snel, Societe Nationale des Distributions d'Eau, 1974 and Institut National de Statistique)

INDUSTRIAL WATER USE, 1969 [In 1,000 m3]

Industry	Withdrawal	Consumption
Coal mines	139,311	10,427
Quarries	41,209	1,972
Food (margerine, oils, etc.)	120,528	14,140
Textile	57,796	4,713
Wood	2,988	551
Paper	96,832	10,138
Leather	5,368	332
Chemical	650,292	39,711
Rubber	7,987	754
Petroleum refineries	335,426	160
Coke plants (gas)	81,029	9,458
Terra cotta	1,476	752
Glass	20,842	1,968
Ceramic	1,051	301
Cement	17,190	5,093
Iron & steel	1,099,867	66,626
Non-ferrous	202,913	39,137
Metallic construction	66,604	4,980
Hydroelectric power	13,257,900	—
Thermoelectric power	3,703,580	11,866
Total	19,910,189	223,079

WITHDRAWAL OF GROUNDWATER, 1970 [In m3]

Department and Province	Industry	Public supply distributions	Dewatering (mines)	Dewatering (other)	Total
Tournai	7,613,829	34,763,286	13,389,090	—	55,766,205
Mons	17,479,426	53,879,377	11,010,751	26,936,064	109,305,618
Charleroi	2,774,947	21,294,734	16,889,515	—	40,959,196
Hainaut	27,868,202	109,937,397	41,289,356	26,936,064	206,031,019
Huy	437,357	42,442,455	24,000	—	42,903,812
Liege	18,740,739	23,332,947	18,783,701	6,813,200	67,670,587
Verviers	3,149,276	6,870,140	—	—	10,019,416
Liege	22,327,372	72,645,542	18,807,701	6,813,200	120,593,815
Arlon	76,950	5,264,700	36,000	—	5,377,650
Marche	104,613	4,479,264	—	—	4,583,877
Neufchateau	361,537	5,385,175	116,724	—	5,863,436
Luxembourg	543,100	15,129,139	152,724	—	15,824,963
Namur	6,620,717	30,629,145	158,548	—	37,408,410
Dinant	750,960	29,262,894	106,000	—	30,119,854
Namur	7,371,677	59,892,039	264,548	—	67,528,264
Bruxelles	8,279,781	4,517,136	—	—	12,796,917
Hal-Vilvorde					
Louvain	6,428,336	11,477,587			17,905,923
Nivelles	14,125,703	32,467,068	1,080,000		47,672,771
Brabant	28,833,820	48,461,791	1,080,000		78,375,611
Hasselt	25,404,484	8,321,955	4,819,069		38,545,508
Tongeren	8,458,486	10,009,851	1,654,660		20,122,997
Limburg	33,862,970	18,331,806	6,473,729		58,668,505
Antwerpen	6,534,808	14,604,516			21,139,324
Mechelen	1,268,151	—			1,268,151
Turnhout	16,290,117	14,212,414			30,502,531
Antwerpen	24,093,076	28,816,930			52,910,006
Brugge	3,695,190	3,798,350			7,493,540
Ieper	853,437	—			853,437
Kortrijk	4,680,012	—			4,680,012
Veurne	256,820	2,240,802			2,497,622
West-Vlaanderen	9,485,459	6,039,152			15,524,611
Gent	8,951,560	4,481,382			13,432,942
Dendermonde	5,308,831	2,805,282			8,114,113
Oudenaarde	1,932,100	689,923			2,622,023
Oost-Vlaanderen	16,192,491	7,976,587			24,169,078
Total Belgium	170,578,167	367,230,383	68,068,058	33,749,264	639,625,872

TABLE 1-13. BELGIUM—AGRICULTURAL WATER DEMAND, 1965-80

(Source: Commissariat royal au probleme de l'eau, 1967)

Nature of demand	Specific water requirement m³/year	1965 Number	1965 Water requirement m³/year	1980 Rate of expansion	1980 Total	Total water requirement m³/year
Irrigation Horticulture:						
in the open	300 per ha	5,234 hectares	4,200,000	600 ha/year	14,000 ha	11,200,000
under glass	5,500 ''	1,499 ''	8,200,000	230 ''	5,000 ''	27,500,000
Agriculture:						
Surface irrigation	5,000 ''	2,100 ''	10,500,000	0	0	10,500,000
Sprinkling	1,500 ''	377 ''	500,000	2,000 ''	30,000 ''	45,000,000
Total irrigation		9,170 hectares	23,400,000	—	49,000 ha	94,200,000
Cattle watering	10m³/head/year	4,600,000 head	48,000,000	+20%	5,520,000 head	57,600,000
Demand of establishments		Number of establishments		Increase in consumption		
'' domestic	12m³/year / establishment	242,000	30,000,000	50%	242,000	45,000,000
'' dairy	2 id.	id.	20,000,000	20%	242,000	24,000,000
'' cleansing	4.8 id.	id.	12,000,000	20%	242,000	14,400,000
Total exploitations			62,000,000			83,400,000
Total agricultural			133,400,000			235,200,000

TABLE 1-14. BELGIUM—MUNICIPAL WATER SUPPLY SYSTEMS
(Source: Internat. Statistical Institute, 1972)

City	Year	Population served	Annual consumption (1,000 m^3) Total	of which domestic
Antwerp (m)[1]	1967	647,187	67,856	40,389
	1969	644,235	82,405	37,865
Liege (m)[1]	1967	—	319,167	59,829
	1969	—	765,311	114,735

1 (m) Metropolitan area
— Not Available

TABLE 1-15. BULGARIA–DISCHARGE OF MARITZA RIVER
(Source: UNESCO, 1971)

River and station	Mean monthly discharge, m3/s												Max flow m3/s	Date	Min flow m3/s	Year
	Jan.	Feb.	Mar.	Apr.	May	Jun.	Jul.	Aug.	Sep.	Oct.	Nov.	Dec.				
Maritza River, Plovdiv	65	54	100	110	175	40	10	6	14	22	24	36	485	May 16	3	1965
	47	52	33	55	65	51	13	14	16	25	77	113	378	Jul. 8	6	1966
	56	61	95	99	105	56	24	17	26	26	32	35	292	May 22	12	1967
	38	49	44	29	20	13	3	8	21	23	35	36	286	Feb. 18	2	1968

TABLE 1-16. BULGARIA—IRRIGATED AREA BY WATER SOURCE, 1970
(Source: Vodproekt Design Institute, 1972)

Water Source	Area under irrigation million Ha	Type of irrigation,%	
		Pump	Gravity
Regulated	498	33.5	66.5
Non-regulated	330	49.0	51.0
Danube River	138	100	—
Ground water	46	100	—
Total	1,014		

TABLE 1-17. BULGARIA—STORAGE RESERVOIRS AND DAMS, 1970
(Source: Vodproekt Design Institute, 1972)

	Existing	Planned
Number of dams	2,015	2,226
Storage capacity (million m^3)		
Total	4,860	17,000
Average per reservoir	2.41	7.64
Volume of dams (million m^3)		
Total	85.70	229.40
Average per dam	42.5	103.0
Surface area of reservoir (Ha)		
Total	45,380	94,400
Average per reservoir	22.5	42.4

TABLE 1-18. BULGARIA—MUNICIPAL WATER SUPPLY SYSTEMS
(Source: Internat. Statistical Institute, 1972)

City	Year	Population served	Total quantity available (1000 m^3)	Total annual consumption (1000 m^3)
Burgas	1967	117,891	6,798	5,704
	1969	131,724	11,675	9,313
Plovdiv	1967	232,466	25,857	16,813
	1969	247,473	29,631	20,826
Ruse	1967	140,492	16,545	15,090
	1969	149,600	20,873	16,908
Sofia	1967	832,177	111,899	84,740
	1969	868,231	139,100	110,351
Varna	1967	195,356	23,634	20,723
	1969	218,988	26,932	23,904

TABLE 1-19. CZECHOSLOVAKIA–DISCHARGE OF SELECTED RIVERS
(Source: UNESCO, 1971)

River and station	Basin area km²	Mean monthly discharge, m³/s												Max flow m³/s	Date	Min flow m³/s	Year or Period of record
		Jan.	Feb.	Mar.	Apr.	May	Jun.	Jul.	Aug.	Sep.	Oct.	Nov.	Dec.				
Elbe (Labe) River, Decin	51,104	287	392	550	496	305	247	249	198	195	218	265	265	–	–	–	1931-60
		294	287	703	781	1,008	1,223	421	233	199	175	185	458	1,940	Jun. 13	132	1965
		354	762	430	590	315	212	578	551	479	290	255	522	1,360	Feb. 11	148	1966
		692	797	728	495	539	494	233	153	296	205	188	389	1,565	Feb. 6	116	1967
		552	511	481	483	289	324	145	150	179	297	279	205	1,207	Jan. 18	121	1968
Morava River, Moravsky Jan	24,129	101	135	216	187	117	84.2	76.0	65.7	66.7	58.8	102	95.4	–	–	–	1931-60
Danube River, Bratislava	131,338	1,419	1,635	2,096	2,385	2,486	2,746	2,725	2,250	1,733	1,525	1,517	1,402	–	–	–	1931-60

TABLE 1-20. CZECHOSLOVAKIA–MUNICIPAL WATER SUPPLY SYSTEMS
(Source: Internat. Statistical Institute, 1972)

City	Year	Population served	Quantity available (1000 m³)			Annual consumption (1000 m³)	
			Production	Purchased from other plants	Total	Total	of which domestic
Bratislava	1967	257,000	45,352	0	45,352	37,908	–
	1969	273,500	52,263	0	52,263	42,802	–
Brno	1967	333,004	41,708	0	41,708	35,081	12,689
	1969	338,965	43,900	0	43,900	36,306	14,353
Kosice	1967	129,782	14,099	0	14,099	11,985	8,070
	1969	139,792	14,638	0	14,638	12,323	8,782
Ostrava	1967	277,003	10,017	25,283	35,300	27,143	27,143
	1969	273,316	11,157	27,741	38,898	29,751	29,751
Plzen	1967	139,505	20,027	0	20,027	16,525	4,295
	1969	140,400	19,149	0	19,149	15,378	5,025
Prague	1961	1,010,000	88,177	0	88,177	88,177	–
	1962	1,013,000	89,779	0	89,779	89,779	38,976

— Not Available

TABLE 1-21. DENMARK–AVAILABLE WATER RESOURCES, BY REGION
(Source: Danish Pollution Council, 1971)

County or region	Area km^2	Population 1965	Mean Annual precipitation 1931-69 mm	Mean Annual evapotranspiration mm	Precipitation less evapotranspiration mm	Available Water Resources	
						Total Million m^3	Estimated recoverable Million m^3
Hjorring	2,886	183,294	660	360	300	864	260
Thisted	1,774	83,270	717	360	360	638	220
Alborg	2,933	248,342	661	360	300	880	350
Nordjylland	7,593	514,906	676	360	310	2,382	830
Viborg	2,992	166,576	705	360	350	1,046	470
Randers	2,509	178,037	603	360	240	603	300
Arhus	2,542	378,094	699	380	320	813	370
Vejle	2,351	226,904	727	360	370	870	390
Ostjylland	10,394	949,611	683	360	320	3,332	1,530
Ringkjobing	4,612	219,952	773	340	430	1,984	1,280
Ribe	3,096	194,681	780	340	440	1,362	890
Vestjylland	7,708	414,633	776	340	430	3,340	2,170
Sonderjylland	3,956	230,220	751	360	360	1,425	850
Jylland	29,651	2,109,370	714	360	350	10,485	5,380
Odense	1,848	278,337	588	380	210	388	160
Svendborg	1,637	146,791	634	380	250	409	160
Fyn	3,485	425,128	612	380	230	797	320
Holbaek	1,754	131,464	536	400	140	246	100
Frederiksborg	1,347	211,449	607	380	230	310	160
Kobenhavn	1,288	1,452,168	576	400	180	232	160
Soro	1,477	135,633	571	400	170	252	110
Praesto	1,683	124,326	586	400	190	320	160
Sjaelland	7,549	2,055,040	574	400	180	1,360	690
Lolland-Falster	1,795	129,315	584	420	160	287	110
Bornholm	588	48,744	596	400	200	117	20
Total Denmark	43,068	4,767,567	673	370	300	13,046	6,520

TABLE 1-22. DENMARK—WATER USE, BY REGION, 1970
(Source: Danish Pollution Council, 1971)

(in million m3/yr)

Region	Domestic	Summer homes	Industrial irrigation and fish farming	Industrial water supplied by public water supply systems	Livestock	Total	Percent of total available water resources presently utilized
Nordjylland	36	0.7	39	12	21	109	13
Ostjylland	55	0.9	51	27	28	162	11
Vestjylland	41	0.5	19	10	19	90	4
Sonderjylland	16	0.2	8.0	5.1	10	39	5
Jylland	148	2.3	117	55	78	400	7
Fyn	30	0.4	9.4	8.5	9.0	57	18
Sjaelland	144	3.6	20	54	13	235	34
Lolland-Falster	9.0	0.3	6.5	2.6	2.3	21	19
Bornholm	3.4	0.1	0.2	0.4	1.3	5.4	25
Total Denmark	334	6.7	153	121	103	718	11

TABLE 1-23. DENMARK—WATER DEMAND, 1970—2000
(Source: Danish Pollution Council, 1971)

(in million m3/yr)

Demand	1970	2000
Domestic (institutions, schools, summer homes)	340	565
Agriculture (livestock, irrigation, gardening)	130	160
Industry	200	320
Fish farms	50	25
Total water demand	720	1,070

TABLE 1-24. DENMARK–MUNICIPAL WATER SUPPLY SYSTEMS

(Source: Assoc. of City and Harbor Engineers and Danish Water Assoc., 1971)

[Data for 1969-70]

City	Population	Population served	Water produced 1000m^3	Water purchased 1000m^3	Water sold to other supply systems 1000m^3	Water distributed 1000m^3	Per capita domestic water use l/day
Copenhagen	640,200	—	97,989	0	34,530	63,459	271 [1]
Arhus	111,306	111,306	15,381	0	745	14,636	360 [1]
Gentofte	79,989	80,000	15,220	0	4,803	10,417	208
Odense	102,698	102,698	12,676	0	405	12,211	326 [1]
Aalborg	153,531	78,701	9,221	0	0	9,221	209
Esbjerg	62,483	56,083	7,184	0	0	7,184	209
Fredericia	35,020	35,020	6,755	0	0	6,755	254
Randers	41,253	41,253	5,994	0	0	5,994	249
Horsens	35,621	36,050	5,510	0	0	5,510	213
Kolding	39,609	39,725	5,290	0	—	5,290	271
Roskilde	40,382	36,317	3,218	861	0	4,079	308 [1]
Frederiksberg	103,689	103,689	2,427	8,355	0	10,782	164
Gladsaxe	74,587	74,587	2,067	5,070	—	7,137	155
Lyngby-Taarbaek	61,034	60,188	1,489	3,777	—	5,266	182
Kastrup-Tarnby	45,868	45,868	7,879	153	933	1,104	201 [1]
Hvidovre	45,611	44,900	901	1,956	0	2,857	174 [1]
Ballerup-Malov	47,144	48,390	563	3,838	0	4,401	170
Rodovre	45,005	45,605	371	3,812	—	4,183	251 [1]

1 Calculated from total water use and includes industrial, commercial and agricultural consumption
— Not Available

TABLE 1-25. FINLAND–DISCHARGE OF SELECTED RIVERS

(Source: ECE, 1970)

[Period of record 1930-40 and 1945-65]

River, station	Mean monthly discharge, m^3/s												Six-monthly average		Yearly average	Mean annual runoff l/s/km^2
	Jan.	Feb.	Mar.	Apr.	May	Jun.	Jul.	Aug.	Sep.	Oct.	Nov.	Dec.	Oct.-Mar.	Apr.-Sep.		
Kymijoki, Kuusankoski	270	260	250	270	334	353	322	283	252	247	252	269	258	302	280	7.8
Vuoksi, Imatra	606	594	525	564	539	531	512	541	559	569	596	600	529	541	567	9.3
Kemijoki, Petajakoski	204	175	159	232	1,557	1,093	563	488	518	513	426	274	292	742	517	10.7

TABLE 1-26. FINLAND–DISCHARGE CHARACTERISTICS OF PRINCIPAL RIVERS

(Source: Jaatinen, Aqua Fennica, 1971)

	River	Basin area km²	Maximum Flow Extreme	Maximum Flow Mean	Monthly discharge, m³/s Mean Flow	Mean Flow Mean	Mean Flow Extreme
Ladoga	Vuoksi	61,560	1,142	700	556	349	56
Gulf of Finland	Kymi	37,235	671	421	288	178	66
	Karjaan	2,010	95	45	19	5	0.1
Gulf of Bothnia	Kokemaen	27,100	881	583	215	72	9.4
	Kyron	4,900	507	307	44	3.7	1.0
	Lapuan	4,110	326	197	31	3.3	1.0
	Ahtavan	2,030	67	39	16	6.9	0.1
	Perhon	2,690	385	252	22	3.7	2.0
	Kala	4,200	518	326	35	1.8	0.1
	Pyha	3,680	425	240	32	6.8	0.9
	Siika	4,440	686	404	34	2.1	0.1
	Oulu	22,925	889	473	248	91	0.1
	Kiiminki	3,880	660	377	44	5.6	1.7
	Ii	14,385	1,397	880	171	43	14
	Simo	3,175	636	417	49	4.8	1.0
	Kemi	51,400	4,400	2,945	536	136	62
	Tornion	40,010	3,180	2,128	376	71	53
Arctic Ocean	Pats	14,575	499	288	152	54	17

TABLE 1-27. FINLAND—MUNICIPAL AND RURAL WATER USE, 1969

(Source: Erkola, Aqua Fennica, 1971)

	Towns	Townships	Total Towns and Townships	Rural districts
Average population	1,940,820	404,900	2,345,720	2,340,000
Average number of consumers	1,754,560	242,890	1,997,450	495,000
Number of consumers (percent of total population)	90.4	60.0	85.2	21.2
Average daily consumption (total m³/day)	556,819	59,866	616,685	131,000
Specific consumption, weighted according to the number of consumers (l/capita/day)	317	246	309	264
Surface water (percent of total consumption)	79	73	79	—
Ground water (percent of total consumption)	21	27	21	—
Total length of distribution network, (km)	5,058	1,739	6,797	—

— Not available.

TABLE 1-28. FINLAND—WATER USE, 1969

(Source: Jaatinen, Aqua Fennica, 1971)

Use category	Quantity, m^3/s
Industry	
Pulp and paper	65
Other.............................	15
Public supplies and domestic	10
Total	90

TABLE 1-29. FINLAND—MUNICIPAL WATER SUPPLY SYSTEMS

(Source: Internat. Statistical Institute, 1972)

City	Year	Population served	Quantity available (1,000 m^3)			Annual consumption (1,000 m^3)	
			Production	Purchased from other plants	Total	Total	of which domestic
Helsinki (m)	1967	682,787	74,763	0	74,763	73,339	62,394
	1969	704,362	81,292	183	81,475	79,523	67,141
Tampere (m)	1967	211,767	14,780	0	14,780	14,780	—
	1969	219,491	16,901	0	16,901	16,901	—
Turku (m)	1967	155,250	17,667	273	17,940	17,198	7,841
	1969	169,000	20,594	493	21,087	20,104	9,209

(m) metropolitan area.
— not available.

TABLE 1-30. FRANCE – DISCHARGE OF SELECTED RIVERS

(Source: Direction du Gaz et de l'Electricite, 1966 and UNESCO, 1967)

River and station	Basin area km2	Mean monthly discharge, m3/s												Year	Period of record
		Jan.	Feb.	Mar.	Apr.	May	Jun.	Jul.	Aug.	Sep.	Oct.	Nov.	Dec.		
Moselle, Hauconcourt	9,400	188	217	159	125	96.5	80.5	56.0	58.0	52.0	57.5	108	156	112	1956-65
Meuse, Chooz	10,120	240	251	172	132	99.0	68.0	47.9	54.5	66.0	78.5	131	218	129	1953-65
Seine,Bazoches-les-Bray	10,240	77.0	85.0	85.0	134	85.0	61.0	26.4	14.3	18.7	23.1	30.4	8.10	60.0	1962-65
Seine, Vitry-sur-Seine	31,300	136	164	223	282	207	138	62.5	48.5	97.5	116	118	345	161	1964-65
Seine, Paris	44,320	505	568	456	314	228	155	114	99	105	131	253	351	273	1927-61
Yonne, Courlon	10,460	122	120	109	130	89.0	59.0	31.3	27.0	34.4	37.7	53.0	119	77.5	1961-65
Marne, Noisiel	12,580	150	158	130	102	78.5	54.5	38.0	36.5	46.4	43.8	72.5	119	86.0	1956-65
Oise, Creil	14,600	125	146	116	124	85.5	60.5	46.5	39.7	46.6	55.0	96.5	155	91.5	1960-65
Allier,Pont-du-Guetin	14,340	212	204	194	178	133	115	58.5	41.5	59.5	80.5	120	162	129	1955-65
Loire, Montjean	110,000	1,530	1,640	1,470	1,100	840	582	375	255	260	381	876	1,180	874	1921-60
Loire, Gien	35,890	552	591	531	442	365	263	135	90	104	187	404	484	345	1921-60
Vienne, Nouatre	19,650	383	250	266	263	185	113	94.0	68.5	91.5	124	193	363	200	1958-65
Dordogne, Bergerac	13,800	453	324	372	365	275	182	106	105	129	211	288	515	278	1958-65
Vezere, Montignac	3,125	96.5	104	81.0	64.5	53.5	36.0	23.9	18.6	20.8	36.2	63.0	88.5	57.0	1921-60
Tarn, Rouby	15,500	328	185	320	198	196	123	60.0	40.0	98.5	274	350	1,035	267	1923-33 1951-59
Garonne,Mas d'Agenais	52,000	861	938	917	822	762	544	279	175	190	274	487	839	590	1921-60
Lot, Cahors	9,170	229	229	186	124	93.2	70.3	34.0	29.7	32.4	60.5	131	203	118	1941-60
Adour,St.Vincent de Paul	7,830	143	136	101	81.6	72.4	64.0	31.0	21.0	26.9	40.1	67.9	109	74.5	1951-60
Rhone, Chateaufort	12,600	219	216	398	415	507	548	653	500	528	296	369	735	451	1965
Rhone, La Mulatiere	50,200	1,250	1,270	1,210	1,070	923	999	904	831	783	803	1,180	1,110	1,030	1921-60

TABLE 1-30. FRANCE—DISCHARGE OF SELECTED RIVERS (continued)

River and station	Basin area km²	Mean monthly discharge, m³/s												Year	Period of record
		Jan.	Feb.	Mar.	Apr.	May	Jun.	Jul.	Aug.	Sep.	Oct.	Nov.	Dec.		
Rhone, Beaucaire	95,590	1,887	1,888	2,005	1,870	1,873	1,899	1,429	1,173	1,199	1,422	2,026	1,863	1,712	1920-60
Ain, Chazey	3,630	132	128	211	157	94.5	74.0	51.0	84.5	95.0	90.5	148	171	120	1960-65
Saone, Le Chatelat-Pouilly	11,660	188	134	244	205	223	187	60.0	43.6	206	166	162	520	195	1965
Ardeche, Sauze Saint Martin	2,240	102	89.0	114	77.0	32.0	48.0	11.0	15.0	22.0	82.0	101	107	66.5	1955-64
Isere, Veurey	9,450	109	101	174	227	475	570	425	269	280	201	188	304	278	1965
Durance, Jouques-Cadarache	11,700	148	161	216	233	302	322	147	96.5	98.5	191	225	198	195	1951-60
Escaut, Conde sur l'Escaut	2,580	20.6	24.1	17.9	18.5	16.3	14.3	11.2	11.2	10.8	12.5	15.9	22.2	16.3	1961-65
Somme, Abbeville	5,560	28.6	28.6	29.6	30.0	25.9	24.5	25.7	25.0	23.2	22.6	25.8	33.7	26.9	1963-65
Charente, Cognac	4,630	80.0	48.3	64.0	46.6	40.7	25.5	21.7	16.3	38.9	45.4	61.0	195	57.5	1965
Aude, Carcassonne	1,794	37.2	36.5	44.4	51.5	51.0	30.6	14.4	9.3	8.5	10.0	13.5	21.1	27.3	1921-30
Orb, Reals	1,148	34.9	33.9	53.0	19.5	14.3	10.5	7.00	4.79	9.55	111	39.6	44.6	31.9	1957-62 1965
Var, La Mescla	1,827	25.5	26.5	35.2	46.5	67.5	61.5	50.5	22.1	25.6	37.9	45.3	35.7	40.0	1921-60
CORSICA															
Golo, Ponte-Leccia	366	6.71	3.91	9.86	5.29	6.04	2.85	0.45	0.23	3.28	4.42	8.37	9.41	5.08	1965
Tavignano, Altiani	489	11.8	5.52	33.5	10.3	17.3	6.61	1.12	0.55	5.65	17.1	23.0	20.1	12.8	1965
Taravo, Guitera	157	5.96	3.96	8.37	5.87	4.66	2.10	1.08	0.84	—	—	6.21	8.13	—	1965

TABLE 1-31. FRANCE—WATER DEMAND, 1955-2050

(Source: Commission de l'eau, Water for Peace, 1967)

[Recirculation of domestic and agricultural water excluded]

Year	1955		1970		2050	
Population (millions)	43		49		60	
Water demand	Billion m³ per year	m³ per person per year	Billion m³ per year	m³ per person per year	Billion m³ per year	m³ per person per year
Drinking water	1.9	44	4	81	7.5	125
Industrial water in circulation	10.4	240	23	470	72	1,200
consumed	6.5	150	14	285	36	600
Agricultural water	10	232	14.5	295	40	670
Inland navigation	1.3	30	1.5	30	2.5	33
Total supplied (excluding consumption)	23.6	546	43	876	122	2,028
Total consumed (or for other needs)	19.7	456	34	691	86	1,628

TABLE 1-32. FRANCE—MUNICIPAL WATER SUPPLY SYSTEMS

(Source: Internat. Statistical Institute, 1972)

City	Year	Population served	Quantity available(1,000 m³) Production	Quantity available(1,000 m³) Purchased from other plants	Quantity available(1,000 m³) Total	Annual consumption (1,000 m³) Total	Annual consumption (1,000 m³) of which domestic
Amiens (m)	1967	—	11,241	0	11,241	7,497	3,504
	1969	—	12,251	0	12,251	7,605	3,813
Avignon	1967	81,927	—	—	—	5,584	4,534
	1969	88,958	—	—	—	5,754	4,712
Bethune (m)	1967	—	2,212	—	2,212	1,932	1,832
	1969	—	2,466	—	2,466	2,153	2,029
Brest (m)	1967	155,000	7,855	0	7,855	5,700	3,620
	1969	167,000	8,751	0	8,751	7,443	4,250
Bruay-en-Artois (m)	1967	—	2,966	—	2,966	2,275	2,199
	1969	—	2,982	—	2,982	2,478	2,175
Caen	1967	95,238	11,002	1,776	12,778	8,092	6,550
	1969	114,398	11,591	1,844	13,435	8,124	6,704
Cannes (m)	1967	59,173	19,000	0	19,000	15,758	15,638
	1969	66,809	17,000	0	17,000	13,858	13,760
Clermont-Ferrand (m)	1967	—	29,072	2,667	31,739	31,739	—
	1969	—	29,190	3,802	32,992	32,992	11,738
Denain (m)	1967	—	—	—	—	2,800	1,756
	1969	—	—	—	—	3,176	1,992
Dijon (m)	1967	192,344	22,780	0	22,780	14,247	—
	1969	192,344	23,909	0	23,909	14,440	—
Douai (m)	1967	—	—	—	—	5,529	3,373
	1969	—	—	—	—	6,223	3,838
Dunkerque (m)	1967	—	10,622	—	10,622	8,757	4,013
	1969	—	12,302	—	12,303	10,186	4,780
Grenoble	1967	—	24,112	86	24,198	24,198	7,063
	1969	165,902	24,551	370	24,921	24,921	7,634
Hagondange-Briey (m)	1967	95,380	5,493	1,671	7,164	4,597	4,041
	1969	95,380	5,204	1,995	7,199	4,666	3,794
Le Havre	1967	—	32,744	213	32,957	20,200	16,075
	1969	—	40,852	231	41,083	24,200	17,200
Lens (m)	1967	—	10,100	—	10,100	7,381	6,477
	1969	—	10,260	—	10,260	7,898	7,015
Lille	1967	—	14,643	3,021	17,664	14,178	7,530
	1969	—	14,482	3,233	17,715	14,173	7,779
Lille (m)	1967	—	—	—	—	43,933	26,331
	1969	—	—	—	—	49,981	29,224
Limoges (m)	1967	145,021	12,158	241	12,399	11,068	8,426
	1969	—	—	—	—	—	—
Lyon	1967	535,000	92,032	0	92,032	74,769	15,441
	1969	535,000	90,259	0	90,259	73,779	13,556
Marseille	1967	879,000	379,801	—	379,801	222,823	77,579
	1969	929,400	382,492	—	382,492	298,829	80,627
Metz (m)	1967	180,000	16,606	77	16,683	11,239	7,867
	1969	210,000	15,512	131	15,643	11,661	6,413
Montpellier (m)	1967	165,016	—	—	—	472,139	462,355
	1969	183,237	—	—	—	479,563	468,734

TABLE 1-32. FRANCE—MUNICIPAL WATER SUPPLY SYSTEMS (continued)

City	Year	Population served	Quantity available(1000m^3) Production	Quantity available(1000m^3) Purchased from other plants	Quantity available(1000m^3) Total	Annual consumption(1000m^3) Total	Annual consumption(1000m^3) of which domestic
Mulhouse (m)	1967	188,729	18,931	647	19,578	18,142	8,465
	1969	193,567	21,518	702	22,220	19,915	8,887
Nancy (m)	1967	259,579	31,988	0	31,988	13,272	2,092
	1969	259,579	32,790	0	32,790	13,357	2,447
Nantes	1967	—	—	—	—	16,860	8,900
	1969	—	—	—	—	18,043	10,315
Nice (m)	1967	294,976	95,450	0	95,450	56,432	28,577
	1969	325,400	98,744	0	98,744	56,997	30,177
Nimes (m)	1967	121,250	68,100	0	68,100	61,160	—
	1969	131,680	79,580	0	79,580	61,580	—
Orleans (m)	1967	—	3,031	0	3,031	2,828	1,832
	1969	—	—	—	—	—	—
Paris	1967	2,635,000	430,515	2,222	430,537	374,478	231,379
	1969	2,546,000	436,202	0	436,202	368,372	228,896
Paris (m)	1967	5,402,631	—	—	711,323	599,107	—
	1969	5,464,619	—	—	709,363	586,901	—
Perpignan (m)	1967	104,530	—	—	78,938	75,633	118,067
	1969	109,474	—	—	102,778	100,207	162,783
Reims (m)	1967	—	14,085	0	14,085	12,041	5,665
	1969	174,011	14,985	0	14,985	12,793	6,387
Rennes (m)	1967	190,000	12,218	0	12,218	10,802	6,095
	1969	200,000	13,148	10	13,158	11,784	7,383
Roubaix	1967	—	7,937	0	7,937	5,457	3,438
	1969	—	8,543	0	8,543	6,558	4,161
Rouen	1967	—	12,660	0	12,660	12,660	5,200
	1969	—	14,946	0	14,946	14,946	5,500
Strasbourg (m)	1967	321,021	35,559	0	35,559	22,154	—
	1969	344,721	39,135	0	39,135	24,140	—
Toulon	1967	172,586	22,513	0	22,513	11,377	—
	1969	173,598	22,821	0	22,821	10,530	—
Toulouse	1967	410,000	51,045	4,850	55,905	—	21,700
	1969	410,000	46,943	8,430	55,373	—	21,260
Toulouse (m)	1967	—	51,946	6,685	58,631	—	24,312
	1969	—	49,930	9,324	59,254	—	24,172
Tours	1967	—	12,300	—	12,300	11,685	5,082
Troyes (m)	1967	—	14,268	0	14,268	7,869	4,320
	1969	76,000	13,898	0	13,898	7,104	3,847
Valenciennes (m)	1967	—	10,900	0	10,900	7,553	3,683
	1969	—	12,000	0	12,000	8,041	3,825
Villeurbanne	1967	—	10,900	0	10,900	8,480	5,000
	1969	122,898	12,000	0	12,000	9,253	5,126

(m) Metropolitan area
— Not available

TABLE 1-33. FRANCE—CONSUMPTION OF BOTTLED MINERAL WATER, 1950-69
(Source: Commission de l'eau, 1971)

| Year | Volume | | | | | Value of Consumption (Millions of francs) | | |
| | Millions of bottles and size | | | Total (millions of bottles) | Total (millions of liters) | Away from home (wholesale price) | At home (wholesale price) | Total |
	90cl.	45cl.	22.5cl.					
1950	404	15	86	505	393	36	97	133
1961	1,100	18	118	1,236	1,024	131	443	574
1962	1,150	19	122	1,291	1,071	137	463	600
1963	1,302	21	133	1,456	1,211	156	523	679
1964	1,465	23	168	1,656	1,364	180	590	770
1965	1,453	23	186	1,632	1,355	182	630	812
1966	1,601	26	183	1,810	1,494	215	710	925
1967	1,731	30	210	1,971	1,593	238	820	1,098
1968	1,758	28	206	1,992	1,650	250	870	1,120
1969	1,891	35	228	2,154	1,775	320	1,000	1,320

FIGURE 1-1. FRANCE — VULNERABILITY OF GROUND WATER TO POLLUTION

(Source : Commission de l'Eau, 1971)

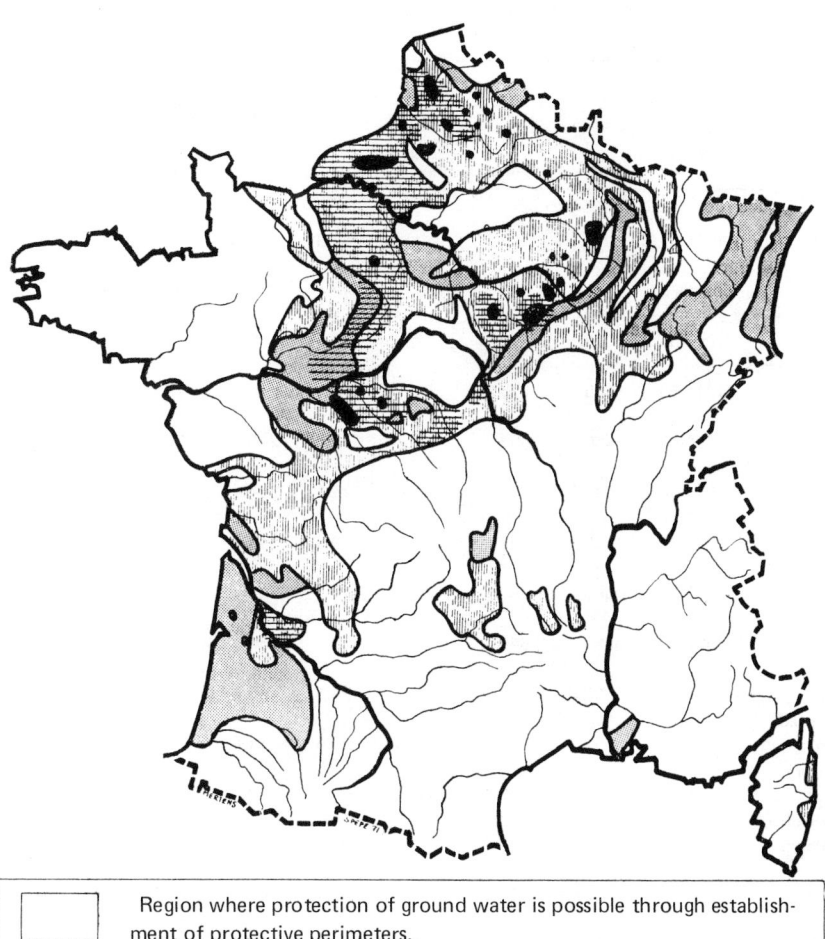

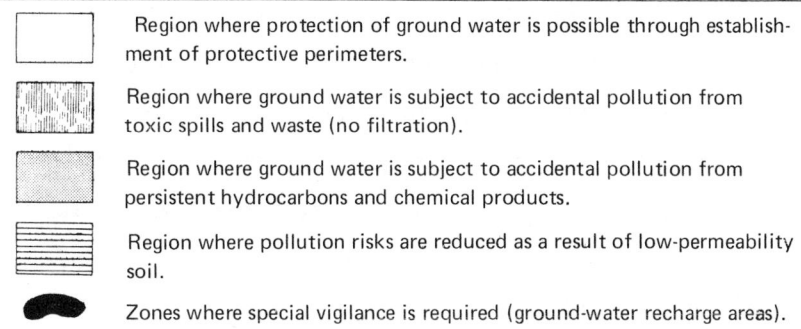

Region where protection of ground water is possible through establishment of protective perimeters.

Region where ground water is subject to accidental pollution from toxic spills and waste (no filtration).

Region where ground water is subject to accidental pollution from persistent hydrocarbons and chemical products.

Region where pollution risks are reduced as a result of low-permeability soil.

Zones where special vigilance is required (ground-water recharge areas).

EAST GERMANY — No tabular data. See index for cross-referenced items.

TABLE 1-34. WEST GERMANY—
(Source: Bundesanstalt fur Gewasserkunde

River	Station and drainage area km^2	Period of record	Mean monthly discharge, m^3/s								
			Jan.	Feb.	Mar.	Apr.	May	Jun.	Jul.	Aug.	Sep.
Treene	Treia 480	1931-60	9.78	9.97	7.84	5.61	3.59	3.85	3.70	4.81	5.15
Elbe	Neu-Darchau 131,951	1931-60	725	843	1,100	1,130	766	591	571	486	448
Ilmenau	Bienenbuttel 1,457	1956-65	12.1	12.0	11.9	9.63	8.34	7.34	7.04	7.60	7.33
Weser	Hann.-Munden 12,444	1931-60	145	173	158	126	80.3	73.7	76.5	64.9	63.3
Weser	Intschede 37,788	1931-60	450	498	478	379	246	206	210	185	171
Leine	Greene 2,920	1931-60	40.0	45.0	43.7	34.1	24.4	21.9	22.9	20.5	18.8
Ems	Versen 8,469	1931-60	136	139	114	79.5	44.7	28.8	35.8	33.6	35.5
Rhein (Rhine)	Kaub 103,730	1931-60	1,550	1,680	1,720	1,610	1,530	1,780	1,880	1,590	1,380
Rhein (Rhine)	Rees 159,680	1936-65	2,560	2,840	2,660	2,450	2,050	2,200	2,170	1,900	1,700
Kinzig	Schwaibach 955	1931-60	33.5	34.5	31.0	24.1	14.1	16.0	16.2	14.7	14.3
Kocher	Neuenstadt 1,762	1931-60	33.7	37.1	34.4	22.6	14.4	14.5	13.2	9.48	10.4
Main	Kemmern 4,251	1931-60	67.3	75.1	73.9	51.5	27.3	25.2	26.3	22.7	20.2
Main	Kleinheubach 21,505	1960-65	134	193	197	190	136	148	81.7	70.1	60.9
Regnitz	Pettstadt 7,005	1931-60	66.6	74.3	79.0	57.1	37.3	38.3	40.4	35.3	32.6
Nidda	Bad Vilbel 1,620	1956-65	15.8	16.2	14.6	12.3	8.65	8.99	7.47	7.81	7.04
Nahe	Grolsheim 4,011	1946-65	48.9	50.5	37.8	24.5	16.6	15.5	10.1	9.37	9.24
Mosel	Cochem 27,100	1931-60	535	555	429	324	201	158	135	125	124
Niers	Goch 1,220	1951-60	11.5	11.8	10.1	7.54	5.66	4.85	4.80	5.34	5.58
Donau (Danube)	Ingolstadt 20,001	1931-60	265	297	347	353	357	380	362	293	262
Donau (Danube)	Hofkirchen 47,496	1931-60	605	687	779	731	678	724	737	605	534
Iller	Kempten 953	1931-60	26.9	28.7	43.1	61.3	72.6	71.2	63.4	50.1	43.8
Naab	Heitzenhofen 5,426	1931-60	57.7	70.7	77.3	56.9	35.1	34.4	40.4	32.6	29.4
Inn	Reisach 9,793	1931-60	109	108	138	219	420	652	625	496	350

DISCHARGE OF SELECTED RIVERS
and Deutsche Forschungsgemeinschaft, 1972)

Oct.	Nov.	Dec.	Year	Maximum discharge, m^3/s			Min. mean daily discharge m^3/s
				Max. mean daily discharge	Peak discharge		
					Discharge	Date	
5.71	6.83	8.29	6.26	—	61.3	Jul. 10,1931	0.23
478	597	652	700	3,620	3,620	Mar.31, Apr. 1,1940	145
7.41	9.35	11.9	9.34	53.5	58.4	Mar. 2,1956	2.76
74.4	103	119	105	1,540	1,540	Feb. 10,1946	17.0
199	277	357	305	3,300	3,500	Feb. 12,1946	59.2
22.3	28.2	33.6	30.0	1,120	1,200	Feb. 9,1946	7.35
49.8	77.6	105	73.5	1,200	1,200	Feb. 12,1946	5.20
1,260	1,370	1,360	1,560	6,050	6,150	Jan. 19,1955	482
1,650	1,970	2,390	2,210	9,440	9,500	Jan.20,21,1955	590
17.3	25.2	25.0	22.2	604	760	Jul. 29,1947	1.22
12.5	21.8	23.4	20.1	483	510	Mar. 4,1956	0.65
26.5	41.3	51.7	42.3	855	952	Dec. 29,1947	3.10
72.2	115	196	133	857	869	Dec. 22,1965	13.8
38.9	46.5	52.5	49.7	660	710	Jan. 26,1941	9.70
8.10	11.0	15.8	11.1	62.3	63.0	Mar. 14,1963	1.65
10.8	26.5	45.8	25.5	463	817	Dec. 21,1952	1.70
168	314	427	291	3,730	3,740	Jan.1,1948	22.0
7.10	8.23	10.4	7.73	44.0	44.9	Dec. 7,1960	2.00
238	252	231	303	1,800	1,860	Jun. 2, 1940	62.0
516	543	530	639	3,830	3,880	Jul. 13,1954	193
34.8	34.3	25.7	46.3	519	595	Nov. 24,1944	4.38
35.9	45.1	48.9	46.9	747	765	Dec. 30,1947	9.60
227	168	127	301	1,550	1,800	May 31,1940	68.0

TABLE 1-35. WEST GERMANY–WATER BALANCE ELEMENTS
(Source: Federal Ministry of Health, 1963)

Element	Volume (km^3)
Precipitation	200
Evapotranspiration	110
Runoff	90
Inflow from other countries	80
Total water use	14

TABLE 1-36. WEST GERMANY–WATER USED BY PUBLIC SUPPLY SYSTEMS, 1969
(Source: Verband der deutschen Gas-und Wasserwerke e.V., 1971)

[by Bundeslander; in 1000 m3]

Land	Number of water supply systems	Total	Self-supplied water of which			Purchased water	Total self-supplied and purchased water
			Ground water	Spring water	Surface water		
Schleswig-Holstein	67	97,950	89,133	79	8,738	7,801	105,751
Hamburg	1	150,275	148,452	–	1,823	30	150,305
Niedersachsen	187	364,228	300,296	25,271	38,661	48,368	412,596
Bremen	2	22,675	14,584	–	8,091	22,634	45,309
Nordrhein-Westfalen	212	1,374,496	1,186,391	20,445	167,660	190,435	1,564,931
Hessen	177	270,079	222,292	47,787	–	98,648	368,727
Rheinland-Pfalz	156	159,987	123,048	27,151	9,788	5,079	165,066
Baden-Wurttemberg	264	427,285	252,765	85,058	89,462	169,400	596,685
Bayern (Bavaria)	179	446,450	300,249	138,930	7,271	8,009	454,459
Saarland	45	58,726	56,240	2,486	–	14,752	73,478
Berlin (West)	1	162,231	162,231	–	–	–	162,231
Total	1,291	3,534,382	2,855,681	347,207	331,494	565,156	4,099,538

TABLE 1-37. WEST GERMANY—MUNICIPAL WATER SUPPLY SYSTEMS
(Source: International Statistical Institute, 1971)

City	Year	Population served	Quantity available(1000m3)			Annual consumption(1000m3)	
			Production	Purchased from other plants	Total	Total	of which domestic
Aachen	1967	177,508	14,034	2,761	16,795	15,175	8,117
	1969	176,694	14,352	2,586	16,938	15,552	8,165
Augsburg	1967	211,070	24,574	0	24,574	17,969	15,774
	1969	212,969	24,251	0	24,251	17,513	15,800
Berlin (West)	1967	2,173,262	149,448	0	149,448	149,448	—
	1969	2,135,057	162,231	0	162,231	162,231	—
Bielefeld	1967	172,091	13,462	0	13,462	9,329	5,308
	1969	170,261	14,997	0	14,997	10,106	5,700
Bochum	1967	358,984	39,835	5,966	45,801	45,025	18,639
	1969	354,884	31,229	13,998	45,227	42,115	22,072
Bonn	1967	134,740	0	10,084	10,084	9,150	6,279
	1969	130,820	0	10,941	10,941	9,891	6,527
Bottrop	1967	111,807	14,883	0	14,883	14,883	5,246
	1969	109,472	12,960	0	12,960	12,960	5,235
Braunschweig	1967	232,930	8,335	6,677	15,012	14,061	9,960
	1969	228,329	8,264	6,932	15,196	14,741	10,781
Bremen 1	1967	602,570	11,503	19,278	30,781	28,649	18,065
	1969	606,265	10,682	22,634	33,316	30,497	18,918
Bremerhaven 1	1967	162,500	12,001	0	12,001	11,539	6,689
	1969	163,300	11,993	0	11,993	11,789	7,610
Darmstadt 1	1967	239,190	16,038	0	16,038	14,374	—
	1969	245,585	17,244	0	17,244	15,697	—
Dortmund	1967	648,912	88,508	6,617	95,125	91,616	23,823
	1969	649,006	91,279	7,915	99,194	92,515	24,778
Duisburg	1967	470,515	82,368	210	82,578	77,842	21,270
	1969	460,658	94,660	202	94,862	89,791	22,802
Dusseldorf	1967	687,800	38,800	41,300	80,100	76,700	36,900
	1969	675,800	39,700	44,300	84,000	80,200	36,900
Essen	1967	705,200	60,497	0	60,497	49,606	—
	1969	696,905	56,247	0	56,247	47,650	—
Frankfurt am Main 1	1967	667,457	34,073	40,114	74,187	62,832	34,139
	1969	666,140	22,219	56,187	78,406	65,677	35,215
Freiburg im Breisgau	1967	160,007	11,774	0	11,774	11,774	7,735
	1969	165,960	12,561	0	12,561	12,561	8,445
Gelsenkirchen	1967	360,721	—	—	—	53,167	16,380
	1969	352,152	—	—	—	56,813	15,520
Gottingen	1967	111,647	6,555	1,144	7,699	7,057	6,516
	1969	113,963	7,223	1,180	8,403	7,330	6,775
Hagen	1967	202,400	15,904	0	15,904	13,670	—
	1969	203,778	18,154	0	18,154	13,600	—
Hamburg	1967	1,832,560	137,258	0	137,258	123,636	97,249
	1969	1,817,122	150,274	31	150,305	135,929	105,663
Hannover (m)	1967	671,018	52,411	288	52,699	49,322	29,340
	1969	673,223	54,421	114	54,535	52,723	31,573

TABLE 1-37. WEST GERMANY—MUNICIPAL WATER SUPPLY SYSTEMS (continued)

City	Year	Population served	Quantity available(1000m³)			Annual consumption(1000m³)	
			Production	Purchased from other plants	Total	Total	of which domestic
Heidelberg	1967	123,446	9,775	1,453	11,228	10,696	4,739
	1969	122,006	10,211	1,270	11,481	10,919	4,873
Herne	1967	106,922	0	5,502	5,502	5,002	3,750
	1969	103,874	0	5,491	5,491	4,992	3,740
Karlsruhe 1	1967	268,526	23,039	0	23,039	21,352	13,045
	1969	274,956	23,253	0	23,253	22,056	13,600
Kassel	1967	211,586	17,851	0	17,851	14,960	9,193
	1969	213,494	18,711	0	18,711	15,685	10,465
Kiel 1	1967	311,000	18,812	0	18,812	17,555	10,371
	1969	315,000	20,932	0	20,932	19,464	11,217
Koblenz	1967	103,879	7,502	20	7,522	5,977	—
	1969	106,454	7,741	17	7,758	6,550	—
Koln 1	1967	854,500	111,747	5	111,752	107,372	32,626
	1969	866,208	113,311	11	113,322	106,459	39,322
Krefeld	1967	223,725	15,700	0	15,700	14,000	8,000
	1969	227,428	17,000	0	17,000	15,100	8,300
Leverkusen	1967	106,092	222,480	2,057	224,537	224,452	4,200
	1969	109,932	258,553	2,182	260,735	260,636	4,200
Ludwigshafen	1967	174,769	16,608	0	16,608	16,448	6,437
	1969	175,738	18,575	0	18,575	17,277	6,576
Lubeck	1967	257,737	10,797	1,234	12,031	11,268	9,214
	1969	258,099	11,700	1,163	12,863	11,961	9,859
Mainz 1	1967	207,137	20,127	337	20,464	19,379	8,175
	1969	215,604	20,871	364	21,235	20,194	8,338
Mannheim	1967	326,625	0	26,721	26,721	26,040	—
	1969	327,418	0	25,895	25,895	25,081	—
Monchengladbach 1	1967	163,000	9,544	0	9,544	8,917	4,007
	1969	164,000	10,141	0	10,141	9,544	4,412
Mulheim a.d. Ruhr	1967	188,892	26,648	0	26,648	26,648	9,532
	1969	190,362	30,913	0	30,913	30,913	10,387
Munich	1967	1,242,190	136,368	0	136,368	117,852	—
	1969	1,303,213	136,842	0	136,842	120,994	—
Munster	1967	202,381	8,449	4,953	13,402	12,622	—
	1969	204,716	9,180	5,136	14,316	13,621	—
Neuss	1967	113,519	8,375	75	8,450	7,748	4,155
	1969	116,783	9,144	221	9,365	8,317	4,153
Nurnberg	1967	465,797	38,367	0	38,367	33,069	—
	1969	477,108	42,475	0	42,475	36,166	—
Oberhausen	1967	255,523	0	22,466	22,466	22,466	10,520
	1969	249,886	0	22,055	22,055	22,055	10,797
Offenbach a. Main	1967	116,830	8,999	0	8,999	7,526	4,419
	1969	117,671	9,285	0	9,285	7,803	4,784
Oldenburg	1967	137,999	6,620	0	6,620	—	—
	1969	139,674	7,357	0	7,357	—	—

TABLE 1-37. WEST GERMANY—MUNICIPAL WATER SUPPLY SYSTEMS (continued)

City	Year	Population served	Quantity available(1000m^3)			Annual consumption(1000m^3)	
			Production	Purchased from other plants	Total	Total	of which domestic
Osnabruck [1]	1967	150,610	9,048	10	9,058	8,206	5,160
	1969	152,629	9,505	84	9,589	8,630	5,394
Recklinghausen	1967	129,218	0	9,361	9,361	9,361	5,964
	1969	128,246	0	9,931	9,931	9,931	5,681
Remscheid	1967	131,782	9,364	1,744	11,108	7,941	7,088
	1969	134,323	9,738	2,437	12,175	9,077	7,938
Rheydt	1967	99,752	5,175	0	5,175	4,917	3,886
	1969	100,340	5,598	0	5,598	5,079	4,000
Saarbrucken	1967	133,888	10,233	586	10,819	9,910	4,493
	1969	131,132	8,039	2,616	10,655	10,160	4,642
Salzgitter	1967	118,865	387	5,356	5,743	5,493	3,797
	1969	118,848	387	5,897	6,284	5,546	4,479
Solingen	1967	174,492	11,466	129	11,595	9,469	7,215
	1969	177,087	12,773	59	12,831	10,494	9,571
Stuttgart	1967	612,907	13,082	45,237	58,319	52,299	21,000
	1969	625,888	16,376	49,412	65,788	59,087	23,000
Trier	1967	102,820	6,947	0	6,947	5,686	3,933
	1969	101,050	7,244	0	7,244	5,716	3,697
Wanne-Eickel [1]	1967	104,556	—	—	—	13,385	4,992
	1969	103,393	0	5,035	5,035	4,577	4,149
Wiesbaden	1967	231,726	15,853	0	15,853	15,853	14,407
	1969	233,559	12,146	4,854	17,000	15,949	15,111
Wilhelmshaven [1]	1967	112,061	11,504	101	11,605	10,710	5,000
	1969	113,465	11,874	89	11,105	11,105	5,000
Wurzburg [1]	1967	119,752	10,782	0	10,782	10,047	8,549
	1969	121,259	10,885	0	10,885	10,088	8,582
Wuppertal [1]	1967	417,203	50,362	2,051	52,413	43,261	—
	1969	413,874	57,995	2,423	60,418	50,751	—

[1] Include some surrounding districts and suburbs
(m) Metropolitan area
— Not available

FIGURE 1-2. WEST GERMANY — WATER SUPPLY IN THE RUHR VALLEY

(Source: Koenig, Water for Peace, 1967)

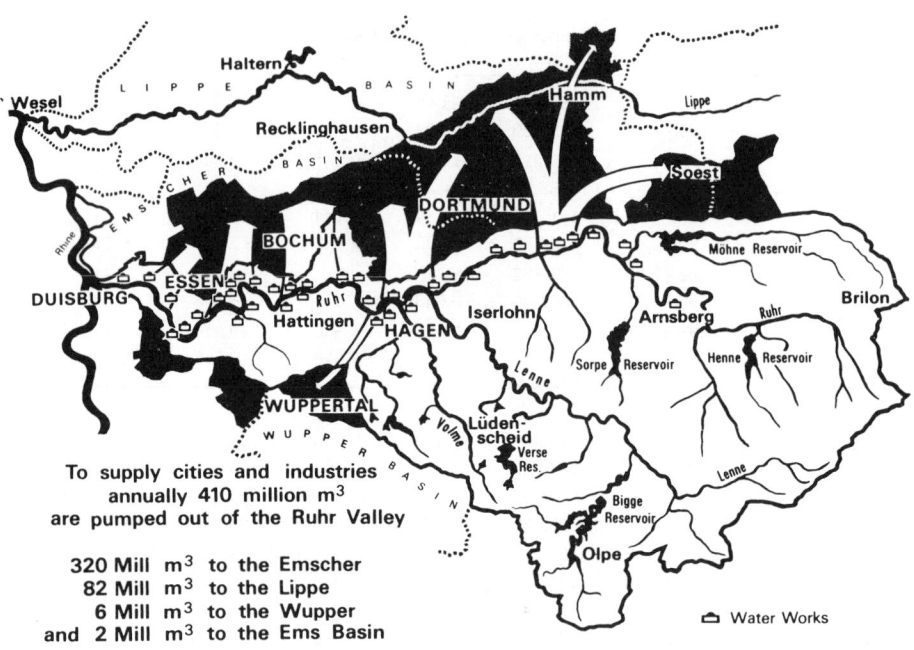

To supply cities and industries
annually 410 million m³
are pumped out of the Ruhr Valley

320 Mill m³ to the Emscher
82 Mill m³ to the Lippe
6 Mill m³ to the Wupper
and 2 Mill m³ to the Ems Basin

FIGURE 1-3. WEST GERMANY — WATER SUPPLY AND WASTE TREATMENT
FACILITIES OF THE RUHR ASSOCIATION (RUHRVERBAND) AND
RUHR RESERVOIR ASSOCIATION (RUHRTALSPERRENVEREIN).

(Source: Koenig, Water for Peace, 1967)

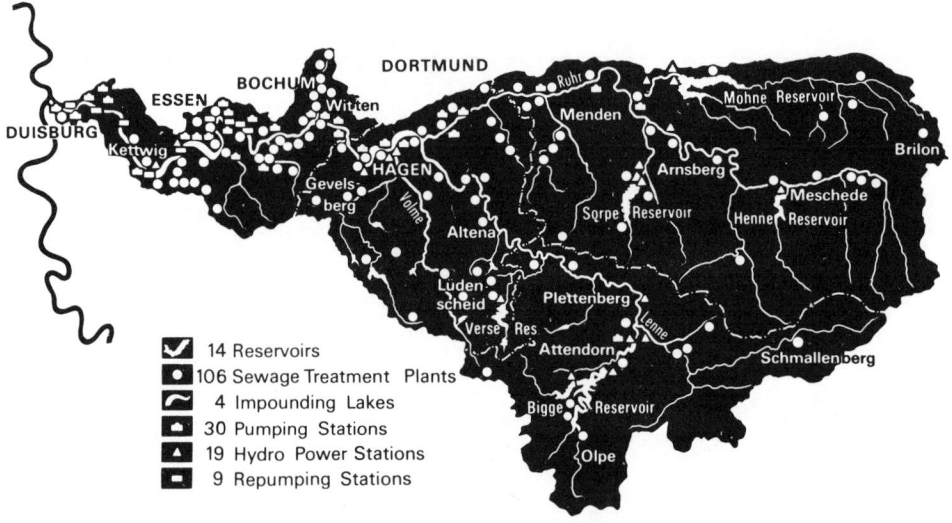

14 Reservoirs
106 Sewage Treatment Plants
4 Impounding Lakes
30 Pumping Stations
19 Hydro Power Stations
9 Repumping Stations

TABLE 1-38. WEST GERMANY—SALT LOAD OF THE RHINE RIVER, 1961-67
(Source: Rijkswaterstaat, The Netherlands, 1968)

Rhine River station	Mean annual chloride-ion load in kg/sec						
	1961	1962	1963	1964	1965	1966	1967
Stein am Rhein	0.9	0.9	1.1	1.0	1.5	1.4	1.5
Kembs	6.3	7.1	7.3	6.7	9.6	9.0	9.0
Seltz	112	112	123	111	155	161	151
Braubach	113	127	136	116	185	183	168
Lobith (Dutch–German border)	265	260	273	264	337	347	337

TABLE 1-39. WEST GERMANY—WATER QUALITY OF THE RHINE RIVER AT DUTCH–GERMAN BORDER
(Source: Rijkswaterstaat, The Netherlands, 1973)

[in mg/l]

Constituent	Natural water quality	1964 (dry year)	1966 (wet year)
Cl^-	20	187	123
SO_4^{--}	35	102	71
HCO_3^-	160	149	151
NO_3^-	2	10.7	9.4
NH_4^+	0.1	3.2	1.5
Na^+	5	98	63
K^+	5	8	6
Ca^{++}	50	91	78
Mg^{++}	10	9.8	10
$KMnO_4$ Consumption	10	61.7	41
Detergents	0	0.52	0.26
Phenols	0	0.038	0.026
Oil	0	—	1–1.5
Oxygen saturation	90-100	50	60

TABLE 1-40. GREECE–DISCHARGE OF ACHELOOS RIVER

(Source: ECE, 1970)

[Period of record 1937-67]

River, station	Mean monthly discharge, m3/s												Six-monthly average		Yearly average	Mean annual runoff l/s/km2
	Jan.	Feb.	Mar.	Apr.	May	Jun.	Jul.	Aug.	Sep.	Oct.	Nov.	Dec.	Oct.-Mar.	Apr.-Sep.		
Acheloos, Kremasta	278	265	242	222	154	92.6	57.1	39.4	42.7	86.3	217	297	230	101	165	46.3

TABLE 1-41. GREECE–DISCHARGE OF RIVERS IN MACEDONIA

(Source: Chorafas, Water for Peace, 1963)

River	Basin Area		Mean monthly discharge, m3/s												Max. known flow m3/s	Min. known flow m3/s
	Total km2	within Greece km2	Jan.	Feb.	Mar.	Apr.	May	Jun.	Jul.	Aug.	Sep.	Oct.	Nov.	Dec.		
Aliakmon	9,455	9,455	100	142	180	140	95	60	33	22	28	38	87	115	1,000	9.0
Axios	24,662	1,818	172	182	228	237	191	129	62	39	38	66	104	170	2,500	12.1
Strymon	16,553	6,027	104	118	135	181	228	156	65	31	27	56	83	135	2,200	5.5
Nestos	6,178	2,524	164	88	178	125	125	124	62	25	19	33	71	137	1,500	9.6

TABLE 1-42. GREECE—POPULATION SERVED BY PUBLIC WATER SUPPLY SYSTEMS AND WELLS

(Source: Georgoulis, Water for Peace, 1963)

[Data as of 1963]

Areas	Population		Water main by tap		Water from wells, carted water and rain water %	Length of distribution network km
	Total	%	within the dwellings %	on street %		
Greater Athens and Saloniki	2,231,053	26.6	77	20	3	4,000
Other urban areas (towns over 10,000 inhabitants)	1,396,952	16.7	60	28	12	
Semi-urban areas (small towns 2,000-10,000 inhabitants)	1,085,856	13.1	27	49	24	20,000
Rural areas	3,674,992	43.6	8	42	50	
Total Greece	8,388,553	100	39	34	27	24,000

TABLE 1-43. GREECE—MUNICIPAL WATER SUPPLY SYSTEMS

(Source: Internat. Statistical Institute, 1971)

City	Year	Population served	Quantity available(1000m^3)			Annual consumption(1000m^3)	
			Production	Purchased from other plants	Total	Total	of which domestic
Athens	1967	2,054,700	—	—	—	108,846	60,960
	1969	2,369,950	—	—	—	122,723	70,192
Patrai (m)	1967	110,886	8,213	0	8,213	7,083	4,191
	1969	116,370	9,103	183	9,286	8,083	4,765
Thessaloniki	1967	—	27,662	0	27,662	17,097	10,426
	1969	—	35,831	0	35,831	21,797	13,006

(m) Metropolitan area
—   Not available

TABLE 1-44. HUNGARY–HYDROLOGIC DATA ON PRINCIPAL RIVERS
(Source: Ballo, Hungarian National Water Authority, 1970)

River	Length		Drainage basin area		Mean width, depth and slope of rivers in Hungary at mean water level			Discharge at mouth, m³/s		
	within the country km	Total km	within the country km²	Total km²	width m	depth m	slope cm/km	Max	Mean	Min
Tisza	763	804	44,619	139,078	150	5.5	6	3,360	190	29.00
Danube	417	1,457	38,936	209,379	450	4.5	15	9,600	2,025	570.00
Drava	225	695	6,242	40,490	150	2.5	23	2,100	600	200.00
Ipoly	193	257	1,518	5,108	20	2.0	30	360	16.2	0.20
Raba	182	283	5,564	10,113	45	3.0	50	560	36.4	6.50
Zagyva	174	174	5,672	5,577	30	3.0	30	150	1.4	0.04
Zala	134	139	2,578	2,578	40	2.0	20	137	5.8	0.25
Sojo	125	229	9,487	12,708	60	2.0	51	520	32.4	2.40
Sio	123	123	8,954	8,954	20	2.5	14	224	21.0	0.60
Hernod	112	282	1,134	5,436	60	2.0	60	552	29.9	2.40
Sarviz	111	111	3,449	3,449	12	1.5	21	45	5.0	1.40
Kapos	111	111	3,242	3,242	25	2.0	30	130	5.1	0.22
Tarna	101	101	2,112	2,116	15	2.0	120	860	2.6	0.015
Marcol	100	100	3,076	3,076	15	2.0	27	122	73	0.33
Hormas Koros	91	363	12,931	27,537	70	3.0	4	1,322	119.0	4.30
Berettyo	78	204	2,649	6,095	40	2.0	18	270	7.9	0.20
Sebes Koros	59	209	3,156	9,199	20	1.5	24	600	25.0	1.90
Szamos	50	415	602	15,881	100	2.5	10	1,350	120.0	15.00
Bodrog	50	267	1,382	13,579	50	3.0	3	1,300	120.0	4.00
Maros	49	754	1,385	30,332	90	3.5	26	1,800	150.0	22.00
Kraszna	46	193	389	3,142	15	1.5	13	83	3.5	0.01
Mura	37	454	1,750	14,138	80	2.0	65	1,050	150	50.00
Kettos Koros	37	273	1,744	10,386	40	2.0	10	687	68.0	0.5
Tur	28	95	112	1,262	30	1.5	26	234	18.0	0.22
Fekete Koros	21	168	151	4,645	20	1.5	14	572	29.0	0.4
Feher Koros	11	236	352	4,275	20	1.5	16	605	23.6	0.001

TABLE 1-45. HUNGARY—DISCHARGE OF DANUBE AND TISZA RIVERS

(Source: UNESCO, 1971)

River, station	Basin area km^2	Mean monthly discharge, m^3/s Jan.	Feb.	Mar.	Apr.	May	Jun.	Jul.	Aug.	Sep.	Oct.	Nov.	Dec.	Year	Period of record
Danube, Nagymaros	183,533	1,880	2,220	2,700	2,950	2,920	3,120	2,920	2,460	1,920	1,690	1,810	1,770	2,360	1931-65
Tisza, Polgar	62,723	463	485	808	1,010	682	489	405	294	301	267	428	441	506	1931-65
Tisza, Szeged	138,408	669	747	1,285	1,465	1,235	885	673	451	384	379	628	683	790	1931-65

TABLE 1-46. HUNGARY—FRESH WATER DEMAND, 1960-85

(Source: Kernacs, Hungarian National Water Authority, 1970)

[In Billion m^3]

	1960		1965		1970		1985	
	Water demand	% increase	Water demand	% increase	Water demand	% increase	Water demand	% increase
Public Supply	0.4	100	0.6	150	0.7	175	1.1	275
Industry	1.6	100	2.1	131	2.6	163	6.6	413
Agriculture	1.4	100	1.8	129	2.3	164	4.8	343
Total	3.4	100	4.5	132	5.6	165	12.5	368
Per capita water use (m^3/yr)	340		440		550		1,100	

TABLE 1-47. HUNGARY—WITHDRAWAL OF WATER BY WATER SUPPLY SYSTEMS, 1965

(Source: Illes and Simo, Aqua, 1970)

Type of water supply systems	No. of systems	Surface Water		Ground Water		Total	
		million m³/d	million m³/yr	million m³/d	million m³/yr	million m³/d	million m³/yr
Municipal and rural	670	0.12	43.0	1.01	370.4	1.13	413.4
Public utility and industrial	6	0.10	32.0	—	—	0.10	32.0
Self-supplied industrial plants	1,020	4.83	1,403.0	0.47	136.0	5.30	1,539.0
Regional	4	0.04	14.6	0.04	14.6	0.08	29.2
Grand total	1,700	5.09	1,492.6	1.52	521.0	6.61	2,013.6

TABLE 1-48. HUNGARY—MUNICIPAL WATER SUPPLY SYSTEMS

(Source: Internat. Statistical Institute, 1971)

City	Year	Population served	Quantity available (1,000 m³)			Annual consumption (1,000 m³)	
			Production	Purchased from other plants	Total	Total	of which domestic
Budapest	1967	1,980,000	254,961	0	254,961	242,165	87,717
	1969	1,930,000	271,227	0	271,227	244,143	96,127
Debrecen	1967	150,000	11,374	0	11,374	11,138	3,691
	1969	150,000	12,636	0	12,636	11,405	4,493
Miskolc	1967	180,000	18,354	0	18,354	17,274	5,714
	1969	170,000	18,841	106	18,947	16,360	5,706
Pecs	1967	140,000	10,317	3,855	14,172	12,592	4,469
	1969	140,000	10,285	5,343	15,628	12,768	5,009
Szeged	1967	120,000	10,598	0	10,598	8,258	3,447
	1969	120,000	11,224	0	11,224	9,656	4,641

TABLE 1-49. ICELAND—DISCHARGE OF PRINCIPAL RIVERS
(Source: UNESCO, 1971)

River and station	Drainage basin km^2	Mean monthly discharge, m^3/s												Year	Period of record
		Jan.	Feb.	Mar.	Apr.	May	Jun.	Jul.	Aug.	Sep.	Oct.	Nov.	Dec.		
Thjorsa, Urridafoss	7,200	245	288	311	302	536	545	509	448	368	338	295	263	371	1947-66
Jokulsa, Dettifoss	7,000	105	110	122	144	218	213	325	345	249	163	128	111	186	1939-66
Olfusa, Selfoss	5,760	386	417	392	394	401	372	358	340	339	387	397	367	379	1950-66
Lagarfljot, Lagarfoss	2,800	70	87	79	94	189	316	215	145	136	138	122	106	141	1949-66

FIGURE 1-4. ICELAND — MINERAL AND THERMAL SPRINGS
(Source: Arnorsson, et al., XXIII Internat. Geol. Congress, 1969)

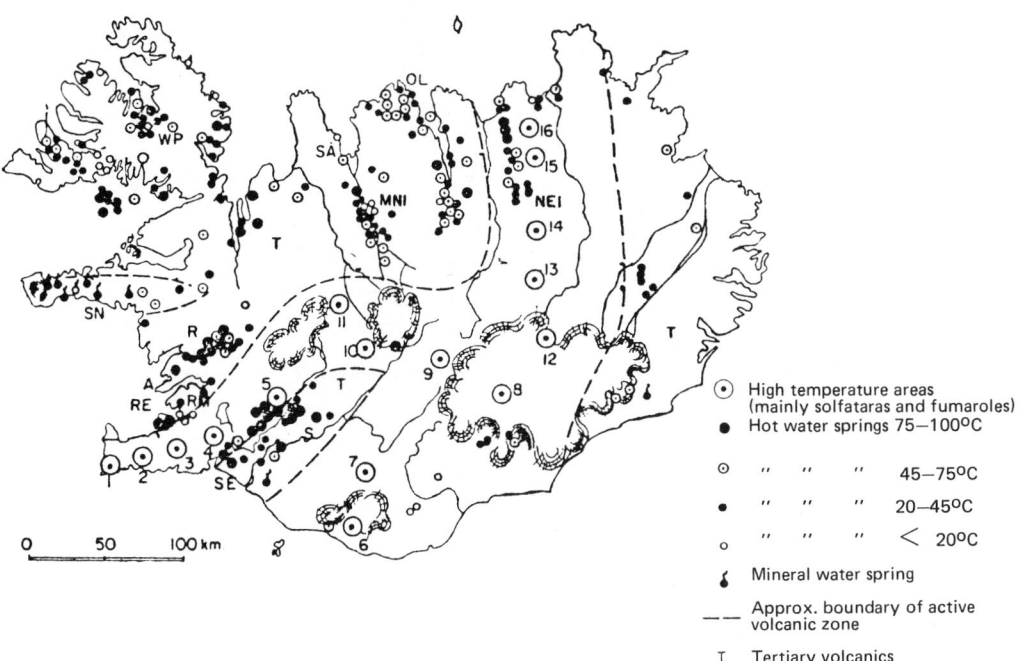

1 — Reykjanes-Eldvorp; 2 — Krysuvik-Trolladyngja; 3 — Brennisteinsfjoll; 4 — Hengill; 5 — Geysir Area; 6 — Sol-heimajokull (sub-glacial); 7 — Torfajokull; 8 — Grimsvotn (sub-glacial); 9 — Vonarskard; 10 — Kerlingarfjoll; 11 — Hveravellir; 12 — Kverkfjoll; 13 — Askja; 14 — Ketildyngja; 15 — Namafjall-Krafla; 16 — Theistareyikir RE — Reykjavik; A — Akranes; R — Reykholtsdalur; SN — Snaefellsnes Peninsula; WP NW — Peninsula; SA — Saudarkrokur; OL — Olafsfjordur; MNI — Middle N. Icleland; NEI — NE Icleland; SE — Selfoss; RM — Reykir Mosfellssveit

TABLE 1-50. ICELAND—DISCHARGE OF THERMAL WELLS

(Source: Arnorsson, et al., XXIII Internat. Geol. Congress, 1969)

Area	Total flow l/sec	Max. temp. °C	Surface manifestation of thermal activity	Remarks
Reykir, Mosfellssveit	290	88	Many surface springs, flow substantial, temp. 83 ° C	70 drillholes, max. depth 1,200 m, mostly natural flow.
Reykjavik	360	146	Surface springs, flow 11 l/sec., temp. 88 ° C	23 deep drillholes, max. depth 2,200 m, numerous shallow drillholes. Most of the water is recovered by pumping.
Saudarkrokur	35	70	Insignificant springs, temp. 35 ° C	12 drillholes, max. depth 480 m, natural flow.
Olafsfjardur	35	57	A single surface spring, small flow, temp. 50 ° C	13 drillholes, max. depth 590 m, natural flow.
Namafjall	60	less than 260	Intense furmarolic activity, temp. 100 ° C	3 drillholes, max. depth 680 m, natural flow.
Selfoss	80	94	Insignificant warm spring that disappeared at the beginning of the century	8 drillholes, max. depth 500 m. Practically all the water is recovered by pumping.
Hengill	600	less than 260	Springs and fumarolic activity, temp. 100 ° C	18 deep drillholes, max. depth 1,200 m, many shallow wells. Natural flow.
Krysuvik	20	220	Intense fumarolic activity, temp. 100 ° C	4 deep drillholes, max. depth 1,200 m, many shallow wells, natural flow.
Total	1,480			

TABLE 1-51. ICELAND—DISCHARGE OF THERMAL SPRINGS

(Source: Arnorsson, et al., XXIII Internat. Geol. Congress, 1969)

[In main low temperature areas]

Area	Total flow l/s	Max. surface temp. °C	Remarks
Southern Lowland	330	100	Many shallow drillholes Total discharge 35 l/s
Reykholtsdalur	300	100	
NW. Peninsula	250	100	One drillhole in the largest group of springs issues 20 l/s
Middle N. Iceland	130	89	One deep drillhole at Laugaland (1,080 m), issues 8 l/s, temp. 93 ° C
NE. Iceland	145	100	
Other areas	50		
Total	1,205		

TABLE 1-52. ICELAND–CHEMICAL ANALYSES OF REPRESENTATIVE THERMAL WATERS

(Source: Arnorsson, et al., XXIII Internat. Geol. Congress, 1969)

[values in ppm]

Location	Akranes drillhole IV	Selfoss drillhole VI	Reykjanes drillhole I	Eyjafjordur Reykhus drillhole I	Reykholtsdalur Skrifta	Namafjall drillhole III	Geysir Geysir-Area	Torfajokull Eyrarhver	Torfajokull Raudihver	Torfajokull Gilshver
Temp. °C	26.6	76	100	76	100	> 260	88	94	94	75
pH	7.1	8.2	6.7	9.8	9.4	9.8	9.0	9.6	2.5	7.2
Eh (volts 25 °C)	+0.12	+0.04			− 0.04		− 0.13	− 0.20		+0.33
SiO_2	60	65	543	106	178	823	501	259	102	177
Na^+	1,360	215	13,800	57.5	70.8	177.0	250.0	370.0	3.8	87.5
K^+	17	4.5	1,920	0.8	3.8	25.5	25.0	19.2	2.9	18.5
Ca^{++}	560	37.2	2,200	3.2	1.42	1.2	0.9	0.0	11.9	60.3
Mg^{++}	12	2.8	45	0.5	0.13	0.0	0.0	0.02	6.6	16.0
Cl^-	3,017	363.0	27,400	13.3	34.2	18.2	127.0	360.0	15.2	5.1
F^-	0.1	0.8	0.4	0.6	2.2	1.1	9.5	25.0	0.4	0.3
OH^-	0.0	0.0	0.0	8.7	0.0	28.2	0.0	1.0	0.0	0.0
CO_3^{--}	0.0	3.0	0.0	26.4	46.6	93.0	70.0	148.2	0.0	0.0
HCO_3^-	12.0	18.0	5.0	0.0	0.0	0.0	133.0	12.8	0.0	503.3
SO_4^{--}	600.0	65.9	1,280.0	48.5	57.4	78.5	108.0	54.8	443.3	5.9
H_2S	< 0.05	0.14	0.2	0.0	1.6		0.7	24.9	0.0	< 0.05
Dissolved solids	6,120	720	47,500	272	429	1,397	1,152	1,351	718	599

TABLE 1-53. IRELAND–DISCHARGE OF SHANNON RIVER
(Source: UNESCO, 1971)

River and station	Drainage basin km2	Mean monthly discharge, m3/s												Year	Period of record
		Jan.	Feb.	Mar.	Apr.	May	Jun.	Jul.	Aug.	Sep.	Oct.	Nov.	Dec.		
Shannon, Killaloe	10,400	326	296	205	152	94.5	71.0	72.5	85.0	130	171	246	321	180	1935-65

TABLE 1-54. IRELAND–FRESH WATER DEMAND, 1961
(Source: R. Common, Queens University, Belfast, 1968)

Northern Ireland	Millions of liters	Millions of gallons	Republic of Ireland	Millions of liters	Millions of gallons
Domestic	70,960	15,606.20	Domestic	140,200	30,821.00
Agricultural	21,140	4,652.05	Agricultural	131,800	29,009.12
Basic Industrial	175,000	38,491.30	Basic Industrial	40,700	8,955.30
1961 Total Estimate	267,100	58,749.55	1961 Total Estimate	312,700	68,785.42

TABLE 1-55. IRELAND–MUNICIPAL WATER SUPPLY SYSTEMS
(Source: Internat. Statistical Institute, 1971)

City	Year	Population served	Quantity available (1000 m3)			Annual consumption (1000 m3)	
			Production	Purchased from other plants	Total	Total	of which domestic
Cork 1	1967	126,000	15,348	0	15,348	14,327	9,204
	1969	130,000	15,348	0	15,348	13,682	8,262
Dublin (m) 1	1967	764,350	183,000	1,000	193,000	66,327	53,100
	1969	802,800	188,800	22,200	211,000	71,873	56,900

(m) Metropolitan area
1 Some surrounding districts and suburbs included

TABLE 1-56. ITALY–DISCHARGE OF PRINCIPAL RIVERS
(Source: UNESCO, 1971)

River and station	Drainage basin km2	Mean monthly discharge, m3/s												Year	Period of record
		Jan.	Feb.	Mar.	Apr.	May	Jun.	Jul.	Aug.	Sep.	Oct.	Nov.	Dec.		
Po, Pontelagoscuro	70,091	1,200	1,220	1,500	1,630	1,900	1,860	1,250	937	1,210	1,600	1,970	1,500	1,480	1918-64
Po, Boretto	55,183	914	953	1,210	1,330	1,470	1,420	881	693	987	1,340	1,690	1,290	1,180	1942-44;1947-64
Po, Piacenza	42,030	652	697	901	962	1,400	1,320	826	610	813	1,010	1,330	879	950	1924-64
Tiber, Rome	16,545	301	343	325	272	242	168	137	123	134	177	264	324	234	1921-64
Adige, Boara Pisani	11,954	149	142	152	187	263	377	274	224	212	146	249	132	223	1951-64

TABLE 1-57. ITALY–AVAILABLE WATER RESOURCES BY RIVER BASIN AND REGION
(Source: Italian Hydrological Service, Water for Peace, 1967)
[Mean annual values]

		INFLOW			OUTFLOW				
	Precipitation mm	Recharge			Surface-water runoff				Coefficient of runoff
		l/sec/km2	109m3	% National recharge	mm	l/sec/km2	109m3	% National discharge	
Po River basin	1,070	33.9	71.8	24.2	663	21.0	46.4	29.2	0.62
Coastal basins in region of:									
Veneto & Venezia Guilia	1,160	36.8	42.8	14.4	908	28.8	33.6	21.1	0.78
Liguria	1,340	42.5	6.4	2.2	990	31.4	4.8	3.1	0.74
Romagna & Marche	940	29.8	20.6	7.0	436	13.8	9.6	6.0	0.46
Toscana	1,010	32.0	20.9	7.1	493	15.6	10.2	6.4	0.49
Lazio	1,020	32.3	24.1	8.1	458	14.5	10.8	6.8	0.45
Abruzzi & Molise	900	28.5	11.9	4.0	477	15.1	6.3	4.0	0.53
Campania	1,200	38.1	23.2	7.8	670	21.2	12.9	8.1	0.56
Puglie	660	20.9	13.2	4.5	150	4.7	2.9	1.8	0.23
Lucania	800	25.4	7.9	2.7	200	6.4	2.0	1.3	0.25
Calabria	1,170	37.1	16.1	5.4	568	18.0	7.9	5.0	0.49
Sicily	730	23.1	18.8	6.4	171	5.4	4.4	2.8	0.23
Sardinia	760	24.1	18.2	6.2	298	9.4	7.2	4.4	0.39
Total Italy	990	31.3	296.0	100.0	528	16.7	159.0	100.0	0.53

TABLE 1-58. ITALY–MUNICIPAL WATER DEMAND, 1967-2015

(Source: Italian Hydrological Service, Water for Peace, 1967)

[by region]

Region	Municipal Water Demand (1967) m³/s				Projected Municipal Water Demand (2015) m³/s			
	Spring water	Surface water	Ground water	Total	Spring water	Surface water	Ground water	Total
Piemonte-Val d'Aosta	4.3	1.9	11.2	17.4	5.6	11.4	14.0	31.0
Liguria	1.2	2.6	4.0	7.8	1.5	6.5	12.7	20.7
Lombardia	3.9	0.2	23.2	27.3	6.2	0.7	41.2	48.1
Trentino-Alto Adige	5.9	–	–	5.9	6.9	–	0.8	7.7
Veneto	3.4	1.7	10.8	15.9	3.9	4.1	14.9	22.9
Friuli-Venezia Giulia	1.2	0.2	2.6	4.0	1.4	0.3	5.6	7.3
Emilia-Romagna	1.3	1.6	7.4	10.3	1.6	6.2	10.3	18.1
Marche	1.9	–	1.3	3.2	4.9	0.2	1.7	6.8
Toscana	1.4	1.5	4.2	7.1	3.0	9.9	4.4	17.3
Umbria	1.3	–	0.2	1.5	2.5	0.7	0.2	3.4
Lazio	23.7	1.8	1.6	27.1	37.1	14.3	3.9	55.3
Campania	11.7	–	1.2	12.9	19.3	6.0	1.0	26.3
Abruzzo	2.7	–	0.6	3.3	4.8	–	0.7	5.5
Molise	0.9	–	–	0.9	1.2	0.2	–	1.4
Puglia	4.8	–	0.8	5.6	6.5	10.1	1.9	18.5
Basilicata	0.8	–	–	0.8	2.7	–	0.1	2.8
Calabria	3.6	–	0.4	4.0	4.3	2.5	1.3	8.1
Sicily	5.0	1.1	1.5	7.6	10.7	11.1	4.8	26.6
Sardinia	1.1	2.6	0.1	3.8	1.5	7.2	0.4	9.1
Total Italy	80.1	15.2	71.1	166.4	125.6	91.4	119.9	336.9
Percentages	48.1	9.1	42.8	100	37.3	27.1	35.6	100

FIGURE 1-5. ITALY – AQUEDUCTS OF ANCIENT ROME

(Source: Da Fabbretti; Martelli; Rome Health Dept., 1968)

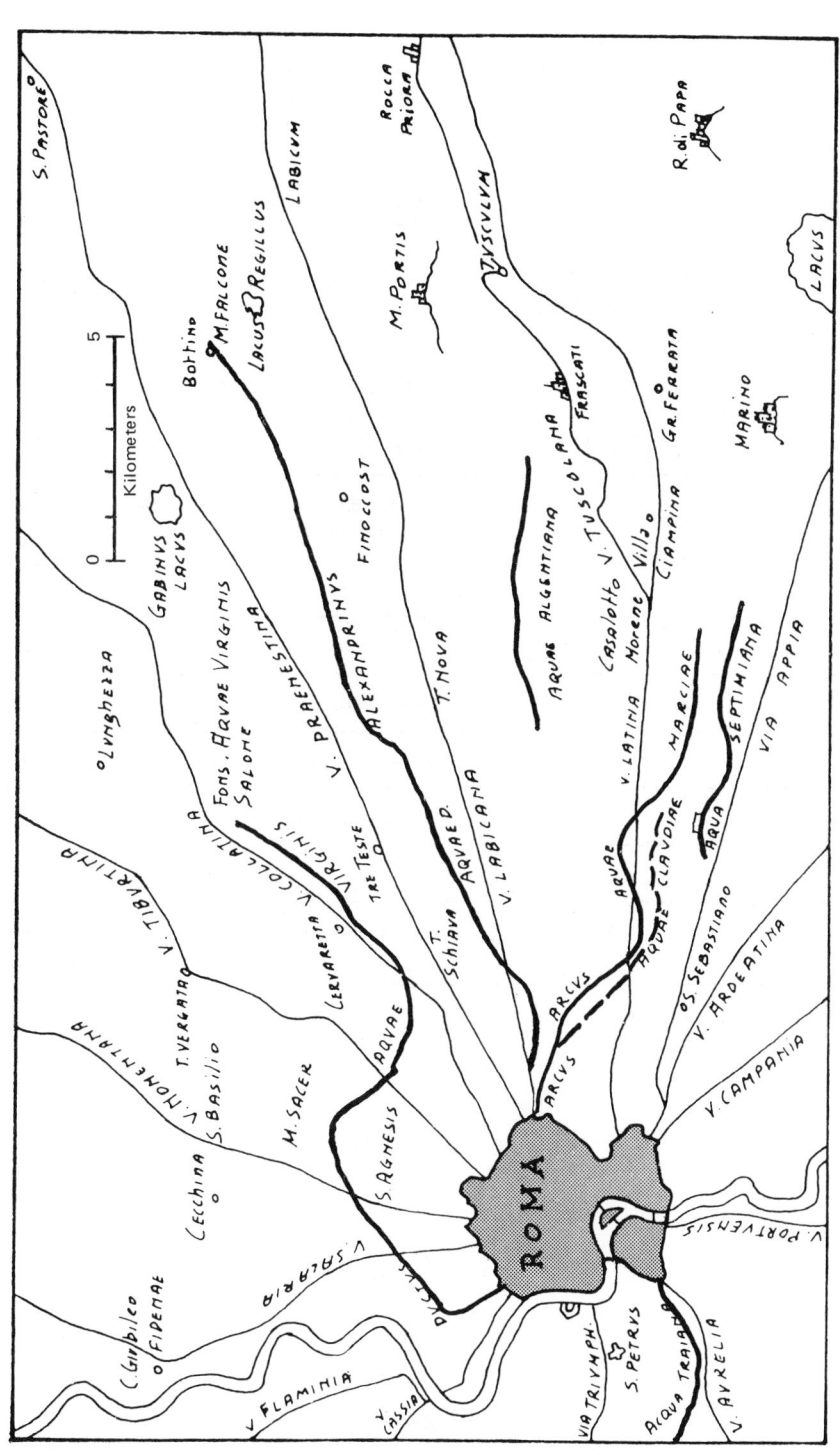

TABLE 1-59. ITALY—AQUEDUCTS OF ANCIENT ROME
(Source: Martelli, Rome Health Dept., 1968)

Name of aqueduct	Date constructed	Name of designer	Carrying capacity	
			Quinariae [1]	l/sec
Appia	312 B.C.	Appio Claudius	1,825	875
Anio Vetus	272-269 "	Manlio Curio Dentato and Flavio Flacco	4,398	2,100
Marcia	144-130 "	Quinto Marcio	4,690	2,250
Tepula	125 "	Servilio Cepione and Cassio Longino	445	214
Julia	33 "	Marcus Agrippa	1,206	577
Virgo	19 "	Marcus Agrippa	2,504	1,200
Alsietina	2 "	Caesar Augustus	392	188
Claudia	38-52 A.D.	Caligula - Claudius	4,607	2,210
Anio Novus	38-52 "	Caligula - Claudius	4,738	2,270
Trajana	109 "	Trajanus	2,848	1,367
Alexandrina	226 "	Alexander Severus	529	254

[1] A Quinaria is equal to 0.47-0.48 l/sec.

TABLE 1-60. LUXEMBOURG—DATA ON PRINCIPAL AQUIFER

(Source: Barthel, Luxembourg Public Health Service, 1965)

[Aquifer yields 90% of national potable water demand] *

Aquifer	Outcrop area km2	Mean Annual		
		Precipitation over outcrop area mm	Recharge million m3	Pumpage million m3
Gres de Luxembourg (Liassic sandstone)	300	750	60-65	22-26

* Editor's note: In 1969 the Esch/Sauer dam and reservoir were completed. This water will gradually replace ground water as source of drinking water.

TABLE 1-61. LUXEMBOURG—WATER DEMAND, 1961-75
(Source: Barthel, Luxembourg Public Health Service, 1965)

[in m3/day]

Category	1961	1975
Domestic	60,500	70,000
Industrial	17,400	50,000
Total potable	77,900	120,000
Industrial (Non-potable)	28,700	—

TABLE 1-62. LUXEMBOURG—MUNICIPAL WATER SUPPLY SYSTEM

(Source: Internat. Statistical Institute, 1971)

City	Year	Quantity available (1000m3)				Annual consumption(1000m3)	
		Population served	Production	Purchased from other plants	Total	Total	of which domestic
Luxembourg-Ville 1	1967	83,621	9,994	0	9,994	8,555	—
	1969	85,092	9,417	287	9,704	9,279	—

1 Some surrounding districts and suburbs included

— Not available

TABLE 1-63. MALTA—AVAILABLE GROUND-WATER RESOURCES AND WATER DEMAND IN 1970

(Source: United Nations, 1964)

Island	Available fresh ground water per year		Estimated water demand 1970	
	Millions of imperial gallons	Cubic meters	Millions of imperial gallons	Cubic meters
Malta	4,180.0	19,000,000	4,202	19,100,000
Gozo	145.9 [1]	663,000	220	1,000,000
Comino	0.35	1,600	33	150,000
Total	4,326.25	19,664,600	4,455	20,250,000

1 An additional source of 26.4 million imperial gallons (120,000 m^3) of fresh ground water per year may eventually be developed.

TABLE 1-64. MALTA—WITHDRAWAL AND USE OF WATER

(Source: United Nations, 1964)

WATER WITHDRAWAL ON MALTA AND GOZO, 1959-62

	Millions of imperial gallons	Cubic meters
1959-60	3,080	14,000,000
1960-61	3,080	14,000,000
1961-62	2,970	13,500,000

PEAK MONTHLY WATER CONSUMPTION

	Millions of imperial gallons	Cubic meters
Malta, July 1961	269.8	1,226,363
Gozo, May 1961	12.3	55,909
Comino, August 1961	0.045	204

PERCENTAGE WATER USE, 1960-61

	Malta	Gozo
Household	72	70
Commercial and industrial	27	15
Irrigation	1	15

TABLE 1-65. NETHERLANDS–DISCHARGE OF SELECTED RIVERS
(Source: Rijkswaterstaat, 1973)

River and station	Basin area km²	Mean monthly discharge, m³/s												Max monthly flow m³/s	Date	Min monthly flow m³/s	Date	Period of record
		Jan.	Feb.	Mar.	Apr.	May	Jun.	Jul.	Aug.	Sep.	Oct.	Nov.	Dec.					
Rhine, Lobith (German border)	—	2,594	2,813	2,642	2,775	2,562	2,517	2,228	2,020	1,910	1,656	1,776	2,569	6,166	Dec. 1965	808	Feb. 1963	1960-71
Maas (Meuse), Lith	28,950	553	591	444	429	304	193	163	151	163	184	323	572	1,500	Dec. 1965	34	Jul. 1964	1960-71
Maas, Borgharen (Belgian border)	—	435	465	324	311	190	115	87	83	91	112	245	453	1,308	Dec. 1965	6	Oct. 1971	1960-71

TABLE 1-66. NETHERLANDS–WATER USE, 1967
(Source: Rijkswaterstaat, 1973)
[in million m3/yr]

Category	Withdrawal
Population........................	540
Industry.........................	840
Agriculture......................	1,000 [1]
Flushing.........................	6,000
Water needed to control salt-water intrusion in the Nieuwe Waterweg (Rotterdam Waterway)........	15,000
Total withdrawal	23,380

[1] Computed for average summer; includes water needed for control of water levels.

TABLE 1-67. NETHERLANDS—AVAILABLE WATER RESOURCES AND PROJECTED WATER DEMAND, 1980-2000

(Source: Rijkswaterstaat, 1964)

[Relative frequency of occurrence 50%]

Source	m³/s	mm	10⁹m³
Average precipitation	950	750	30
Discharge River Rhine	2,200	1,750	70
Discharge River Meuse	220	175	7
Discharge small rivers	105	83	3.3
Total	3,475	2,758	110.3

Needs and losses	m³/s		mm		10⁹m³	
	1980	2000	1980	2000	1980	2000
Average evaporation	630	630	500	500	20	20
Domestic water use	19	33.5	15	27	0.6	1.1
Industrial use	51	86.5	40	69	1.6	2.7
Flushing of "boezems", control of water levels and necessary discharge to main rivers to combat salt-water intrusion	1,680	1,680	1,335	1,335	53	53
Additional flushing due to supplementary irrigation	35	44	28	35	1.1	1.4
Excess	1,060	1,001	840	792	34	32.1
Total	3,475	3,475	2,758	2,758	110.3	110.3

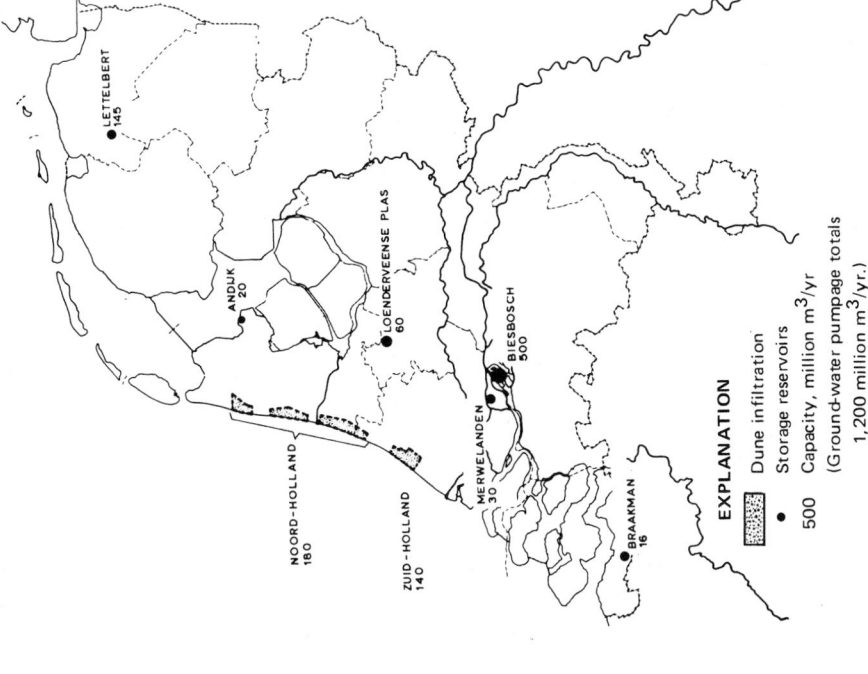

FIGURE 1-7. NETHERLANDS – WATER MANAGEMENT SYSTEMS IN 1972

(Source: Rijksinstituut voor Drinkwatervoorziening, 1973)

LETTELBERT
145

ANDIJK
20

LOENDERVEENSE PLAS
60

BIESBOSCH
500

MERWELANDEN
30

NOORD-HOLLAND
180

ZUID-HOLLAND
140

BRAAKMAN
16

EXPLANATION

Dune infiltration

Storage reservoirs

500   Capacity, million m³/yr
(Ground-water pumpage totals
1,200 million m³/yr.)

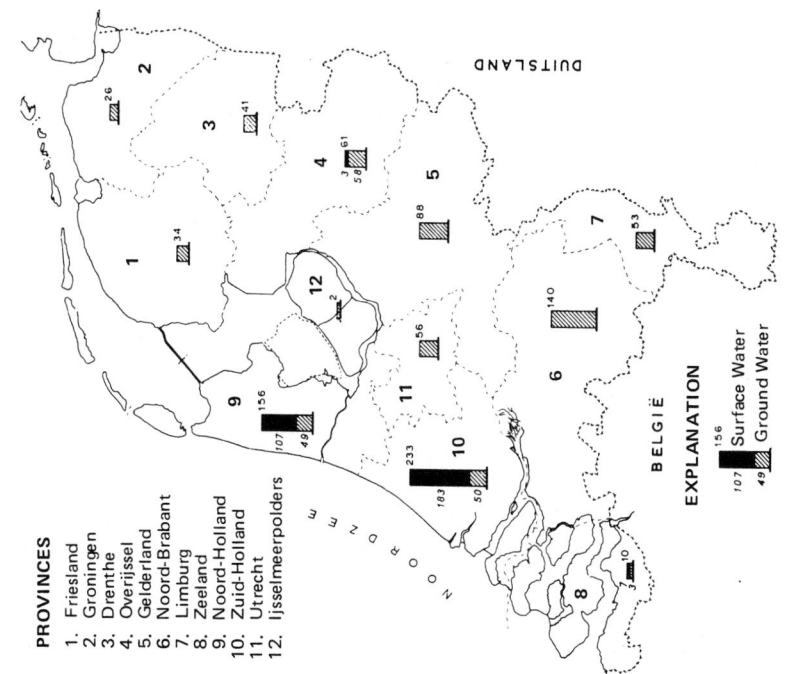

FIGURE 1-6. NETHERLANDS – WITHDRAWAL OF SURFACE AND GROUND WATER BY WATER-SUPPLY SYSTEMS IN 1971

(Source: Rijksinstituut voor Drinkwatervoorziening, 1973)

[withdrawals in million m³, by province]

PROVINCES

1. Friesland
2. Groningen
3. Drenthe
4. Overijssel
5. Gelderland
6. Noord-Brabant
7. Limburg
8. Zeeland
9. Noord-Holland
10. Zuid-Holland
11. Utrecht
12. Ijsselmeerpolders

DUITSLAND

NOORDZEE

BELGIË

EXPLANATION

156 Surface Water
107 49 Ground Water

2   26

3   41

4   3 58 61

5   88

7   53

1   34

12   2

11   56

6   140

9   156 107 49

10   233 183 50

8   3 10

TABLE 1-68. NETHERLANDS—DATA ON PUBLIC AND INDUSTRIAL
WATER SUPPLY SYSTEMS, 1952-67

(Source: Huisman, J. Inst. Water Engineers, Nov. 1970)

	1952		1957		1962		1967
Population, in millions	10.42		11.09		11.88		12.66
Increase, % per year		1.2		1.4		1.3	
Percentage served	83.5		90.8		96.2		98.8
Dwellings, in millions	2.35		2.66		3.01		3.48
Increase, % per year		2.5		2.5		2.9	
Percentage served	85.0		91.8		96.7		98.9
Percentage metered	54.6		58.1		63.7		70.5
Number of public water supply systems:							
Practicing production and distribution ..	131		125		113		106
Practicing distribution only	64		61		58		41
Total	195		186		171		147
Decrease, % per year		0.9		1.7		3.1	
Public supply, in mgd, of:							
Ground water	106		144		190		266
Dune water	47		29		26		32
Artificial ground and dune water.......	2		28		50		64
Surface water	47		57		63		81
Total	202		258		329		443
Increase, % per year................		5.0		5.0		6.1	
Public supply, in mgd, for:							
Industrial purposes	43		65		93		137
Non-industrial purposes	159		193		236		306
Financial return, in million pound sterling	7.3		11.8		17.9		33.4
Increase, % per year................		9.9		8.7		12.6	
Per capita per day consumption, in gal., for:							
Industrial purposes	5.1		6.5		8.2		10.9
Non-industrial purposes	18.2		19.2		20.6		24.5
Total	23.3		25.7		28.8		35.4
Increase, % per year................		2.0		2.1		4.3	
Industrial water supply, in mgd, for:							
Ground water			222		262		285
Surface water			502		871		1,257
Total			724		1,133		1,542
Increase, % per year................				9.4		6.4	
Industrial water supply, in mgd, for cooling:							
With ground water.................					122		166
With surface water.................					762		1,142
Total					884		1,308
Cooling water, in mgd, for public electricity supply..................			1,770		2,340		3,360

TABLE 1-69. NETHERLANDS—DISCHARGE INTO THE NORTH SEA OF HEAVY METALS AND METALLOIDS FROM THE RHINE RIVER

(Source: Weichart, AMBIO, April 1973)

Metal	Discharge from Rhine River (metric tons per year)	Concentration in Rhine Water/Concentration in Oceanic Water
Chromium	1,000	20
Manganese	6,000	30
Iron	80,000	300
Nickel	2,000	10
Copper	2,000	10
Zinc	20,000	40
Cadmium	200	40
Mercury	100	20
Lead	2,000	700
Arsenic	1,000	5

TABLE 1-70. NETHERLANDS—MUNICIPAL WATER SUPPLY SYSTEMS

(Source: Internat. Statistical Institute, 1971)

City	Year	Population served	Quantity available (1000m^3) Production	Quantity available (1000m^3) Purchased from other plants	Quantity available (1000m^3) Total	Annual consumption (1000m^3) Total	Annual consumption (1000m^3) of which domestic
Amsterdam [1]	1967	962,000	78,629	0	78,629	78,629	67,576
	1969	943,000	82,744	0	82,744	82,744	72,614
Apeldoorn	1967	118,694	5,910	66	5,976	5,147	5,147
	1969	123,628	6,988	99	7,087	5,758	5,758
Arnheim	1967	135,090	7,669	0	7,669	6,721	3,730
	1969	132,531	8,846	0	8,846	7,720	—
Breda [1]	1969	131,000	5,506	227	5,733	5,253	3,681
Eindhoven [1]	1967	232,606	14,119	60	14,179	14,109	5,857
	1969	238,782	16,634	59	16,693	16,196	6,254
Enschede	1967	136,242	5,703	0	5,703	5,504	3,302
	1969	137,834	6,309	0	6,309	6,198	3,720
Groningen [1]	1967	180,503	11,760	232	11,992	11,992	7,579
	1969	193,716	12,656	174	12,830	12,830	7,943
Haarlem	1967	172,953	7,067	13	7,080	6,581	5,061
	1969	172,475	7,376	16	7,392	7,052	5,476
Hilversum	1967	101,983	7,632	0	7,632	—	—
	1969	99,811	7,911	0	7,911	—	—
Leiden [1]	1967	183,443	9,911	0	9,911	8,507	—
	1969	210,041	11,430	0	11,430	10,443	—
Nijmegen	1967	145,455	10,372	0	10,372	7,592	—
	1969	148,790	11,017	0	11,017	8,611	—
The Hague [1]	1967	582,714	37,910	0	37,910	24,211	—
	1969	558,364	42,751	0	42,751	24,520	—
Tilburg	1967	161,142	10,718	0	10,718	10,718	4,120
	1969	164,013	11,804	0	11,804	11,804	6,190
Utrecht	1967	274,329	—	—	—	14,894	8,978
	1969	277,673	—	—	—	16,004	9,474

[1] Includes some surrounding districts or suburbs.
— Not available

TABLE 1-71. NORWAY–DISCHARGE OF GLOMMA RIVER

(Source: ECE, 1970)

[Period of record 1930-40 and 1945-65]

River, station	Mean monthly discharge, m^3/s												Six-monthly average		Yearly average	Mean annual runoff l/s/km^2
	Jan.	Feb.	Mar.	Apr.	May	Jun.	Jul.	Aug.	Sep.	Oct.	Nov.	Dec.	Oct.-Mar.	Apr.-Sep.		
Glomma, Langnes	209	158	178	526	1,677	1,476	1,126	874	787	652	447	314	331	1,078	704	17.5

TABLE 1-72. NORWAY–ESTIMATED WATER USE, 1972

(Source: Hjelm-Hansen, Norwegian Water Resources and Electricity Board, 1973)

Category	Water use (Million m^3)
Municipal water supply (300 water systems)	200
Industrial	1,200
Agriculture	30
Total	1,430

TABLE 1-73. NORWAY—MAJOR RESERVOIRS FOR POWER GENERATION

(Source: Norwegian Water Resources and Electricity Board, 1973)

[Data as of 1973]

	Name of Reservoir	Reservoir Capacity Billion m^3
Existing	Rosvatn	2.36
	Mjosa	1.31
	Akersvatn	1.28
	Mosvatn	1.06
	Altevatn	1.03
	Roskreppfijorden	0.70
	Songavatn	0.64
	Strandevatn	0.62
	Aursjoen	0.56
	Namsvatn	0.46
	Randsfjorden	0.41
Planned or Under Construction	Blasjo	3.36
	Svartevatn	1.40

TABLE 1-74. NORWAY—MUNICIPAL WATER SUPPLY SYSTEMS

(Source: Internat. Statistical Institute, 1971)

City	Year	Population supplied	Annual consumption (1000 m^3)	
			Total	of which domestic
Bergen (m)	1967	116,939	23,777	17,763
	1969	115,718	24,876	19,049
Oslo	1967	486,750	75,600 *	26,200 *
	1969	488,500	83,700 *	26,700 *
Trondheim	1967	120,900	21,700	9,750 *
	1969	126,100	22,250	10,000 *

(m)   Metropolitan area
*   Provisional

TABLE 1-75. POLAND–DISCHARGE OF PRINCIPAL RIVERS
(Source: UNESCO, 1971)

River and station	Drainage basin area km²	Mean monthly discharge, m³/s												Year	Period of record
		Jan.	Feb.	Mar.	Apr.	May	Jun.	Jul.	Aug.	Sep.	Oct.	Nov.	Dec.		
Oder, Gozdowice	109,360	538	591	721	728	556	454	398	365	332	363	451	481	498	1921-37;1946-65
Oder, Gozdowice		576	611	579	787	590	455	405	377	368	380	448	515	522	1901-65
Vistula, Tczew	193,870	816	911	1,480	1,890	1,080	854	793	766	686	715	870	832	974	1921-37;1946-65
Vistula, Tczew		895	982	1,570	1,920	1,160	857	835	828	710	710	827	856	1,010	1901-65
Vistula, Warsaw	84,695	436	504	856	932	544	520	533	446	395	397	482	420	539	1921-37;1946-65
Bug, Wyszkow	38,665	125	127	218	336	152	98.0	83.2	85.3	78.9	94.2	122	126	137	1921-37;1951-65

TABLE 1-76. POLAND–WATER BALANCE OF PRINCIPAL RIVER BASINS
(Source: Mikulski, An Outline of Poland's Hydrography, 1968)

Basin and area within Poland km²	Units	Winter (November through April)				Summer (May through October)				Year (November through October)			
		Precipi-tation	Runoff	Evapo-transpi-ration	Runoff Coeffi-cient	Precipi-tation	Runoff	Evapo-transpi-ration	Runoff Coeffi-cient	Precipi-tation	Runoff	Evapo-transpi-ration	Runoff Coeffi-cient
Vistula 173,066	10⁶ m3	36.9	17.7	19.2	0.48	67.6	12.2	55.4	0.18	104.5	29.9	74.6	0.29
	mm	213.2	102.4	110.8		390.6	70.6	320.0		603.8	173.0	430.8	
Odra 107,586	10⁶ m3	24.7	9.9	14.8	0.36	38.1	6.2	31.9	0.16	62.8	16.1	46.7	0.26
	mm	229.1	91.9	137.2		354.4	57.6	296.8		583.5	149.5	434.0	
Coastal Region 31,078	10⁶ m3	7.8	4.6	3.2	0.59	12.2	2.8	9.4	0.23	20.0	7.4	12.6	0.37
	mm	249.5	145.6	103.9		393.7	91.2	302.5		643.2	236.8	406.4	
Poland 311,730	10⁶ m3	69.4	32.2	37.2	0.46	117.9	21.2	96.7	0.18	187.3	53.4	133.9	0.29
	mm	233.3	103.1	120.2		378.5	68.2	310.3		601.8	171.3	430.5	

TABLE 1-77. POLAND—WATER BALANCE
(Source: Mikulski, An Outline of Poland's Hydrography, 1968)

Available water	km^3	%	Losses of water	km^3	%
Precipitation	187.2	97.3	Runoff through river channels to the sea:		
			a) surface-water runoff	24.6	12.9
			b) ground-water runoff	32.7	17.1
			c) pumped water	1.3	0.6
River inflow from adjacent countries	5.2	2.7	Evaporation, transpiration and economic use:		
			a) evaporation from land surface	133.5	69.2
			b) consumption of pumped water	0.3	0.2
Total	192.4	100.0		192.4	100.0

TABLE 1-78. POLAND—MUNICIPAL WATER SUPPLY SYSTEMS
(Source: Internat. Statistical Institute, 1971)

City	Year	Population served	Quantity available (1,000m^3) Production	Quantity available (1,000m^3) Purchased from other plants	Quantity available (1,000m^3) Total	Annual consumption (1,000m^3) Total	Annual consumption (1,000m^3) of which domestic
Bialystok [1]	1967	146,600	11,025	0	11,025	10,186	3,849
	1969	163,000	12,399	0	12,399	11,618	4,518
Bielsko-Biala [1]	1967	—	19,803	0	19,803	16,105	5,586
	1969	103,000	24,405	0	24,405	21,501	7,190
Bydgoszcz [1]	1967	262,500	18,146	0	18,146	16,863	9,406
	1969	279,000	22,900	0	22,900	21,089	12,206
Bytom	1967	191,200	0	30,057	30,057	27,156	10,858
	1969	187,000	0	30,089	30,089	27,220	12,296
Chorzow	1967	154,000	0	23,496	23,496	22,563	6,080
	1969	151,000	0	23,596	23,596	23,047	6,328
Czestochowa [1]	1967	178,300	18,430	326	18,756	16,390	7,387
	1969	186,000	19,097	95	19,192	16,798	7,559
Gdansk	1967	329,900	30,365	231	30,596	27,537	16,591
	1969	370,000	34,530	82	34,612	30,750	18,394
Gdynia	1967	170,400	14,783	0	14,783	12,487	7,339
	1969	182,000	16,482	0	16,482	13,011	7,663
Gliwice [1]	1967	164,300	15,639	8,422	24,061	22,202	9,630
	1969	168,000	16,763	8,950	15,715	23,800	11,664

TABLE 1-78. POLAND—MUNICIPAL WATER SUPPLY SYSTEMS (continued)

City	Year	Population served	Quantity available(1000m³)			Annual consumption(1000m³)	
			Production	Purchased from other plants	Total	Total	of which domestic
Gliwice [1]	1967	164,300	15,639	8,422	24,061	22,202	9,630
	1969	168,000	16,763	8,950	25,715	23,800	11,664
Katowice	1967	289,700	0	38,767	38,767	33,514	16,653
	1969	296,000	0	41,078	41,078	35,505	18,365
Kielce	1967	111,700	7,234	0	7,234	8,002	4,349
	1969	121,000	8,953	0	8,953	8,438	4,339
Krakow	1967	554,800	52,500	0	52,500	45,800	22,700
	1969	576,500	56,000	0	56,000	48,800	25,300
Lublin	1967	209,700	15,635	0	15,635	14,451	6,860
	1969	238,000	18,089	0	18,089	16,074	7,957
Lodz [1]	1967	745,400	56,400	0	56,400	51,400	26,000
	1969	752,700	62,800	0	62,800	57,300	30,500
Poznan	1967	453,300	38,100	0	38,100	34,800	16,500
	1969	462,100	43,300	0	43,300	39,400	18,700
Radom	1967	147,100	10,072	0	10,072	9,518	5,850
	1969	154,000	10,825	0	10,825	10,252	6,723
Ruda Slaska	1967	142,600	0	15,754	15,754	12,873	5,679
	1969	141,000	0	15,239	15,239	12,769	5,852
Sosnowiec	1967	142,500	0	19,602	19,602	18,388	9,585
	1969	144,000	0	22,277	22,277	20,080	9,653
Szczecin	1967	319,800	29,248	0	29,248	25,691	16,067
	1969	335,000	31,904	0	31,904	28,093	17,668
Torun	1967	117,200	8,618	0	8,618	7,967	3,893
	1969	126,000	10,100	0	10,100	9,185	4,420
Walbrzych [1]	1967	126,500	13,152	0	13,152	11,217	6,201
	1969	126,000	12,735	0	12,735	10,905	5,729
Warsaw	1967	1,266,700	143,000	0	143,000	130,400	80,400
	1969	1,288,400	160,700	0	160,700	144,700	90,800
Wroclaw	1967	506,100	42,400	0	42,400	37,300	21,200
	1969	517,400	43,600	0	43,600	39,700	23,100
Zabrze	1967	196,900	0	20,978	20,978	18,903	9,758
	1969	200,000	0	21,512	21,512	18,913	9,819

1 Including some surrounding districts and suburbs.
— Not available

TABLE 1-79. PORTUGAL–DISCHARGE OF SELECTED RIVERS

(Source: UNESCO, 1971)

River and station	Basin area km2	Mean monthly discharge, m3/s												Year	Period of record
		Jan.	Feb.	Mar.	Apr.	May	Jun.	Jul.	Aug.	Sep.	Oct.	Nov.	Dec.		
Douro, Regua	91,491	998	1,070	1,020	786	516	297	151	114	131	204	468	707	496	1932-66
Tagus, V.V. de Rodao	59,167	540	694	656	392	217	110	44	28	40	118	307	535	305	1905-66
Guadiana, Pulo do Lobo	60,883	350	530	511	234	90	40	17	14	15	51	135	293	202	1947-66
Mondego, Coimbra	4,957	173	193	134	126	72	35	11	3	6	26	78	143	82	1921-66

TABLE 1-80. PORTUGAL. PORTUGAL–WATER BALANCE OF SELECTED RIVER BASINS

(Source: Quintela, Recursos de Aguas Superficiais em Portugal Continental, 1967)

River	Mean Annual					Period of record
	Precipitation mm	Runoff mm	Deficit of runoff mm	Coefficient of runoff	Temperature °C	
Rabagao ar Venda Nova	1,687	1,085	602	0.64	10.5	1944-52
Tamega at Marco de Canaveses	1,614	841	773	0.52	12.0	1955-63
Tua	1,112	509	603	0.46	12.5	1955-63
Sabor at Quinta das Laranjeiras	791	247	544	0.31	12.5	1938-60
Paiva at Castro Daire	1,482	768	714	0.51	10.5	1944-61
Mondego at Ponte de Tabua	1,054	391	663	0.37	12.0	1937-60
Alva at Ponte da Mucela	1,183	628	555	0.53	10.5	1945-60
Ponsul at Cabeco Monteiro	822	311	511	0.38	15.0	1934-45
Zezere at Cabril	1,304	692	612	0.53	11.5	1934-50
Ocreza at Almourao	910	404	506	0.44	13.0	1941-60
Raia at Cabecao	670	142	528	0.21	15.5	1935-54
Degebe at Amieira	561	92	469	0.16	16.0	1944-60
Odivelas at Odivelas	594	129	465	0.22	16.0	1940-60
Sado at Moinho da Gamitinha	581	97	484	0.17	16.0	1934-51
Arade at Casa Queimada	752	251	501	0.33	17.0	1935-51

TABLE 1-81. PORTUGAL—MUNICIPAL WATER SUPPLY SYSTEMS

(Source: Internat. Statistical Institute, 1971)

City	Year	Population served	Annual consumption (1000m³) Total	of which domestic
Lisboa	1967	824,800	60,350	17,856
	1969	828,000	63,289	21,759
Porto	1967	322,800	12,663	7,285
	1969	324,000	13,318	7,946

TABLE 1-82. ROMANIA—DISCHARGE OF SELECTED RIVERS

(Source: UNESCO, 1971)

River and station	Drainage basin area km2	Jan.	Feb.	Mar.	Apr.	May	Jun.	Jul.	Aug.	Sep.	Oct.	Nov.	Dec.	Year	Period of record
Danube, Orsova	576,232	4,464	4,714	6,403	7,698	7,632	6,627	5,533	4,518	3,911	3,910	4,762	4,953	5,427	1838-65
Danube, Ceatal Izmail	807,000	5,713	5,768	6,246	8,437	8,798	8,330	7,026	5,356	4,405	4,147	5,017	5,758	6,250	1928-65
Somes, Satu Mare	15,155	91.8	143	220	232	167	116	93.1	77.0	50.6	57.8	76.4	94.1	115	1933-64
Mures, Arad	27,061	104	131	209	280	302	218	150	105	80.1	76.6	96.1	104	155	1933-65
Olt, Izbiceni	24,203	101	104	238	251	300	236	122	93.4	68.3	97.2	108	96.9	151	1961-65
Siret, Lungoci	36,122	81.9	102	160	308	298	259	153	157	105	84.3	83.5	84.8	156	1951-65

TABLE 1-83. ROMANIA—MUNICIPAL WATER SUPPLY SYSTEMS
(Source: Internat. Statistical Institute, 1971)

City	Year	Population served	Quantity available(1000m³)			Annual consumption(1000m³)	
			Production	Purchased from other plants	Total	Total	of which domestic
Arad	1967	128,427	7,396	0	7,396	6,784	2,208
	1969	135,181	9,295	0	9,295	8,664	3,072
Braila	1967	140,390	9,989	0	9,989	9,031	4,266
	1969	149,686	12,498	0	12,498	11,404	5,891
Brasov	1967	168,582	19,060	0	19,060	18,257	6,891
	1969	179,316	20,919	0	20,919	20,306	8,356
Bucharest	1967	1,390,496	202,972	160	203,132	188,093	71,881
	1969	1,457,802	236,438	1,785	238,223	213,605	83,140
Cluj	1967	188,610	18,604	0	18,604	16,384	5,763
	1969	197,902	21,634	0	21,634	19,066	6,560
Constanta	1967	156,821	24,287	0	24,287	22,184	5,691
	1969	170,026	28,952	0	28,952	26,678	6,919
Craiova	1967	154,156	20,937	0	20,937	18,041	4,822
	1969	171,676	31,556	0	31,556	28,318	9,336
Galati (m)	1967	154,390	20,737	0	20,737	18,864	7,239
	1969	172,687	23,934	0	23,934	21,699	8,707
Iasi	1967	166,169	27,570	0	27,570	25,256	7,561
	1969	179,405	32,447	0	32,447	29,365	8,335
Oradea	1967	127,120	15,515	0	15,515	14,055	3,564
	1969	135,361	18,516	0	18,516	15,988	4,072
Ploesti	1967	150,897	13,431	302	13,733	12,580	4,651
	1969	160,011	16,062	459	16,521	15,746	6,171
Sibiu	1967	112,662	6,065	0	6,065	6,065	2,401
	1969	118,893	9,444	0	9,444	9,444	3,177
Timisoara	1967	178,787	16,450	0	16,450	14,759	5,039
	1969	189,264	20,211	0	20,211	18,408	6,641

(m) Metropolitan area

TABLE 1-84. SPAIN–DISCHARGE OF SELECTED RIVERS
(Source: UNESCO, 1971)

River and station	Basin area km2	Jan.	Feb.	Mar.	Apr.	May	Jun.	Jul.	Aug.	Sep.	Oct.	Nov.	Dec.	Year	Period of record
Duero, Villachica	41,856	696	267	271	225	160	100	50	28	35	56	111	195	183	1920-65
Guadalquivir, Alcala del Rio	46,995	359	441	454	273	139	57	34	30	40	75	154	209	189	1931-65
Jucar, Masia del Mompo	17,876	56	68	68	58	56	50	41	37	41	44	49	52	52	1911-65
Ebro, Zaragoza	40,434	412	444	437	337	247	151	57	28	45	117	225	386	240	1913-65
Ebro, Tortosa	84,230	723	776	888	741	700	586	266	149	193	334	552	720	552	1913-65

Mean monthly discharge, m3/s

TABLE 1-85. SPAIN–AVAILABLE WATER RESOURCES, BY BASIN
(Source: M.R. Llamas and Direccion de Obras Hidraulicas, 1969)

Basin		North	Duero	Tajo	Guadiana	Guadal-quivir	South	Segura	Jucar	Ebro	Eastern Pyrenees	Total
Area	km2	53,430	79,330	56,750	60,270	61,060	20,860	16,160	43,090	86,000	16,500	493,500
1. Rainfall	hm3	72,900	48,100	36,500	33,400	36,500	10,200	7,050	22,100	52,100	12,300	331,050
	m	1.36	0.61	0.64	0.53	0.60	0.50	0.43	0.51	0.60	0.74	0.67
2. Water resources												
2.1 Surface runoff	hm3	32,000	8,150	7,525	3,895	4,800	1,900	580	2,250	14,000	1,200	76,300
	m	0.60	0.10	0.13	0.06	0.08	0.09	0.04	0.05	0.16	0.07	0.15
2.2 Ground-water discharge into streams	hm3	5,500	3,000	1,700	1,000	1,800	250	300	700	3,400	500	18,150
	m	0.10	0.04	0.03	0.02	0.03	0.015	0.02	0.02	0.04	0.03	0.04
2.3 Ground-water discharge into sea	hm3	1,100	—	—	60	600	300	160	800	80	250	3,350
	m	0.02	—	—	0.001	0.01	0.015	0.01	0.02	0.001	0.015	0.01
2.4 Total stream run-off (2.1+2.2)	hm3	37,500	11,150	9,225	4,895	6,600	2,150	880	2,950	17,400	1,700	94,450
	m	0.70	0.14	0.16	0.08	0.11	0.10	0.05	0.07	0.02	0.10	0.19
2.5 Total ground-water discharge (2.2+2.3)	hm3	6,600	3,000	1,700	1,060	2,400	550	460	1,500	3,480	750	21,500
	m	0.72	0.04	0.03	0.02	0.04	0.03	0.03	0.04	0.04	0.05	0.04
2.6 Gross water supply (2.1+2.2+2.3)	hm3	38,600	11,150	9,225	4,995	7,200	2,450	1,040	3,750	17,480	1,950	97,800
	m	0.72	0.14	0.16	0.08	0.12	0.12	0.06	0.09	0.20	0.12	0.20
3. Actual evapo-transpiration	hm3	34,300	36,950	27,275	28,445	29,300	7,750	6,010	18,350	34,520	10,350	233,250
	m	0.64	0.47	0.48	0.45	0.48	0.38	0.37	0.42	0.40	0.62	0.47
4. Estimated ground-water reserves assuming a drawdown of 50 m	m	25,000	60,000	30,000	40,000	50,000	10,000	15,000	40,000	30,000	8,000	358,000

FIGURE 1-8. SPAIN – WATER RESOURCES BY RIVER BASIN

(Source: Llamas and Direccion de Obras Hidraulicas, 1969)

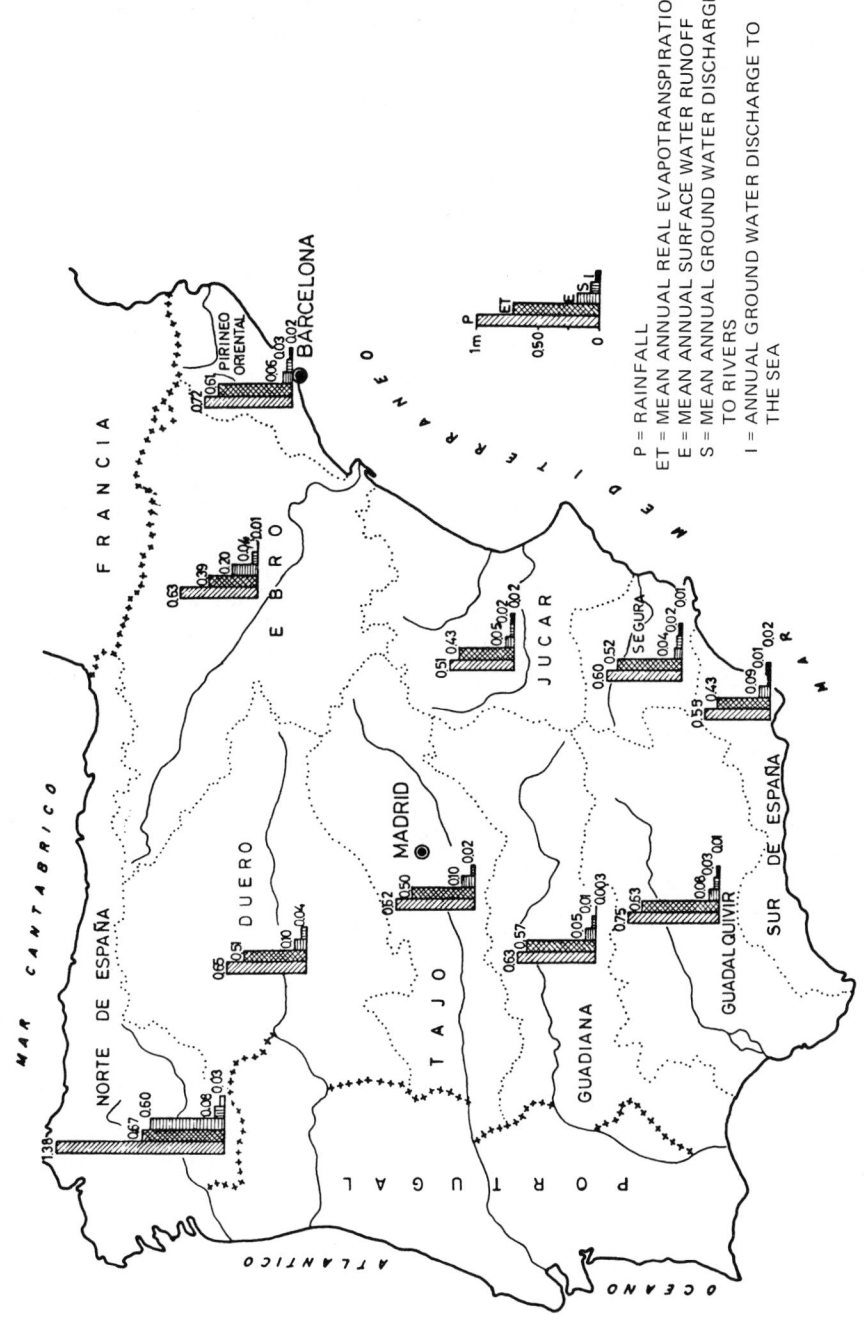

TABLE 1-86. SPAIN—SURFACE WATER BALANCE, 1972
(Source: S. Martin and Comision de Recursos Hidraulicos, 1969)

[Estimated; by river basin]

River basin	Supply hm³/yr	Demand, hm³/yr			Surplus (+) or Deficit (−) hm³/yr
		Urban (including Industry)	Irrigation	Total	
North	10,780	316	253	569	(+) 10,211
Duero	7,367	170	2,045	2,215	(+) 5,152
Tajo	7,582	681	1,219	1,900	(+) 5,682
Guadiana	2,261	167	1,176	1,343	(+) 918
Guadalquivir	4,234	366	3,110	3,476	(+) 758
South	814	160	808	968	(−) 154
Segura	670	83	976	1,059	(−) 389
Jucar	2,400	289	1,674	1,963	(+) 437
Ebro	13,253	220	6,277	6,497	(+) 6,756
Eastern Pyrenees	947	580	415	995	(−) 48
Total	50,308	3,032	17,953	20,985	(+) 29,323

TABLE 1-87. SPAIN—WATER USED BY AGRICULTURE, INDUSTRY AND PUBLIC SUPPLIES, 1968-69

(Source: Servicio Sindical de Estadistica, 1971)

Category and economic activity	Water Consumption			
	1968		1969	
	10^6m^3	%	10^6m^3	%
Agriculture				
Irrigation	14,271.0	57.49	14,753.0	55.66
Livestock	327.0	1.32	359.0	1.35
Total	14,598.0	58.81	15,112.0	57.01
Industry				
Mining	155.8	0.63	155.9	0.59
Food	873.5	3.52	942.5	3.55
Beverages and alcoholic drinks	84.8	0.34	99.5	0.37
Tobacco	0.3	0.01	0.3	0.01
Textile	104.2	0.42	113.6	0.43
Leather, tanning, shoes	17.9	0.07	20.1	0.08
Wood, cork	41.0	0.17	47.1	0.18
Paper, printing, publishing	265.8	1.07	294.0	1.11
Chemicals	949.8	3.83	1,104.6	4.17
Products derived from petroleum and coal	313.6	1.26	344.7	1.30
Ceramics, stone and cement	255.8	1.02	278.0	1.05
Base metals	513.5	2.07	599.3	2.26
Other metals	171.1	0.69	199.8	0.75
Construction (buildings and public works)	399.7	1.61	448.9	1.69
Water, gas and electric (incl. thermal energy production)	3,642.9	14.67	4,172.2	15.74
Total	7,788.9	31.38	8,820.5	33.28
Services				
Transportation	420.6	1.70	452.9	1.71
Business, banking and insurance	24.1	0.98	26.3	0.99
Other	822.7	3.32	885.9	3.35
Total	1,267.4	5.10	1,365.1	5.15
Public Water Supply	1,168.4	4.71	1,209.6	4.56
TOTAL SPAIN	24,822.7	100.00	26,507.2	100.00

TABLE 1-88. SPAIN—MUNICIPAL WATER SUPPLY SYSTEMS, 1969
(Source: Servicio Sindical de Estadistica, 1971)

Provincial Capital	Domestic	Industrial	Public fountains	Municipal services	Other	Total	Per capita consumption m³
Alava	6,020,000	2,320,000	1,700,000	2,978,000	2,750,000	15,768,000	121.0
Albacete	2,240,000	285,000	275,000	510,000	90,000	3,400,000	40.0
Alicante	9,234,183	4,076,330	49,780	1,586,607	—	14,946,900	90.0
Almeria	4,006,786	660,209	62,403	790,000	50,000	5,569,398	52.0
Avila	1,659,858	120,610	134,042	498,264	—	2,412,774	79.0
Badajoz	1,725,963	540,600	183,417	—	—	2,449,980	23.4
Baleares	12,575,561	—	578,000	152,819	200,000	13,506,380	63.2
Barcelona	137,633,830	23,105,118	17,394,356	—	7,453,471	185,586,325	105.0
Burgos	3,594,308	1,740,839	—	420,940	369,982	6,126,069	55.1
Caceres	1,993,034	—	—	168,699	—	2,161,733	38.0
Cadiz	5,421,019	3,718,899	—	—	—	9,139,918	65.4
Castellon	3,960,990	1,990,990	—	960,020	—	6,912,000	77.0
Ciudad Real	1,390,000	550,900	53,990	120,710	—	2,115,600	52.0
Cordoba	15,363,517	—	—	1,592,346	—	16,955,863	71.2
Coruna, La	6,113,677	4,427,282	806,076	450,000	227,008	11,574,043	58.0
Cuenca	1,400,000	300,000	350,000	450,000	—	2,500,000	75.0
Gerona	1,432,780	—	558,340	1,432,780	—	3,423,900	69.0
Granada	9,940,000	525,000	325,000	710,000	—	11,500,000	67.0
Guadalajara	990,500	—	406,000	265,000	50,240	1,711,740	61.0
Guipuzcoa	11,078,987	3,744,975	2,687,914	3,162,240	—	20,674,116	126.1
Huelva	2,750,600	850,000	—	561,720	—	4,162,320	42.3
Huesca	950,800	169,680	—	—	—	1,120,480	38.0
Jaen	2,349,484	1,119,708	100,000	650,000	50,000	4,269,192	56.4
Leon	5,682,831	174,914	—	420,260	—	6,278,005	60.0
Lerida	4,600,343	499,942	—	—	—	5,100,285	60.0
Logrono	4,910,210	1,456,710	868,600	1,019,700	849,200	9,104,420	113.0
Lugo	3,213,760	—	—	—	3,763,318	6,977,078	98.1
Madrid	168,518,027	81,313,465	371,465	999,492	21,954,458	273,156,907	88.0
Malaga	14,529,600	6,400,584	500,809	1,521,850	—	22,952,843	64.0
Murcia	8,250,500	1,900,000	60,500	1,933,200	1,600,800	13,745,000	50.0
Navarra	7,793,487	3,133,787	21,280	711,369	433,044	12,092,967	86.0
Orense	1,850,550	300,500	500,500	400,800	684,590	3,736,940	76.0
Oviedo	6,022,500	2,737,500	547,500	1,095,000	547,500	10,950,000	75.0
Palencia	1,013,540	659,910	130,000	1,100,000	1,350,500	4,253,950	74.5
Palmas, Las	7,257,867	571,184	53,291	1,063,855	52,304	8,998,501	33.2

Water Consumption, m³

Pontevedra	2,337,240	300,500	260,900	500,000	550,000	3,948,640	59.2
Salamanca	5,561,320	675,000	—	3,500,960	—	9,737,280	78.4
S.C. de Tenerife	3,459,681	2,290,500	900,980	169,268	—	6,820,429	37.0
Santander	4,509,889	5,154,124	290,586	809,732	—	10,764,331	73.4
Segovia	952,315	856,218	—	—	296,320	2,104,853	51.5
Sevilla	36,241,601	6,483,200	350,500	1,964,819	—	45,040,120	71.2
Soria	1,184,458	270,967	—	—	8,803	1,464,228	62.0
Tarragona	4,555,400	—	1,000,000	1,000,000	—	6,555,400	90.0
Teruel	758,756	758,756	175,200	84,000	—	1,776,712	79.3
Toledo	1,784,256	450,624	1,540,900	—	95,900	3,871,680	87.2
Valencia	28,587,732	5,087,559	558,641	1,208,629	2,058,924	37,501,485	59.0
Valladolid	10,310,000	10,220,420	990,620	1,980,100	620,220	24,121,360	109.5
Vizcaya	13,489,100	5,531,600	850,950	4,730,900	2,700,810	27,303,360	67.2
Zamora	1,639,500	625,000	255,200	314,840	—	2,834,540	59.3
Zaragoza	35,421,200	6,542,120	1,420,500	2,340,300	1,430,200	47,154,320	104.4
TOTAL	628,261,090	194,641,224	37,313,240	45,879,219	50,237,592	956,332,365	79.5

TABLE 1-89. SWEDEN—DISCHARGE OF PRINCIPAL RIVERS
(Source: UNESCO, 1971)

River and station	Basin area km²	Mean monthly discharge, m³/s													Period of record	
		Jan.	Feb.	Mar.	Apr.	May	Jun.	Jul.	Aug.	Sep.	Oct.	Nov.	Dec.	Year		
Vanern-Gota, Vanersborg	46,830	530	520	510	512	549	585	582	567	552	541	539	536	544	1807-1937	unregulated
		500	500	498	504	526	532	522	506	497	490	494	503	506	1938-66	calculated [unregulated
		604	645	641	574	447	332	304	392	471	506	566	583	505	1938-66	regulated
Angerman, Solleftea	30,640	161	136	126	199	892	1,370	803	484	449	459	346	229	472	1909-48	unregulated
		179	144	134	213	1,020	1,270	752	535	423	386	351	240	472	1949-66	calculated [unregulated
		328	326	309	383	993	1,350	750	410	381	355	367	350	526	1949-53	regulated
		420	438	409	353	561	586	473	504	398	378	419	423	448	1954-66	regulated
Lule, Boden Waterworks	24,490	134	104	20	107	426	1,060	1,370	1,030	695	463	308	192	501	1900-22	unregulated
		141	117	101	101	439	1,190	1,320	831	650	503	322	200	500	1923-66	calculated [unregulated
		163	152	147	148	502	1,100	1,340	938	646	517	307	212	516	1923-40	regulated
		284	288	276	259	507	911	870	731	609	456	337	284	486	1941-66	regulated

TABLE 1-90. SWEDEN—REGIONAL DISTRIBUTION OF WATER RESOURCES
(Source: Falkenmark, Swedish IHD Committee, 1973)

Region	Area km²	Annual runoff mm	Population Million 1970	Per capita runoff m³/year
Norrland	248,700	450	1.2	93,000
Svealand	87,700	325	3.0	9,500
Gotaland	113,900	300	3.8	9,000
Country	450,300	390	8.0	22,000

TABLE 1-91. SWEDEN—CHARACTERISTICS OF PRINCIPAL LAKES AND THE BALTIC SEA

(Source: Falkenmark, Swedish IHD Committee, 1973)

Parameter	Unit	Lake Vanern	Lake Vattern	Lake Malaren	Baltic Sea
Area	km2	5,550	1,910	1,140	385,800
Volume	km3	140	74	14	20,500
Basin area	km2	46,830	6,360	22,600	1,649,600
Basin area in relation to lake volume	km2/km3	335	85	1,600	80
Mean depth	m	25	39	13	50-60
Maximum depth	m	100	120	61	459
Through-flow	m3/s	544	35	168	14,100
Turnover time	years	8.2	67	2.7	35-40
Population in the basin	million	0.7	0.175	1.3	20
Annual load per year: Phosphorus	kg/ha	3	1	7	0.36
Nitrogen	kg/ha	22	10	90	—

TABLE 1-92. SWEDEN—WATER DEMAND, 1969-2000

(Source: Falkenmark, Swedish IHD Committee, 1973)

[million m3/year]

Year	Industry	Urban	Total
1969	4,000	1,000	5,000
2000	6,700	2,000	8,700

TABLE 1-93. SWEDEN—WATER WITHDRAWN FOR PUBLIC SUPPLIES, 1968
(Source: Falkenmark, Swedish IHD Committee, 1973)

Region	Million m3/year			Per capita withdrawal m3/year	% of potential resources
	Total	Ground water	Surface water		
Norrland	130	75	55	108	0.11
Svealand	374	165	210	125	1.30
Gotaland	367	168	198	97	1.04
Country	871	408	463	109	0.47

TABLE 1-94. SWEDEN—MUNICIPAL WATER SUPPLY SYSTEMS
(Source: Internat. Statistical Institute, 1971)

City	Year	Population served	Annual consumption (1000m3)	
			Total	of which domestic
Goteborg 1	1967	482,800	49,902	28,177
	1969	489,800	56,158	29,674
Malmo	1967	242,000	30,100	21,000
	1969	247,000	30,600	20,200
Stockholm	1967	775,646	124,590	66,810
	1969	756,126	130,036	67,850
Vasteras	1967	100,000	12,612	12,612
	1969	105,000	13,676	13,676

1 Including some surrounding districts or suburbs.

TABLE 1-95. SWITZERLAND–DISCHARGE OF PRINCIPAL RIVERS
(Source: UNESCO, 1971)

River and station	Basin area km²	Mean monthly discharge, m³/s												Year	Period of record
		Jan.	Feb.	Mar.	Apr.	May	Jun.	Jul.	Aug.	Sep.	Oct.	Nov.	Dec.		
Rhine, Basel (St. Alban)	35,925	708	692	801	992	1,250	1,530	1,500	1,310	1,100	903	812	745	1,030	1808-1965
Rhone, Chancy	10,299	223	241	254	275	336	522	555	496	369	253	258	222	334	1935-64

TABLE 1-96. SWITZERLAND–DISCHARGE CHARACTERISTICS OF PRINCIPAL RIVERS
(Source: Walser, Swiss Federal Office of Water Management, 1961)

River and station	Rhine at Rheinfelden	Rhone at Chancy	Ticino at Bellinzona	Inn at Martinsbruck
Area of drainage basin, km²	34,550	10,299	1,515	1,945
Period of record	1935-60	1935-60	1918-60	1904-60
Minimum mean monthly discharge,				
Month	January	Feb.-Jan.	February	February
m³/s	740	221	25.6	14.0
Mean discharge,				
m³/s	1,026	338	71.2	58.1
Maximum mean monthly discharge,				
Month	July	July	June	June
m³/s	1,489	567	146	149
Maximum instantaneous discharge,				
m³/s	3,670	1,700	1,500	580
Month, Year	June, 1953	Nov., 1944	Sep., 1927	Sep., 1960
Maximum instantaneous discharge, since start of observations,				
m³/s	5,600 [1]	1,700	1,500	580
Month, Year	June, 1876	Nov., 1944	Sep., 1927	Sep., 1960
Period	(1808-1960)	(1905-1960)	(1918-1960)	(1904-1960)
Minimum mean daily discharge,				
m³/s	337	103	12.5	9.60
Month, Year	Jan., 1954	Mar., 1949	Jan.-Feb., 1922	Mar., 1905

[1] Calculated according to peak discharge of 5,700 m³/s observed at Basel.

TABLE 1-97. SWITZERLAND—WATER USE, 1962

(Source: Swiss Federal Office of Water Management, 1963)

Category	Estimated Water use 1962 million m^3
Public water supplies (including commercial and industrial uses)	920
Self-supplied industrial	1,000
Agriculture (livestock; irrigation not included)	85
Total water use	2,005

TABLE 1-98. SWITZERLAND—MUNICIPAL WATER SUPPLY SYSTEMS

(Source: Societe Suisse de l'Industrie du Gaz et des Eaux, 1972)

[Data for year 1971]

City	Population served	Water Withdrawn-1000 m^3					Water Delivered-1000 m^3				
		Spring water	Ground water	Lake water	Purchased water	Total	Commercial & Industry	Domestic	Public Services	Losses	Total 1
Zurich	423,660	9,335	16,196	51,583	60	77,174	17,878	33,508	5,751	8,649	77,174
Geneva	283,170	–	3,692	56,770	–	60,462	501	46,576	4,160	5,487	60,462
Basel	252,751	1,962	27,770	–	22,915	52,647	17,848	22,059	1,179	7,102	52,647
Lausanne	199,031	8,603	–	33,683	55	42,341	5,147	23,612	3,000	4,672	42,341
Bern	177,000	6,000	24,024	–	–	30,024	650	24,952	1,067	1,474	30,024
Winterthur	94,736	423	16,423	–	–	16,846	7,580	6,881	900	1,200	16,846
St. Gallen	79,063	152	333	9,582	11	10,078	–	6,598	413	1,555	10,074
Luzern	72,060	6,369	3,684	3,866	–	13,919	1,264	8,477	1,794	1,126	13,919
Biel	63,725	7,045	4,540	–	–	11,585	38	7,522	415	–	11,585
Vevey-Montreux	53,000	9,398	37	1,381	276	11,092	3,223	3,805	1,334	2,013	11,092

1 Includes water consumed by water-supply services and water sold to other systems.
– Not applicable.

TABLE 1-99. UNION OF SOVIET SOCIALIST REPUBLICS—DISCHARGE OF SELECTED RIVERS

(Source: UNESCO, 1971)

River and station	Basin area km²	Mean monthly discharge, m³/s												Year	Period of record
		Jan.	Feb.	Mar.	Apr.	May	Jun.	Jul.	Aug.	Sep.	Oct.	Nov.	Dec.		
EUROPE															
Byelorussian S.S.R.															
Pripiat, Mozyr	97,200	234	230	414	1,140	744	369	236	198	177	190	232	235	367	1881-1917
Ukrainian S.S.R.															
Southern Bug, Aleksandrovka	46,200	54.0	88.0	252	223	72.0	55.5	47.5	41.0	37.0	40.0	49.5	53.5	84.5	1914-64
Dnieper, Dnieper Hydroelectric Plant	463,000	974	1,200	1,480	2,580	3,410	1,780	932	781	764	839	848	862	1,370	1952-64
Desna, Chernigov	81,400	147	140	269	1,140	906	283	174	149	137	147	185	163	320	1884-1964
Pechora, Ust-Tsilma	248,000	653	501	426	812	9,130	13,700	4,330	2,220	2,750	3,270	1,590	960	3,360	1932-64
Usa, Adzva	54,700	116	89.0	88.0	192	1,940	4,230	1,350	793	958	829	385	180	929	1931-64
Mezen, Malonisogorskaya	56,400	169	145	135	390	2,720	1,370	567	421	493	652	464	250	648	1920-64
Northern Dvina, Ust-Pinega	348,000	1,020	820	718	2,300	13,900	7,200	3,020	2,260	2,440	3,040	2,450	1,380	3,380	1882-1964
Vychegda, Malaya Kushba	26,500	65.0	58.5	57.5	165	1,040	459	181	125	149	202	169	89.0	230	1930-64
Vaga, Filaievskaya	13,200	25.5	22.5	22.0	224	497	167	76.0	49.0	65.0	97.0	73.0	35.0	113	1938-64
Onega, Porog	55,700	162	137	124	442	1,980	836	434	308	358	452	359	208	483	1943-64
Kem, Podoushemie	27,600	147	122	97.5	132	492	502	366	268	252	268	251	186	257	1917-20; 1925-64
Kola, 1,429th km of Octiabrsky Railway	3,780	11.5	9.4	8.2	13.0	126	128	43.5	34.5	40.5	40.0	25.5	15.0	41.5	1928-64
Neva, Novosaratovka	281,000	1,760	1,750	1,850	2,480	3,050	3,140	3,080	2,980	2,890	2,860	2,600	2,060	2,540	1859-1941; 1943-64
Olonka, Olonets	2,120	13.5	9.0	7.3	62.5	95.0	24.5	13.5	15.0	21.5	29.0	33.0	21.0	28.5	1949-64
Piarnu, Orekula	5,180	32.5	28.0	34.5	136	61.5	21.5	23.5	30.5	35.0	51.5	53.5	60.0	47.5	1942-64
Daugava, Daugavpils (Western Dvina)	64,600	187	193	393	1,640	992	344	255	262	271	337	413	289	465	1881-1916; 1921-64
Neman, Smalininkai	81,200	447	459	745	1,360	632	401	366	387	372	404	484	490	546	1812-1943; 1946-64
Dniester, Bendery	66,100	168	211	471	578	380	339	357	301	234	210	222	209	307	1881-1915; 1945-64
Don, Razdorskaya	378,000	295	449	918	2,410	3,050	861	422	344	307	308	340	274	832	1891-1964
Vorona, Chutanovka	5,560	5.4	4.9	15.5	151	23.0	8.1	6.5	6.0	5.6	6.0	6.5	5.2	20.5	1915-19; 1922-64
Medveditsa, Archedinskaya	33,700	17.5	23.5	61.0	418	166	39.5	26.5	23.0	18.0	19.5	19.5	16.0	70.5	1927-42; 1945-64
Kalaus, Svetlograd	4,540	0.7	3.6	9.5	5.9	1.7	3.0	1.8	0.7	0.3	0.3	0.4	0.8	2.4	1930-41; 1943-64
Kuban, Tikhovsky	48,100	241	297	374	445	633	714	603	423	266	230	223	258	392	1911-12; 1927-41; 1945-64

TABLE 1-99. UNION OF SOVIET SOCIALIST REPUBLICS—DISCHARGE OF SELECTED RIVERS (continued)

River and station	Basin area km2	Mean monthly discharge, m3/s												Year	Period of record
		Jan.	Feb.	Mar.	Apr.	May	Jun.	Jul.	Aug.	Sep.	Oct.	Nov.	Dec.		
Ukrainian S.S.R.(cont.)															
Rioni, Sakochakidze	13,300	278	354	451	639	667	560	410	275	245	311	302	312	400	1928-64
Kura, Surra	178,000	345	379	521	1,000	1,480	1,150	531	272	278	387	411	370	594	1930-52
	—	545	540	574	765	828	693	392	320	391	421	453	554	540	1953-64
Vorotan, Eivazlar	2,020	10.5	11.0	14.0	34.5	63.0	38.5	15.5	11.0	10.5	11.5	12.0	11.0	20.0	1927-28; 1930-64
Kara-Samur, Luchek	481	1.8	1.7	2.1	6.4	17.0	22.5	16.0	7.3	5.4	5.0	3.2	2.1	7.5	1932-63
Terek, Ordzhonikidze	1,490	12.0	11.0	11.5	18.5	44.0	71.5	83.5	66.0	42.0	27.0	19.0	14.5	35.0	1912-13; 1925-64
Volga, Volgogradskaia Hydroelectric Plant	1,350,000	3,040	2,990	3,080	7,300	26,000	24,500	9,120	5,610	4,940	5,380	5,410	3,170	8,380	1879-1955
	—	4,960	5,860	6,030	9,330	20,500	13,200	6,310	5,630	5,410	5,440	5,370	4,790	7,740	1956-58; 1961-64
Unzha, Makariev	18,500	40.0	34.0	35.5	360	676	173	101	69.0	92.5	123	110	65.0	157	1896-1964
Oka, Kostomarovo	4,900	6.3	7.3	45.0	106	13.0	7.9	7.3	6.2	5.2	6.5	7.5	7.2	19.0	1881-1917; 1922-41; 1948-64
Protva, Spas-Zagorie	3,640	6.6	6.4	13.5	118	23.0	10.0	8.0	8.6	9.2	10.5	11.5	8.5	19.5	1937-64
Belaya, Birsk	121,000	302	247	235	1,780	3,630	1,030	662	504	471	560	481	353	855	1881-1964
Viatka, Kirov	48,300	118	104	104	680	1,870	518	269	211	237	278	228	162	398	1878-1964
Maly Uzen, Aleksashkino	1,910	nil	0.0[1]	7.4	7.5	0.1	0.0[1]	0.0[1]	0.0[1]	nil	nil	0.0[1]	0.0[1]	1.2	1934-35; 1952-55; 1959-61
ASIA															
Ural, Kushum	190,000	58.5	52.0	61.5	944	1,540	448	211	140	110	101	91.0	60.0	318	1915-17; 1921-64
Ilek, Aktubinsk	11,000	0.8	0.6	4.0	173	31.0	5.9	2.5	1.6	1.5	2.0	2.2	1.3	18.9	1938-64
Amur, Komsomolsk	1,730,000	1,930	1,160	858	2,910	14,400	16,900	16,700	20,100	21,700	17,900	6,860	2,570	10,300	1933-64
Selemdhza, Ust-Ulma	67,000	33.5	18.5	15.0	160	1,290	1,430	1,450	1,700	1,350	613	153	68.5	690	1940-64
Ussuri, Kirovsky	24,400	18.0	11.0	15.0	413	588	384	247	231	328	215	123	41.5	218	1952-64
Tym, Ado-Tymovo	3,420	12.0	9.9	9.4	26.5	204	126	43.0	33.5	51.0	65.0	29.0	17.0	52.0	1937-64
Khasyn, Kolyma Rd.,79th km	682	0.3	0.1	0.1	0.2	16.0	35.5	16.5	13.5	14.5	6.9	1.7	0.8	0.8	1941-64
Penzhina, Kamenskoe	71,600	30.0	25.0	21.0	22.5	694	4,680	1,140	992	574	278	78.0	47.0	715	1957-64
Kamchatka, Kluchi	45,600	392	374	376	429	806	1,560	1,690	1,020	772	662	458	398	745	1931-64
Anadyr, Novy Eropol	47,300	15.0	11.5	11.0	10.5	109	3,260	1,040	704	336	132	47.5	25.0	475	1958-64
Amguema, mouth of the brook Shumny, Chuckchee Sea	26,700	0.3	0.0[1]	nil	0.0[1]	15.5	1,460	898	609	288	66.5	19.5	3.5	280	1944-64
Kolyma, Srede-Kolymsk	361,000	113	75.0	61.0	54.0	2,110	10,300	5,280	4,110	3,290	950	301	198	2,240	1927-31; 1934-57; 1964

Station	Area	I	II	III	IV	V	VI	VII	VIII	IX	X	XI	XII	Annual	Period
Sugov, 3.2 km below the mouth of the brook Omchikchan	5,880	1.7	0.9	0.7	0.6	55.5	305	134	94.0	59.0	22.5	7.8	3.8	57.0	1941-64
Indigirka, Vorontsovo	305,000	39.5	20.0	11.5	8.0	291	5,580	5,680	4,090	2,130	480	131	81.0	1,550	1937-64
Nera, Ala-Chubuk	22,300	0.0¹	nil	nil	nil	62.5	560	386	271	105	17.0	3.1	0.4	117	1944-64
Yana, Dzhangky, Laptev Sea	216,000	2.8	1.7	1.6	1.4	446	3,800	3,040	2,300	1,150	171	39.5	9.9	914	1938-64
Lena, Kusur	2,430,000	2,610	1,900	1,370	1,120	4,490	73,800	39,100	27,000	24,400	14,400	3,270	2,710	16,300	1935-64
Kirenga, Shorokhovo	46,500	180	159	147	205	1,250	2,460	1,070	735	681	494	236	211	652	1927-64
Vitim, Bodaibo	186,000	118	92.5	77.5	92.5	1,370	4,930	4,110	3,260	3,040	1,020	231	164	1,540	1912-64
Maya, Chabda	165,000	87.0	63.0	52.5	56.0	2,900	3,860	2,100	1,720	1,560	776	244	141	1,130	1935-64
Amga, Buyaga	23,900	17.5	15.5	14.0	20.5	636	291	119	104	126	60.5	30.5	21.5	121	1933-35; 1937-64
Markha, Malykai	89,600	1.9	1.6	1.3	2.8	902	2,100	636	439	474	126	10.5	3.4	392	1938; 1940-64
Olenek, 8 km upstream of Pur river	181,000	7.8	4.0	2.2	1.8	161	5,610	1,740	939	903	250	61.5	19.0	808	1952-63
Anabar, Saskylakh	78,500	1.3	nil	nil	nil	36.0	3,040	711	459	249	100	47.0	12.5	388	1954-64
Yenisei, Igarka	2,440,000	4,840	4,510	4,190	3,980	30,200	76,000	27,900	18,900	17,900	14,700	6,120	4,880	17,800	1936-64
Tuba, Bugurtak, Yenisei	31,400	126	102	94.5	563	2,210	2,610	1,030	722	722	490	257	149	756	1911-23; 1927-28; 1930-64
Iya, Tulun	14,500	21.0	16.0	14.5	63.0	212	351	356	319	238	122	44.5	27.5	149	1920-22; 1927-33; 1936-64
Khilok, Maleta	25,700	4.5	2.3	2.4	19.5	186	116	141	124	158	92.5	24.5	10.5	73.5	1936-64
Podkamennaya Tunguska, Yenisei	218,000	332	272	244	253	6,070	6,850	1,390	1,210	1,270	944	481	434	1,650	1954-64
Gerasimova, Bolshoy Porog, Nizhnaya Tunguska	9,100	24.0	18.5	16.5	17.0	250	1,040	126	108	138	89.0	53.5	29.5	159	1949-62
Pur, Samburg	95,100	307	258	235	231	736	3,780	1,610	965	889	730	489	388	885	1939-64
Ob, Salekhard	2,430,000	4,370	3,610	3,110	3,210	14,500	32,100	29,100	22,000	13,600	10,300	6,070	4,970	12,200	1930-64
Tom, Tomsk	57,000	201	138	133	1,970	4,890	2,400	852	538	657	746	472	256	1,100	1918-64
Tym, Napas	24,500	60.5	56.0	55.0	78.5	454	564	254	144	157	147	97.0	73.0	178	1937-64
Bolshoy Yugan, Ugut	22,100	28.5	22.5	20.0	43.0	472	441	167	107	98.0	109	85.0	44.0	136	1945-64
Ulba, Ulba-Perevalochnaya	4,900	17.0	17.0	24.5	224	400	204	83.0	50.5	54.0	52.5	37.0	20.0	98.5	1930-64
Ishim, Petropavlovsk	106,000	3.5	3.0	3.1	129	361	84.0	25.5	12.5	9.0	8.3	6.8	4.8	54.0	1932-64
Lobva, Lobva	2,940	2.5	2.1	2.1	26.0	80.5	36.0	26.0	16.0	16.5	11.5	6.3	3.4	19.0	1938-64
Kargat, Gavrilovsky	3,910	0.1	0.0¹	0.1	12.5	36.0	12.5	3.3	1.6	1.0	1.0	0.8	0.1	5.8	1948-64
Northern Sosva, Sosvinskaya Kultbasa	65,200	72.0	57.0	48.5	137	1,660	2,100	960	644	653	475	204	107	593	1937-64
Amu-Darya, Chatly	450,000	616	597	546	873	1,800	2,590	3,260	2,800	1,700	1,050	834	702	1,450	1931-64
Gunt, Khorog	13,700	31.0	28.5	27.0	30.0	67.0	217	327	250	127	65.0	45.0	36.5	104	1940-64
Zerafshan, Dupuli	10,200	38.0	35.0	35.0	50.5	141	338	466	373	191	85.5	57.0	44.5	155	1914-21; 1923-64
Chu, Chapaevo (Tash-Utkul)	26,700	73.5	89.5	118	107	68.0	69.0	22.5	36.0	55.0	80.5	94.0	94.0	75.5	1926-64
Syr-Darya, Tyumen-Aryk	219,000	494	559	633	805	1,040	1,190	949	536	380	446	550	532	676	1930-44; 1947-64
Arys, Arys	13,100	47.0	57.5	89.0	104	57.5	28.5	15.5	9.8	19.0	31.5	38.5	44.0	45.0	1927-64
Kara-Turgay, Akutkul	14,700	0.1	0.1	9.8	80.0	16.0	4.0	1.8	1.0	0.8	0.8	0.6	0.4	9.5	1942-47; 1950-64
Nura, Sergiopolskoye	12,300	0.1	0.1	0.9	43.0	9.5	2.9	2.3	0.8	0.8	0.9	0.7	0.2	5.2	1934-64
Karatal, Ush-Tobe	13,200	35.0	38.0	59.0	81.0	110	135	106	70.5	45.5	47.5	48.5	40.0	68.0	1915-18; 1923-64

1 River frozen

FIGURE 1-9. UNION OF SOVIET SOCIALIST
(Source: Fox, I.K., 1971, Water Resources
The University of Wisconsin Press; Copyright

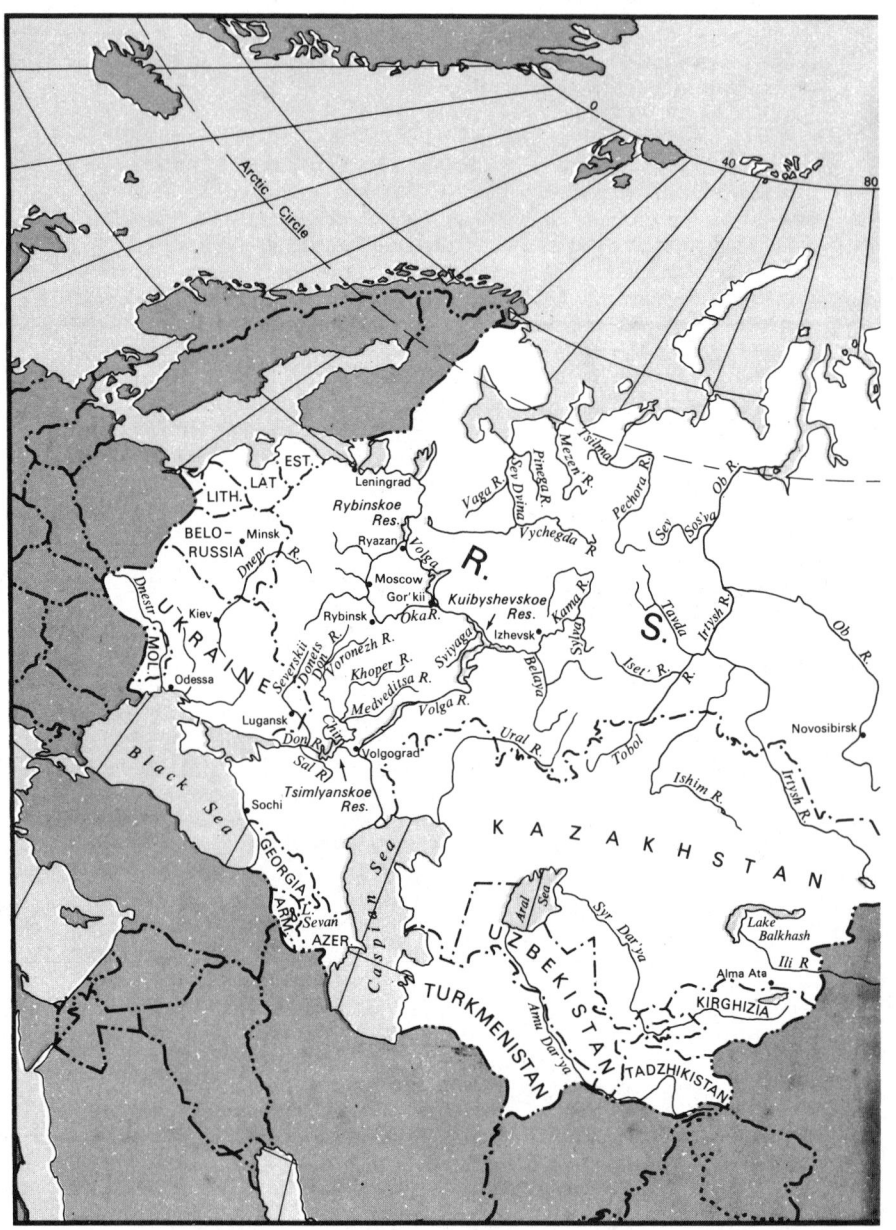

REPUBLICS — PRINCIPAL RIVER SYSTEMS

Law and Policy in the Soviet Union,
The Regents of the University of Wisconsin)

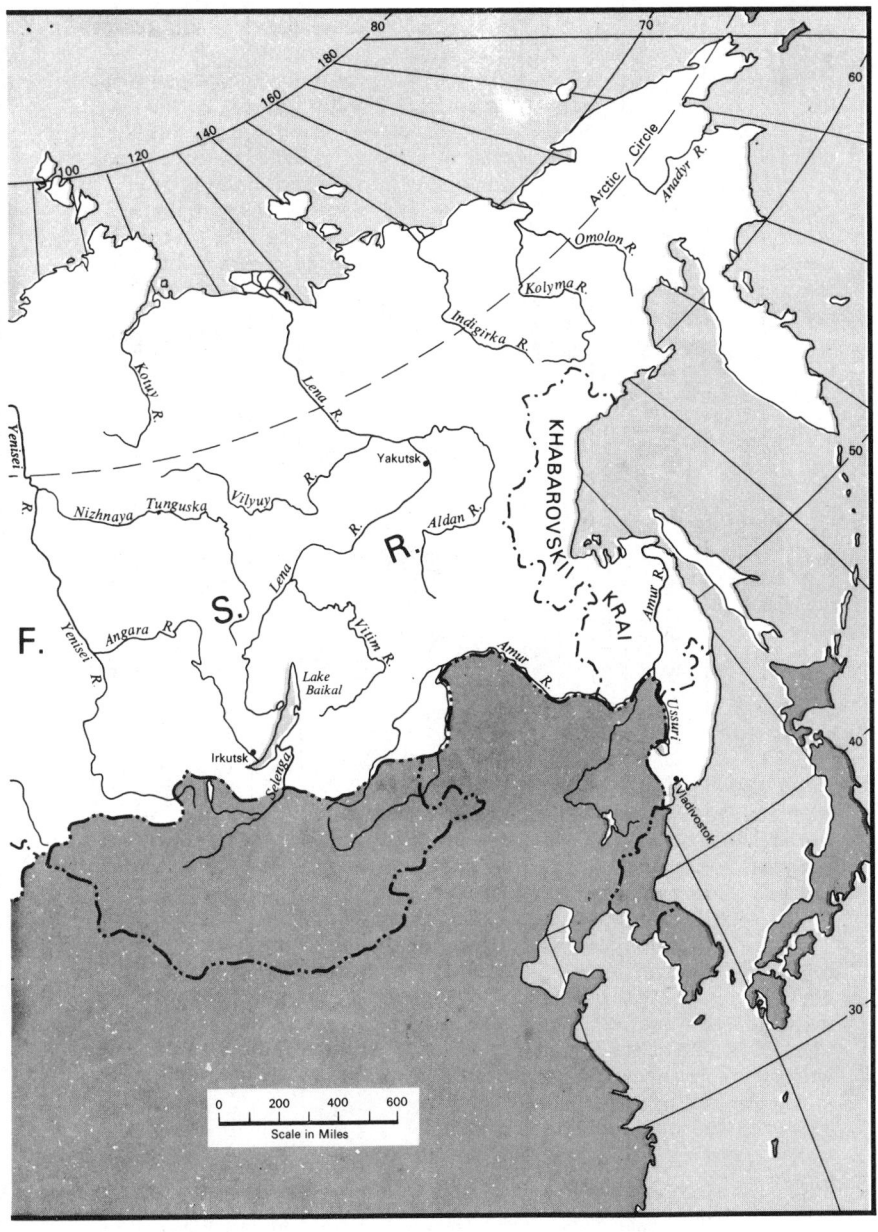

TABLE 1-100. UNION OF SOVIET SOCIALIST REPUBLICS—SURFACE WATER RUNOFF
(Source: Guerardy, United Nations, WRS 32, 1967)

Ocean and sea basins	Basin area 1,000 km²	Surface-water runoff	
		Mean annual discharge m³/s	Volume in mean water year 10⁹ m³
1	**2**	**3**	**4**
Arctic Ocean	12,612	89,500	2,811
Pacific Ocean	3,508	29,900	938
Atlantic Ocean	1,969	10,500	334
including:			
Black and Azov Seas	1,329	5,300	166
Aral-Caspian basin	5,450	13,100	415
Total	23,539	143,000	4,498

TABLE 1-101. UNION OF SOVIET SOCIALIST REPUBLICS—WATER BALANCE
(Source: Kudelin, et al., Proc. Reading Symposium, IASH—UNESCO—WMO, 1972)

Basins	Area 1,000 km²	Water balance elements amount in km³					Coefficients of underground alimentation of rivers in per cent	Coefficients of underground flow in per cent
		Precipitation	Runoff			Evapotranspiration and deep percolation		
			Total	Surface	Base			
White and Barents Seas	1,170	834	399	321	77.8	435	19	9
Baltic Sea	653	500	169	131	37.6	331	20	8
Black Sea and Sea of Azov	1,322	875	152	100	52.1	723	34	6
Caspian Sea	2,832	1,411	290	193	96.6	1,121	33	7
Kara Sea	6,251	3,350	1,283	995	288	2,267	22	8
Laptev, East Siberian and Chukotsk Seas	5,048	2,135	1,038	901	137	1,097	13	6
Bering, Okhotsk Seas and Sea of Japan	2,549	1,736	789	597	192	974	24	11
Closed regions of the Middle Asia and Kazakhstan	2,188	653	88 [2]	—	57.3 [3]	565	2	—
Total	22,013 [1]	11,694	4,208 [2]	—	939 [4]	7,486	22 [5]	

1 Without large islands of the Arctic Ocean and the area of large water reservoirs without outflow (Aral Sea, Balkhash and Issyk-Kul Lakes).
2 Without account of runoff losses on evapotranspiration, infiltration and unconsumptive use in arid regions (estimated losses for the Middle Asia and Kazakhstan are equal to 112 cu km, (or the USSR territory about 150 cu km.).
3 For mountainous part of the territory with an area of 591,000 sq km.
4 In addition, underground flow in the arid zones equal to 80 cu km (The total ground-water flow of the USSR territory equals 1,019 cu km).
5 The total value of underground flow (1,019 cu km) is equal to 23.4% of the total amount of water resources (4,358 cu km) formed within the USSR territory.

TABLE 1-102. UNION OF SOVIET SOCIALIST REPUBLICS—WATER BALANCE IN THE UKRAINIAN S.S.R.

(Source: Perekhrest, Nature and Resources, UNESCO, 1971)

Zone	Area	Precipitation		Average annual runoff		Evaporation, transpiration, seepage, population needs	
	1,000 km^2	mm	million m^3	mm	million m^3	mm	million m^3
Ukraine as a whole	601,000 [1]	512	308,123	84	50,137	428	257,986
Polesye	108,427	600	65,050	110	11,926	490	53,124
Forest steppe	209,971	500	104,985	75	15,748	425	89,237
Central steppe	91,225	450	41,051	48	4,379	402	36,672
Donets basin (Ukrainian part)	79,360	450	35,717	52	4,126	398	31,591
Southern steppe	72,463	400	29,389	12	882	388	28,507
Upland Carpathians	30,649	900	27,584	405	12,412	495	15,772
Upland Crimea	7,905	550	4,347	84	664	466	3,683

[1] Excluding sea.

TABLE 1-103. UNION OF SOVIET SOCIALIST REPUBLICS—WATER BALANCE IN DIFFERENT GEOGRAPHICAL ZONES

(Source: Kuznetsov and L'vovich, in Fox, I.K., Water Resources Law and Policy in the Soviet Union, Univ. Wisconsin Press, 1971)

[values in mm]

Zone and subzone	Basin	Precipitation P	FLOW			Evaporation E	Gross Wetting W	Coefficients		
			Total R	Underground U	Surface S			K U	K E	K R [a]
Tundra	Shchuch'ya	450	340	34	306	110	144	0.24	0.76	0.76
	Amguema	400	296	15	281	104	119	0.14	0.86	0.74
Permafrost taiga	Vilyui	300	117	13	104	183	196	0.07	0.93	0.39
	Olenek	350	177	14	163	173	187	0.08	0.82	0.50
Taiga without permafrost	Pinega	490	302	106	196	188	294	0.36	0.64	0.62
	Vym'	510	350	140	210	160	300	0.47	0.53	0.69
Mixed forest	Berezina	600	183	73	110	417	490	0.15	0.85	0.31
	Klyaz'ma	500	158	63	95	342	405	0.16	0.84	0.32
Forest-steppe	Psel	500	82	16	66	418	434	0.04	0.96	0.16
	Oka	530	167	50	117	363	413	0.12	0.88	0.32
	Medveditsa	370	66	20	46	304	324	0.06	0.94	0.18
Steppe	Ingulets	435	28	3	25	407	410	0.01	0.99	0.06
	Sal	370	19	3	16	351	354	0.01	0.99	0.05
	Malyi Uzen'	250	42	0.4	42	208	208	0.00	1.00	0.17
Semidesert	Turgai	175	10	0	10	165	165	0.00	1.00	0.06
	Sary-Su	175	5	0	5	170	170	0.00	1.00	0.03

[a] Coefficient of flow—relation of total flow to precipitation.

TABLE 1-104. UNION OF SOVIET SOCIALIST REPUBLICS—WATER BALANCE BY REPUBLIC
(Source: USSR National Committee for IHD, 1969)

[Values in mm]

Union Republic	Precipitation	Runoff	Evaporation	Coefficient of runoff
Russian Soviet Federative Socialist Republic ...	571	235	336	0.41
Ukrainian SSR	609	83	526	0.14
Moldavian SSR	520	24	496	0.05
Byelorussian SSR	745	175	570	0.24
Estonian SSR	746	259	487	0.35
Latvian SSR	820	268	552	0.33
Lithuanian SSR	805	235	570	0.29
Georgian SSR	1,420	769	651	0.54
Azerbaijan SSR	616	101	515	0.16
Armenian SSR	678	218	460	0.32
Kazakh SSR	308	24	284	0.08
Uzbek SSR	232	27	205	0.12
Kirghiz SSR	470	274	196	0.58
Tadjik SSR	596	358	238	0.60
Turkmen SSR	205	2	203	0.01
Total area of the USSR	531	198	333	0.37

TABLE 1-105. UNION OF SOVIET SOCIALIST REPUBLICS—PRINCIPAL RESERVOIRS
(Source: LaMothe, U.S.Army Medical Intelligence Office, 1971)

[Reservoirs with surface area of 1,000 km^2 or more]

River	Name of Reservoir	Date of Construction	Area km^2
Volga	V.I. Lenin	1955	6,500
Volga	Rybinsk	1941	4,550
Volga	Cheboksary	U/C	3,780
Volga	22nd Congress (Volgograd)	1958	3,160
Volga	Saratov	1965	1,950
Volga	Gorky	1955	1,570
Irtysh	Bukhtarma	1960	5,500
Angara	Bratsk	1961	5,426
Kama	Nishne Kamskaya	U/C	5,400
Kama	Kama	1954	1,720
Kama	Votkinsk	1962	1,120
Ob	Kamenskoye	Completed	4,500
Ob	Novosibirsk	1957	1,070
Don	Tsimlyansk	1952	2,700
Dnepr	Kakhovka	1955	2,155
Dnepr	Kremenchug	1961	2,500
Yenisey	Krasnoyarsk	1966	2,130

U/C   Under construction

TABLE 1-106. UNION OF SOVIET SOCIALIST REPUBLICS—ICE STORAGE IN GLACIERS

(Source: Kotlyakov, Proc. Reading Symposium, IASH—UNESCO—WMO, 1972)

Glacier provinces	Area of glaciers km^2	Average depth of ice, m	Volume of ice, km^3	Ice mass, 10^{17}g of water
Atlantic-Arctic (islands of the Soviet Arctic)	61,000	210	13,100	113.5
Atlantic-Eurasian (mountains of the Caucasus, Altay and Central Asia)	19,000	140	2,700	23.0
Pacific-Asian (mountains of East Siberia and Far East)	1,900	70	130	1.1

TABLE 1-107. UNION OF SOVIET SOCIALIST REPUBLICS—DISTRIBUTION OF HYDROELECTRIC POTENTIAL

(Source: Guerady, United Nations, WRS 32, 1967)

[Unit: 10^{12} kWh]

	Republic and economic regions	Power resources	
		technically possible	economically effective
1.	RSFSR	1,670	851
	incl. the economic regions		
	a) Northwest	55	42
	b) Povolzhie	41	39
	c) North Caucasus	53	25
	d) Ural	62	42
	e) Eastern Siberia	580	390
	f) Western Siberia	93	46
	g) Far East	768	254
2.	Ukrainian SSR	21	17
3.	Transcaucasian economic region	93	45
4.	Middle-Asian economic region	248	146
5.	Kazakhstan economic region	62	27
6.	Baltic economic region	7.5	6
7.	Other economic regions	8.5	2
A.	Within the European part of the USSR (with the Ural area)	355	230
B.	Within the Asian part of the USSR ...	1,751	864
	Total for the USSR	2,106	1,094

TABLE 1-108. UNION OF SOVIET SOCIALIST REPUBLICS—WATER USE, 1970

(Source: Papisov, et al., UN Water Resources Series No. 44, 1973)

Use category	Total water use		Consumption	
	km^3/year	percentage	km^3/year	percentage
Population	12.6	4.4	3.0	1.8
Industry, including thermal power generation	89.6	31.1	5.0	3.1
Agricultural, total	147.7	51.1	123.0	75.2
(Irrigation only)	(130.5)	(45.0)	(117.0)	(72.5)
Fishery	9.2	3.2	2.0	1.2
Free water storage surface evaporation losses	29.5	10.2	29.5	18.1
Total	288.6	100	162.5	100

TABLE 1-109. UNION OF SOVIET SOCIALIST REPUBLICS—HYDROPOWER PRODUCTION, INLAND NAVIGATION AND RECLAMATION

(Source: Kuznetsov and L'vovich, in Fox I.K., Water Resources Law and Policy in the Soviet Union, Univ. Wisconsin Press, 1971)　*

Use	Years				
	1913	1940	1955	1958	1965[1]
Hydro-energy Production of hydro-energy, billion kilowatt hours	0.04	5.1	23.1	46.5	100.0
Significance of hydro-energy, in total production of electrical energy, percentage	2.0	10.5	13.6	20.0	20.0
Navigation Length of inland waterways in operation, thousand km	64.6	107.3	132.0	133.4	152.0
Length of artificial waterways, thousand km	3.1	4.6	5.6	9.7	15.1
Reclamation Actual watered lands, million hectares, approx.	3.5	6.1	7.0	8.0[2]	10.6
Drained lands, million hectares, approx.	2.8	5.9	8.2	8.4[3]	12.4

* Original source: Vendrov, S.L., and Kalinin, G.P. Resursy poverkhnostnykh vod SSSR, ikh ispol'zovanie i izuchenie [Surface water resources of the USSR, their utilization and study]. Leningrad, 1959. [Papers presented at the third conference of the Geographical Society of the USSR; speeches on "The role of geography in the study, utilization, conservation and renewal of the natural resources of the USSR"].

[1] As planned
[2] In 1959
[3] In 1956

TABLE 1-110. UNION OF SOVIET SOCIALIST REPUBLICS—WATER REQUIREMENTS FOR IRRIGATION

(Source: Guerardy, United Nations, WRS 32, 1967)

Republic and regions	Irrigation areas provided with water without flow diversion of Northern and Siberian rivers, million ha	Water intake at heads of irrigation system 10^9 m3	Flow deficit 10^9 m3	Proposed irrigation areas million ha	Water intake at heads of irrigation system 10^9 m3	Flow deficit 10^9 m3	Sources for meeting water deficits
I. RSFSR							
1. Povolzhie	2.1 - 3.0	21.4 - 30.6	—	8	51.4 - 60.6	30	Northern rivers
2. The Northern Caucasus and the Don River basin	3.2	22.7	—	12	67.7	45	Northern rivers
3. Western Siberia	1.6	7.4	—	2	10.4	—	
4. Eastern Siberia	0.5	3.1	—	3	13.3	—	
II. Ukrainian and Moldavian Republics	4.7 - 5.7	24.3 - 29.4	—	13	64.3 - 69.4	33 / 7	The Danube, North-western rivers
III. Kazakh Republic excluding regions of Syr-Darya basin	3.1	33.5	—	20	117.5	28 / 116	Northern rivers / Siberian rivers
IV. Republics of the Middle Asia and regions of Kazakh Republic in Syr-Darya basin	10.0	143.0	30	14	208	65	Siberian rivers
V. Republics of the Transcaucasus	2.9	17.8	—	3	18.5	—	
Total for the USSR	28.1 - 30.0	273.2 - 287.5	30	75	611.1 - 625.4	324 / 110 / 181 / 33	Including: Northern rivers & Northwestern rivers / Siberian rivers / The Danube

TABLE 1-111. UNION OF SOVIET SOCIALIST REPUBLICS—WATER SUPPLY AND WASTE DISPOSAL

(Source: Guerardy, United Nations, WRS 32, 1967)

Description	1965	1966	1967 [in million m3/day]
Total water consumption	176	197	211
Industrial	149	170	182
Domestic	27	27	29
Total sewage water discharged	151	174	188
Industrial waste water	64	69	74
Total industrial waste water treated	31	39	45
Percentage (%)	48.5	56.5	60.8
Municipal per capita consumption (l/day)	150	156	164

TABLE 1-112. UNITED KINGDOM—DISCHARGE OF PRINCIPAL RIVERS
(Source: Willis, Water Resources Board, 1973 and UNESCO, 1971)

River and station	Mean monthly discharge, m3/s													Period of record
	Oct.	Nov.	Dec.	Jan.	Feb.	Mar.	Apr.	May	Jun.	Jul.	Aug.	Sep.	Year	
Spey, Boat o Brig	66.3	71.6	81.4	75.3	69.3	75.5	69.2	62.5	42.1	38.4	61.1	53.3	63.9	1959-70
Tweed, Norham	93.2	122.4	115.3	103.3	100.5	108.6	71.6	69.1	40.7	37.8	68.5	80.7	84.3	1962-70
Clyde, Blairston	51.0	58.6	64.0	53.1	44.0	39.9	30.7	27.2	17.1	17.2	28.4	38.1	39.1	1958-70
Tyne, Bywell	51.8	63.6	70.5	69.0	58.5	54.5	44.2	28.7	17.7	23.0	36.4	42.1	46.7	1956-70
Trent, Colwick	77.5	92.1	140.6	136.8	111.9	90.9	83.7	66.4	43.5	44.7	43.1	60.3	82.6	1958-68
Nene, Orton	4.3	9.7	13.1	17.3	18.7	16.3	9.8	6.4	3.8	3.6	3.3	3.2	9.1	1940-70
Avon, Bath	15.1	25.3	32.8	36.3	31.5	24.5	20.2	13.1	11.5	8.1	7.9	9.6	19.6	1953-66
Wye, Cadora	61.7	112.2	166.0	132.6	121.9	89.1	63.2	46.3	36.6	27.6	31.9	46.6	78.0	1937-69
Usk, Chain Bridge	35.2	41.9	57.9	54.1	43.9	31.1	27.4	21.2	11.8	9.9	11.1	18.1	30.3	1957-70
Lune, Halton	45.4	48.0	55.5	44.8	32.9	34.1	36.5	27.1	17.6	20.3	32.7	40.9	36.3	1960-70
Thames, Teddington	45.5	94.1	114	131	142	117	85.0	57.8	43.6	30.8	29.3	32.6	76.9	1937-64
Severn, Bewdley	52.4	92.6	99.0	110	107	69.9	51.5	34.7	28.2	22.4	29.1	42.0	61.5	1937-64

TABLE 1-113. UNITED KINGDOM—AVAILABILITY OF WATER IN NORMAL AND DRY YEARS
(Source: Institution of Civil Engineers, 1963)

Area	Average rainfall	Coefficient of variation	Rainfall in dry year (1 in 10)		Average evaporation	Water available in:		
						average year	dry year	dry year as % of average year
	in.	%	%	in.	in.	in.	in.	
Northern Ireland	42	11	85.9	36.1	14	28	22.1	78.9
Scotland	52	14	82.1	42.7	15	37	27.7	74.9
Pennines	42	15	80.8	33.9	16	26	17.9	68.8
Wales and West Midlands	43.4	15	80.8	35.1	19	24.4	16.1	66.0
South West England	38	16	79.5	30.2	19	19	11.2	58.9
South East England	27	16	79.5	21.5	18	9	3.5	38.9

TABLE 1-114. UNITED KINGDOM—AVAILABILITY OF WATER IN 1990

(Source: Pugh, Inst. of Civil Engineers, 1963)

Area	Estimated average daily demand 1990 mgd	Estimated sustained max. demand to be provided for mgd	Average intensity of rainfall inches	Total water available mgd	Average net quantity available surface and underground mgd	Dry year quantity available mgd	Surplus resources for all other water uses including industry & agriculture mgd	Estimated population 1990 millions	Surplus resources expressed as gal. per head of population in 1990
Northern Ireland	100	110	42	8,700	4,800	4,400	4,300	1.65	2,600
Scotland	640	700	52	61,500	50,000	45,000	44,300	6.09	7,270
Pennines	950	1,045	42	20,775	13,100	11,790	10,745	13.25	810
Wales and West Midlands	560	615	43.4	17,972	12,700	11,430	10,815	8.42	1,275
South West England	200	220	38	9,550	5,200	4,680	4,460	2.86	1,575
South East England	1,900	2,090	27	26,945	9,800	8,820	6,730	28.30	235
Totals:	4,350	4,780	—	145,442	95,600	86,120	81,350	60.57	—

TABLE 1-115. UNITED KINGDOM—WATER WITHDRAWN IN ENGLAND AND WALES

(Source: U.K. Water Resources Board, 1971)

[Year ending Sep. 30, 1969; in million m3]

	Surface Water			Ground Water			Total		
	Quantity withdrawn	Quantity licensed to be withdrawn	%	Quantity withdrawn	Quantity licensed to be withdrawn	%	Quantity withdrawn	Quantity licensed to be withdrawn	%
Public Water Supply	3,241	4,911	65	1,737	2,714	65	4,978	7,625	65
C.E.G.B. 1	12,595	21,136	60	10	21	50	12,605	21,157	60
Industry:									
Process water and cooling (other than C.E.G.B.)	3,633	5,927	60	593	937	65	4,226	6,864	60
Agriculture:									
Spray irrigation	12	65	20	8	29	25	20	94	20
Other than spray irrigation	2	5	50	29	60	50	31	65	50
Miscellaneous	262	437	60	15	24	60	277	461	60
Total	19,745	32,481	60	2,392	3,785	65	22,137	36,266	60

1 Central Electricity Generating Board (Cooling Water).

Note: Percentages have been given to the nearest 5 per cent.

TABLE 1-116. UNITED KINGDOM—WITHDRAWAL OF SURFACE WATER IN ENGLAND AND WALES, BY RIVER AUTHORITY

(Source: Water Resources Board, 1971)

[Period Oct. 1, 1968 to Sep. 30, 1969; in thousand m3]

River Authority	Public and Private water supply	C.E.G.B. [1] Cooling and Hydro-power	Industry Process water and cooling (other than C.E.G.B.)	Agriculture Spray Irrigation	Agriculture Other than Spray Irrigation	Miscellaneous	Total (rounded to nearest 100x10³m3)
Northumbrian	294,435	1,528,154	431,112	19	597	0	2,255,300
Yorkshire Ouse and Hull	326,178	2,157,370	491,371	257	292	0	2,975,500
Trent	189,017	2,500,068	247,673	844	*	221,890	3,159,500
Lincolnshire	16,507	13,074	34,446	223	*	9	64,300
Welland and Nene	31,575	100,311	10,146	183	16	*	142,200
Great Ouse	13,000	—	—	5,100	200	*	18,300 2
East Suffolk and Norfolk	30,202	503,364	5,350	842	0	10	539,800
Essex	81,249	600,027	269,762	1,129	0	46	952,200
Kent	13,926	25,573	353,705	786	5	246	394,200
Sussex	7,880	431,281	581 2	361	42	11	440,200
Hampshire	19,157	40,000 E	473 2	268	0	0	59,900
Isle of Wight	2,239	18,359	130	12	0	*	20,700
Avon and Dorset	41,323	0	2,367 2	—	*	0	43,700 2
Devon	49,256	8,600 E	9,986 E	342 E	*	*	68,200 E
Cornwall	66,485	—	72,318 E	195 E	0	0	139,000 E
Somerset	48,030	1,122	1,163	216	*	0	50,500
Bristol Avon	43,616	0	99,652	29	0	62	143,400
Severn	236,569	1,108,244	25,587	128	834 E	913	1,372,300
Wye	138,958	0	509	55	*	0	139,500
Usk	107,931	649,573	37,891	14	27	9,683	805,100
Glamorgan	95,650	59,546	383,412	18	*	*	538,600
South West Wales	50,788	47,849 2	15,490	70	*	*	114,200
Gwynedd	32,656	893,172 / 766,000 3	65,000 E	*	6	460	1,757,300
Dee and Clwyd	131,870	4,683	99,163	14	*	0	235,700
Mersey and Weaver	168,625	620,256	619,292	73	*	0	1,408,200
Lancashire	147,563	199,303	248,275	14	210	28,801	624,200
Cumberland	214,238	54,483	43,646	*	*	*	312,400
Thames Conservancy	521,349	134,548	52,715	1,073	27	227	709,900
Lee Conservancy	121,108	130,122	10,554	18	53	*	261,800
Total (rounded to nearest 100x103 m3)	3,241,300	12,595,100	3,632,800	12,300	2,300	262,400	19,746,200
Licensed quantity to which above totals relate (rounded)	4,911,200	21,136,200	5,850,700	64,600	4,800	437,300	32,404,700

— No details available
* Possible individual withdrawals of less than 8,000 m3/yr
1 Central Electricity Generating Board
2 Partial returns only
3 Hydropower
E Estimate

FIGURE 1-10. UNITED KINGDOM – HYDROMETRIC AREAS
AND RIVER AUTHORITIES IN GREAT BRITAIN

(Source: Water Resources Board, 1964)

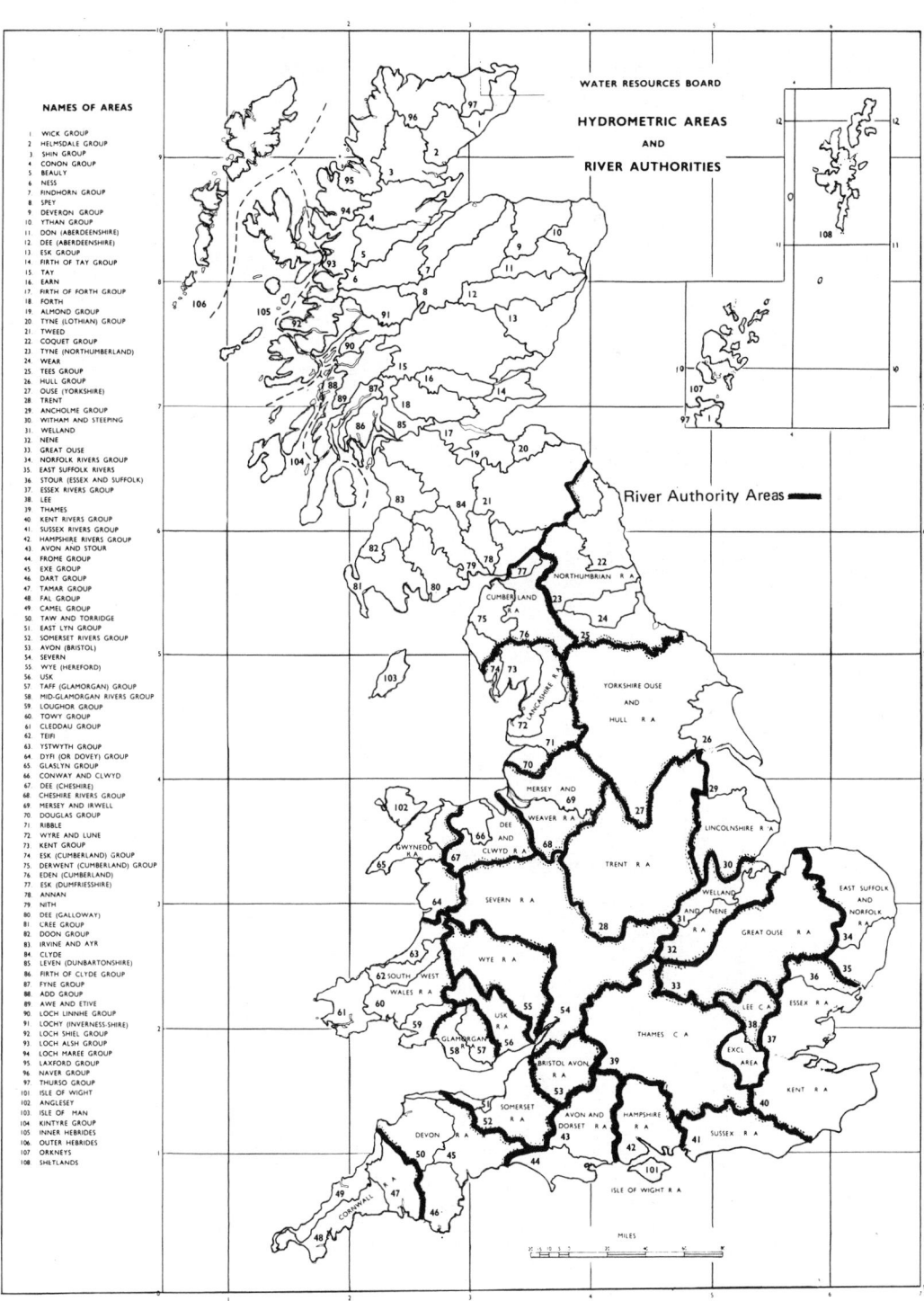

NAMES OF AREAS

1. WICK GROUP
2. HELMSDALE GROUP
3. SHIN GROUP
4. CONON GROUP
5. BEAULY
6. NESS
7. FINDHORN GROUP
8. SPEY
9. DEVERON GROUP
10. YTHAN GROUP
11. DON (ABERDEENSHIRE)
12. DEE (ABERDEENSHIRE)
13. ESK GROUP
14. FIRTH OF TAY GROUP
15. TAY
16. EARN
17. FIRTH OF FORTH GROUP
18. FORTH
19. ALMOND GROUP
20. TYNE (LOTHIAN) GROUP
21. TWEED
22. COQUET GROUP
23. TYNE (NORTHUMBERLAND)
24. WEAR
25. TEES GROUP
26. HULL GROUP
27. OUSE (YORKSHIRE)
28. TRENT
29. ANCHOLME GROUP
30. WITHAM AND STEEPING
31. WELLAND
32. NENE
33. GREAT OUSE
34. NORFOLK RIVERS GROUP
35. EAST SUFFOLK RIVERS
36. STOUR (ESSEX AND SUFFOLK)
37. ESSEX RIVERS GROUP
38. LEE
39. THAMES
40. KENT RIVERS GROUP
41. SUSSEX RIVERS GROUP
42. HAMPSHIRE RIVERS GROUP
43. AVON AND STOUR
44. FROME GROUP
45. EXE GROUP
46. DART GROUP
47. TAMAR GROUP
48. FAL GROUP
49. CAMEL GROUP
50. TAW AND TORRIDGE
51. EAST LYN GROUP
52. SOMERSET RIVERS GROUP
53. AVON (BRISTOL)
54. SEVERN
55. WYE (HEREFORD)
56. USK
57. TAFF (GLAMORGAN) GROUP
58. MID-GLAMORGAN RIVERS GROUP
59. LOUGHOR GROUP
60. TOWY GROUP
61. CLEDDAU GROUP
62. TEIFI
63. YSTWYTH GROUP
64. DYFI (OR DOVEY) GROUP
65. GLASLYN GROUP
66. CONWAY AND CLWYD
67. DEE (CHESHIRE)
68. CHESHIRE RIVERS GROUP
69. MERSEY AND IRWELL
70. DOUGLAS GROUP
71. RIBBLE
72. WYRE AND LUNE
73. KENT GROUP
74. ESK (CUMBERLAND) GROUP
75. DERWENT (CUMBERLAND) GROUP
76. EDEN (CUMBERLAND)
77. ESK (DUMFRIESSHIRE)
78. ANNAN
79. NITH
80. DEE (GALLOWAY)
81. CREE GROUP
82. DOON GROUP
83. IRVINE AND AYR
84. CLYDE
85. LEVEN (DUNBARTONSHIRE)
86. FIRTH OF CLYDE GROUP
87. FYNE GROUP
88. ADD GROUP
89. AWE AND ETIVE
90. LOCH LINNHE GROUP
91. LOCHY (INVERNESS-SHIRE)
92. LOCH SHIEL GROUP
93. LOCH ALSH GROUP
94. LOCH MAREE GROUP
95. LAXFORD GROUP
96. NAVER GROUP
97. THURSO GROUP
101. ISLE OF WIGHT
102. ANGLESEY
103. ISLE OF MAN
104. KINTYRE GROUP
105. INNER HEBRIDES
106. OUTER HEBRIDES
107. ORKNEYS
108. SHETLANDS

TABLE 1-117. UNITED KINGDOM—WITHDRAWAL OF GROUNDWATER IN ENGLAND AND WALES, BY RIVER AUTHORITY

(Source: Water Resources Board, 1971)

[Period Oct.1,1968 to Sep. 30, 1969; in thousand m3]

River Authority	Public and Private water supply	C.E.G.B.[1] Cooling and Hydro-power	Industry Process Water and Cooling	Agriculture Spray Irrigation	Other	Miscellaneous	Total	Total withdrawal of surface and ground water (rounded to nearest 100x10³m3)
Northumbrian	21,609	5,732	16,247	12	38	*	43,600	2,298,900
Yorkshire Ouse and Hull	71,329	1,259	28,815	394	113	*	101,900	3,077,400
Trent	191,742	528	91,205	190	268	186	284,600	3,444,100
Lincolnshire	69,267	686	19,884	221	3,287	14	93,400	157,700
Welland and Nene	25,651	0	6,119	223	395	*	32,400	174,600
Great Ouse	76,000	0	—	2,700	3,000	*	81,700	100,000
East Suffolk and Norfolk	32,488	0	15,917	650	*	222	49,300	589,100
Essex	41,717	0	24,848	378	143	501	67,600	1,019,800
Kent	178,175	300	99,764	111	846	244	279,400	673,600
Sussex	97,089	0	5,184	184	423	*	102,900	543,100
Hampshire	86,524	0	5,773 2	73	*	*	92,400	152,300
Isle of Wight	9,715	0	14	34	*	*	9,800	30,500
Avon and Dorset	40,549	0	2,995	—	—	0	43,500	87,200
Devon	16,570	0	240 E	*	*	138	17,000	85,200
Cornwall	3,755	0	6,896 E	10 E	1,864 E	*	12,500	151,500
Somerset	23,091	77	911	28	*	0	24,100	74,600
Bristol Avon	24,613	0	2,331	*	*	91	27,000	170,400
Severn	119,871	0	16,677	32	4,273 E	499	141,300	1,513,600
Wye	2,719	0	1,127	64	*	*	3,900	143,400
Usk	251	0	4,637	*	*	0	4,900	810,000
Glamorgan	12,624	0	5,967	*	*	29	18,600	557,200
South West Wales	3,059	0	1,160	*	*	0	4,200	118,400
Gwynedd	0	0	0	0	0	0	0	1,757,300
Dee and Clwyd	8,125	0	21,854	*	0	*	30,000	265,700
Mersey and Weaver	108,054	45	94,684	32	*	0	202,800	1,611,000
Lancashire	21,598	0	19,630	*	*	*	41,200	665,400
Cumberland	6,642	0	1,608	*	*	*	8,200	320,600
Thames Conservancy	367,426	5	76,023	1,427	13,670	12,192	470,800	1,180,700
Lee Conservancy	76,764	1,251	22,407	760	698	*	101,900	363,700
Total (rounded to nearest 100x103m3)	1,737,000	9,900	592,900	7,500	29,000	14,600	2,390,900	22,137,100
Licensed quantity to which above totals relate (rounded)	2,714,100	21,000	937,200	28,900	60,500	24,400	3,824,400	36,229,100

* Possible individual withdrawals of less than 16,000 m3/yr
E Estimate
— No data available

1 Central Electricity Generating Board
2 Partial return only

TABLE 1-118. UNITED KINGDOM—DEMAND FOR PUBLIC SUPPLIES IN
NORTHERN ENGLAND AND NORTHERN WALES, 1967—2001

(Source: Water Resources Board, 1971)

[in 1000 m^3/d]

River Authority	Average Daily Demand			Authorized Resources (including imports)	Deficiency (nearest 5,000 m^3/d)	
	1967	1981	2001		1981	2001
Northumbrian	840	1,335	1,890	1,138	205	750
Yorkshire Ouse & Hull	1,089	1,495	2,140	1,272	250	865
Dee & Clwyd	105	187	264	152	45	115
Mersey & Weaver	1,545	2,230	3,000	1,997	295	1,025
Lancashire	464	716	1,018	546	180	475
Cumberland	123	296	423	201	90	230
Total	4,166	6,295	8,735		1,065	3,550
Lincolnshire (Humberside)	—	135	455		135	455
Overall Total	4,166	6,394	9,190	5,306	1,200	4,005

TABLE 1-119. UNITED KINGDOM—WATER USED FOR PUBLIC SUPPLIES IN SCOTLAND, 1971

(Source: Scottish Development Department, 1973)

Water Board	Census population (1,000)	Water delivered 1,000 m3/d			Per capita consumption liters/day			Estimated population not receiving public supply	
		Total	Domestic	Commercial and Industrial	Total	Domestic	Commercial and Industrial	Number (1,000)	%
Boards in Central Area									
Ayrshire and Bute	375	185.7	126.3	59.4	495	336	159	1-3	0.3-0.8
East of Scotland	429	138.0	100.2	37.8	322	233	89	12.5	2.9
Fife and Kinross	328	108.0	75.0	33.0	329	229	100	1.2	0.4
Lanarkshire	562	204.3	140.0	64.3	364	249	115	1-2	0.2-0.4
Lower Clyde	1,501	722.9	479.3	243.6	482	319	163	2	0.1
Mid-Scotland 1	319	203	87.6	41.7 (73.7)	636	275	130 (231)	0.7	0.2
South-East of Scotland	854	273.9	181.4	92.5	321	213	108	2-4	0.2-0.5
Totals (and averages) for Central Area	4,368	1,835.8	1,189.8	646.0	(420)	(272)	(148)	20.4-25.4	0.5-0.6
Boards in Country Area									
Argyll	59	22.0	21.6	0.4	373	366	7	5	8
Inverness-shire	89	28.6	21.3	7.3	321	239	82	9	10
North of Scotland	76	28.5	23.3	5.2	375	307	68	3	4
North East of Scotland	434	127.5	80.2	47.3	335	211	124	39	9
Ross and Cromarty	59	26.3	19.0	7.3	446	322	124	5	8
South West of Scotland	143	61.8	36.4	25.4	432	255	177	9.6	7
Totals (and averages) for Country Area	860	294.7	201.8	92.9	(343)	(235)	(108)	70.5	8
Totals (and averages) for all areas	5,228	2,130.5	1,391.6	738.9	(407)	(266)	(141)	90.9-95.9	1.7-1.8

1 Special industrial supplies at Grangemouth shown separately in brackets.

TABLE 1-120. UNITED KINGDOM—WITHDRAWAL OF WATER EXCLUDING PUBLIC SUPPLIES IN SCOTLAND, 1971

(Source: Scottish Development Department, 1973)

[in 1,000 m³/d]

| Water Board Region | General Withdrawals | | | | | | Electricity Boards | | | | Canals | | Total withdrawals | |
| | Reservoirs | | River Intakes | | Water Wells | | Hydro-electric reservoirs and river intakes | | Thermal cooling water-river intakes | | User withdrawals | | | |
	Gross 1	Net 2	Gross	Net	Gross	Net	Gross	Net	Gross	Net	Gross	Net	Gross	Net
Ayrshire and Bute	–	–	5.0	5.0	–	–	400	–	5	–	–	–	410	5.0
East of Scotland	–	–	2.5	2.5	–	–	6,700	–	–	–	–	–	6,702.5	2.5
Fife and Kinross	–	–	19.5	10.1	15.9	15.9	–	–	–	–	–	–	35.4	26.0
Lanarkshire	–	–	–	–	–	–	540	–	–	–	11.4	2.9	551.4	2.9
Lower Clyde	–	–	148.3	40.6	–	–	540	–	910	–	145	4.1	1,743.3	44.7
Mid-Scotland	6.4	6.4	4.7	4.7	3.3	3.3	–	–	–	–	66.9	0.2	81.3	14.6
South-East of Scotland	–	–	900.2	22.9	1.0	1.0	–	–	–	–	16.7	12.9	917.9	36.8
Totals for Central Area	6.4	6.4	1,080.2	85.8	20.2	20.2	8,180	–	915	–	240.0	20.1	10,441.8	132.5

1 Total withdrawal including water returned locally such as water circulated.
2 Water withdrawn and not returned locally to the source.

TABLE 1-121. UNITED KINGDOM—MUNICIPAL WATER SUPPLY SYSTEMS IN ENGLAND, WALES AND SCOTLAND

(Source: Internat. Statistical Institute, 1971)

City	Year	Population served	Quantity available(1000m^3)			Annual consumption(1000m^3)	
			Production	Purchased from other plants	Total	Total	of which domestic
England and Wales							
Birmingham	1967	1,288,000	157,221	879	158,100	119,040	61,710
	1969	1,303,000	156,392	879	157,271	125,329	66,871
Blackburn 1	1967	119,890	—	0	—	17,290	9,000
	1969	119,820	—	0	—	18,497	11,100
Bolton 1	1967	466,931	28,180	12,227	40,407	40,394	28,442
	1969	474,784	28,180	12,227	40,407	43,528	31,351
Bournemouth 1	1967	278,600 *	1,440	0	1,440	17,863	12,558
	1969	283,100 *	1,520	0	1,520	18,706	13,335
Brighton 1	1968	318,000	27,166	1,065	28,231	28,159	19,638
	1970	320,000	28,826	1,244	30,070	29,862	20,816
Bristol	1967	899,000	99,114	18,496	117,610	97,664	58,493
	1969	927,000	113,345	26,544	139,889	110,686	65,726
Coventry	1969	382,324	—	—	—	31,802	16,674
Derby	1967	127,910	—	—	—	12,413 *	7,380 *
	1969	221,240	—	—	—	22,774 *	13,014 *
Greater London	1967	7,880,760	—	—	—	744,871	—
	1969	7,703,410	—	—	—	777,934	—
Leeds	1967	608,566	63,600	0	63,600	51,800	33,700
	1969	608,801	76,300	0	76,300	56,800	38,200
Luton	1967	—	—	—	—	—	—
	1969	160,000	—	—	—	14,222	9,240
Oxford	1968	362,603	28,107	1,058	29,165	29,165	15,431
	1969	369,800	28,929	1,136	30,065	30,065	15,430
Plymouth 1	1967	268,000	86,400	0	86,400	89,400	60,100
	1969	268,000	86,400	0	86,400	94,700	64,800
St. Helens 1	1967	145,400	11,379	5,461	16,440	14,859	7,617
	1969	147,270	13,678	4,964	18,642	16,824	8,840
Southampton 1	1967	445,000	43,405	11	43,416	43,416	23,444
	1969	459,000	48,494	7	48,501	48,501	26,304
Southend on Sea	1967	165,760	—	—	—	11,057	—
	1969	167,000	—	—	—	11,500	—
Stoke on Trent 1	1967	480,000	38,246	329	38,575	40,664	26,568
	1969	480,000	39,836	269	40,105	43,990	28,992
Sunderland	1967	580,000	40,332	6,393	46,725	42,324	29,257
	1969	590,000	40,664	6,343	47,007	42,573	29,821
Teesside 1	1967	485,325	126,114	0	126,114	126,114	26,162
	1969	512,085	139,084	0	139,084	139,084	31,072
Scotland							
Aberdeen 1	1967	183,000	23,381	0	23,381	18,687	11,251
	1969	180,000	24,404	0	24,404	18,927	10,465
Edinburgh	1967	468,361	63,000	0	63,000	54,000	32,000
	1969	468,361	75,000	0	75,000	57,000	35,000
Glasgow 1	1967	1,085,730	176,530	38	176,568	160,000	104,000
	1969	1,109,603	179,800	38	179,838	166,300	117,700

1   Including some surrounding districts or suburbs.
—   Not available
*   Provisional

TABLE 1-122. UNITED KINGDOM—SOURCES AND CONSUMPTION OF DRINKING WATER IN GIBRALTAR

(Source: Gonzalez, City Engineer, 1966)

[in millions of gallons]

Year	Water Produced						Total water consumed
	Ground water	Rain water	Distilled water	Imported water	Other sources	Total	
1951	23	23	0	0	2	48	42
1953	22	23	1	3	0	49	46
1955	23	29	4	0	0	56	49
1957	29	18	3	10	0	60	54
1959	28	17	2	3	0	50	56 *
1961	31	23	3	15	2	74	58
1963	39	24	0	0	1	64	66 *
1965	34	18	10	15	0	77	76

* water deficit supplied from storage.

TABLE 1-123. YUGOSLAVIA–DISCHARGE OF PRINCIPAL RIVERS
(Source: UNESCO, 1971)

River and station	Basin area km2	Mean monthly discharge, m3/s												Year	Period of record
		Jan.	Feb.	Mar.	Apr.	May	Jun.	Jul.	Aug.	Sep.	Oct.	Nov.	Dec.		
Drava, Donji Miholjac	8,880	391	396	482	558	775	846	691	589	502	487	578	472	561	1921-66
Sava, Sremska Mitrovica	7,222	1,830	1,980	2,320	2,440	2,080	1,460	977	630	589	900	1,760	2,070	1,590	1941-66
V. Morava, Ljubicevski most	7,342	262	396	450	438	362	246	133	82.7	78.4	93.0	147	201	240	1941-66
Danube, Veliko Gradiste	6,217	5,460	6,000	7,120	8,100	7,780	6,730	5,720	4,670	4,080	4,000	2,250	5,500	5,870	1921-66

TABLE 1-124. YUGOSLAVIA–RUNOFF OF PRINCIPAL RIVER BASINS
(Source: Pecinar, Water Power of Yugoslavia, Serbian Ac. of Sci., 1968)

Drainage basin	River		Basin area km2	Mean elevation above sea-level m	Elevation at point of entry or departure m	Mean annual precipitation mm	Mean discharge at mouth m3/s	Runoff Coefficient	Mean annual runoff l/km2
Black Sea	Dunav (Danube)	a)1	15,200	119	27.8	620	5,750	0.20	4.0
		b)2	585,220			855		0.37	9.8
	Tisa	a)	8,880	85	69.8	570	945	0.22	4.0
		b)	157,220			715		0.22	6.4
	Drava	a)	12,120	282	79.4	930	620	0.33	9.8
		b)	40,150			1,080		0.45	15.5
	C.D. Sava	a)	26,810	172	66.7	810	1,690	0.28	7.2
		b)	93,719			1,090		0.53	17.5
	Upper Sava		10,250	590	132.1	1,490	330	0.68	32.0
	Kupa		9,800	378	90.6	1,360	283	0.67	29.0
	Una		9,690	600	87.0	1,210	202	0.52	21.0
	Vrbac		5,570	687	86.0	1,100	102	0.52	18.2
	Bosna		10,460	640	80.0	960	174	0.54	16.5
	Drina		19,570	934	75.4	1,050	371	0.58	19.0

Kolubara	3,620	206	67.4	760	24	0.27	6.6
Morava	37,444	672	66.6	700	232	0.28	6.2
Mlava	1,860	353	66.0	670	13	0.31	7.0
Pek	1,240	420	62.4	750	13	0.44	10.5
Porechka	520	370	49.2	770	5	0.39	9.6
Timok	4,630	481	27.8	720	39	0.57	8.4
Total to Black Sea	177,614	473	27.8	900	5,750	0.47	12.4
Adriatic Sea							
Socha	2,550	767	55.0	2,030	147	0.87	58.0
Pechina	480	530	0.0	2,380	24	0.82	50.0
Reka	1,000	742	0.0	1,480	17	0.65	17.0
Mirna	560	282	0.0	1,080	16	0.80	29.0
Rasha	420	318	0.0	1,360	12	0.65	29.0
Lika-Gatska	2,400	763	0.0	1,530	42	0.45	17.5
Krka	2,250	475	0.0	1,260	50	0.55	22.0
Zrmanja	780	403	0.0	1,390	23	0.66	29.5
Cetina	5,800	820	0.0	1,330	127	0.59	22.0
Neretva	12,750	795	0.0	1,580	378	0.65	22.0
Moracha-Zeta	3,200	838	3.0	2,040	152	0.73	47.5
Beli Drim	4,300	870	270.0	920	56	0.45	13.0
Crni-Drim	3,385	1,190	485.8	810	56	0.64	16.5
Coastal Rivers	14,383	400	0.0	1,180	—	0.55	—
Total to Adriatic Sea	54,258	695	—	1,380	—	0.51	25.1
Aegean Sea							
Dragovishtica	690	990	660.0	1,010	8	0.34	11.6
Strumica	1,460	640	180.8	670	9	0.29	6.2
Vardar	21,815	810	43.6	690	146	0.30	6.7
Total to Aegean Sea	23,968	807	—	700	—	0.30	6.8
Total Yugoslavia	255,840	555	—	930	—	0.48	14.50

1 Basin area within Yugoslavia
2 Total basin area

TABLE 1-125. YUGOSLAVIA—MUNICIPAL WATER SUPPLY SYSTEMS

(Source: Internat. Statistical Institute, 1971)

City	Year	Population served	Total quantity available (1000m^3)	Annual consumption (1000m^3)	
				Total	of which domestic
Beograd (Belgrade)	1967	747,000	82,234	65,795	32,700
	1969	797,000	100,756	74,475	35,547
Novi Sad	1967	129,306	10,912	9,094	3,175
	1969	136,064	10,412	8,026	4,068
Sarajevo	1967	233,000	27,820	17,768	9,178
	1969	—	31,456	19,048	9,758
Skopje	1967	264,000	18,015	12,614	7,271
	1969	—	20,308	13,466	7,459
Split	1967	101,423	28,836	28,836	8,032
	1969	108,098	29,731	29,731	8,451
Zagreb	1967	463,372	49,018	44,219	17,766
	1969	480,699	56,671	49,502	20,177

— Not available

EUROPE
REFERENCES CITED

References cited in more than one country:

UNESCO, Discharge of Selected Rivers of the World; Vol. I. General and Regime Characteristics of Stations Selected (1969); Vol. II. Monthly and Annual Discharges Recorded at Various Selected Stations (1971); Vol. III. Mean Monthly and Extreme Discharges 1965-1969 (1971). Copyright by UNESCO/IASH/WMO. Reprinted by permission.

International Statistical Institute, 1970, International Statistical Yearbook of Large Towns. Vol. V., The Hague, The Netherlands.

Proceedings of the International Conference on Water for Peace, 1967, Superintendent of Documents, Washington, D.C. 20402.

Economic Commission for Europe, 1970, Bulletin on Conditions of Hydraulicity in Europe, Vol. I, 1968-69.

EUROPE GENERAL

Economic Commission for Europe, 1967, Ad Hoc Group of Experts for the Study of Concepts and Methods Required for Analyzing the Situation and Development of Water Resources in ECE Countries, Water Util./Meth/13/ Corr. 1,9 May 1967.

Verband der Deutschen Gas-und Wasserwerke e.V., 1971,81. Wasserstatistik-Berichtsjahr 1969, VGW, Frankfurt am Main.

U.S. Federal Power Commission, 1971, World Power Data-1968, Superintendent of Documents, Washington, D.C. FPC P-40.

AUSTRIA

Hydrographischer Dienst in Osterreich, 1972, Hydrographisches Jahrbuch von Osterreich-1967, Hydrographisches Zentralburo, Wien.

Verlag fur Jugend und Volk, 1967, die Wiener Wasserversorgung, Vienna.

BELGIUM

Interministeriele Commissie voor het Waterbeleid, 1971, Hydrologisch Jaarboek van Belgie-1969, Bruxelles.

Commissariat Royal au Probleme de l'Eau, 1967, Le Probleme de l'Eau en Belgique, Vol. III. Bruxelles.

Snel, M.J., 1974, L'Industrie et son Approvisionnement en Eau en Belgique, Hydrographica, Vol. I., No. 1, Societe Nationale des Distributions d'Eau, Bruxelles.

BULGARIA

Vodproekt Design Institute, 1972, Irrigation in Bulgaria, Eighth International Congress on Irrigation and Drainage - Varna, May, 1972.

DENMARK

Forureningsradet, 1971, Vandressource, Publ. Nr. 14, Copenhagen.

Stads-og Havneingenior Foreningen i Danmark and Dansk Vandteknisk Forening, 1971, Oversigt over Indretning og Drift af Vandvaerker, 1969-1970, Herning.

FINLAND

Jaatinen, S., Water Resources in Finland, Aqua Fennica-1971, Water Association of Finland, Helsinki.

Erkola, P., Water Supply and Sewerage in Finland, Aqua Fennica-1971, Water Association of Finland, Helsinki; also private communication (1972).

FRANCE

Direction du Gaz et de l'Electricite, 1966, Stations de Jaugeage, Annee 1965, Imprimerie Nationale, Paris.

Commission de l'Eau, 1971, Rapports des Commissions du 6e Plan 1971-1975, la Documentation Francaise, Paris.

WEST GERMANY

Bundesanstalt fur Gewasserkunde/Deutsche Forschungsgemeinschaft, 1972, International Hydrological Decade - 1969 Yearbook, Koblenz.

HUNGARY

Kernacs, S., National Water Authority, 1970, Short History of Water Management in Hungary, Hydraulic Documentation and Information Centre, Budapest.

Ballo, I.Z., National Water Authority, 1970, Water Resources Administration in Hungary, Interregional Seminar on Current Issues of Water Resources Administration, United Nations, New Delhi, 1973.

Illes, G. and Simo, J., 1970, Water Supply in Hungary, Aqua, No. 4, 1970.

ICELAND

Arnorsson, et al., 1969, General Aspects of Thermal Activity in Iceland, XXIII International Geological Congress Proc. Symp. II - Mineral and Thermal Waters of the World-1968; Vol. 18, Academia, Prague.

IRELAND

Common, R., 1967?, Land Drainage and Water Use in Ireland; Irish Geographical Studies, Queen's University, Belfast.

ITALY

Martelli, T., et al., 1968, Gli Acquedotti di Roma dall'Eta Imperiale ad Oggi, Ufficio d'Igiene e Sanita del Comune di Roma.

LUXEMBOURG

Barthel, J., 1966, Politique de l'Eau au Luxembourg, Techniques et Sciences Municipales 61e Anne. No. 3.

MALTA

United Nations, 1964, Water Desalination in Developing Countries, Publ. No. 64, II. B. 5., ST/ECA/82.

NETHERLANDS

Rijkswaterstaat, 1968, De Waterhuishouding van Nederland, The Hague; also river flow and water use data (private communication-1972).

Rijkswaterstaat, 1964, Policy in the Development of the Water Resources of the Netherlands, in Proc. Sixth Regional Conference on Water Resources Development in Asia and the Far East, UN/ECAFE Water Resources Series No. 28.

Huisman, L., 1970, Water Use Statistics of the Netherlands, Journal of the Institution of Water Engineers, Vol. 24, No. 8, Nov.

Weichart, G., The North Sea, Ambio, Vol. 2, No. 4, Stockholm; also in Environment, Vol. 16, No. 1, 1974.

POLAND

Mikulski, Z., 1968, An Outline of Poland's Hydrography, U.S. Dept. of Commerce, National Technical Information Center, Springfield, Virginia.

PORTUGAL

Quintela, A., 1967, Recursos de Aguas Superficiais em Portugal Continental, Lisboa.

SPAIN

Direccion General de Obras Hidraulicas, 1971, El Inventario de Recursos Hidraulicos y los Balances Hidraulicos de Caracter Nacional, Centro de Estudios Hidrograficos, Madrid.

Llamas, M.R., Sobre el Papel de las Aguas Subterraneas en Espana, Agua, July-August 1967.

Martin, S., 1969, Water Resources Administration in Spain, UN Seminar on Current Issues of Water Administration, New Delhi, 1973.

Sindicato Nacional de Agua, Gas y Electricidad, 1971, Datos Estadisticos, Tecnicos y Laborales de las Industrias de Abastecimientos de Agua para Usos Domesticos e Industriales y para Regadios en Espana, Madrid.

SWEDEN

Falkenmark, M., Report on Water Resources Activities in Sweden; First World Congress on Water Resources, International Water Resources Association, Chicago, 1973.

SWITZERLAND

Walser, M.E., 1961, Sur le Regime des Eaux en Suisse, Office Federal de l'Economie Hydraulique.

Office Federal de l'Economie Hydraulique, 1963, l'Utilisation des Ressources Hydraulique de la Suisse.

Societe Suisse de l'Industrie du Gaz et des Eaux, 1972, Distributions d'Eau en Suisse-Resultats Statistiques-1971.

UNION OF SOVIET SOCIALIST REPUBLICS

Guerardy, I.A., Experience in Planning for Utilization Development of Water Resources in the Soviet Union, Proc. Seventh Regional Conference on Water Resources Development in Asia and the Far East, UN Water Resources Series No. 32, 1967.

Papisov, V.K., et al., Multipurpose Water Resources Use in the USSR, Proc. Tenth Session of the Regional Conference on Water Resources Development in Asia and the Far East, UN Water Resources Series No. 44, 1973.

Kudelin, B.I., et al., 1970, The Role of the Underground Flow in the Water Balance of the USSR, Proc. Reading Symposium on World Water Balance IASH-UNESCO-WMO 1972.

Perekhrest, S.M., 1971, Water Resources in the Ukrainian S.S.R., Present and Future, Nature and Resources, Vol. 7, No. 3, Sept.

Kuznetsov, N.T., and L'vovich, M.I., 1971, Problems of the Complex Use and Conservation of Water Resources: Theoretical Prerequisites, in Fox, Water Resources Law and Policy in the Soviet Union, University of Wisc. Press. Translated from "Problemy kompleksnogo ispol 'zovaniya i okhrana vodnykh resursov: Teoreticheskie predposylki," in Academy of Sciences of the USSR, Institute of Geography, Prirodnye resursy Sovetskogo Soyuza, ikh ispol 'zovanie i vosproizvodstvo [Natural resources of the Soviet Union, their use and renewal] . Moscow: Academy of Sciences, 1963.

USSR National Committee for International Hydrological Decade, 1969, Summary of the USSR Activities on the I.H.D. Programme for Five Years (1965-69), Leningrad.

LaMothe, J.D., 1971, Water Quality of the Soviet Union, Medical Intelligence Office, U.S. Army, Washington, D.C. NTIS AD-728 516.

Kotlyakov, V.M., 1970, Land Glaciation Part in the Earth's Water Balance, Proc. Reading Symposium on World Water Balance, IASH-UNESCO-WMO, 1972.

UNITED KINGDOM

Willis, A., 1973, Data on Discharge of Principal Rivers, Water Resources Board, Reading (private communication).

Pugh, N.J., 1963, Water Supply, Proc. Symposium Conservation of Water Resources in the United Kingdom, 30 Oct-1Nov. 1962, Institution of Civil Engineers, London.

Water Resources Board, 1971, Seventh Annual Report for Year Ending 30 Sept., 1970, HMSO, London.

Scottish Development Board, 1973, A Measure of Plenty-Water Resources in Scotland-A General Survey, HMSO, Edinburgh. (Data reproduced with the permission of the Controller of Her Britannic Majesty's Stationery Office).

Gonzalez, F. J., The Water Supply in Gibraltar, Aqua, 1966, No. 2.

YUGOSLAVIA

Pecinar, M., 1968, Water Power of Yugoslavia, Serbian Academy of Science (in Serbian).

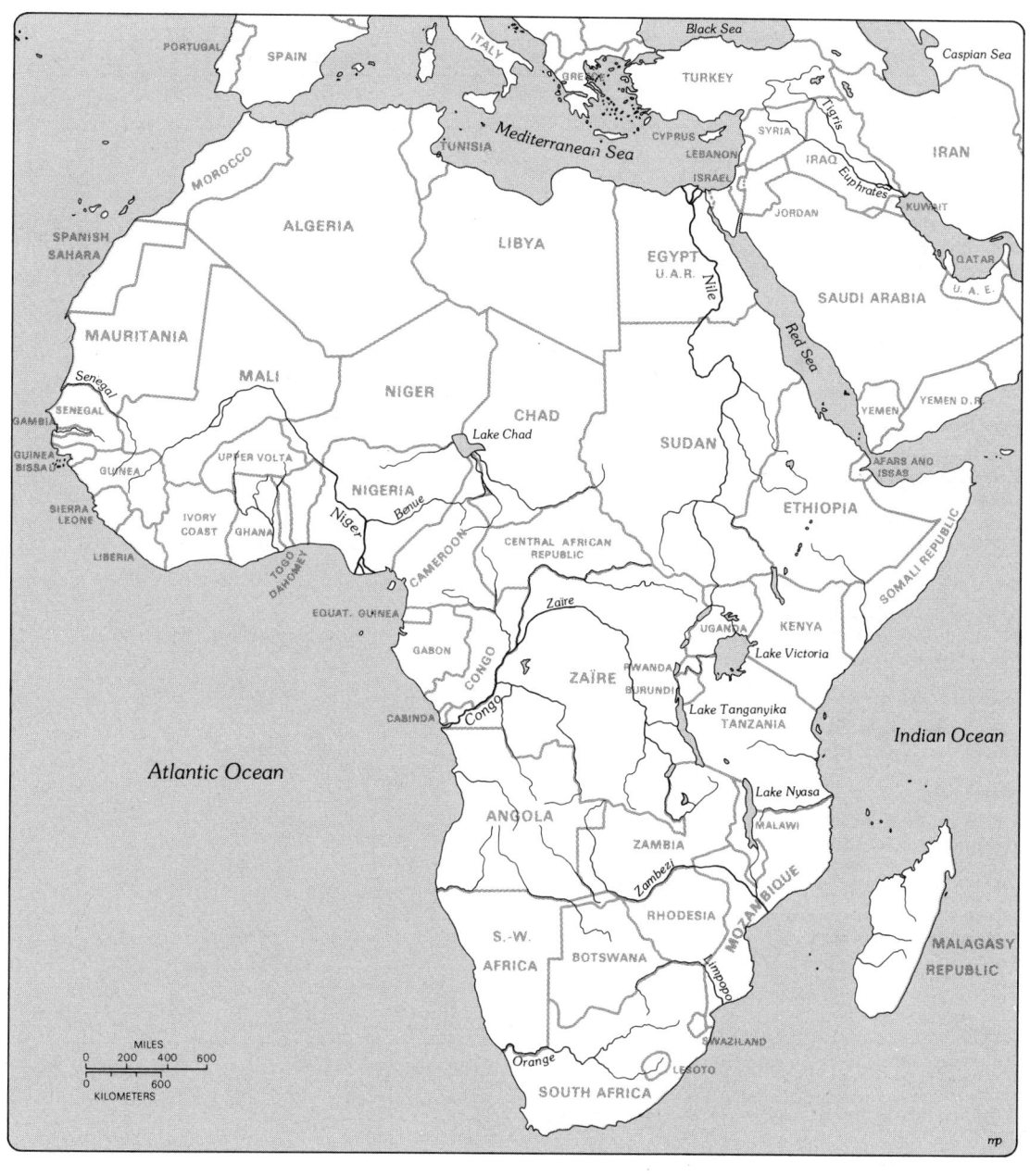

PORTUGAL SPAIN

Black Sea

Caspian Sea

ITALY GRE. TURKEY

Mediterranean Sea

TUNISIA CYPRUS SYRIA IRAN
LEBANON
ISRAEL IRAQ KUWAIT

MOROCCO JORDAN

SPANISH
SAHARA ALGERIA LIBYA EGYPT
U.A.R. SAUDI ARABIA QATAR U.A.E.

Nile

MAURITANIA Red Sea

Senegal MALI NIGER CHAD SUDAN YEMEN YEMEN D.R

SENEGAL Lake Chad AFARS AND
GAMBIA ISSAS
GUINEA UPPER VOLTA NIGERIA ETHIOPIA
BISSAU GUINEA Niger Benue CENTRAL AFRICAN
SIERRA IVORY REPUBLIC SOMALI REPUBLIC
LEONE COAST GHANA CAMEROON
LIBERIA TOGO Zaire
DAHOMEY UGANDA KENYA
EQUAT. GUINEA Lake Victoria
GABON CONGO ZAÏRE RWANDA
BURUNDI
CABINDA Congo Lake Tanganyika
TANZANIA Indian Ocean

Lake Nyasa

Atlantic Ocean

ANGOLA ZAMBIA MALAWI

Zambezi MOZAMBIQUE MALAGASY
RHODESIA REPUBLIC
S.-W.
AFRICA BOTSWANA
Limpopo
SWAZILAND

MILES
0 200 400 600 Orange LESOTO
0 600 SOUTH AFRICA
KILOMETERS

rrp

TABLE 2-1. AFRICA—

(Source: ECA

[Information

Country	Basuto-land	Central African Rep.	Chad	Congo, Dem. Rep. of	Ethiopia	French Islands		
Area, km^2	30,000	660,000	1,284,000	2,344,116	907,230	12,815		
Reference No.	1	2	3	4	5	6a	6 6b	6c
Precipitation								
Total number of stations in 1964 (1)	71	131	240	500	76	127	6	31
Stations equipped with recorders (2)	4	7	10	54	16	12	2	5
Percentage of (2) to (1)	5.6	5.35	4.2	10.8	21	9.5	33	16.2
Density of stations/1,000 km^2	2.35	0.2	0.18	0.21	.084	51	.74	14.1
Average length of records	40	33	35	56	—	48	13	43
Evaporation								
Total number of stations in 1964	4	21	18	181	3	13	4	12
Percentage of evaporation to precipitation	5.6	16	7.5	36	3.95	10.2	66	39
Average length of records	7	23	7	31	—	16	13	8
Water levels								
Total number of stations in 1964 (1)	11	43	114	97	64	—	—	—
Stations equipped with recorders (2)	11	1	1	3	22	—	—	—
Percentage of (2) to (1)	100	2.33	0.88	3.1	34.4	—	—	—
Density of stations/1,000 km^2	0.36	0.65	0.09	.04	0.07	—	—	—
Average length of records	8	13	24	72	12	13	—	—
Discharge								
Total number of stations in 1964	—	43	58	3	55	14	—	—
Density of stations/1,000 km^2	—	0.65	.045	.013	0.06	5.6	—	—
Average length of records	—	13	24	49	12	13	—	—
Sediment Transportation								
Longest continuous record of any station	—	—	—	—	3	—	—	—
Year first station established	—	—	—	—	1960	—	—	—
Ground water								
Longest continuous record of any station	—	—	5	—	—	—	—	—
Year first station established	—	—	1958	—	—	—	—	—
Quality of water								
Longest continuous record of any station	—	—	—	—	1	—	—	—
Year first station established	—	—	—	—	1960	—	—	—

Data are supplied by the Governments unless otherwise stated; length of records in years.
6a — Reunion area — 2,512 km^2
6b — Islands New Amsterdam, Kerguelen and Crozet together area — 8,104 km^2
6c — Grand Comore, Maheli, Anjouan and Mayotte together area — 2,199 km^2

AVAILABILITY OF HYDROLOGIC DATA

and WMO, 1966)

as of 1964]

Terr. of Afars and Issas	Gambia	Ghana	Guinea	Kenya	Liberia	Malawi	Mauri-tania	Mauri-tius	Morocco
23,000	10,400	238,000	255,000	583,000	111,300	118,485	1,085,000	2,090	500,000
7	8	9	10	11	12	13	14	15	16
20	5	493	49	999	17	219	39	271	626
2	–	44	6	56	–	–	6	12	24
10	–	8.9	12.25	5.6	0	0	15.4	4.5	3.83
0.87	0.48	2.07	0.19	1.71	.153	1.85	0.04	129	1.25
63	15	63	61	73	36	69	42	101	46
5	2	133	15	136	–	38	33	9	128
25	40	30.4	30.6	13.6	–	17.3	85	3.33	20.4
20	12	11	32	18	6	11	5	5	–
–	–	92	–	371	25	190	125	7	99
–	–	1	–	32	4	24	23	–	–
–	–	1.08	–	8.6	16	12.6	18.4	0	0
–	–	0.39	–	0.64	0.225	1.6	.115	3.35	0.20
–	–	34	1	42	17	69	62	21	21
–	–	39	–	364	145	130	4	18	57
–	–	0.16	–	0.62	1.3	1.1	.004	8.6	0.11
10	–	13	1	42	5	7	14	21	21
–	–	–	–	7	–	11	–	–	–
–	–	–	–	1957	–	1954	–	–	–
10	–	–	–	8	–	–	4	6	–
1900	–	–	–	1956	–	–	1960	1958	–
60	–	–	–	–	–	–	3	–	–
1900	–	–	–	–	–	–	1961	–	–

TABLE 2-1. AFRICA —

(Source: ECA

[Information

Country	Portuguese Provinces				Rhodesia	Rwanda	Senegal	Spanish	
Area, km^2	1,288,571				389,000	26,338	201,000	302,824	
Reference No.	17				18	19	20	21	
	17a	17b	17c	17d				21a	21b
Precipitation									
Total number of stations in 1964 (1)	148	98	39	560	1,150	10	73	285	3
Stations equipped with recorders (2)	14	4	16	32	39	1	8	6	1
Percentage of (2) to (1)	9.5	4.1	41	5.7	3.4	10	11	2.1	33
Density of stations/1,000 km^2	87	24.3	1.1	0.45	2.96	0.38	0.36	39.2	2
Average length of records	99	72	40	107	75	—	30	45	20
Evaporation									
Total number of stations in 1964	13	21	25	294	53	10	11	6	1
Percentage of evaporation to precipitation	8.8	21.4	64	52.5	4.6	100	15	2.1	33
Average length of records	99	72	40	83	8	—	30	44	20
Water levels									
Total number of stations in 1964 (1)	14	—	14	124	387	25	15	1	—
Stations equipped with recorders (2)	—	—	2	—	268	8	8	—	—
Percentage of (2) to (1)	0	—	15	0	69	32	53.4	0	—
Density of stations/1,000 km^2	8.20	—	0.39	0.10	1	0.95	.075	.014	—
Average length of records	10	—	8	4	40	—	—	1	—
Discharge									
Total number of stations in 1964	—	—	1	—	349	8	4	1	—
Density of stations/1,000 km^2	—	—	.028	—	0.90	0.34	0.02	.014	—
Average length of records	—	—	7	11	—	—	—	—	—
Sediment Transportation									
Longest continuous record of any station	—	—	—	—	—	—	—	—	—
Year first station established	—	—	1957	1957	—	—	—	—	—
Ground water									
Longest continuous record of any station	—	—	—	—	10	—	—	—	—
Year first station established	—	—	—	1923	1946	—	—	—	—
Quality of water									
Longest continuous record of any station	—	—	—	—	—	—	—	—	—
Year first station established	—	—	1942	—	—	—	—	—	—

17a — Madeira, Porto Santo, S. Tome and Principe together area - 1,713 km^2
17b — Cape Verde Islands area - 4,033 km^2
17c — Guinea area - 36,125 km^2
17d — Angola area - 1,246,700 km^2
21a — Canary Islands, area - 7,273 km^2
21b — Ifni, area - 1,500 km^2

AVAILABILITY OF HYDROLOGIC DATA (continued)

and WMO, 1966)

as of 1964]

Territories		Tanzania	Togo	Tunisia	Uganda	United Arab Rep.	Zambia
		833,000	56,000	167,000	236,037	1,002,000	752,000
21c	21d	22	23	24	25	26	27
32	15	978	—	519	443	118	660
2	1	36	—	49	19	27	57
6.4	6.6	3.7	—	9.4	4.3	21.8	8.64
1.14	.056	1.17	—	3.1	1.88	0.12	0.88
11	17	40	—	80	68	63	60
2	—	74	—	52	56	74	38
6.4	—	7.6	—	20	12.6	63	5.75
11	6	—	5	20	35	63	9
—	—	179	46	86	118	111	72
—	—	12	—	65	13	—	1
—	—	6.7	0	75.6	11	0	1.4
—	—	.215	0.82	.515	0.50	0.11	.096
—	—	13	20	38	67	1,100	58
—	—	133	27	75	78	66	118
—	—	0.15	0.48	0.45	0.33	.065	0.16
—	—	13	20	38	33	62	15
—	—	—	2	16	—	30	1
—	—	—	1961	1948	—	1918	1962
—	—	—	—	12	—	11	10
—	—	—	—	1952	—	1927	1932
—	—	—	—	16	—	11	4
—	—	—	—	1948	—	1927	1959

21c — Guinea, area - 28,051 km2
21d — Sahara, area - 266,000 km2

TABLE 2-2. AFRICA—HYDROGRAPHIC DATA ON SELECTED RIVERS

(Source: Welcomme, Dept. of Fisheries, FAO, 1972)

River	Cavally	CHARI–LOGONE SYSTEM — Chari (Shari)	CHARI–LOGONE SYSTEM — Logone	Comoe (Komoe)	Congo (Zaire)	CONGO (ZAIRE) SYSTEM — Kasai-Kwa	CONGO (ZAIRE) SYSTEM — Luapula	CONGO (ZAIRE) SYSTEM — Ubangi (Oubangui)
Source	Nimba Mts., Guinea	Confluence of rivers Salamat Bahr Aouk and Ouham	Adamaoua Mts., Cameroon	Upper Volta	Many sources [1]	Angola, near Villa Luso	L. Bangweulu, Zambia	Zaire, Highlands north of Lake Albert
Altitude (m. above sea level)	—	—	—	420	Lualaba 1,540	1,500	1,140	—
Drainage area (km2)	22,400	600,000	73,700 (at Bongor)	76,500	4,104,500 (incl. Ubangi)	—	—	772,800
Total length (km)	700	950	550	1,160	4,700	1,735 (Kasai)	560	1,060
Countries traversed	Guinea, Ivory Coast, Liberia	Cameroon, Central African Rep., Chad, Sudan	Cameroon, Chad	Ivory Coast, Upper Volta	Angola, Burundi, Cameroon, Central African Rep., Congo, Rwanda, Tanzania, Zaire, Zambia	Angola, Zaire	Zambia, Zaire	Congo, Central African Rep., Zaire
Major tributaries	—	Bahr Aouk, Lagone	—	—	Alima, Aruwimi, Elila, Itimbiri, Kwa, Lomami, Lowa, Lufira, Lukuga, Lulonga, Luvua, Mongala, Sangha, Ruki, Oubangui	Kasai (1,735 km), Lubilash - Sankuru (1,150 km), Lukenie-Fimi (1,060 km)	—	Kotto, Ouaka, Mbori, Chinko
Discharge to	Atlantic Ocean	Lake Chad	Chari River at Fort Lamy	Atlantic Ocean	Atlantic Ocean	Congo River	L. Mweru, thence to Lualaba and Congo Rivers through Luvua R.	Congo River at Liranga
Volume of discharge at mouth (m3/s)	—	1,011-1,181	40-3,000 (mean 1,500)	500 (Oct.), 15 (Feb.)	22,000-67,000 (mean 39,160)	—	—	—
Suspended silt load (tons/yr)	—	—	—	—	50,500,000	—	—	—
Flood regime	Aug.-Nov.	Jul. to Feb. Max. in Oct.	Jun.-Jan. Max. in Sep.-Oct.	Jul. to Oct.	Peak floods May and Dec.	—	—	Aug.-Dec. Max. in Oct.
Special features	—	The Chari and Logone Rivers together flood the Yaeres, a plain of some 6,000 km2. Lake Iro		—	Lakes: Kivu, Tanganyika, Bangweulu, Mweru, Maji Ndombe, Pool Malebo, Upemba, Tumba. Waterfalls: Portes d'Enfer, Wagenia, numerous cataracts; dams on nearly all tributaries	Maji Ndombe lake discharges into Congo/Zaire through Fimi and Kwa Rivers	Flows through extensive swamp system in Kifukula depression (1,500 km2)	—
Conductivity K20 (micromhos)	—	42-73 (K24)	41-82 (K24)	—	44-108 (Luapula)	—	—	—

TABLE 2-2. AFRICA–HYDROGRAPHIC DATA ON SELECTED RIVERS (continued)

River	Corubal (Komba)	Ganale Dorya/Juba	Gambia	Geba	Konkoure	Limpopo	Moa	Mono
Source	Fouta Djallon, Guinea	Mendebo Mts., Ethiopia	Fouta Djallon, Guinea	Senegal	Fouta Djallon, Guinea	Witwatersrand, South Africa	Guinea	Togo
Altitude (m. above sea level)	650	—	1,100	—	900	—	—	—
Drainage area (km²)	—	—	77,000	8,000	—	358,000	17,900	22,000
Total length (km)	600	1,600	1,120	—	365	1,680	300	360
Countries traversed	Guinea, Portuguese Guinea	Ethiopia, Somalia, Kenya	Guinea, Senegal, Gambia	Portuguese Guinea, Senegal	Guinea	Botswana, Rhodesia, South Africa, Mozambique	Guinea, Liberia, Sierra Leone	Togo, Dahomey
Major tributaries	—	Webbe Schibeli, Lagh Bor	—	—	—	Marico, Shashi, Olifants (790 km)	—	—
Discharges to	Atlantic Ocean	Indian Ocean	Atlantic Ocean	Atlantic Ocean	Atlantic Ocean	Indian Ocean	Atlantic Ocean	Atlantic Ocean
Volume of discharge at mouth (m³/s)	—	—	—	—	—	—	—	—
Suspended silt load (tons/yr)	—	—	—	—	—	—	—	—
Flood regime	—	—	—	—	—	—	—	Jul.-Oct.
Special features	—	—	—	—	—	—	—	—
Conductivity K20 (micromhos)	—	—	—	—	—	—	36	—
Dissolved solids (mg/l)	—	—	—	—	—	—	—	—
pH	6.1	—	6.3	—	5.9 (upper) 6.2 (middle) 6.2 (lower)	—	6.6	—
Temperature Range	22	—	20	—	19-25 (upper) 23-27 (middle) 25-29 (lower)	—	—	—
Dissolved solids (mg/l)	—	—	—	—	—	—	—	—
pH	6.9-7.7	—	—	—	5.6-6.5	—	—	7.0-7.2
Temperature Range (°C)	19.4-30.4	—	—	—	23-26 (upper course)	—	—	—

TABLE 2-2. AFRICA—HYDROGRAPHIC DATA ON SELECTED RIVERS (continued)

River	NIGER–BENUE SYSTEM		NILE SYSTEM				Okavango/ Cubango	Orange	Ouema
	Niger	Benue (Benoue)	Nile	White Nile 3)	Blue Nile	Kagera			
Source	Fouta Djallon, Guinea	Adamowa Mts., Cameroon	2)	Owen Falls Dam, Uganda	Lake Tana Ethiopia	Rwanda, Burundi	Bie Plateau, Angola	Drakensberg Mts., Lesotho	Atakora Massif, Dahomey
Altitude (m. above sea level)	1,000	—	—	1,136	1,829	—	—	3,300	600
Drainage area (km2)	1,125,000 (incl. Benue)	—	3,000,000	—	325,000	—	—	640,000	40,150
Total length (km)	4,183	1,400	6,669	2,084	1,460 (1,000 in Ethiopia)	785	1,600	2,160	700
Countries traversed	Guinea, Mali, Ivory Coast, Niger, Upper Volta, Dahomey, Nigeria	Cameroon, Nigeria	Burundi, UAR, Ethiopia, Kenya, Rwanda, Sudan, Tanzania, Uganda	Uganda, Sudan	Ethiopia, Sudan	Burundi, Rwanda, Tanzania, Uganda	Angola, Namibia, Botswana	Lesotho, South Africa, Namibia	Dahomey, Nigeria
Major tributaries	Bani(Mali), Alibori, Mekrou, Sota(Dahomey), Sokoto, Benue (Nigeria)	Mayo-Kebbi (Cameroon), Faro and Gongola (Nigeria)	Atbara, Blue Nile, White Nile	Kagera, Semliki, Aswa, Bahr el Ghazal, Atbara	Dinder, Rahad	Akanyaru	Cuito	Vaal (1,200 km)	Okpara, Zou
Discharges to	Atlantic Ocean	Niger	Mediterranean Sea	Nile at Khartoum	Nile at Khartoum	Lake Victoria	Okavango Swamps	South Atlantic Ocean	Lake Nokoue and Porto Novo lagoon
Volume of discharge at mouth (m3/s)	6,100	—	2,832	71 x 106m3/d4)	1,640 or 145 x 106m3/d6)	—	254	—	—
Suspended silt load (tons/yr)	5,050,000	—	103,200	—	—	—	—	—	—
Flood regime	Guinean: Jan., Feb. Dahomean: Aug., Sep., Oct.	Aug.-Sep.	—	—	—	—	—	Nov.-Mar.	Jul.-Nov.; Peak in Aug.
Special features	Internal delta and Mali Lakes, Kainju Reservoir, Delta at mouth	—	The cataracts, Lake Nasser/ Nubia, Aswan dam	Lake Kyoga, Murchison Falls, L. Albert (Uganda), Sudd, Jebel Aulia Reservoir, (Sudan)	Tissisat Falls (Ethiopia), Roseires Reservoir, Sennar Reservoir, (Sudan)	Lakes Rugwero and Cyohoha (Rwanda/ Burundi) and other lakes (Rwanda), Rusumo Falls (Rwanda)	Okavango Swamps reach 20,000 km2 in floods (May-Aug.), Lake Ngami	Aughrabies Falls	Lakes Azilli and Cele extensive delta (1,000 km2) and flood plain at mouth

TABLE 2-2. AFRICA—HYDROGRAPHIC DATA ON SELECTED RIVERS (continued)

River	Ruvuma	Ruzizi	Senegal	Sassandra	Volta	Black Volta	Pendjari/Oti	Zambezi
					VOLTA SYSTEM 7)			
Source	Lugenda R. and Rovuma R., Mozambique	Lake Kivu	Fouta Djallon, Guinea	Ivory Coast	Upper Volta	8)	Atakora Mts., Dahomey	N.W. Zambia
Altitude (m. above sea level)	—	—	1,200	—	—	300	—	1,600
Drainage area (km2)	140,000	—	338,000	75,000	390,000	149,600	72,900	1,300,000
Total length (km)	640	—	1,641	650	1,270	—	900	2,574
Countries traversed	Mozambique, Tanzania	Rwanda, Burundi, Zaire	Guinea, Mali, Mauritania, Senegal	Ivory Coast	Upper Volta Ghana, Ivory Coast, (Red Volta only)	—	Dahomey, Togo, Ghana	Angola, Malawi, Mozambique, Rhodesia, Zambia, Namibia
Major tributaries	Lugenda, Rovuma	—	Bafing, Baoule, Bakoye, Gorgol Faleme, Kolombine, Karakoro	Nzo	Oti (Pendjari)	—	—	Kafue, Luangwa, Shire, Cuango
Discharges to	Indian Ocean	Lake Tanganyika	Atlantic Ocean	Atlantic Ocean	Atlantic Ocean	—	Lake Volta	Indian Ocean
Volume of discharge at mouth (m3/s)	—	—	5,000 during 100 year flood	425	—	243	500	7,070
Suspended silt load (tons/yr)	—	—	—	—	—	—	—	—
Flood regime	—	—	Aug.-Nov.	Jul.-Nov.; Max. in Sep.	—	—	—	Dec.-Jul.; Max. in Mar.
Special features	—	—	—	—	Volta Lake (Akosombo Dam)	—	—	Barotse flood plain, Kafue flats, Lukanga Swamps. Lake Kariba (Kariba Dam). Victoria Falls
Conductivity K20 (micromhos)	31 (upper)	—	150-240 5) 550 in Sudd	—	99	—	159 (K25)	—
Dissolved solids (mg/l)	—	—	—	—	—	—	—	—
pH	7.2 (upper) 6.7-6.8 (middle)	—	7.2-7.9	—	—	—	7.7	—
Temperature Range (°C)	19-30 (upper)	21-32 (upper)	24-29	—	—	—	—	—

[continued on next page]

TABLE 2-2. AFRICA—HYDROGRAPHIC DATA ON SELECTED RIVERS (continued)

River	Ruvuma	Ruzizi	Senegal	Sassandra	Volta	Black Volta	Pendjari/Oti	Zambezi
					VOLTA SYSTEM			
Conductivity K_{20}	–	828-1,190(K_{18})	72	–	–	41-124	–	50 (flood) 96 (low water)
Dissolved solids (mg/l)	–	–	–	–	–	–	–	– } Above Lake Kariba
pH	–	–	6.8-7.1	–	76 White Volta 6.5 Red Volta	6.5-7.3	6.4-6.7	7.4
Temperature Range (°C)	–	–	18.6-26.8	–	–	–	–	30 (flood) 17 (low water) }

1) Longest continuous stream: Chambezi River, Lake Bangweulu (Zambia) through Luapula River into Lake Mweru, through Luvua River into Lualaba River. The Lualaba River changes its name to Congo (Zaire) at Kisangani, 2,000 km from the mouth.

2) The Nile takes its name at the confluence of the Blue Nile and White Nile at Khartoum 3,800 km (2,380 mi) from the mouth. The longest continuous stream which measures 6,669 km (4,160 mi) is the Nile River, White Nile River, Lake Victoria, Kagera River and Akanyaru River.

3) The White Nile changes its name several times in its course. From Lake Victoria to Lake Albert it is known as the Victoria Nile. From Lake Albert to the Uganda border it is known as the Albert Nile, and from the Sudan to its confluence with the Blue Nile it is called the White Nile or Bahr el Abiad.

4) White Nile contributes 10% of total Nile flow in summer floods, 83% of the total Nile flow at low water.

5) Varies with distance from Lake Victoria as follows: 800 km (180-230), 1,200 km (190-240), 2,000 km (150-180).

6) Contributes 68 percent of summer flood (July to October) to Nile.

7) Consists of Black, Red and White Voltas all of which now feed Lake Volta.

8) The major tributary (1,300 km long) arises in Upper Volta southwest of Bazo at an altitude of 300 m and flows for 650 km in Upper Volta.

TABLE 2-3. AFRICA—HYDROGRAPHIC DATA ON SELECTED LAKES AND RESERVOIRS
(Source: Welcomme, Dept. of Fisheries, FAO, 1972)

Lake or Reservoir	Abaya (Margherita)	Abiyata (Abiata)	Albert (Mobutu)	Awasa	Bangweulu	Baringo	Bunyoni	Chad	Chilwa	Edward
Country (ies)	Ethiopia	Ethiopia	Uganda, Zaire	Ethiopia	Zambia	Kenya	Uganda	Cameroon, Chad, Niger, Nigeria	Malawi, Mozambique	Uganda, Zaire
Altitude (m)	1,285	1,573	619	1,708	1,160	965	1,973	282	654	914
Depth (m)	13(max) 7(mean)	14.2(max) 7.6(mean)	58(max) 25(mean)	21.6(max) 10.7(mean)	10(max) 4(mean)	5.6(mean)	39.3(max)	9.5(max) 3.9(mean)	5(max) 2.0(mean)	117(max) 34(mean)
Surface area (km2)	1,161	205	5,600	130	9,850 1)	130	57	13,000-26,000 16,317(mean)	Variable 2)	2,300
Volume (km3)	8.2	1.6	140	1.3	11.2	0.7	—	75	—	78.2
Max. length (km)	70	20.9	180	17	72	21	—	224	45	90
Max. width (km)	28	12.4	40	11	39	11	—	144	32	40
Shoreline length (km)	225	61.9	—	52	490	—	—	1,000	—	—
Drainage basin area (km2)	17,300	1,630	—	1,250	100,800	—	—	2,500,000	—	—
Major inflowing rivers	Bilate	—	Victoria, Nile, Semliki	—	Chambezi	Molo, Tangulbei	—	Chari, Yobe, Ngadda, Yedseram	—	Rutshuru 4)
Major outflowing rivers	—	—	Albert Nile	—	Luapula	—	—	—	—	Semliki
Annual water level fluctuation (m)	—	—	—	—	1.2(mean)	—	—	1(approx.)	1.3 3)	—
Conductivity K20 (micromhos)	670-766	10,700-30,000	730	1,050	14-38(open water) 27-52(swamps)	416	260-262	50(open water) 1,000 (vegetated areas)	1,200-2,000	925
Dissolved solids (mg/l)	517	8,358	565	650	—	—	—	75-1,000	—	521
pH	—	—	8.4-9.5	—	6.1-7.0	8.7-8.9	8.0-9.3	7.1-8.3	8.4-9.6	8.5-9.3
Surface temperature range (°C)	22-28	23-26	26-29	23-26	18-26	—	21-22	18.7-32.3	21-37	25.7

TABLE 2-3. AFRICA—HYDROGRAPHIC DATA ON SELECTED LAKES AND RESERVOIRS (continued)

Lake or Reservoir	George	Guiers	Jebel Aulia Reservoir	Jipe	Kainji	Kariba	Kitangiri	Kivu	Kyoga
Country (ies)	Uganda	Senegal	Sudan	Kenya, Tanzania	Nigeria	Rhodesia, Zambia	Tanzania	Rwanda, Zaire	Uganda
Altitude (m)	914	Near sea level	377	700	142 (when full)	485	800	1,463	1,100
Depth (m)	7(max) 2.5(mean)	3.5(max) 2(mean)	12(max) 6(mean)	—	60(max) 11(mean)	93(max) 29.2(mean)	5(max)	489(max) 240(mean)	8(max) 6(mean)
Surface area (km2)	270	170	600	39	1,270	5,364	1,200	2,699	2,700
Volume (km3)	0.5	.19	3.5	—	14	156	—	—	—
Max. length (km)	23	70	—	12	136	277	40	96	90
Max. width (km)	18	8	—	3	24	40	12	48	25
Shoreline length (km)	—	—	—	—	716	2,164	—	—	—
Drainage basin area (km2)	9,000	—	—	—	—	—	—	—	—
Major inflowing rivers	Semliki Channel to Lake Edward	Connected to Senegal R. by Tawey Channel	White Nile	Lumi	Niger	Zambezi, Umiali	Wembere, Manonga	—	Victoria Nile
Major outflowing rivers	—	—	White Nile	—	Niger	Zambezi	Sibiti to Lake Eyasi	Ruzizi	Victoria Nile
Annual water level fluctuation (m)	—	—	6	—	10-11	3-4	—	—	—
Conductivity K20 (micromhos)	165-207	45-75	—	—	46.6-99.6	88-115	785	1,240	245-365
Dissolved solids (mg/l)	264	—	120-180	—	50	40-70	404-432	975-1,020	—
pH	8.5-9.5	6.6-8.1	7.5-9.2	—	6.0-7.6	7.5-8.9	8.0-8.9	9.1-9.5	7.6-9.0
Surface temperature range (°C)	25-35	15-35	21.0-29.5	—	23-31	17-32 5)	—	24-26	28

TABLE 2-3. AFRICA—HYDROGRAPHIC DATA ON SELECTED LAKES AND RESERVOIRS (continued)

Lake or Reservoir	Langano (Langana)	Leopold II (Maji Ndombe)	Malawi	Malomba	Mweru	Naivasha	Nasser (Nubia)	Rkiz	Rudolf
Country (ies)	Ethiopia	Zaire	Malawi, Mozambique, Tanzania	Malawi	Zaire, Zambia	Kenya	Egypt, Sudan	Mauritania	Ethiopia, Kenya
Altitude (m)	1,585	200	471	470	927	1,890	185	–	406
Depth (m)	46.2(max) 17(mean)	5(mean)	758(max) 426(mean)	6(max) 4(mean)	37(max) 10(mean N.end) 3(mean S. end)	–	97(max) 6) 25(mean) 6)	2.5(max)	73
Surface area (km2)	230	2,300	30,800	390	4,580	189	3,330 7) (half full) 5,000 6)	120	7,200
Volume (km3)	3.8	–	8,400	–	36.6	–	130 6)	–	–
Max. length (km)	23.2	–	603	29	124	20	490	–	296
Max. width (km)	16	–	87	17	51	13	20	–	59
Shoreline length (km)	77.5	–	1,500	–	340	–	–	–	–
Drainage basin area (km2)	1,600	–	65,000	–	–	–	–	–	–
Major inflowing rivers	–	Lotoi, Lokoro, Olongo-Lule, Olongo-Nsongo, Bowele	Ruhuhu, Songwe	Shire	Luapula, Kalungwishi	Melawa	Nile	–	Omo, Suam-Turkwell
Major outflowing rivers	–	Fimi, into Kwa and Congo/Zaire	Shire	Shire	Luvua (into Lualaba)	–	Nile	Connected to Senegal R. by Laouwaja, Sakan, Sebereim and Kamlach Rivers	–
Annual water level fluctuation (m)	–	–	6	–	–	–	10-16(1970) 7-10 6)	2.5 9)	–
Conductivity K20 (micromhos)	1,900	–	220	225	70-125	318-400	210-250	–	3,300
Dissolved solids (mg/l)	1,644	–	–	–	76	–	260	–	–
pH	–	>4	7.7-8.6	–	6.4-9.3	8.8-9.0	7.6(mean)	–	9.3-9.7
Surface temperature (°C)	22-26	–	23-25	–	19-30	–	17-32 8)	–	29.6(mean)

TABLE 2-3. AFRICA—HYDROGRAPHIC DATA ON SELECTED LAKES AND RESERVOIRS (continued)

Lake or Reservoir	Sennar	Shala	Shamo	Tana	Tanganyika	Tumba	Upemba	Victoria	Volta	Ziway (Zwei)
Country (ies)	Sudan	Ethiopia	Ethiopia	Ethiopia	Burundi, Tanzania, Zaire, Zambia	Zaire	Zaire	Kenya, Tanzania, Uganda	Ghana	Ethiopia
Altitude (m)	422	1,567	1,282	1,829	773	350	1,000	1,136	85	1,848
Depth (m)	26(max)	266(max) 86(mean)	12.7(max)	14(max) 8(mean)	1,435(max) 700(mean)	3-5	3.5(max) 0.3(mean)	84(max) 40(mean)	74(max) 19(mean)	7(max) 2.5(mean)
Surface area (km2)	140	409	551	3,500	32,900	720	530	68,800	8,482	434
Volume (km3)	–	37	–	28	18,940	–	0.9	2,750	165	1.1
Max. length (km)	–	27	36	80	673	–	40	400	400	32
Max. width (km)	–	17	23	64	48	–	20	240	24	20
Shoreline length (km)	–	110	118	–	1,500	–	–	3,440	4,828	102
Drainage basin area (km2)	–	3,920	–	–	249,000	–	–	263,000	–	7,025
Major inflowing rivers	Blue Nile	–	–	Little Abai, Reb, Gumura	Malagarasi, Ruzizi	Nganga, Lobambo, Butuka, Lolo, Modala, Membe	Lualaba, Lufira	Kagera, Nzoia, Yala	White Volta, Black Volta, Oti	–
Major outflowing rivers	Blue Nile	–	–	Blue Nile	Lukuga	Congo/Zaire through Irebu Channel	Lualaba	Victoria, Nile	Volta	–
Annual water level fluctuation (m)	17	–	–	2.2	–	3 10)	–	–	3-4	–
Conductivity K20 (micromhos)	–	20,400	927	–	520-610	24-32	145-255	91-98	63-172(K18)	372-427
Dissolved solids (mg/l)	–	–	651	151-174	–	72-90	–	97	70	354
pH	–	–	–	7.5-8.2	7.3-7.8	–	6.4-8.0	8.0-9.0	6.7-7.5	–
Surface temperature range (°C)	–	–	25-28	–	–	28-33	24-33	23-28	27-32	22-27

1) Includes adjoining Bangweulu swamps.
2) Varies from 259 - 2,590 km2; mean 750 km2 with a surrounding swampy area of 1,000 km2 .
3) Lake dries up completely on occasions.
4) Also connected to Lake George by Kasinga Channel.
5) Lake overturns once a year between March and July; monothermy about 22 oC.
6) When full.
7) Lake Nasser 2,500 km2 and Lake Nubia in Sudan 830 km2 (1972).
8) At winter overturn: 18 oC.
9) Completely dry for five months of year due to dams and water control.
10) Maximum in Nov.-Dec.; minimum in Apr.-Jun.

FIGURE 2-1. AFRICA — DIAGRAM SHOWING MAJOR NILE RIVER PROJECTS
(Source: Simaika, Water for Peace, 1967)

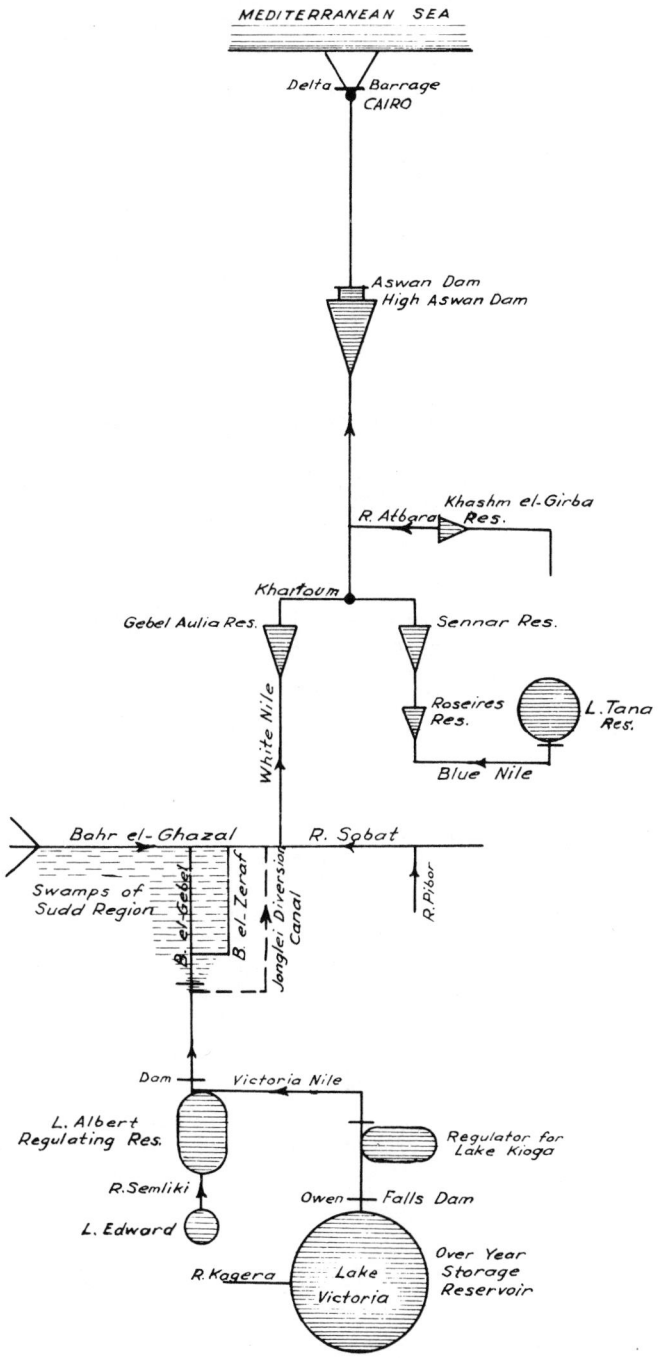

TABLE 2-4. AFRICA—BALANCE OF WATER RESOURCES AND WATER NEEDS OF THE NILE BASIN

(Source: Dekker, Water Resources Res., V. 8, p. 824-825, 1972, Copyright Am. Geophysical Union

[in millions of cubic meters]

INFLOW		OUTFLOW	
Sources		**Ultimate requirements**	
Bahr el Jebel at Mongalla	26,500	Egypt, losses in river	
Sobat at junction with White Nile	13,300	below Aswan	6,000
Blue Nile at Khartoum	51,400	Egypt, losses in canals	10,000
Atbara at Atbara	11,800	Egypt, application	
Bahr el Ghazal	15,000	on fields	46,000
Total	118,000	Total	62,000
		Sudan	23,800
		Ethiopia, Blue Nile	5,310
		Ethiopia, Baro plains	2,000 [1]
		Total	7,310
Losses			
Bahr el Ghazal	15,000	Kenya	480
Sudd region, Bahr el Jebel	12,300	Uganda	1,500
Transmission losses	7,600	Tanzania	480
Total	34,900	Total	2,460
Average flow at Aswan	83,100		
Mean Benefits			
Jonglei bypass canal	7,820	Congo	N.A.
Baro reservoir	1,710	Burundi	N.A.
Balas tunnel	140	Rwanda	N.A.
Total	9,670	Total	1,000 [1]
		Grand Total	96,570
		Reservoir Losses	
		Upper Blue Nile (450)	1,280
		Roseires (280)	
		Sennar (60)	
		Khashm el Girba	
		Lake Albert (or Nimule)	
		Baro	60
		Jebel Aulia	
Other Potential Increases		Sabaloka	470
		Fifth cataract	810
Reclamation in the Bahr el		Fourth cataract	820
Ghazal swamps	N.A.	Senna	
Baro diversion scheme and	2,300	Aswan	
marsh reclamation		Sadd el Aali	10,000
		Additional Upper Blue	
		Nile reservoirs	
Total	2,300	Total	13,440
Balance (shortage)	14,940		
Grand Total	110,010	**Grand Total**	110,010

N.A. — Not available
[1] Guess

TABLE 2-5. AFRICA—CONTRIBUTION TO MAIN NILE RIVER FLOW
FROM TRIBUTARY RIVERS

(Source: Sudan Almanac, 1968)

Contribution from	Main Nile River Stage	
	High Flood %	Low River %
Blue Nile	68	17
White Nile.......................	10	83
River Atbara	22	—

TABLE 2-6. AFRICA—WATER– LEVEL ELEVATIONS ALONG THE NILE RIVER SYSTEM

(Source: Sudan Almanac, 1966)

River	Mean Height of Low Water above the Sea		Distance from the Mediterranean Sea	
	Meters	Feet	Kilometers	Miles
Main Nile:				
Rosetta Mouth	0	0	0	0
Aswan Dam (downstream)	87	285	1,180	737
Wadi Halfa	115	377	1,531	957
Merowe (4th cataract)	239	784	2,270	1,419
Atbara	340	1,115	2,699	1,687
Khartoum	370	1,214	3,025	1,891
Blue Nile:				
Rahad Mouth	386	1,266	3,219	2,012
Wad Medani	388	1,273	3,227	2,017
Dinder Mouth	393	1,289	3,280	2,050
Sennar Dam (downstream)	405	1,328	3,379	2,112
Roseires	438	1,437	3,645	2,278
Sudan Frontier	—	—	3,758	2,349
Lake Tana (exit)	1,780	5,840	4,588	2,867
White Nile:				
J. Aulia Dam (downstream)	371	1,217	3,065	1,916
Dueim	372	1,220	3,224	2,015
Kosti Bridge	374	1,227	3,342	2,089
Malakal	382	1,253	3,832	2,395
Sobat Mouth	383	1,256	3,855	2,409
Lake No	386	1,266	3,978	2,486
Bahr El Jebel:				
Shambe	405	1,328	4,399	2,749
Bot	419	1,374	4,605	2,878
Juba	452	1,483	4,787	2,992
Nimule	612	2,007	4,955	3,097
Victoria Nile:				
Lake Albert (entrance)	617	2,024	5,391	3,369
Namasagali	1,030	3,378	5,530	3,456
Lake Victoria (exit)	1,133	3,716	5,611	3,507

TABLE 2-7. AFRICA—EXISTING AND POSSIBLE DAMS IN THE NILE BASIN
(Source: Dekker, Water Resources Res., V.8, p. 823, 1972,(Copyright Am. Geophysical Union))

Dam	River	Purpose	Reservoir Surface km2	Estimated Annual Evaporation Losses 106 m3	Storage Capacity 109 m3	Installed Capacity kw	Present Production kw hour	Ultimate Capacity kw	Ultimate Capacity kw hour	Remarks
Owen Falls	Victoria Nile	Energy and over-year storage	67,000	?	200 2)	120,000	430X106	150,000		Existing
Bujagali	Victoria Nile	Energy		?	Run of the river			180,000		Possible
Busowoko	Victoria Nile	Energy		?	Run of the river			150,000		Possible
Kalagala	Victoria Nile	Energy		?	Run of the river			125,000		Possible
Kamdinia	Victoria Nile	Energy		?	Run of the river			234,000		Possible
Aingo	Victoria Nile	Energy		?	Run of the river			490,000		Possible
Murchison	Victoria Nile	Energy		?	Run of the river			664,000		Possible
Nimule or Mutir	Bahr El Jebel	Overyear storage and energy		?	Lake Albert Reservoir: 100			?		Proposed for century storage project
Balancing reservoir	Bahr El Jebel	Balancing		?						
Gambela	Baro (Eth.)	Multipurpose		60	5,5 1)			?		Proposed
Jebel Aulia	Bahr El Jebel	Storage for Egypt	3,100	2,800	3.575			0		Existing
Lake Tana	Lake Tana (Blue Nile)	Storage and energy		?	15			?		Proposed
Tis Abbay	Blue Nile	Energy		—	Run of the river			9,600		Existing
Series of dams	Blue Nile			?						
Roseires	Blue Nile	Storage and energy	315	450	2.7 (net) Ultimate:7.3			175,000	109	Existing
Sennar	Blue Nile	Storage and energy	100	280	0.93	15,000		?		
Sabaloka	Main Nile	Energy		—	Run of the river			?		
Khashm El Girba	Atbara	Energy and storage		60	1.1	7,000		?		Existing
5th cataract	Main Nile	Energy		—	Run of the river			?		Proposed
4th cataract	Main Nile	Energy		—	Run of the river			?		Proposed
High Aswan	Main Nile	Energy and over-year storage 2)	5,000	10,000	90 (net)	2,100,000		10,000X106		Existing
Aswan	Main Nile	Energy and storage		?	5	352,000	1,860,000	?	10,000X106	Existing
Isna	Main Nile	Diversion for irrigation		—	0					Existing
Nag Hammadi	Main Nile	Diversion for irrigation		—	0					Existing
Assyut	Main Nile	Diversion for irrigation		—	0					Existing
Delta	Main Nile	Diversion for irrigation		—	0					Existing
Zifta	Main Nile	Diversion for irrigation		—	0					Existing
Edfina	Main Nile	Diversion for irrigation		—	0					Existing

1) Potential storage capacity on the Baro in Ethiopia may be larger, up to 10X109 m3.
2) At present operated as a run-of-the-river power plant. Lake Victoria is not being used for storage.

TABLE 2-8. AFRICA—WATER BUDGET OF LAKE VICTORIA

(Source: Afifi, ECA and WMO, 1971)

[Period 1959-67]

Year	Inflow Million m^3	Rainfall mm	Rainfall Million m^3	Outflow Million m^3	Change in Storage Million m^3	Evaporation mm	Evaporation Million m^3
1959	15,180	1,320	91,080	18,430	− 6,900	1,375	94,730
1960	17,740	1,374	94,806	20,420	1,035	1,320	91,090
1961	24,130	1,680	115,920	20,580	74,520	650	44,950
1962	25,550	1,506	103,914	38,720	31,050	865	59,695
1963	28,900	1,549	106,881	44,820	35,190	810	55,770
1964	25,850	1,398	96,462	50,480	− 345	1,045	72,177
1965	18,080	1,245	85,905	46,880	− 28,290	1,238	85,395
1966	20,140	1,383	95,427	42,950	− 8,970	1,182	81,585
1967	21,310	1,356	93,546	37,760	− 4,485	1,185	81,581

TABLE 2-9. AFRICA—ANNUAL INFLOW INTO LAKE VICTORIA

(Source: Afifi, ECA and WMO, 1971)

Name of river	Annual inflow into Lake Victoria in million m^3								
	1959	1960	1961	1962	1963	1964	1965	1966	1967
Gaged streams									
Kagera	5,030	6,450	5,200	9,410	11,790	11,360	8,060	8,250	6,730
Gucha	630	730	1,230	1,195	975	845	675	975	825
Sondu	830	1,250	2,100	1,975	1,960	1,900	745	1,190	1,845
Nyando	540	620	980	830	835	685	470	625	770
Nzoia	2,010	2,060	4,720	3,670	4,450	3,740	1,450	2,125	3,360
Sio	250	360	510	630	480	450	240	350	520
Total	9,290	11,470	14,740	17,710	20,490	18,980	11,640	13,515	14,050
Ungaged streams									
Isanga	174	180	290	174	250	174	189	180	184
Magogo	162	166	288	260	202	209	166	137	212
Simiyu	550	560	890	590	890	600	940	790	735
Mbalagati	200	200	200	200	200	200	200	200	200
Rwana	760	780	985	800	840	625	687	625	750
Suguti	92	100	119	110	99	93	92	103	115
Mori	127	145	242	218	208	165	131	156	188
Mara	1,325	1,295	2,025	1,960	2,125	1,565	1,255	1,280	1,750
Yala	700	875	1,225	1,030	1,160	910	770	860	940
Ruizi	57	50	72	68	76	58	49	78	37
Katonga	98	104	120	99	120	102	89	104	55
North Shore	44	45	59	46	53	49	38	42	44
South Shore	756	875	1,490	1,110	1,147	1,100	944	1,040	1,140
Kavirondo gulf streams	846	895	1,385	1,175	1,040	1,020	890	1,030	910
Total	5,890	6,270	9,390	7,840	8,410	6,870	6,440	6,625	7,260
Grand Total	15,180	17,740	24,130	25,550	28,900	25,850	18,080	20,140	21,310

FIGURE 2-2. AFRICA — RIVERS AND RAIN GAGING STATIONS IN
LAKE VICTORIA REGION

(Source: Afifi, ECA and WMO, 1971)

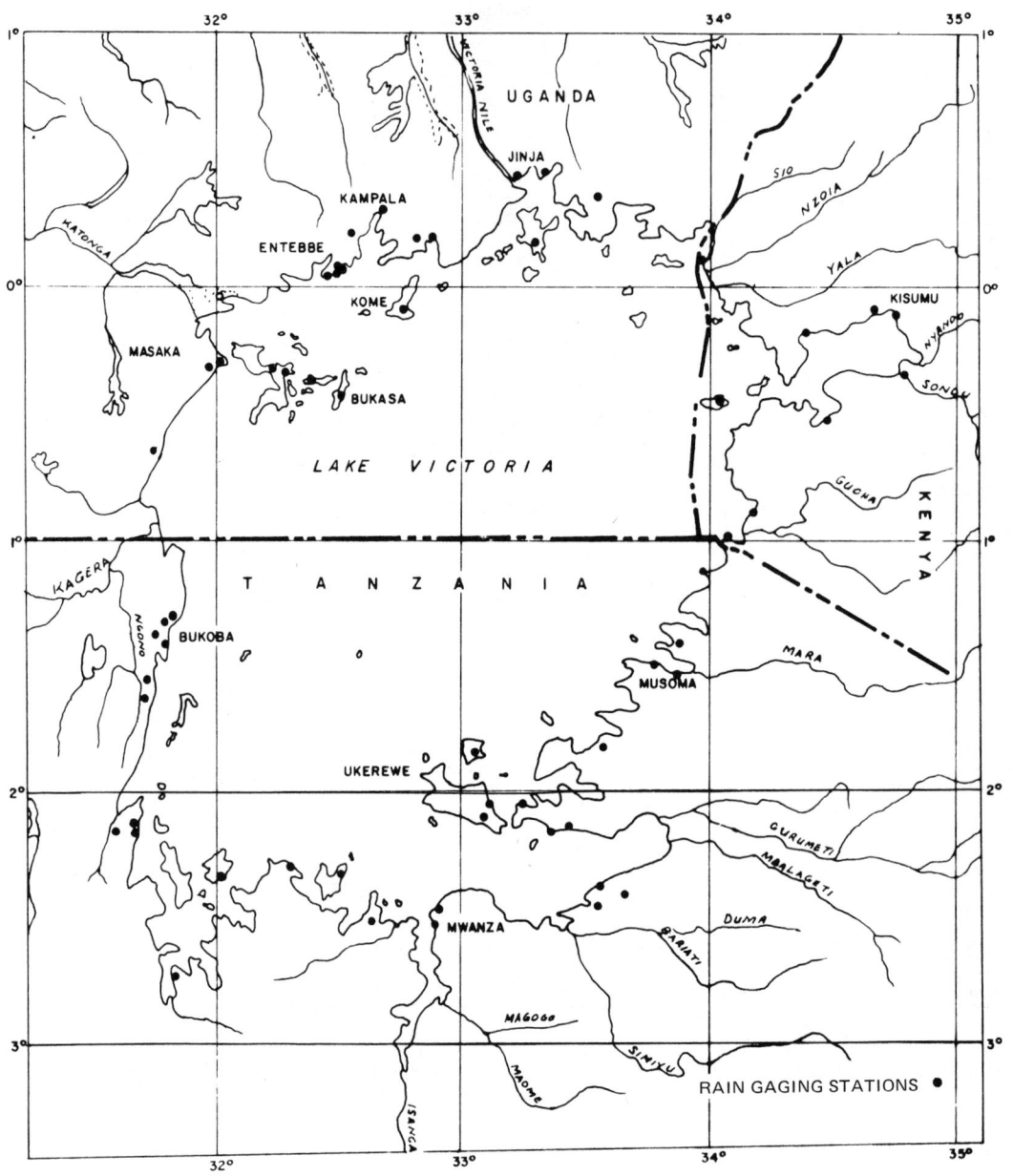

TABLE 2-10. AFRICA—WATER QUALITY OF LAKE VICTORIA

(Source: Welcomme, R.L., FAO, 1972)

Constituent	Concentration
Na	10.4 mg/l
K	3.8 mg/l
Ca	5.6 mg/l
Mg	2.6 mg/l
$HCO_3 + CO_3$	0.9 meq/l
Cl	3.9 mg/l
SiO_2	4.2 mg/l
$NO_2.N$	11 μg/l
$PO_4.P$	13 μg/l
P total	47 μg/l
Dissolved solids	97 mg/l
Conductivity K_{20}	91-98 μmhos
pH Kavirondo Gulf	8.2
pH Open lake	8.0
Surface temperature	23-28 $^{\circ}$C

TABLE 2-11. AFRICA—WATER QUALITY OF LAKE TANGANYIKA

(Source: Welcomme, R.L., FAO, 1972) [1]

Constituent	Concentration
Na	57 mg/l
K	35 mg/l
Ca	9.8 mg/l
Mg	43.3 mg/l
$HCO_3 + CO_3$	6.71 meq/l
Cl	26.5 mg/l
SO_4	5 mg/l
SiO_3	0.38 mg/l
Conductivity K_{20}	520-610 μmhos
pH	7.3-7.8

[1] Original analyses from Talling J. and Talling, I.B., The Chemical Composition of African Lake Waters, Int. Rev. Gesamten Hydrobiol., Vol. 50, No. 3, 1965.

FIGURE 2-3. AFRICA — MAJOR RIVERS AND NUMBER OF RAIN GAGES,
EVAPORATION PANS, AND STREAM GAGES IN SOUTHERN AFRICAN COUNTRIES
(Source : SARCUSS, 1970)

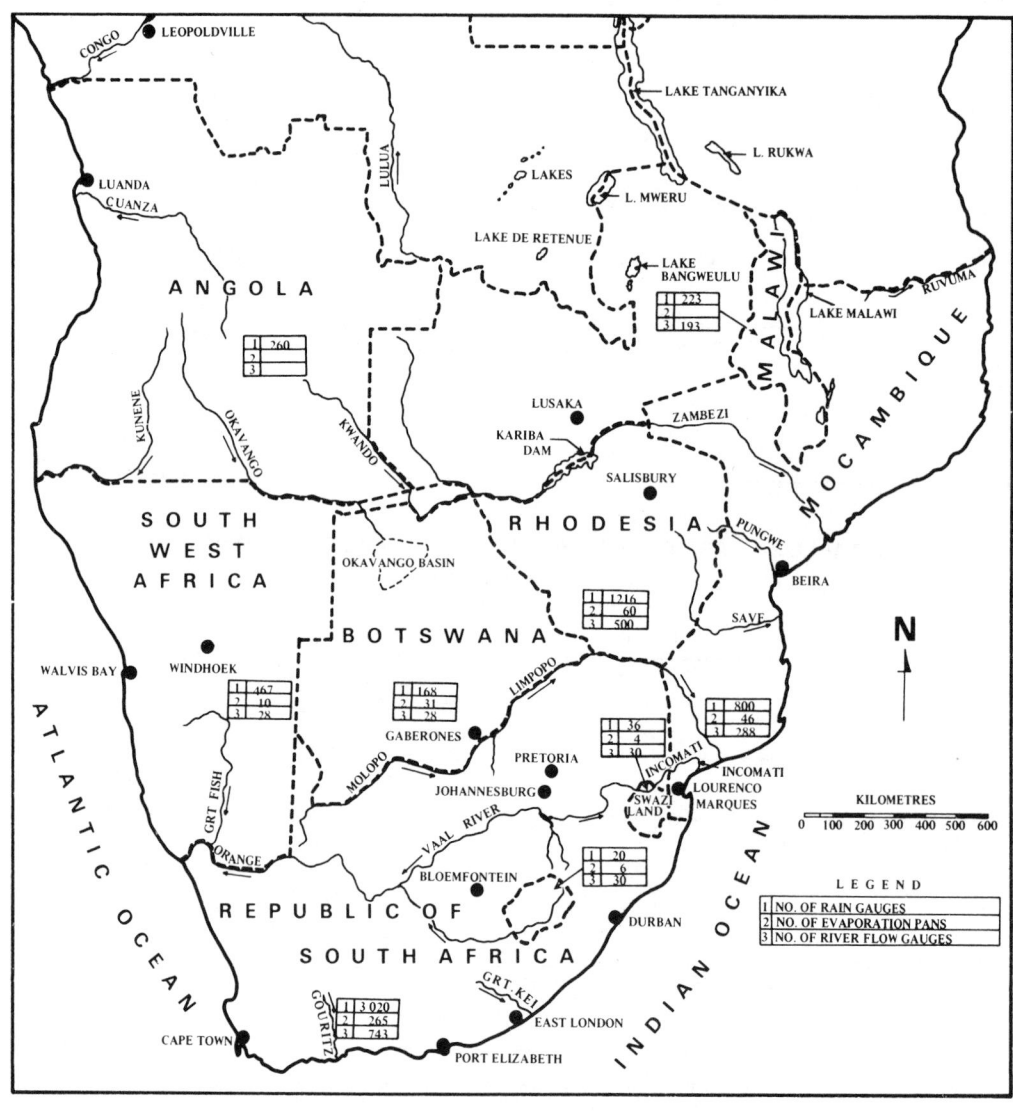

TABLE 2-12. AFRICA—RAIN GAGE DENSITY AND PRECIPITATION
IN SOUTHERN AFRICAN COUNTRIES

(Source: SARCCUS, 1970)

Country	Area km^2	Official raingages	Density km^2/gage	Overall mean annual rainfall in mm	Total mean precipitation in 10^6 m^3
Angola	1,242,847	260	4,800	1,068	1,325,000
Botswana	587,399	168	3,500	440	258,000
Lesotho	30,458	20	1,523	815	24,800
Malawi	96,524 (excluding Lake)	223	433	11,077	104,000
Mozambique	776,696	800	971	962	746,000
Rhodesia	395,100	1,216	325	668	264,000
South Africa	1,219,549	3,020	404	483	590,000
South West Africa	833,213	467	1,780	290	242,000
Swaziland	17,327	36	482	867	15,000
Total:	5,199,113				3,568,800

TABLE 2-13. AFRICA—RUNOFF IN PRINCIPAL RIVER BASINS AND REGIONS
OF SOUTHERN AFRICAN COUNTRIES

(Source: SARCCUS, 1970)

River or region	Drainage area, km^2	Mean annual runoff, 10^6 m^3
Zambezi	1,200,000 *	90,000
Orange	650,000	11,000
Limpopo, Incomati	440,000	12,000
Save, Rungwe, Ruvuma, etc.	611,500	85,000
East Coast rivers South of Incomati	166,700	26,000
Kei to Gouritz	175,000	3,700
South West and West Cape	118,300	5,000
Kunene and West Coast	508,300	6,000
Okavango, Kwando (Cuando) and Kalahari	1,200,000	12,000
Western Angola	395,500	54,000
Northeastern Angola, Lulua-Congo	263,500	80,000
Total:	5,728,800	384,700

* Includes drainage area within Zambia.

TABLE 2-14. AFRICA—WATER WELLS AND WITHDRAWAL OF GROUND WATER IN SOUTHERN AFRICAN COUNTRIES

(Source: SARCCUS, 1970)

Country	Number of water wells	Number of water wells drilled per year	Annual withdrawal of ground water million m3	Observation wells measured	Observation wells equipped with water-level recorders	Springs and fountains gauged
South Africa	385,000	7,000	1,110	952	330	60
South West Africa	22,000	500	103	0	29	0
Botswana	4,000	250	45 [1]	60	20	N.A.
Malawi	2,000	400	4.5 [2]	0	0	0
Lesotho	N.A.	N.A.	N.A.	0	0	0

N.A. — Not available
1   4,000 wells at average yield of 1,250 gallons per hour and 6 hours pumping per day.
2   2,000 wells at average yield of 950 gallons per hour and 6 hours pumping per day.

TABLE 2-15. AFRICA—HYDROELECTRIC AND THERMAL POWER GENERATING CAPACITY AND PRODUCTION, 1968

(Source: U.S. Federal Power Commission, 1971)

Country	Installed capacity [1] (MW.) [2]			Energy production [1] (GWh.) [3]			Population (1,000)	Kwh. per capita
	Hydro	Thermal	Total	Hydro	Thermal	Total		
Algeria	340	423	763	557	1,013	1,570	12,943	121
Angola	214	75	289	382	69	451	5,362	84
Cameroon	155	17	172	973	34	1,007	5,562	181
Central African Republic	7	3	10	34	0	34	1,488	23
Chad	0	16	16	0	30	30	3,460	9
Congo (Brazzaville)	15	7	22	34	20	54	870	62
Congo (Kinshasa)	810	90	900	2,581	145	2,726	16,730	163
Dahomey	0	10	10	0	24	24	2,571	9
Ethiopia	92	57	149	231	123	354	24,212	15
French Terr. Afar and Issas. [4]	0	15	15	0	33	33	45	733
Gabon	0	35	35	0	71	71	480	148
Gambia [5]	0	6	6	6	12	12	350	34
Ghana	512	119	631	2,521	63	2,584	8,376	309
Guinea	20	72	92	22	190	212	3,795	56
Ivory Coast	50	86	136	254	113	367	4,100	90
Kenya	66	87	153	247	147	394	10,209	39
Lesotho	0	3	3	0	6	6	876	7
Liberia	38	175	213	36	522	558	1,130	494
Libya	0	168	168	0	263	263	1,803	146
Madagascar	35	55	90	101	90	191	6,500	29
Malawi	28	21	49	105	12	117	4,270	27
Mali	1	15	16	1	33	34	4,787	7
Mauritania	0	25	25	0	42	42	1,120	38
Morocco	355	168	523	1,043	501	1,544	14,580	106
Mozambique	66	219	285	129	277	406	7,274	56
Niger	0	11	11	0	25	25	3,806	7
Nigeria	29	456	485	125	934	1,059	62,650	17
Rhodesia	705	486	1,191	4,833	743	5,576	4,940	1,129
Rwanda	21	1	22	67	2	69	3,405	20
Senegal	0	95	95	0	270	270	3,685	73
Sierra Leone [6]	2	74	76	6	133	139	2,475	56
Somalia	0	6	6	0	16	16	2,670	6
South Africa	5	8,272	8,277	20	40,094	40,114	19,167	2,093
Sudan	27	70	97	104	219	323	14,770	22
Swaziland [6]	16	35	51	56	94	150	395	380
Tanzania [7]	41	56	97	238	72	310	12,590	25
Togo	2	18	20	6	47	53	1,769	30
Tunisia	32	230	262	33	613	646	4,920	131
Uganda	157	23	180	747	18	765	8,133	94
United Arab Rep.	1,051	1,674	2,725	2,922	3,600	6,522	31,693	206
Upper Volta	0	11	11	0	22	22	5,175	4
Zambia	50	212	262	294	365	659	4,080	162
Total	4,942	13,697	18,639	18,702	51,100	69,802	329,216	212

[1] As of end of 1968.
[2] MW. - Megawatts (thousand kilowatts).
[3] GWh. - Gigawatt-hours (million kilowatt-hours).
[4] Former French Somaliland.
[5] As of June 30.
[6] Fiscal year beginning April 1.
[7] Includes Zanzibar.

TABLE 2-16. ALGERIA – DISCHARGE OF PRINCIPAL RIVERS
(Source: Algerian Hydrological Service, 1962)

River and station	Basin area km2	Mean monthly discharge, hm3 (million m3)												Total	Period of record
		Sep.	Oct.	Nov.	Dec.	Jan.	Feb.	Mar.	Apr.	May	Jun.	Jul.	Aug.		
Oued Tafna, Beni-Bahdel	1,016	2.393	3.006	3.922	6.760	10.395	11.626	11.468	10.593	7.474	3.419	2.326	2.089	75.471	33 years
Oued Tafna, Pierre du Chat	6,900	6.777	11.588	18.804	29.866	41.948	34.384	31.560	26.924	14.835	9.147	6.145	5.534	237.512	1954-61
Oued Meffrouch, Tlemcen	90	.409	.983	.717	2.268	3.291	2.394	2.757	2.649	1.717	.632	.491	.418	18.726	1943-61
Oued Ysser, Montagnac	1,860	3.203	7.971	9.379	18.907	24.099	21.480	20.284	26.439	10.720	4.857	2.940	2.280	152.559	18 years
Oued Tlelat, Saint-Lucien	122	.042	.044	.140	.475	.827	.457	.353	.160	.120	.080	.074	.055	2.827	1959-61
Oued Chouly, Pont de la R.N. 7	178	.444	.770	1.080	1.889	3.686	3.846	3.344	2.399	2.203	.806	.542	.520	21.529	23 years
Oued Mekerra, Chanzy	1,850	2.690	2.672	1.732	2.011	1.803	1.692	1.926	2.860	2.467	1.799	1.574	1.665	24.891	1949-61
Oued Mekerra, Sidi-bel-Abbes	3,050	2.736	3.095	2.702	3.201	3.617	3.617	3.947	3.700	3.371	2.535	1.956	1.743	36.220	28 years
Oued Sarno, Sarno	264	.009	.018	.443	.865	1.103	1.507	1.000	3.216	.269	.044	.013	0	8.487	1953-61
Oued Mekerra, Sig	4,240	2.266	3.408	4.326	5.801	3.866	3.965	5.463	5.914	8.218	5.009	2.244	2.419	52.899	1959-61
Oued El-Hamman, Bou-Hanifia	7,850	8.750	14.380	12.617	16.895	16.326	13.051	13.674	13.324	11.246	7.056	5.338	4.954	137.611	32 years
Oued El-Hamman, Trois Rivieres	7,610	10.093	17.534	11.660	15.693	20.346	16.181	18.528	16.005	11.447	8.234	6.356	5.404	157.481	1949-61
Oued Saida, Saida	101	.156	.200	.248	.778	1.320	.751	.421	.330	.183	.215	.150	.094	4.846	1960-61
Oued Mina, d'Uzes-le-Duc	5,460	10.350	10.174	9.738	11.485	26.842	25.706	23.226	24.440	13.486	8.288	8.430	9.800	181.965	1953-61
Oued Mina, Bakhadda	1,300	3.492	3.766	4.049	8.954	10.325	10.689	9.627	7.407	5.736	3.345	2.542	2.321	72.253	29 years
Oued Mina, Relizane	6,830	10.615	16.383	20.722	28.766	36.570	11.018	14.999	10.575	8.422	6.628	6.497	7.114	178.489	1956-61
Oued Chelif, Ponteba	29,300	19.916	33.737	32.941	56.671	97.579	92.686	67.189	57.374	40.112	16.217	16.748	20.139	551.309	26 years
Oued Riou, d'Ammi-Moussa	1,900	1.551	1.540	9.198	39.673	55.262	17.631	18.040	3.895	4.537	1.460	.885	.834	154.506	1958-61
Oued Fodda, Lamartine	800	1.098	3.110	4.420	9.007	20.876	16.649	13.155	10.024	5.607	1.497	.363	.284	86.090	19 years
Oued Chelif, Ghrib	23,300	8.785	12.702	6.129	11.312	28.088	26.112	18.469	17.696	11.115	3.322	1.315	1.469	146.514	24 years
Oued Mazafran, Fer a Cheval	1,850	4.053	21.037	13.430	37.167	93.487	58.118	37.723	61.331	28.726	6.471	3.305	2.641	367.489	12 years
Oued Hamiz, Hamiz	139	.574	2.567	6.994	8.789	10.689	6.546	5.158	5.454	3.056	1.256	.319	.108	51.510	43 years
Oued Boudouaou, Pont de la R.N.29	94	.697	1.222	2.702	4.132	9.271	2.993	1.818	4.841	4.531	.630	.523	.423	33.783	1959-61
Oued Leham, d'Ain el Hadjel	2,770	.683	2.096	1.175	2.613	7.188	3.204	3.397	3.434	.672	.209	.009	.051	24.731	1960-61
Oued Leham, Pont de la Rocade	5,460	4.264	7.119	1.889	7.841	12.920	4.697	5.356	4.518	2.238	2.570	.762	1.141	55.315	1952-61
Oued Ksob, Ksob	1,480	3.578	4.520	1.958	4.373	5.397	4.097	3.877	3.457	3.695	4.511	1.431	1.224	42.118	18 years
Oued Azerou, Portes de Fer	560	.085	.182	.104	.499	1.499	.559	.359	.173	.100	.052	.054	.035	3.701	1960-61
Oued El Abjod, Foum-el-Gherza	1,300	2.340	2.016	2.173	1.827	1.663	1.890	2.502	2.112	2.409	1.401	.388	.497	21.218	1950-61
Oued Saf-Saf, Zardezas	345	.291	2.327	9.564	7.142	17.362	11.563	6.138	5.597	4.055	1.209	.260	.147	65.655	1953-61
Oued Saf-Saf, Khemakem	325	.064	.071	.087	.189	3.319	1.044	.408	.251	.172	.167	.080	.080	5.932	1960-61
Oued Kebir-Ouest, Pont de la R.N.44	1,120	3.183	11.701	25.513	28.478	47.351	38.109	20.770	16.921	10.204	4.983	2.782	2.188	212.183	1953-61
Oued Ressoul, Pont du C.D. 13	105	.055	.030	.573	1.436	2.921	1.054	.547	.992	1.547	.514	.220	.103	9.992	1959-61
Oued Gueiss, Foum el Gueiss	156	.480	.580	.663	.868	1.050	.970	2.545	1.824	1.308	.428	.206	.299	11.221	31 years
Oued Mellah, Duvivier	545	1.366	5.065	9.356	17.427	29.411	23.264	19.278	15.979	9.222	8.430	1.532	1.151	141.481	1949-61
Oued Medjerdah, Souk-Ahras	220	.428	.939	4.195	9.273	14.448	11.612	9.238	5.816	4.289	1.362	.562	.400	62.562	1959-61
Oued Kebir Est, Yusuf	665	.578	10.084	31.836	50.245	75.149	47.091	40.239	20.467	8.070	1.750	.734	.450	286.693	9 years

FIGURE 2-4. ALGERIA – WATER RESOURCES REGIONS OF NORTHERN ALGERIA
(Source: Ministry of Industry and Energy, 1973)

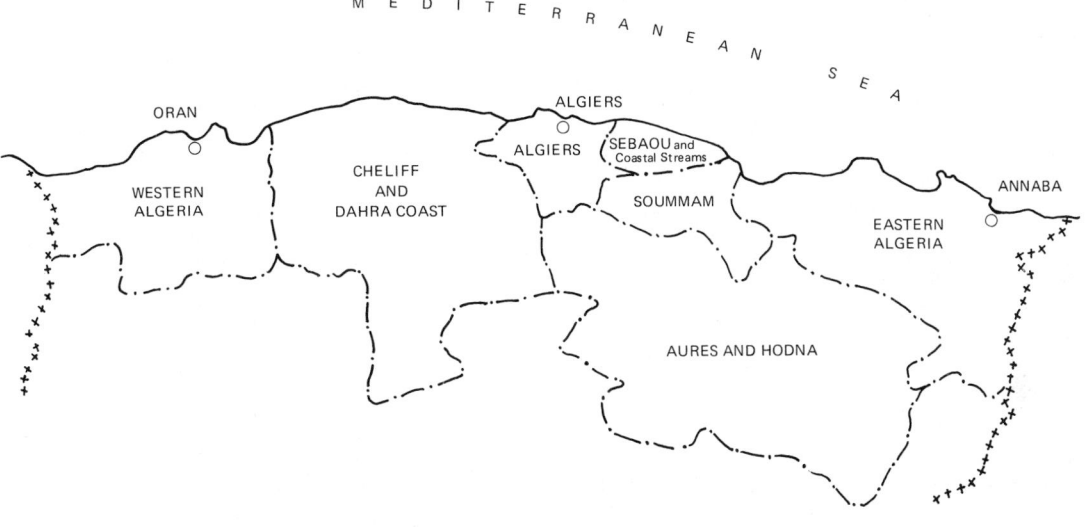

TABLE 2-17. ALGERIA—MEAN ANNUAL RUNOFF BY REGION
(Source: Ministry of Industry and Energy, 1973)

Region	Total drainage area km²	Mean annual flow volume-hm³ [1]	Mean annual runoff	
			mm	% of average Northern Algeria
Western Algeria	27,690	600	22	34
Cheliff & Dahra Coast	48,600	1,720	35	54
Algiers	8,720	1,500	172	274
Sebaou & Coastal Streams	3,900	1,600	410	631
Soummam Basin	9,200	750	82	126
Aures & Hodna	61,000	900	15	23
Eastern Algeria	34,800	5,500	158	243
Northern Algeria	193,910	12,570	65	100

[1] Million m³

TABLE 2-18. ALGERIA—DAMS, RESERVOIRS AND WATER USE
(Source: Ministry of Industry and Energy, 1973)

BREAKDOWN BY REGION:

Region Oued	Dam	Height m	Estimated average reservoir inflow hm^3/yr	Reported storage capacity hm^3	Reported regulated flow hm^3/yr	Principal use
WESTERN ALGERIA						
Tafna	Beni Bahdel	55	73	42	48	Municipal
Meffrouch	Meffrouch	26	18	15	14	Municipal
Sarno	Sarno	27	9	22	15	Irrigation
Mekerra	Cheurfas	29	65	8	15	Irrigation
El Hammam	Bou Hamidia	50	160	52	100	Irrigation
El Hammam	Fergoug	50		18		
CHELIFF & DAHRA COAST						
Fodda	Fodda	85	85	228	70	Irrigation
Cheliff	Ghrib	65	150	280	105	Irrigation
Mina	Bakhadda	45	73	45	43	Irrigation
ALGER						
Hamiz	Hamiz	45	50	15	11	Irrigation
Bou Djabroun	Merad	23	N.A.	1	1	Irrigation
SEBAOU			none			
SOUMMAN			none			
AURES & HODNA						
Gueiss	Foum El Gueiss	23	11	3	7	Irrigation
El Abiod	Foum El Gherza	65	23	43	18	Irrigation
Ksob	Ksob	32	50	8	15	Irrigation
EASTERN ALGERIA						
Agrioun	Irhil-Emda	85	186	127	180	Power
Djendjen	Erraguene	76	116	200	120	Power
Safsaf	Zardezas	37	54	10	12	Irrigation
B. Namoussa	La Cheffia	51	147	170	98	Irrig/Munic

SUMMARY INCLUDING DATA ON POWER GENERATION AND WATER USE:

Region	No. of storage dams	Reported total flow regulating capability hm^3/yr	% of total mean runoff	Present utilization Power GWh/yr	Irrigation hm^3/yr	Urban water supply hm^3/yr
Western Algeria	6	192	32	20	117	69
Cheliff & Dahra Coast	3	218	15	29	213	4
Algiers	2	12	1	1	12	nil
Sebaou & Coastal Streams	—	—	0	—	—	—
Soummam	—	—	0	—	—	—
Aures & Hodna	3	40	4	—	39	1
Eastern Algeria	4	410	7	330	83	26
Northern Algeria	18	872	7	380	464	100

N.A.—Not available.

TABLE 2-19. BOTSWANA—DATA ON PRINCIPAL DRAINAGE SYSTEMS

(Source: Jennings, South African J. of Science, 1971)

Basin or River	Basin area mi^2	Mean annual flow acre-feet
Limpopo	78,000 (30,000 within Botswana)	N.A.
Okavango-Makarikari	53,000	6,290,000 (Okavango at Shakawe)
Molopo-Nossop (Border Rivers)	N.A.	Mostly dry
Chobe (Border River)	56,000 (above junction with Zambezi River)	2,500,000—3,000,000

N.A. — Not available

TABLE 2-20. BOTSWANA—WITHDRAWAL OF GROUND WATER

(Source: UNDP/FAO, 1972)

[Includes pumpage from valley-fill aquifers]

Use	Estimated annual consumption million gallons	million m^3	Remarks
Rhodesian Railway	370	1.7	From Railway records.
Township, village and domestic	480	2.2	From water use records for Francistown and Lobatse; based on 6 gpd per person for 12 large villages with reticulated supply; and based on 1.5 gpd per person for 75% of remainder of population.
Agricultural	3,129	14.2	Based on Min. of Agric. livestock census figures and estimated water consumption for types of animals. Does not include irrigation usage which is minimal.
Total	3,979	18.1	

TABLE 2-21. BOTSWANA—WATER WITHDRAWN FROM VALLEY - FILL AQUIFERS

(Source: UNDP/FAO, 1972)

River valley	Use	Estimated annual quantity 10^6 gal	m^3	Remarks
Mahalatswe	Rhodesian Railways Mahalapye Township Supply	29.2 14.6	132,495 66,430	At Mahalapye
Motloutse	Selebi-Pikwe Mining Complex	36.5	165,710	
Shashe	Rhodesian Railways	91.2	414,275	At Shashe Siding
Ramokgwebana	Rhodesian Railways	12.8	58,035	At Tsamaya
Ntshe	Rhodesian Railways	14.6	66,430	At Tshesebe
Thune Lotsane Metsemotlhaba	—	—		No known pumpage
Limpopo	—	—		Relatively large pumpage for agricultural use. Quantities unknown.
Total		198.9	903,375	

TABLE 2-22. CAMEROON–DISCHARGE OF SELECTED RIVERS
(Source: ORSTOM, 1969)

River and station	Basin area km²	Mean monthly discharge, m³/s												Max flow m³/s	Year	Period of record
		Mar.	Apr.	May	Jun.	Jul.	Aug.	Sep.	Oct.	Nov.	Dec.	Jan.	Feb.			
Wouri, Yabassi	8,250	96.0	131	165	263	447	573	819	687	371	188	116	88.0	1,820	1960	1951-66
Sanaga, Edea	135,000	492	651	952	1,340	2,020	2,660	4,450	5,760	3,530	1,480	881	571	7,620	1955	1944-66
Mbaum, Bac de Goura	43,000	131	213	339	535	894	1,060	1,670	2,140	1,240	4,550	223	143	3,040	1954	1951-66
Djerem, Mbakaou	20,390	41.0	83.0	168	243	558	849	1,250	1,150	496	214	128	75.0	2,090	1964	1959-66
Lom, Betare Oya	10,680	53.0	62.0	80.0	118	193	250	374	425	246	140	92.0	66.0	689	1954	1951-66
Nyong, M'Balmayo	13,750	54.0	83.0	116	142	124	97.0	125	241	330	217	110	59.0	557	1964	1940-66
Lokoundje, Lolodorf	1,177	17.8	31.3	33.8	35.5	18.2	11.6	35.1	64.2	62.3	27.8	94.0	86.0	219	1954	1951-66
Lobe, Bac Kribi-Campo	1,940	43.0	105	158	119	34.0	23.2	116	295	236	94.0	34.0	23.8	542	1951	1950-66
Ntem, Bac de Ngoazik	18,060	168	300	362	304	155	82.0	228	550	576	323	143	122	1,010	1965	1953-66
Benoue, Garoua	64,000	5.30	3.20	15.2	79.0	336	1,110	1,920	846	171	62.0	25.0	12.6	6,130	1948	1930-66
Riao	31,000	1.30	0.40	4.30	35.0	226	795	1,430	612	82.0	19.0	7.80	3.30	3,180	1960	1950-66
Mayo-Kebi, Cossi	26,000	2.30	1.20	10.5	53.8	136	272	355	154	84.3	52.6	18.3	6.60	1,190	1960	1950-66
Vina du Nord, Touboro	12,200	46.9	30.8	36.5	72.5	209	437	520	314	101	45.0	26.6	158	1,010	1964	1964-66

TABLE 2-23. CAMEROON–MUNICIPAL WATER SUPPLY SYSTEM OF JAOUNDE
(Source: Internat. Statistical Institute, 1971)

Year	Population served	Total quantity available (1,000 m³)	Annual consumption (1,000 m³)	
			Total	of which domestic
1968	135,800	4,826	3,392	2,295
1969	160,000	5,221	3,907	2,601

TABLE 2-24. CENTRAL AFRICAN REPUBLIC—DISCHARGE OF SELECTED RIVERS
(Source: ORSTOM, 1969)

River and station	Basin area km^2	Mean monthly discharge, m^3/s												Max flow m^3/s	Year	Period of record
		Apr.	May	Jun.	Jul.	Aug.	Sep.	Oct.	Nov.	Dec.	Jan.	Feb.	Mar.			
Oubangui, Bangui	500,000	1,270	1,930	3,010	4,210	6,200	8,300	9,640	8,600	4,620	2,400	1,380	1,070	17,300	1916	1910-66
Lobaye, M'Bata	30,000	278	287	297	320	356	419	452	427	340	293	270	271	740	1955	1950-66
Tomi, Sibut	2,380	4.20	5.30	7.20	16.8	30.8	37.9	43.3	23.6	10.5	6.60	3.80	3.20	150	1955	1951-66
Kotto, Kembe	78,400	146	174	252	431	600	880	928	622	301	219	151	136	1,500	1962	1948-66
Chinko, Rafai	52,500	71.0	103	228	409	642	981	1,120	677	273	122	75.0	53.0	1,880	1955	1952-66

TABLE 2-25. CHAD—DISCHARGE OF SELECTED RIVERS
(Source: ORSTOM, 1969)

River and station	Basin area km^2	Mean monthly discharge, m^3/s												Max flow m^3/s	Year	Period of record
		May	Jun.	Jul.	Aug.	Sep.	Oct.	Nov.	Dec.	Jan.	Feb.	Mar.	Apr.			
Chari River, Fort-Lamy	600,000	199	293	573	1,290	2,440	3,330	3,390	1,920	837	468	271	183	5,160	1961	1936-66
Bousso	450,000	189	204	392	924	1,940	2,760	2,350	1,230	637	378	249	191	3,980	1961	1952-66
Fort-Archambault	193,000	56.0	66.0	117	272	614	1,050	801	409	208	112	73.0	60.0	2,090	1961	1938-66
Bahr-Sara, Moissala	67,600	130	186	394	916	1,530	1,420	923	459	263	171	97.0	78.0	3,680	1955	1951-66
Logone River, Bongor	73,700	94.0	142	468	1,090	1,750	1,780	735	239	137	93.0	66.0	63.0	2,630	1955	1948-66
Lai (Mission)	56,700	97.0	163	501	1,110	1,950	1,570	527	204	125	87.0	62.0	65.0	3,770	1955	1948-66
Moundou	33,970	85.0	143	381	983	1,410	1,000	310	135	85.0	59.0	40.0	49.0	3,640	1956	1935-66
Nya, Argao	2,840	0.050	0.323	13.8	85.5	105.5	38.7	6.25	0.650	0.300	13.6	1.53	0.200	268	1964	1964-66
Pende, Doba	14,300	13.0	20.0	100	323	561	462	136	48.0	23.0	13.0	8.00	8.90	928	1963	1950-66
M'Bere, M'Bere	7,430	45.0	84.0	145	254	333	265	91.0	51.0	34.0	22.0	16.0	26.0	1,940	1965	1951-66
Ba-Tha, Ati	46,000	0.000	0.000	4.50	124	163	9.10	0.100	0.000	0.000	0.000	0.000	0.000	790	1961	1955-66

TABLE 2-26. CONGO (BRAZZAVILLE)–DISCHARGE OF SELECTED RIVERS

(Source: ORSTOM, 1969)

River and station	Basin area km2	Mean monthly discharge, m3/s												Max flow m3/s	Year	Period of record
		Sep.	Oct.	Nov.	Dec.	Jan.	Feb.	Mar.	Apr.	May	Jun.	Jul.	Aug.			
Kouyou, Linnegue	10,750	169	255	290	282	236	245	261	279	302	248	165	148	658	1961	1956-65
N'Keni, Gamboma	6,200	194	212	214	210	207	207	211	213	214	197	189	185	265	1963	1951-65
Foulakari, Kimpanzou	2,813	14.0	21.0	80.0	95.0	66.0	68.0	78.0	106	92.0	30.0	21.0	17.0	365	1951	1947-65
Kouilou, Sounda	55,010	324	434	1,060	1,390	1,130	1,180	1,300	1,590	1,660	799	526	407	4,100	1950	1952-65
Loudima, Station de IFAC	3,990	12.0	14.0	33.0	42.0	36.0	36.0	40.0	57.0	51.0	22.0	17.0	14.0	192	1955	1953-65
Nyanga, Donguila	5,800	66.0	126	288	379	306	269	286	320	296	147	95.0	75.0	760	1962	1954-65

TABLE 2-27. DAHOMEY–DISCHARGE OF SELECTED RIVERS

(Source: ORSTOM, 1969)

River and station	Basin area km2	Mean monthly discharge, m3/s												Max flow m3/s	Year	Period of record
		May	Jun.	Jul.	Aug.	Sep.	Oct.	Nov.	Dec.	Jan.	Feb.	Mar.	Apr.			
Alibori River, Route Kandi-Banikoara	8,150	7.45	7.08	30.1	114	236	78.0	5.39	1.22	0.400	0.110	0.010	0.040	685	1962	1952-66
Pendjari, Porga	22,280	2.23	9.40	26.9	170	405	307	41.5	6.89	3.53	1.67	0.760	0.630	776	1952	1952-66
Oueme, Pont de Save	23,600	0.726	28.0	154	474	758	490	114	11.6	1.98	0.216	0.253	0.460	2,650	1949	1942-66
Zou, Atcherigbe	6,950	5.21	25.3	60.9	58.2	73.5	76.9	11.7	0.530	0.010	0.020	0.420	0.990	581	1963	1951-66
Okpara, Kaboua	9,600	0.524	6.32	32.9	94.9	170	177	47.0	6.13	1.09	0.210	0.190	0.126	535	1957	1951-66

TABLE 2-28. ETHIOPIA—DISCHARGE OF SELECTED RIVERS IN THE BLUE NILE BASIN, 1970
(Source: Dept. of Water Resources, 1971)

River and station	Basin area km2	Mean monthly discharge, m3/s												Max. mean daily flow m3/s	Date
		Jan.	Feb.	Mar.	Apr.	May	Jun.	Jul.	Aug.	Sep.	Oct.	Nov.	Dec.		
Blue Nile (Abbay), Bahir Dar (Lake Tana)	15,243	69.8	43.4	25.5	11.8	*	*	*	117	259	240	173	111	285	Sep. 17
Blue Nile (Abbay), Sudan border	173,813	227	165	114	*	*	*	*	6,760	4,652	2,681	1,022	456	9,741	Aug. 23
Dinder, Abu Mendi	3,110	0	0	0	0	0.05	3.7	36.2	145	98.1	22.0	2.3	0.03	311	Aug. 16
Diddessa, Arjo	9,486	11.8	*	*	*	16.5	111	336	*	*	*	71.8	23.9	–	–
Dabus, Assosa	10,100	52.3	31.6	22.0	14.5	15.1	58.0	139	253	462	409	218	103	535	Sep. 28
Angar, Lekemt	4,349	9.5	5.4	5.2	3.0	4.1	2.3	91.0	262	244	120	39.7	19.7	477	Aug. 28
Beles, Metekel	3,520	0.3	0.1	0.09	0.03	0.03	2.3	18.7	99.7	81.4	26.8	4.6	0.4	240	Aug. 24

* No record.

TABLE 2-29. FRENCH AFRICAN DEPENDENCIES : REUNION—DISCHARGE OF SELECTED RIVERS
(Source: ORSTOM, 1968)

River and station	Basin area km2	Mean monthly discharge, m3/s												Year	Period of record
		Jan.	Feb.	Mar.	Apr.	May	Jun.	Jul.	Aug.	Sep.	Oct.	Nov.	Dec.		
R. du Mat, Route Nationale	145	7.38	14.0	14.7	8.15	7.10	7.65	6.03	7.94	7.56	6.63	10.5	8.46	9.42	1959-65
R. des Roches, Grand Bras	24.4	6.57	5.66	11.8	3.85	2.61	2.38	2.90	3.22	2.69	2.02	2.39	4.03	4.18	1947-65
R. des Marsouins, Cascade Gingembre	27.5	4.47	6.85	11.5	6.23	5.14	4.65	4.31	4.39	4.21	4.11	3.21	4.48	5.29	1950-65
R. Langevin, Passerelle	36	3.62	2.92	3.40	2.89	2.42	2.78	2.47	2.25	1.99	1.80	1.48	1.55	2.46	1950-65
Bras de la Plaine, Passerelle de l'Entre-Deux	83	7.38	7.27	7.21	6.25	5.94	5.95	5.98	5.71	5.55	5.36	5.23	5.36	6.09	1950-65

TABLE 2-30. FRENCH AFRICAN DEPENDENCIES: TERRITORY OF THE AFARS AND THE ISSAS—
MUNICIPAL WATER SUPPLY SYSTEM OF DJIBOUTI
(Source: United Nations, 1964)

Year	Annual water consumption, m3			Per capita consumption l/d	Source of water	Quality of water
	Town	Port	Total			
1958	1,229,903	104,027	1,333,930	–	Infiltration galleries and wells in Wadi Hambouli	Total Dissolved Solids contents 1,300-2,000 ppm
1959	1,147,944	127,529	1,275,473	–		
1960	1,255,000	131,000	1,386,000	100		

TABLE 2-31. GABON–DISCHARGE CHARACTERISTICS OF PRINCIPAL RIVERS

(Source: Libizargomo-Joumas, Water for Peace, 1963)

[Discharge in m3/s; runoff in l/s/km2]

River and gaging station	Drainage basin km2	Long-term average discharge			Absolute low flow		Median minimum monthly flow	Maximum streamflow			Unit
		Median	10-year dry period	10-year rainy period	Median	10-year dry period		2-year	10-year	Exceptional	
Ogooue											
Mingara (near Franceville)	8,800	255	190	300	110	—	135	500	700	1,500	m3/s
		29.0	21.6	34.1	12.5	—	15.3	57	80	170	l/s/km2
Mafoula-Matato	19,800	600	450	725	300	—	350	1,000	1,500	2,500	m3/s
		30.5	22.8	36.7	15.2	—	17.5	51	76	127	l/s/km2
Lastourville (Boutemba)	47,700	1,250	950	1,700	500	(350)	615	2,250	3,250	5,500	m3/s
		26.2	20	35.7	10.5	7.3	12.9	47	68	115	l/s/km2
Booue	129,600	2,785	2,250	3,400	1,000	650	1,150	5,500	6,500	12,000	m3/s
		21.5	17.4	26.2	7.7	5.0	8.9	42.5	50	92.5	l/s/km2
Pte Okanda	140,000	2,870	2,320	3,500	1,050	680	1,230	5,750	7,150	13,000	m3/s
		20.5	16.6	25.1	7.5	4.8	8.8	41	51	93	l/s/km2
Ndjole	158,100	3,085	2,500	3,800	1,150	750	1,365	6,250	8,250	15,000	m3/s
		19.5	15.8	24	7.2	4.7	8.6	39.5	52	95	l/s/km2
Lambarene	205,000	4,670	—	—	1,600	1,250	1,940	9,150	12,200	20,000	m3/s
		22.8	—	—	7.3	6.1	9.5	44	60	97.5	l/s/km2
Mpassa											
Bac Okondja	6,400	235	—	—	125	—	150	600	850	—	m3/s
		36.7	—	—	19.5	—	23.5	94	133	—	l/s/km2
Ivindo											
Makokou	35,800	545	360	730	80	55	165	1,350	1,800	3,500	m3/s
		15.2	10	20.4	2.2	1.5	4.5	38	50	98	l/s/km2
Tsengue-Leledi	62,700	1,005	750	1,200	200	125	300	2,500	3,500	6,000	m3/s
		16	12	19.2	3.2	2.0	4.8	40	56	96	l/s/km2
Ngounie											
Fougamou (Ch. Imperatrice)	22,400	745	530	950	160	135	190	1,800	2,000	4,200	m3/s
		34	24	43	7.3	6.1	8.6	82	90	190	l/s/km2
Nyanga											
Pont Dounguila	5,600	222	160	260	60	45	66	560	750	1,900	m3/s
		39.6	28.6	46.5	10.7	8.0	11.8	100	134	340	l/s/km2
Mitoungou	8,200	290			70	55	86	750	1,000	2,500	m3/s
		35.4			8.5	6.7	10.5	91	120	300	l/s/km2
Ouyama	20,800	650			80	60	100	1,350	1,900	4,000	m3/s
		31.2			3.8	2.9	4.8	65	90	190	l/s/km2

TABLE 2-32. GABON–DISCHARGE OF SELECTED RIVERS
(Source: ORSTOM, 1969)

River and station	Basin area km2	Mean monthly discharge, m3/s												Max flow m3/s	Year	Period of record
		Jan.	Feb.	Mar.	Apr.	May	Jun.	Jul.	Aug.	Sep.	Oct.	Nov.	Dec.			
Ogooue, Lambarene	203,500	4,880	4,370	4,840	5,940	6,530	4,740	2,840	2,000	1,950	4,180	7,340	7,000	13,500	1961	1929-65
Ogooue, Franceville	9,000	269	308	332	345	351	236	172	150	143	197	300	300	688	1961	1953-65
Ivindo, Makokou	35,800	415	256	331	630	791	625	294	145	204	731	1,320	937	1,980	1962	1954-65
N'Gounie, Fougamou	22,000	924	874	909	1,070	1,010	472	304	218	186	464	1,190	1,260	2,620	1962	1953-65

TABLE 2-33. GABON–MUNICIPAL WATER SUPPLY SYSTEMS
(Source: Libizangomo-Joumas, Water for Peace, 1968)

City	Population	Capacity of water-supply system m3/d	Source of water
Libreville	73,000	10,000	N'Toum River, 40 km from Libreville
Port Gentil	—	5,000	106 wells in 5 well fields

TABLE 2-34. GHANA–DISCHARGE OF SELECTED RIVERS
(Source: UNESCO, 1971)

River and station	Basin area km2	Mean monthly discharge, m3/s												Year	Period of record
		Jan.	Feb.	Mar.	Apr.	May	Jun.	Jul.	Aug.	Sep.	Oct.	Nov.	Dec.		
Pra, Daboasi	22,710	80	135	235	530	630	285	260	570	370	180	75	50	283	1954-65
Tano, Alenda	15,800	35	65	120	310	360	145	100	245	195	90	30	23	143	1956-65
Volta, Yeji	261,000	59	35	30	34	59	160	370	910	2,420	2,380	480	90	585	1955-64
Volta, Senchi	394,100	100	55	43	48	110	380	920	1,870	4,930	5,120	1,310	240	1,260	1936-63
Black Volta, Bui	125,100	52	31	20	19	35	84	165	390	905	789	228	99	234	1953-64
White Volta, Nawuni	92,880	4	2	1	1	3	28	170	690	1,320	590	65	11	240	1959-65

TABLE 2-35. GHANA—WITHDRAWAL OF GROUND WATER, 1963

(Source: Gill, 1964 U.S.Geol. Survey)

Geohydrologic province or subprovince	Geologic unit	Geographic area or political subdivision	Approximate pumpage (gpd)
Lower Precambrian	Dahomeyan	Accra Plains	124,000
Coastal Plain	Cretaceous to lowerTertiary	Half Assini area	1,058,000
Middle Precambrian	Upper Birrimian	Western and Central Regions	401,000
Coastal Block Fault	Devonian	Central Region	31,000
Middle Precambrian	Lower Birrimian	Western and Central Regions	1,272,000
Precambrian	Granite	Northern and Upper Regions	4,024,000
Voltaian	Voltaian	do	298,000
Precambrian	Togo and Buem	Volta Region	1,301,000
Voltaian	Voltaian	do	281,000
Precambrian	Dahomaeyan and Buem	Ho area	245,000
Middle Precambrian	Lower Birrimian	Brong-Ahafo Region	190,000
Precambrian	Granite	Ashanti Region	224,000
Coastal Plain	Cretaceous to lower Tertiary	Keta area	1,506,000
Precambrian	Granite	Central Region	87,000
Total Ghana	. . . .		11,042,000

TABLE 2-36. GHANA—MUNICIPAL WATER SUPPLY SYSTEMS

(Source: Internat. Statistical Institute, 1971)

City	Year	Population served	Total quantity available (1,000 m3)	Annual consumption (1,000 m3)	
				Total	of which domestic
Accra City	1967	523,800	24,545	22,273	N.A.
	1969	591,000	27,273	24,545	3,864
Accra-Tema City Council	1967	616,000	34,455	32,636	N.A.
	1969	705,800	40,909	38,864	18,200

N.A. — Not available

TABLE 2-37. IVORY COAST–DISCHARGE OF SELECTED RIVERS
(Source: ORSTOM, 1969)

River and station	Basin area km²	Mean monthly discharge, m³/s												Max flow		Period of record
		May	Jun.	Jul.	Aug.	Sep.	Oct.	Nov.	Dec.	Jan.	Feb.	Mar.	Apr.	m³/s	Year	
Bagoe, Guinguerini	1,042	0.04	3.21	17.8	48.0	65.0	32.0	10.0	37.6	1.77	0.75	0.09	0.01	431	1964	1955-66
Sassandra, Guessabo	35,400	77.0	145	309	581	1,080	794	358	156	78.0	51.0	49.0	64.0	1,830	1957	1953-66
Bandama River, Brimbo	60,300	48.3	117	232	467	1,090	1,100	386	135	54.0	28.1	20.3	32.9	2,270	1957	1953-66
Beoumi	26,200	17.7	46.0	102	269	650	563	183	72.0	29.9	14.4	8.90	13.3	1,210	1964	1954-66
N'Zi, Zienoa	33,150	35.0	111	152	113	204	317	141	27.4	8.60	3.30	3.90	14.3	768	1957	1953-66
Comoe, Aniassue	66,500	39.4	81.5	227	451	1,040	943	301	92.4	28.5	13.1	13.4	21.9	2,340	1954	1953-66
Agneby, Agboville	4,600	8.70	29.8	45.0	9.10	10.9	16.8	8.10	2.50	6.10	0.35	1.02	3.06	171	1955	1955-66

TABLE 2-38. KENYA–DISCHARGE OF SELECTED RIVERS
(Source: UNESCO, 1971)

River and station	Basin area km²	Mean monthly discharge, m³/s												Year	Period of record
		Jan.	Feb.	Mar.	Apr.	May	Jun.	Jul.	Aug.	Sep.	Oct.	Nov.	Dec.		
Nzoia, Ukwala	11,851	47.8	33.8	42.2	84.4	158	91.9	90.8	134	120	91.6	73.9	75.6	91.0	1963-65
Tana, Garissa	42,217	105	67.9	69.5	219	372	184	102	82.2	70.4	90.4	241	204	151	1933-65
Ewaso N'giro, Archers Post	15,022	15.2	6.6	9.6	40.4	24.9	11.9	11.2	17.8	17.3	15.9	63.0	33.2	22.4	1949-65

TABLE 2-39. KENYA—RUNOFF AND HYDROELECTRIC POWER GENERATION
(Source: Ojany, F.F., 1973)

Runoff		Existing Hydroelectric Power Plants				
River	Mean annual runoff million m³	River	Location of plant	Available Head meters	Minimum Flow m³/s	Output Kw
Nzoia	1,920					
Yala	965					
Nyando..................	500	Maragua	Mesco	40	1.4	380
Sondu....................	1,235					
Kuja-Migori...............	870	Maragua	Wanjii	68	1.7	2,000
Others	1,800 1)					
Total	7,290	Mathioya	Wanjii	110	2.5	5,400
Melawa	184					
Gilgil	28	Maragua	Tana	73	1.7	6,400
Molo	39					
Perkerra	125	Tana	Tana	54	4.5	8,000
Others	430 1)					
Total	806	Sagana	Sagana Falls	37	1.1	1,500
Athi	720	Thika	Ndula	29	2.0	2,250
Tsavo	138					
Njoro-Lumi...............	293	Liki	Nanyuki	26	0.1	100
Springs	—					
Others	113 1)	Sosiani	Selby Falls	43	0.04	360
Total	1,294	Kuja	Macalder Mines	24	0.8	900
Tana (Garissa)	4,700					
Ewaso Ngiro .... (Archer's Post)	740	Tana	Kindaruma	36	15	40,000
Estimated total runoff of whole country	14,830					

1) Estimated.

FIGURE 2-5. LIBYA — GROUND-WATER PROVINCES
(Source: Jones, U.S. Geol. Survey, 1971)

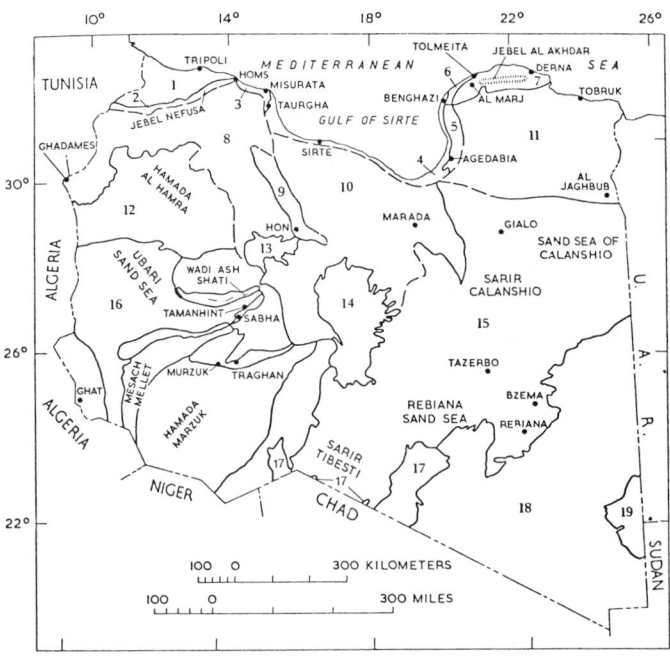

1. Jefara; 2. Jebel Nefusa; 3. Misurata; 4. Sirtic coast; 5. Agedabia; 6. Sahel; 7. Jebel Akhdar; 8. Ghibla; 9. Hon graben; 10. Sirtica; 11. Balta; 12. Hamada al Hamra; 13. Jebel es Soda; 14. Haruj al Aswad; 15. Calanshio; 16. Murzuk basin; 17. Tibesti; 18. Kufra basin; 19. 'Uweinat.

TABLE 2-40. LIBYA–GROUND-WATER PROVINCES AND PUMPAGE

(Source: Jones, U.S. Geol. Survey, 1971)

[For location of ground-water provinces see adjacent map; for explanation of water quality terms see footnotes]

Ground-water province	Area km²	Principal aquifer	Depth to water	Discharge or pumpage	Quality of water [1]
1. Jefara	18,500	Quaternary sediments Tertiary sandstones	Regional water table <20 m below land surface (l.s.) Artesian, 40-50 m above l.s.	Pumpage 250X106m3/yr	Fair to good (western part) Poor or brackish
2. Jebel Nefusa	—	Mesozoic sediments	Perched water conditions; regional water table at considerable depth	Yields of wells small	Fair; hard
3. Misurata	1,800	Quaternary sediments Tertiary sediments	Regional water table <20 m below l.s. —	Pumpage 100X106m3/yr; spring discharge - 240 X 106m3/yr; little developed	Poor
4. Sirtic coast	—	Sand dunes and alluvium Miocene sediments	— —	Total withdrawal very small	Fair Brackish to salty
5. Agedabia	—	Miocene sediments	—	Total withdrawal very small	Brackish to salty
6. Sahel	1,300	Miocene limestone	Water levels slightly above sea level	Pumpage (chiefly for Benghazi) is 6X106m3/yr	Shallow water poor to brackish in south. Fair in northern part
7. Jebel Akhdar	—	Cretaceous to Miocene sediments	Perched water conditions; regional water table >200 m below l.s.	Spring discharge - 24-48X106m3/yr	Good to fair
8. Ghibla	—	Ain Tobi limestone (Cret.) and Chicla sand and sandstone (Cret.)	Perched water conditions; regional water table 100-200 m below l.s.	Yields of wells small	Mostly brackish
9. Hon Graben	—	Quaternary and Tertiary sediments	Flowing artesian wells in south	—	Mostly brackish
10. Sirtica	133,000	Tertiary sediments and local Quaternary deposits	Water table 50 m below l.s. some deep flowing wells in south	—	Salty to brackish
11. Balta	—	Miocene sediments	Water table >100 m below l.s. in northern half and <50 m in south and west; water table below sea level in Jaghbub trough	Pumpage very small	Poor to salty

TABLE 2-40. LIBYA—GROUND-WATER PROVINCES AND PUMPAGE (continued)

Ground-water province	Area km2	Principal aquifer	Depth to water	Discharge or pumpage	Quality of water 1)
12. Hamada al Hamra	—	U. Cretaceous limestone and Chicla sandstone	Water table 50-100 m below l.s.; in places <50 m or >200 m	Undeveloped	Predominately brackish (east of long. 12°); fair to the west
13. Jebel es Soda and al Haruj al Aswad	—	Basalt, Tertiary and Cretaceous sediments	Unknown	Undeveloped	Poor to saline
14. Calanshio	375,000	Post-Miocene sand	Regional water table varies from <1->200 m below l.s. 20 m at oases; <50 m south of lat. 26°N	Undeveloped	Brackish to salty north of lat. 28°15'; fair south of line in east-central part. Poor water along east border and east of Tazerbo. Good quality water in Rebiana and Bzema areas.
		Older aquifers	Confined conditions		
15. Murzuk basin	320,000	Paleozoic and Mesozoic rocks including Nubian sandstone and Quaternary sediments	Water table<50 m below l.s. in general; 100 m under Mesak Mellet and western edge of Sand Sea; <20 m at oases. Flowing wells at some localities.	Discharge from wells and springs in Wadi Shati area 30X10⁶m3/yr; Yields of wells range from 1-190 m3/hr	Fair to good
16. Kufra basin	245,000	Nubian sandstone and Quaternary sands	Shallow at oases	Total withdrawal 2X10⁶m3/yr	Poor to mildly brackish in shallow wells; fair to good in Nubian sandstone and older rocks.
17. Tibesti and Jebel al Uweinat	—	Granite, metamorphics, volcanics and sandstones. Alluvial fill in wadis.	Unknown	Undeveloped	Unknown

1)

CLASSIFICATION OF GROUND WATER IN LIBYA

Class	Chloride (Cl)	Sulfate (SO_4)	Total dissolved solids (ppm)	Specific Conductance (Micromhos at 25°C)
Good	less than 250	less than 250	less than 1,000	less than 1,600
Fair	251 to 800	251 to 600	1,010 to 1,500	1,610 to 2,500
Poor	801 to 1,500	601 to 1,200	1,510 to 3,500	2,510 to 6,000
Brackish	1,510 to 3,500	1,210 to 2,500	3,510 to 6,000	6,010 to 10,000
Salty	more than 3,500	more than 3,500	more than 6,000	more than 10,000

TABLE 2-41. LIBYA—DOMESTIC AND INDUSTRIAL WATER REQUIREMENTS
(Source: General Water Authority, 1974)

Category	Estimated withdrawal million m^3 per year	Source of water	Quality of water
Domestic (Coastal zone)	100	Ground water from shallow aquifers	Total dissolved solids content 500-2,500 ppm
Industrial	Small, to rise to 40 in near future	—	—

TABLE 2-42. LIBYA—IRRIGATED AREA AND WATER REQUIREMENTS FOR IRRIGATION
(Source: General Water Authority, 1974)

Region	Irrigated Area 1974 ha	Near future ha	Near future Water requirement million m^3/yr	Water quality Total dissolved solids ppm	Aquifer Pumping Depths m	Aquifer Drilling Depths m
SOUTH Fezzan	5,000	15,000	100	700 to 1,500	up to 50	100 to 400 (Nubian Sandstone)
Kufra-Sarir	10,000	20,000 50,000	400 1,000	700 to 1,000	50 to 100	100 to 300 (Nubian and Continental Post Eocene)
NORTH Northwest	57,000	114,000	700	500 to 3,000	10 to 100	10 to 100 (Shallow aquifer)
Northeast	15,000	21,000	100	500 to 2,000	10 to 200	up to 300 (Shallow and perched aquifers)
North-Central	5,000	40,000	400 [1]	1,500 to salty	Shallow and deep aquifer	
Total	92,000	260,000	2,800			

[1] Most of this water will be transferred to the central coastal region.

TABLE 2-43. MADAGASCAR—DISCHARGE OF SELECTED RIVERS
(Source: ORSTOM, 1969)

River and station	Basin area km2	Mean monthly discharge, m3/s												Max flow m3/s	Year	Period of record
		Nov.	Dec.	Jan.	Feb.	Mar.	Apr.	May	Jun.	Jul.	Aug.	Sep.	Oct.			
Sambirano, Ambanja	2,980	33.8	106	219	290	300	190	92.4	52.9	36.4	28.8	23.0	21.6	6,700	1959	1951-65
Ikopa River, Antsatrana	18,550	238	613	952	1,030	1,060	534	295	227	188	155	125	125	2,820	1958	1948-65
Bac de Fiadana	9,450	102	313	415	381	448	248	128	98.0	86.0	75.0	56.0	48.0	1,840	1964	1958-65
Bevomanga	4,250	41.9	102	149	146	163	109	51.8	40.8	38.9	33.0	26.6	22.9	550	1959	1947-65
Sisaony, Andramasina	318	4.42	9.81	15.6	11.5	12.4	5.45	3.12	2.99	3.10	2.87	2.30	1.87	172	1963	1959-65
Betsiboka, Ambodiroka	11,800	163	445	678	620	666	313	169	144	124	107	86.2	74.3	12,000	1959	1957-65
Isinko, Ambodiroka	600	14.7	34.0	56.9	47.5	60.3	26.8	12.7	8.80	6.80	5.00	3.90	3.10	1,550	1963	1957-65
Andromba, Tsinjony	350	5.09	12.1	16.7	16.4	21.5	10.3	5.05	4.15	3.91	3.62	3.26	3.00	202	1959	1959-65
Vohitra, Rogez	1,825	44.4	67.0	97.0	109	155	78.0	57.0	59.0	62.0	61.0	48.3	37.4	3,950	1959	1948-65
Ivondro, Ringaringa	2,775	69.8	94.7	126	141	189	132	96.3	92.4	93.8	95.8	80.9	67.2	1,940	1959	1952-65
Mananjary, Antsindra	2,260	56.7	119	198	216	216	146	98.6	82.3	80.1	83.5	68.6	50.9	1,330	1963	1955-65
Ivoanana, Fatihita	835	24.7	42.8	77.6	69.5	106	68.0	39.0	29.4	31.7	25.4	19.9	46.8	710	1959	1955-65
Namorona, Vohiparara	445	6.88	11.9	23.1	20.8	27.1	14.8	10.3	10.1	9.40	9.50	7.00	5.10	542	1959	1951-65
Mangoky, Banian	50,000	175	768	1,410	1,100	911	378	183	158	135	118	89.5	79.6	14,800	1956	1951-65
Mananantanana River, Tsitondroina	6,510	46.2	220	278	198	190	70.5	27.6	28.0	20.7	20.3	15.2	12.3	2,150	1964	1952-65
Ihosy, Ihosy	1,500	6.20	26.8	55.3	37.6	23.1	13.3	7.20	5.10	5.30	5.10	4.50	4.30	580	1954	1952-65
Mania, Fasimena	6,675	88.7	205	303	259	219	133	92.8	81.0	82.3	76.0	65.0	62.3	765	1963	1956-65
Mandrare River, Amboasary	12,435	30.6	200	292	202	190	53.3	23.4	17.2	12.2	14.4	14.0	11.3	6,350	1960	1951-65
Bevia	1,137	2.09	6.80	18.9	9.57	15.2	3.22	1.75	2.34	2.08	1.68	1.55	1.30	700	1961	1951-65

TABLE 2-44. MALAWI—DISCHARGE AND IRRIGATION POTENTIAL OF MAJOR RIVERS
(Source: Framji and Mahajan, ICID, 1969)

River	Estimated maximum flow m³/s	Estimated minimum flow m³/s	Potential reservoir site
Songwe (Internat. boundary)	Very high	4.25	No
Lufira	113.27	0.057 - 0.141	No
N. Rukuru	679.60	0.085 - 1.13	Yes
Waye and Chiwondo Lagoon	Moderate	Nil	Yes
Nyungwe	Moderate	Nil	Yes
Wovwe	56.63	0.85 - 1.13	No
Chananga and Hara	56.63 - 84.95	0.142	No
Runyina	50.97	0.85	No
Kasitu and Henga Valley	Flood control with irrigation and power possible		
Upper S. Rukuru Valley	Irrigation from reservoirs possible		
Limpasa	254.85	0.28 - 0.42	No
Luweya	665.44	0.57 - 0.85	Yes
Dwambazi	Moderate	1.416 - 1.7	No
Dwanga	High	0.85 - 1.13	Yes
Bua	184.06	0.141	Yes
Kaombe	High	Very low	Yes
Linthipe/Lilongwe	962.77	Very low	Possible
Lifisi	56.63 - 113.26	0.42	No
Shire (Internat. boundary in South)	Very high	141.58	No
Rou and tributaries (Internat. boundary)	Irrigation of tea areas with river water supported by storage possible		

TABLE 2-45. MALAWI—DISCHARGE OF THE SHIRE RIVER
(Source: Pike and Rimmington, Malawi, A Geographical Study, Oxford Univ. Press, 1965)

[Mean annual flow in cfs]

Year	Gaging Station				
	Liwonde	Matope	Chikwawa	Chiromo	Port Herald
1948-49	10,557	10,265	N.A.	10,798	9,459
1949-50	10,392	10,804	N.A.	11,480	9,677
1950-51	10,474	10,754	N.A.	11,089	9,534
1951-52	12,090	12,808	14,243	14,743	11,022
1952-53	10,468	10,754	11,165	11,545	10,023
1953-54	7,855	8,179	8,215	7,729	7,349
1954-55	7,179	7,828	8,309	9,108	8,529
1955-56	8,106	8,616	9,493	9,633	8,475
1956-57 [*]	2,667	2,810	3,424	3,091	4,932
1957-58	14,070	14,317	14,674	15,168	10,875
1958-59	11,422	11,497	11,644	12,300	10,106
1959-60	9,942	10,064	10,186	9,983	9,819
1960-61	9,342	9,613	10,388	10,374	10,191

[*] Flow from Lake Nyasa ceased owing to river obstruction.
N.A. — Not available

TABLE 2-46. MALI–DISCHARGE OF SELECTED RIVERS
(Source: ORSTOM, 1969)

River and station	Basin area km2	Mean monthly discharge, m3/s												Max flow m3/s	Year	Period of record
		May	Jun.	Jul.	Aug.	Sep.	Oct.	Nov.	Dec.	Jan.	Feb.	Mar.	Apr.			
Senegal, Galougo	128,400	10.6	135	600	2,080	2,660	1,350	484	232	131	74.5	37.0	15.0	6,880	1958	1905-66
Niger River, Dire	330,000	143	82.0	273	863	1,500	1,890	2,150	2,290	2,050	1,540	959	434	2,680	1926	1924-66
Mopti	281,600	67.0	143	699	1,760	2,580	2,820	2,680	2,020	998	406	176	98.0	3,100	1924	1922-66
Koulikoro	120,000	97.0	356	1,240	3,190	5,310	4,600	2,090	868	403	196	101	67.0	9,700	1925	1907-66
Bani, Douna	101,600	31.0	46.0	181	1,210	2,510	2,520	1,200	414	171	102	67.0	40.0	3,550	1964	1922-66
Sankarani, Gouala	35,300	41.0	103	319	792	1,420	1,240	509	212	106	65.0	38.0	32.0	2,200	1962	1954-66

TABLE 2-47. MAURITANIA–DISCHARGE OF OUED GHORFA
(Source: ORSTOM, 1969)

River and station	Basin area km2	Mean monthly discharge, m3/s												Max flow m3/s	Year	Period of record
		May	Jun.	Jul.	Aug.	Sep.	Oct.	Nov.	Dec.	Jan.	Feb.	Mar.	Apr.			
Oued Ghorfa, Ghorfa	5,020	0.000	0.000	8.38	56.4	56.0	2.25	0.000	0.000	0.000	0.000	0.000	0.000	240	1965	1964-66

TABLE 2-48. MAURITIUS—MAJOR RESERVOIRS

(Source: Sentenac, Mauritius Sugar Industry Res. Inst., 1962)

[Multiple-purpose storage reservoirs]

Name of reservoir	Maximum capacity million cu.ft.	Maximum capacity million m³	Maximum surface area acres	Maximum surface area hectares
Mare aux Vacoas	597	16.9	1,000	404.7
	975	27.6	1,300	526
Mare Longue	222	6.28	238	96.5
Tamarin Falls	72	2.4	177	71.5
Piton du Milieu	112	3.17	170	68.8
La Ferme	416	11.8	561	227
La Nicoliere	204	5.78	232	94
Eau Bleue	216	6.1	222	90
Midlands (under construction)	550	15.6	620	251

TABLE 2-49. MAURITIUS—MUNICIPAL WATER SUPPLY SYSTEMS

(Source: Internat. Statistical Institute, 1971)

City	Year	Population served	Annual consumption 1,000 m³
Port Louis	1967	125,800	16,593
	1969	130,000	21,571

TABLE 2-50. MOROCCO—DISCHARGE OF SELECTED RIVERS

(Source: UNESCO, 1969)

River and station	Basin area km2	Mean monthly discharge, m3/s Jan.	Feb.	Mar.	Apr.	May	Jun.	Jul.	Aug.	Sep.	Oct.	Nov.	Dec.	Year	Period of record
Mediterranean Sea															
Moulouya, Mechra Homadi	51,950	55.4	56.6	74.8	92.6	102	62.2	18.5	14.1	26.2	21.9	33.5	44.7	50.2	1951-64
Atlantic Ocean															
Sebou, Azib El Soltane	17,250	200	191	178	133	68.6	49.5	27.4	22.3	16.2	21.6	54.3	124	90.6	1959-64
Oum Er Rebia, Im Fout	30,000	114	150	187	191	138	72.8	41.9	36.4	42.4	52.9	74.8	106	101	1941-64
Souss, Ait Melloul	16,150	5.3	5.4	3.8	6.3	1.3	0.9	0.7	0.7	0.9	0.6	0.8	54.1	6.8	1963-64
Dra, Zagora	20,130	3.6	4.2	0.3	5.0	1.2	0.3	0.2	0.2	0.0	0.0	0.0	3.0	1.5	1963-64

TABLE 2-51. MOROCCO—DISCHARGE CHARACTERISTICS OF MAJOR RIVERS

(Source: Combe, Direction de l'Hydraulique, 1968)

River basin	Basin area km^2	Annual precipi- tation mm	Period of record	Station or location	Mean annual flow m^3/s	Mean flow during driest month m^3/s
Sebou	39,000	710	1932-1964	Mouth	200.0	20.0
Oum Er Bia	34,400	480	1918-1966	Mouth	130.0	30.0
Moulouya	53,700	300	1952-1966	Mouth	44.0	4.0
Loukkos	3,800	950	1961-1967	Mouth	42.0	0.5
Tensift	20,100	365	Calculated	Mouth	29.0	0.3
Bou Regreg	7,800	600	1930-1968	Mouth	20.0	0.4
Martine	1,220	935	Calculated (1963-1967)	Mouth	14.5	0.2
Draa	15,100	190	1936-1967	Zaouia-N'Ourbaz	13.5	1.0
Lau	920	980	Calculated (1963-1967)	Mouth	13.0	2.0
Neckor	960	590	Calculated	Mouth	9.0	1.0
Kerte	3,080	355	Calculated	Mouth	8.0	0.5
Rhis	800	595	Calculated	Mouth	7.5	0.8
El Had	600	825	Calculated	Mouth	7.2	0.8
Hachef	650	905	Calculated	Mouth	6.3	0.0
Ziz	4,600	290	1948-1966	Ksar es Souk	6.0	0.5
	186,730	Oueds with discharge greater than 6 m^3/s			550.0	62.0
	113,270	Other oueds			110.0	3.0
Total General	300,000				660.0	65.0

TABLE 2-52. MOROCCO—DAMS AND RESERVOIRS

(Source: Direction de l'Hydraulique, 1972)

| Name of dam | Year of comp- letion | Location | | Length of crest m | Volume of dam 10^3m^3 | Gross capacity of reservoir 10^3m^3 | Pur- pose [2] | Maximum discharge capacity of spillways m^3/s |
		River	Nearest city					
Sidi Maachou	1929	Oum-er-R'bia	El-Jadida	150	32	2,000	H/S	4,500
Oued Mellah	1931	Mellah	Casablanca	138.50	25	18,000	I/S	300
Ali Thelat	1934	Laou	Chaouen	320	52.4	25,000	I/H	500
El Kansera [1]	1935	Beht	Sidi-Slimane	177.50	200	297,000	I/H	1,750
Lalla Takerkoust	1935	N'Fis	Marrakech	357	150	52,000	I/H	2,000
Ouezzane	1937	Bou Droua	Ouezzane	235	75	400	S	
Imfout	1944	Oum-er-R'bia	Settat	200	130	83,000	I/H	3,500
Zemrane	1950	Mellah	Khouribga	112.50	7.8	600	S	
Daourat	1950	Oum-er-R'bia	Settat	125	50	24,000	H	3,500
Bin-El-Ouidane	1953	El Abid	Beni-Mellal	290	365	1,500,000	I/H	2,500
Ait Ouarda	1954	El Abid	Beni-Mellal	120	28	3,800	I/H	2,750
Mechra Homadi	1955	Moulouya	Oujda	215	125	42,000	I/H	7,000
Taghdout	1956	Taghdout	Ouarzazate	21.8	0.93	3,000	I	
Nakhla	1961	Nakla	Tetouan	240	159	9,200	I/S	700
Safi	1965	Sahim	Safi	328	63	2,100	I/S	200
Mohammed V (Mechra Klila)	1967	Moulouya	Nador	305	323	730,000	I/H	7,000
Grou	1968	Grou	Rabat	500	300	18,000	S	2,300
Moulay Youssef	1970	Tessaout	Marrakech	725	5,300	200,000	I/H	3,000
Hassan Addakhil	1971	Ziz	Ksar-es-Souk	785	5,800	380,000	I	1,700
Mansour Eddahbi	1972	Draa	Ouarzazate	285	160	560,000	I/H	7,200
Youssef Ben Tachfine	1973	Massa	Tiznit	670	3,700	310,000	I	3,400
Idriss 1er	1973	Inaouene	Fes	447	450	1,270,000	I/H	2,400
Bou-Regreg	1974	Bou-Regreg	Rabat	340	3,000	570,000	S	5,000

[1] Raised 5 m. in 1969

[2] H Hydropower
S Water supply
I Irrigation

TABLE 2-53. MOROCCO—ANNUAL WITHDRAWAL AND LOSSES OF GROUND WATER

(Source: Combe, Direction de l'Hydraulique, 1968)

Hydrogeologic basin	Ground water withdrawn		Ground water lost		Type of losses
	Rate m^3/s	Volume 10^6 m^3	Rate m^3/s	Volume 10^6 m^3	
Couloir sud-rifain and recharge basin of Moyen Atlas	14.0	441.0	1.0	31.5	Evaporation following drainage.
Souss, Chtoukas, Massa	10.0	315.0	2.0	63.0	Losses to ocean and evaporation.
Haouz Mejjate	9.0	283.5	—	—	—
Haut-Atlas central and piedmont	6.0	189.0	—	—	—
Tadla	4.5	142.0	1.0	31.5	Evaporation from drains and canals.
Rharb, Mamora, Dradere	4.0	126.0	6.0	189.0	Evaporation following drainage and losses to ocean.
Tafilalet - Maidere	3.2	100.0	5.0	157.5	Evaporation and losses to desert.
Coastal zone Rabat-Azemmour	2.0	63.0	—	—	—
Mediterranean coastal basins (Rif, Bareg, Bou Areg, Neckor)	2.0	63.0	3.0	94.5	Losses to sea and evaporation.
Foums of Anti Atlas	2.0	63.0	1.0	31.5	Evaporation
Bahira	2.0	63.0	1.5	47.0	Evaporation
Couloir Taourirt-Oujda	1.6	50.0	1.0	31.5	Losses across border and evaporation.
Berguent	1.5	47.0	1.0	31.5	Decompression of artesian aquifer and evapotranspiration of seeps.
Moyenne and haute Moulouya	1.2	38.0	1.0	31.5	Evaporation of seeps.
Sillon Ouarzazate-Boudenib	1.0	31.5	1.0	31.5	Losses across borders and to desert.
Plain of Berrechid	1.0	31.5	1.0	31.5	Evaporation
Basin Essaouira-Chichaoua	1.0	31.5	2.0	63.0	Losses to sea.
Plains of Loukkos and R'Mel	1.0	31.5	0.5	16.0	Losses to sea.
Doukkala and coastal Sahal	—	—	4.0	126.0	Losses to sea.
Total major aquifers	67.0	2,109.5	32.0	1,000.0	
Other aquifers	10.0	315.0	1.0	31.5	
					Withdrawn and lost
Grand Total	77.0	2,424.5	33.0	1,039.5	110 m^3/s or 3,500 x10^{6}m^3/yr

TABLE 2-54. MOROCCO—WATER USE, 1972-2000

(Source: Combe, Direction de l'Hydraulique, 1968)

Category	Annual water requirements, million m^3	
	1972	2000
Public water supply	300	1,300
Industry	200	700
Agriculture	7,500	10,000
Total	8,000	12,000
Available water resources.......... 16,000		16,000
Percent utilized	50	75

FIGURE 2-6. MOROCCO — WATER BALANCE DIAGRAM
(Source: Direction de l'Hydraulique, 1972)

(in 10^9 m^3)

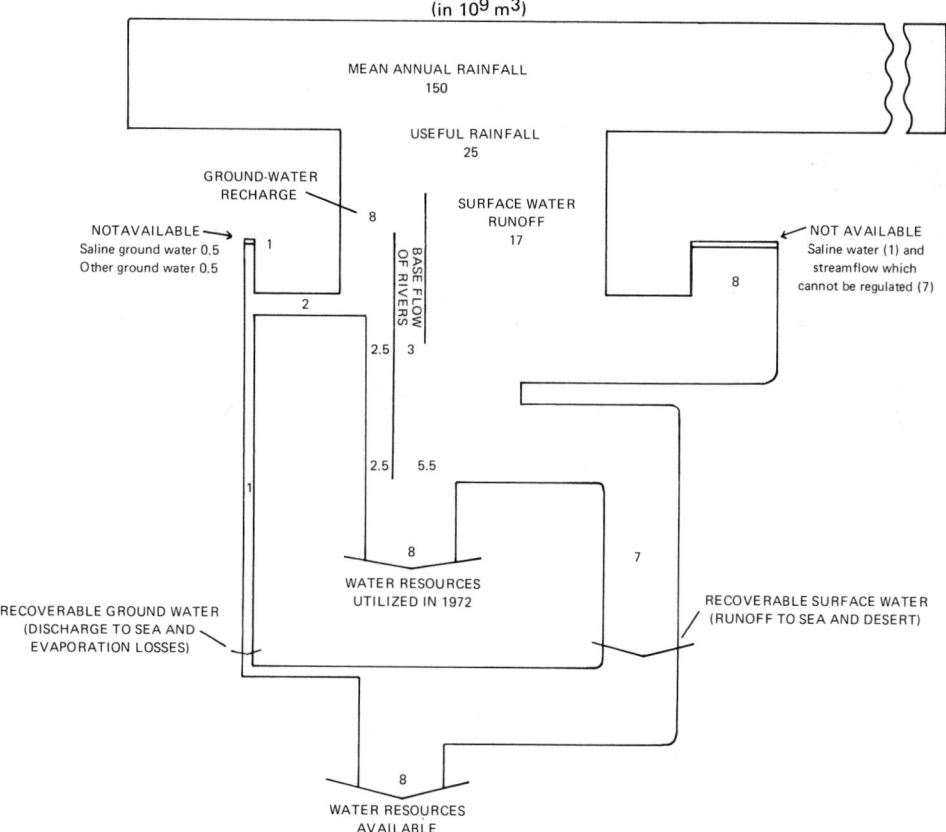

TABLE 2-55. MOROCCO–MUNICIPAL WATER SUPPLY SYSTEM

(Source: Monition, Wasserwirtschaft in Afrika, 1963)

City or town	Population in 1952	Quantity of water distributed in 1960 (l/s)			Aquifer
		Surface water	Ground water	Total	
North Flank Atlas Mountains					
Oujda	115,446	–	150	150	Liasic
Berkane	18,830	–	15	15	Liasic
Nador	26,300	–	30	30	Andesite and Quaternary
Jerada	17,400	–	120	120	Liasic
Taza	24,000	–	85	85	Liasic
Fes	235,000	–	680	680	Liasic
Meknes	190,000	–	415	415	Liasic and Quaternary
Ouezzane	24,500	8	8	16	Oligo-Miocene
Sidi Kacem	18,500	–	18	18	Quaternary
Larache	43,000	–	35	35	Plio-Quaternary
El-Ksar-El-Kbir	36,000	–	50	50	Plio-Quaternary
Tanger	140,000	100	200	200	Plio-Quaternary
Tetouane	100,000	40	25	65	Plio-Quaternary
Casablanca **	750,000	1,350	250	1,600	
Mohammedia *	30,000				
Bou-Znika *	–				
Temara *	–	{A. Reboula 160 and Bab Tamesna}			
Rabat *	180,000		543	543	
Sale *	100,000		160	160	
Kenitra *	100,000			50	
Settate	27,000	–	40	40	Cenomanian-Turonian
Khouribga	50,000	12	58	70	Precambrian
Beni-Mellal	27,000	–	30	30	Lias
Safi	76,000	–	85	85	Hauterivian and Plio-Quaternary
Marrakech	220,000	20	410	430	Quaternary
El-Jadida	35,000	115	–	115	–
Es Saouira	21,000	–	36	36	Quaternary
Agadir	32,800	–	130	130	Cenomanian-Turonian
Total	2,638,176	1,645	4,473	5,168	
South Flank Atlas Mountains					
Ouarzazate	2,856	3	–	3	Quaternary
Ksar es Souk	5,500	–	10	10	Quaternary

* Coastal communities supplied by Fouarate pipeline.

** Partially supplied by Fouarate pipeline.

TABLE 2-56. NIGER—DISCHARGE OF SELECTED RIVERS
(Source: ORSTOM, 1969)

River and station	Basin area km2	Mean monthly discharge, m3/s												Max flow m3/s	Year	Period of record
		May	Jun.	Jul.	Aug.	Sep.	Oct.	Nov.	Dec.	Jan.	Feb.	Mar.	Apr.			
Niger, Niamey	700,000	339	126	117	501	1,100	1,270	1,420	1,600	1,750	1,750	1,430	8,540	2,150	1956	1922-66
Gorouol, Dolbel	7,500	0.000	4.01	13.7	38.8	68.9	0.980	0.000	0.000	0.000	0.000	0.000	0.000	117	1961	1964-66
Goulbi de Maradi, Madarounfa-Pont	5,400	0.000	0.660	13.9	42.4	28.4	0.373	0.000	0.000	0.000	0.000	0.000	0.000	400	1961	1964-66

TABLE 2-57. NIGERIA—DISCHARGE OF SELECTED RIVERS
(Source: Framji and Mahajan (ICID) and van Blommestein (FAO), 1969)

[Estimated value]

River	Station	Basin area km2	Mean annual flow million m3	Maximum flow m3/s	Minimum flow m3/s
Benue River Basin					
Benue	Yola	106,200	23,400	—	7
Benue	Ibi	257,700	49,300	—	14
Benue	Confluence Niger R.	336,000 1)	100,000	10,000-14,000	100
Gongola	Bare	55,690	8,000	—	6
Taraba	Gassol	21,340	12,300	—	1
Chad Basin					
Yobe	Yau	97,130	430	—	3
Ebeji	Gamboru	16,210	1,700	—	3
Yedseram	Bama	4,740	290	—	0
Ngadda	Maiduguri	6,530	250	—	0
Niger River Basin					
Niger	Confluence Benue R.	1,215,000 3)	86,300 2)	—	1,840 2)
Niger	Onitsha	—	200,000	22,500	—
Kaduna	Niger R.	66,400	24,000	—	—
Rima	Niger R.	—	3,200 4)	—	—

1) Within Nigeria 230,000 km2
2) Prior to construction of Kainji dam; guaranteed minimum regulated flow is 1,500 m3/s
3) Within Nigeria 440,000 km2
4) Dry-year flow

TABLE 2-58. PORTUGUESE AFRICAN DEPENDENCIES: ANGOLA—DISCHARGE CHARACTERISTICS OF PRINCIPAL RIVERS

(Source: Quintela, Proc. Reading Symp. IASH-UNESCO-WMO, 1972)

[Period 1958-68]

River	Section	Drainage area km2	Precipitation mm	Flow million m3	Flow mm	Deficit mm
Bengo	Lalama	7,370	1,168	1,194	162	1,006
Cuanza	Cambambe	121,470	1,285	26,355	216	1,069
Queve	J. Amboiva	17,231	1,404	5,838	338	1,066
Cunene	Matala	28,037	1,217	5,449	194	1,023
Cunene	Ruacana	83,000	958	6,774	82	876
Cubango	Runtu	115,000	948	7,447	76	883

TABLE 2-59. PORTUGUESE AFRICAN DEPENDENCIES: MOZAMBIQUE—DISCHARGE OF PRINCIPAL RIVERS

(Source: de Ataída, Service Hydraulique, 1972)

River basin	Basin area km2 Mozambique	Basin area km2 Foreign	Basin area km2 Total	Length of river km	Gaging station	Effective basin area km2	Mean annual flow million m3
Maputo	1,570	28,230	29,800	565	Madubula	28,500	2,800
Umbeluzi	2,356	3,244	5,600	314	Goba	3,100	315
Incomati	14,925	31,321	46,246	714	R. Garcia	21,200	2,300
Limpopo	79,600	332,400	412,000	1,461	Trigo de Morais	340,000	5,330
Save	4,550	83,845	88,395	735	V. Franca do Save	100,885	5,000
Gorongosa	13,150	—	13,150	135	—	—	—
Buzi	25,600	3,200	28,800	360	Estaquinha	26,314	1,450
Pungoe	28,000	1,500	29,500	372	Bue Maria	15,046	3,080
Zambeze	140,000	920,000	1,200,000 1)	2,700	D. Ana	1,035,900	103,380
Licungo	27,726	—	27,726	336	Errego	5,800	1,210
Molocue	6,500	—	6,500	325	Gile	2,900	865
Ligonha	16,299	—	16,299	295	E.N. 232	5,410	820
Meluli	9,700	—	9,700	255	E.N. 495	9,607	1,915
Nonapo	8,800	—	8,800	240	Entete	8,000	1,005
Mecuburi	8,900	—	8,900	295	E.N. 106	4,053	460
Lurio	60,800	—	60,800	605	Namapa	56,200	7,330
Montepuez	9,500	—	9,500	315	E.N. 509	2,415	195
Messalo	24,000	—	24,000	530	Nairoto	10,000	1,030
Ruvuma	101,160	54,240	155,400	800	—	—	—

1) 140,000 km2 in Angola.

TABLE 2-60. PORTUGUESE AFRICAN DEPENDENCIES: MOZAMBIQUE—EXISTING DAMS AND RESERVOIRS

(Source: de Ataida, Service Hydraulique, 1972)

Name of dam	River	Nearest town	Height m	Reservoir capacity 10⁶ m³	Irrigated area ha	Installed power capacity MW	Use 1)
Trigo de Morais	Limpopo	Trigo Morais	15	15	30,000	—	I
Massingir	Elefantes	Trigo Morais	48	2,844	97,000	45-60	I
Oliveira Salazar	Revue	Vila Pery	75	1,920	—	40	P
Mavuzi	Revue	Vila Pery	17	1.5	—	46	P
Vila Pery	Mezingaze	Vila Pery	15	0.3	—	—	WS
Nampula	Monapo	Nampula	17.5	4.3	—	—	WS
Nacala	Moecula	Nacala	17.4	4.4	—	—	WS
Cabora-Bassa	Zambezi	Tete	160	63,000	—	{5X400 4X400	P

1) I–Irrigation, P–Power generation, WS–Public water supply.

TABLE 2-61. PORTUGUESE AFRICAN DEPENDENCIES: MOZAMBIQUE—WATER USE

(Source: de Ataida, Service Hydraulique, 1972)

Category	Water use million m³
Agriculture	1,000
Domestic, industry and other	200
Total Mozambique	1,200

TABLE 2-62. RHODESIA—EXISTING DAMS AND RESERVOIRS

(Source: Fessler, Österreichische Wasserwirtschaft, 1972)

[Data as of November, 1971]

Name	Region	River	Basin area km²	Date constructed	Height m	Length of crest m	Volume content of dam m³	Storage capacity 10⁶ m³	Surface area of reservoir km²
Kariba	Border	Zambezi	—	1956-60	128	620	1,050,000	160,000	5,200
Kyle	South	Several rivers near Fort Victoria	4,000	1958-61	63	309	55,000	1,330	91
Bangala	South	Downstream Kyle (Lovveld)	5,800	1961-63	50	396	88,000	130	11
Manjirenji	South	Chiredzi	1,540	1964-66	52	315	660,000	284	20
Tokwe Mokorsi	South	Tokwe	7,100	—	97	378	500,000	2,100	113

TABLE 2-63. RHODESIA—WATER BALANCE OF LAKE KARIBA

(Source: Haviland, J. Inst. Water Engineers, 1969)

[Data for year 1968-69]

Inflow	Million acre-ft.
From upper drainage basin	55.9
From lower drainage basin	6.8
From rainfall on lake	3.7
Total inflow	66.4

Disposal	Million acre-ft.
Turbine discharge	16.9
Spillway discharge	32.5
Gross evaporation	6.7
Plus increase in storage	10.6
Apparent gain	− 0.3
Total disposal	66.4

TABLE 2-64. SENEGAL—DISCHARGE OF SELECTED RIVERS

(Source: ORSTOM, 1969)

River and station	Basin area km²	Mean monthly discharge, m³/s												Max flow m³/s	Year	Period of record
		May	Jun.	Jul.	Aug.	Sep.	Oct.	Nov.	Dec.	Jan.	Feb.	Mar.	Apr.			
Senegal, Dagana	268,000	17.5	33.0	414	1,220	2,000	2,310	1,420	438	201	113	62.0	31.0	3,570	1936	1903-66
Senegal, Bakel	218,000	9.60	113	590	2,350	3,430	1,650	568	255	142	83.0	46.0	20.0	9,340	1906	1903-66
Faleme, Kidira	28,900	0.900	26.1	145	707	953	401	110	41.0	18.8	9.20	4.40	1.60	3,120	1961	1930-66

FIGURE 2-7. SOUTH AFRICA – RIVER BASINS AND RUNOFF

(Source: South African Water Commission, 1970)

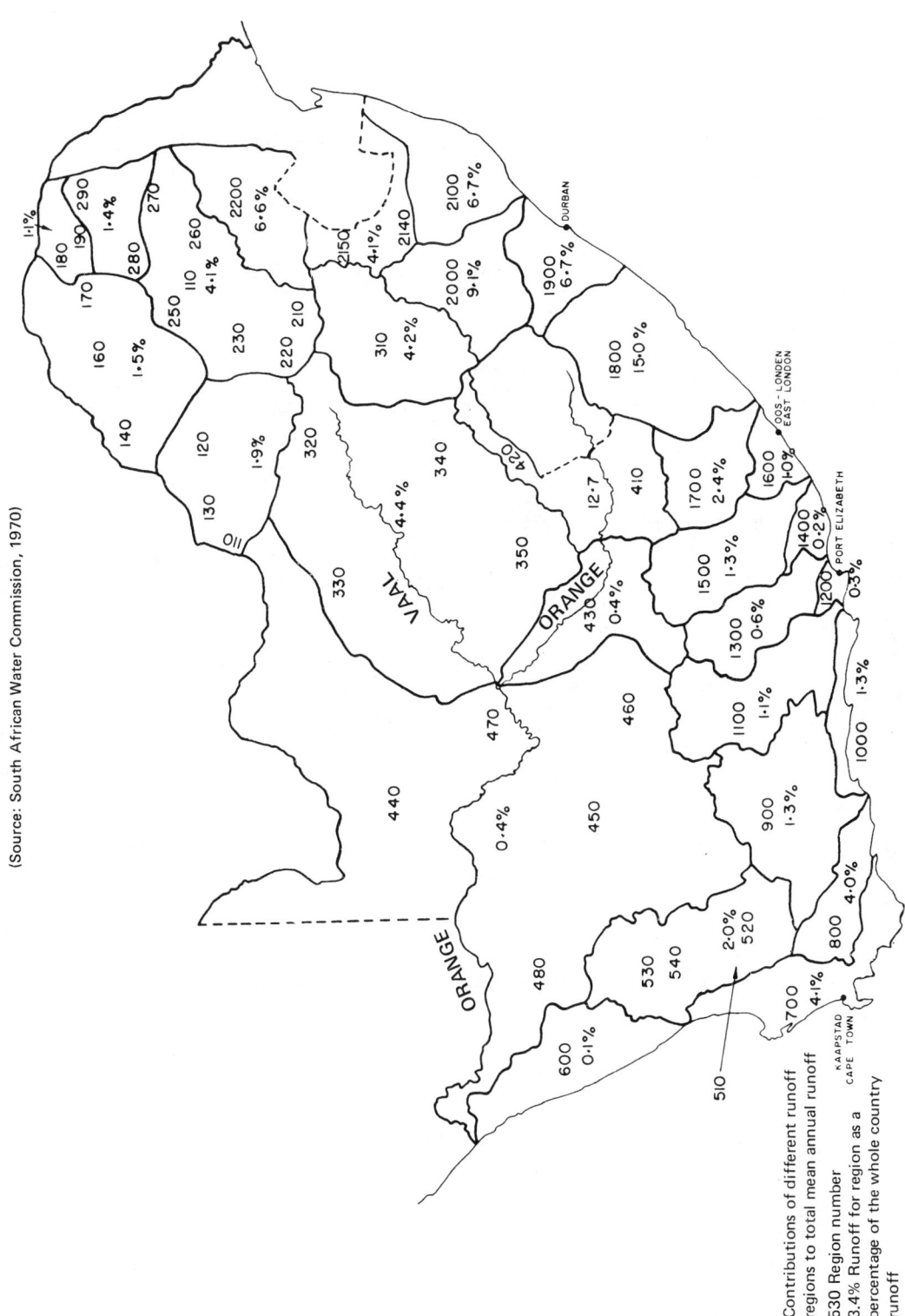

Contributions of different runoff regions to total mean annual runoff

530 Region number

3.4% Runoff for region as a percentage of the whole country runoff

TABLE 2-65. SOUTH AFRICA—ANNUAL PRECIPITATION AND RUNOFF

(Source: South African Water Commission, 1970)

[Basin regions shown on page 155]

Basin region number	Basin or region	Total basin area mi2	Mean annual runoff morgen ft. [1]	Average rainfall over basin mm	Runoff as percentage of rainfall
110-130	North Western Transvaal	16,820	370,000	633	3.5
140-170	North Western Transvaal	21,560	287,000	506	2.7
180-and 190	Soutpansberg	3,940	212,000	609	8.9
210-270	Olifants (Transvaal)	21,060	803,000	676	5.7
280-290	Letaba and Shingwedzi	7,310	280,000	592	6.5
310 and 380	Vaal River down to Vaal Dam	14,860	825,000	725	7.7
320-370 and 390	Vaal River between Vaal Dam and Orange River	59,985	875,000	537	2.7
410 and 420	Orange and Caledon down to Bethulie R.S.A.	13,310			
	Lesotho	11,760	2,502,000	717	14.0
430	Orange between Bethulie and confluence with Vaal	12,975	75,200	363	1.6
440-480	Lower Orange	121,090	78,000	225	0.3
510	Olifants Western Cape Province	1,090	219,000	523	38.7
520-540	Doorn and Sout	17,670	172,000	188	5.2
600	Western Coastal Region (Namaqualand)	11,155	27,000	130	1.9
700	South Western Region	9,800	801,000	482	17.1
800	Breede River	5,955	775,800	651	20.2
900	Gouritz	17,490	258,000	249	6.0
1,000	Outeniqua	2,780	260,000	772	12.2
1,100	Gamtoos	13,320	217,000	277	5.9
1,200	Algoa	1,015	58,000	625	9.2
1,300	Sundays	8,150	114,000	340	4.2
1,400	Great Bushmans	2,175	31,800	549	2.7
1,500	Great Fish	11,690	222,200	423	4.5
1,600	Amatola	3,055	205,000	716	9.5
1,700	Great Kei	7,910	467,000	613	9.7
1,800	Bashee to Umzimkulu	17,995	2,955,000	842	19.7
1,900	Southern Natal	11,230	1,330,000	1,002	19.0
2,000	Tugela	7,050	1,800,000	901	17.9
2,110-2,130 and 2,170	Zululand	11,230	1,318,000	896	13.8
2,140-2,150	Pongola and Usutu in R.S.A.	10,710 / 6,675	820,000	871	14.2
2,160	Black Imbuluzi in R.S.A.	40	2,000	880	5.7
2,200	Komati in R.S.A., Crocodile and Sabie	11,005	1,300,000	813	14.7
Totals: Republic		470,870	19,660,000	483	8.1
Lesotho		11,760	—		
Grand Total		482,630	19,660,000		

1) 1 morgen foot = 2.11653 acre feet, 2,610.71 m3 or 574.275 Imp. gal.

TABLE 2-66. SOUTH AFRICA—DAMS AND RESERVOIRS

(Source: South African Water Commission, 1970)

Description	No. of dams	Capacity morgen ft.
(a) Completed dams		
Farm dams for storage of surface water, as derived from data in the Agricultural Census of 1960	170,000	670,000
Dams of capacity between 100 mg. ft. and 1,999 mg. ft.	181	97,242
Dams of capacity from 2,000 to 499,999 mg. ft.	82	1,877,363
Large dams, 500,000 mg. ft. and more (Vaal Dam)	1	895,000
Sub-total for completed dams	170,264	3,539,605
N.B. A considerable percentage of these dams were not built by the State.		
(b) Raising of existing dams now in hand (increase of capacity only indicated)		
By Department of Water Affairs	11	113,495
By other organizations	1	34,321
Sub-total for dams presently being raised	12	147,816
(c) New dams under construction		
Capacities from 100 to 500,000 mg. ft. by Department of Water Affairs	30	421,331
Capacities from 100 to 500,000 mg. ft. by Municipalities	3	33,900
Capacities of 500,000 mg. ft. and more, by Department of Water Affairs (Hendrik Verwoerd Dam, J.G. Strijdom Dam and Oppermansdrift Dam)	3	3,665,000
Sub-totals for dams under construction	36	4,120,231
(d) Dams already approved by Parliament but not yet commenced by Department of Water Affairs		
Capacities from 100 to 500,000 mg. ft.	9	270,855
Capacities 500,000 mg. ft. and more (P.K. le Roux Dam)	1	1,220,000
Sub-total	10	1,490,855
e) Total for all dams completed, currently being raised, under construction, and already approved		9,298,507

TABLE 2-67. SOUTH AFRICA—GROUND-WATER USE, 1961

(Source: Enslin, 1964, Dept. of Water Affairs)

Category	Unit	Per capita use gpd	Total use million gpd
RURAL AREAS			
Drinking water	3,450,000 persons	8.3	28.5
Stock watering			
large stock	5,500,000 head	10	55.0
small stock	26,000,000 head	1.5	39.0
Irrigation	100,000 morgen 1)	24 inches per year	314.5
URBAN AREAS			
Drinking water	300,000 persons	46	13.0
Gardening	50,000 erven	500	25.0
Rand Water Board	—	—	5.0
TRANSPORT AND MINES			
Railways - Power and domestic	—	—	8.0
Gold mines - Mine processes,			
domestic, irrigation	—	—	117.0
Unusable water	—	—	23.0
Other mines - General	—	—	12.0
Total			640.0

1) 1 morgen = 2.11653 acres

TABLE 2-68. SOUTH AFRICA—WATER USE, 1965

(Source: South African Water Commission, 1970)

Use category	Surface water mg. ft. p.a. 1)	Ground water mg. ft. p.a.	Total mg. ft. p.a.	Total mgd
Irrigation	2,640,000	256,000	2,896,000	4,561
Stockwatering	17,000	18,000	35,000	55
Total agricultural consumption	2,657,000	274,000	2,931,000	4,616
Domestic, municipal, industrial and mining purposes	521,700	55,000	576,700	909
Water pumped from mines and only partly beneficially utilized	—	97,000	97,000	153
	521,700	152,000	673,700	1,062
Total consumption, excluding water pumped from mines	3,178,700 (5,006 mgd)	329,000 (519 mgd)	3,507,700 (5,525 mgd)	5,525

1) morgen feet per year

TABLE 2-69. SOUTH AFRICA—PROJECTED WATER DEMAND IN 2000
(Source: South African Water Commission, 1970)

	Agricultural purposes		Municipal and industrial purposes		Total	
	morgen ft. p.a.	mgd	morgen ft. p.a.	mgd	morgen ft. p.a.	mgd
Groundwater	310,000	490	600,000	950	910,000	1,440
Intermittent floodwaters	61,000	90	—	—	61,000	90
Permanent surface water sources	4,500,000 *	7,050	5,700,000	9,050	10,200,000	16,100
Total..........................	4,870,000	7,630	6,300,000	10,000	11,170,000	17,630

* Including water for stockwatering.

TABLE 2-70. SOUTH AFRICA—WATER USE IN VAAL RIVER BASIN, 1968 [1]
(Source: South African Water Commission, 1970)

Sector	Vaal Dam to Oppermans-drift mg. ft.	Oppermans-drift to Douglas mg. ft.	Total mg. ft.	Total mgd average
Urban and industrial	185,801	5,231	191,032	300.6
Power generation	23,716	—	23,716	37.3
Irrigation	11,765	180,204	191,969	302.1
Losses	39,483	55,547	95,030	149.5
Total	260,725	240,942	501,747	789.5

[1] Water withdrawn from Vaal River below Vaal dam.

TABLE 2-71. SOUTH AFRICA—COST OF RECLAIMING POTABLE WATER FROM RAW SEWAGE AT WINDHOEK, SOUTH WEST AFRICA
(Source: Clayton and Pybus, Civil Eng. ASCE, Sept. 1972; reprinted with permission)

	South African cents/kl *	U.S. cents/ 1,000 U.S. gals.
Primary and secondary treatment, including maturation ponds and all capital and labor costs	3.54	17.69
Capital costs of treatment works including those portions of the existing works that were utilized	3.88	19.36
Proportional cost of labor	0.89	4.45
Chemicals	2.87	14.33
Activated Carbon	2.40	11.98
Supervision by Bio-Chemists' Dept.	1.50	7.50
Total Unit Cost of obtaining potable water from raw sewage	15.08 **	75.31

* 1 Rand S.A. = 1.32 U.S. Dollars.

** The cost of water from the Goreangab Dam — conventional water treatment — is about 9 S.A. ¢/kl or 45 ¢/1,000 U.S. gal. That is, water from raw sewage costs about 67% more.

TABLE 2-72. SPANISH AFRICAN DEPENDENCIES—GROUND WATER DEVELOPMENT IN THE CANARY ISLANDS

(Source: Dingman and Nunez, U.S. Geol. Survey, 1969)

Province and island	Source of water and yield m3/s								Total yield m3/s
	Wells		Galleries		Springs		Desalinization plants		
	Number	Yield	Number	Yield	Number	Yield	Number	Yield	
Las Palmas:									
Gran Canaria	2,830 1	4-5.2 2	360	1.0	10 3	0.2	4)	0.23	6.43
Lanzarote	3	0.003	2	.006	None		1	.027	.036
Fuerteventura	650 5	.227	2	0	6	.006	None		.233
Santa Cruz de Tenerife:									
Tenerife	73	.85	485	7.4	54	.18	None		8.43
La Palma	30 6	.46 6	102	1.2	125	.65	None		2.31
Gomera	36	.17	3	.07	36	.38	None		.62
Hierro	3	.12	8	None	None		None		.12

1) Includes 510 shallow wells in the alluvium of Aldea Valley.
2) Yield reached 6.2 m3/s during the drought of the middle 1960's.
3) More than 1,000 small springs with yields of a fraction of a liter per second are used for domestic water or for irrigation of a very small plot of land.
4) A desalinization plant is planned for Las Palmas.
5) All wells produce brackish water, salinity 2,000 mg/l or higher.
6) Incomplete data.

TABLE 2-73. SPANISH AFRICAN DEPENDENCIES—SURFACE WATER DEVELOPMENT IN THE CANARY ISLANDS

(Source: Dingman and Nunez, U.S. Geol. Survey, 1969)

[As of May 1967; Dams tabulated are more than 15 meters high]

Province and island	Dams existing		Dams planned or under construction		Total storage 106 m3
	Number	Storage capacity 106 m3	Number	Additional storage 106 m3	
Las Palmas:					
Gran Canaria 1	54	52	12	30	82
Lanzarote	0	0	1	.2	.2
Fuerteventura	3	2.5	2	3	5.5
Santa Cruz de Tenerife:					
Tenerife	9	1.8	3	.7	2.5
La Palma	1	.045	0	0	.045
Gomera	9	1.1	3	1.2	2.3
Hierro	0	0	1	.045	.045

1) Thousands of small reservoirs, with estimated storage capacity of 20X106 m3, exist in Gran Canaria.

TABLE 2-74. SPANISH AFRICAN DEPENDENCIES—PRECIPITATION AND WATER USE IN THE CANARY ISLANDS

(Source: Dingman and Nunez, U.S. Geol. Survey, 1969)

[Data as of 1966]

Island	Area km²	Population	Average annual precipitation			Water use, m³/s					Percentage of annual precipitation used
			mm/yr	106m³/yr	m³/s	Agriculture	Municipal	Industrial	Harbor	Total	
Gran Canaria	1,551	500,000	400 1)	620 1)	19.66	5.64	0.5	0.3	0.03	6.5	34
Lanzarote	796	40,000	160	130	4.1	None	.006	.003	None	.009 2)	.22
Fuerteventura	2,017	17,000	100	200	6.3	.229	.004	None	None	.233	3.7
Tenerife	2,058	400,000	445	915	29	—	—	—	—	8.43	29
La Palma	750	75,000	530	400	12.7	—	—	—	—	2.31	18
Gomera	378	10,000	—	—	—	—	—	—	—	.62	—
Hierro	278	6,000	—	—	—	—	—	—	—	.12	—

1) Ranges from an average of 130 mm/yr (200×106m3/yr) at low altitudes to 800 mm/yr (1,200×106m3/yr) at altitudes of more than 1,500 meters.
2) Desalinization plant yield of 0.027 m3/s not included.
— Data not obtained.

TABLE 2-75. SPANISH AFRICAN DEPENDENCIES—MUNICIPAL WATER SUPPLY SYSTEM OF CEUTA

(Source: Troyano Lobaton, Agua, 1972)

[Population 75,000 in 1972]

Source of water	Yield, l/s			Mean annual yield m³
	Maximum	Minimum	Mean	
Benzu (11 springs)	200	40	80	—
Arroyo de las Bombas (well)	40	10	20	—
Arroyo del Renegado (surface water)	?	0	15	400,000
Arroyo del Infierno (surface water)	?	0	10	300,000
Desalinization plant	—		45	—
Tankers		in emergency		

TABLE 2-76. SUDAN—WATER LEVEL FLUCTUATION OF THE NILE RIVER
(Source: Sudan Almanac, 1966)

	Normal Year			Extreme River Gages				
	Lowest level	Highest level	Range meters	Lowest	Date	Highest	Date	Period of record
Atbara (Main Nile)	May	Early Sep.	6	9.42	23-25 May 1922	16.52	27-28 Aug. 1946	1907-50
Khartoum	May	Early Sep.	5½	9.21	20 May 1922	17.14	2 Sep. 1946	1899-1950
Khashm el Girba (River Atbara)	(dry)	Aug.	5½	(dry)		17.80	13 Aug. 1916	1903-50
Er Roseires (Blue Nile)	April	Aug.	8	10.53	22 May 1914	22.68	24 Aug. 1946	1903-50
Rabak (White Nile)	April	Sep.	2½	9.27	13 May 1922	14.15 [1]	4-5 Jan. 1948	1906-50
Malakal (White Nile)	April	Oct. & Nov.	2¼	9.16	15 April 1922	13.35	11-15 Mar. 1918	1905-50
Juba (Bahr el Jebel)	March	Sep.	0.60	12.36	12-20 Apr. 1945	15.45	11 Aug. 1942	1925-50

[1] Now influenced by Jebel Aulia reservoir.

TABLE 2-77. SUDAN—MUNICIPAL WATER SUPPLY SYSTEM OF KHARTOUM
(Source: Internat. Statistical Institute, 1971)

Year	Households served (No. of connections)	Annual water consumption 1,000 m^3
1967	21,955	14,308
1969	22,055	16,452

TABLE 2-78. TANZANIA—DISCHARGE OF SELECTED RIVERS
(Source: Tanzania Hydrological Yearbook, 1967)

River and station	Basin area km²	Nov.	Dec.	Jan.	Feb.	Mar.	Apr.	May	Jun.	Jul.	Aug.	Sep.	Oct.	Max flow m³/s	Date	Period of record
Pangani, Nyumba ya Mungu	9,139	41.60	41.07	36.99	24.22	23.83	58.12	79.05	43.34	33.06	25.86	21.67	20.84	256.01	May 5,1964	1959-64
Pangani, Kirua	16,155	18.91	25.49	30.26	25.22	20.56	23.78	33.50	42.65	29.16	23.45	18.61	16.47	74.11	Jan. 5,1962	1959-64
Pangani, Korogwe	25,110	29.11	24.36	43.46	29.94	22.67	32.20	39.61	42.97	35.20	25.21	19.86	18.01	167.54	Nov. 14,1962	1959-65
Mkomazi, Gomba	3,340	2.52	3.53	13.59	5.38	3.05	7.27	11.20	4.45	2.75	1.18	0.73	0.67	69.38	Jan. 4,1964	1962-65
Ruvu, Tanga Road Bridge	3,024	5.75	7.59	8.28	5.23	4.75	7.54	7.16	5.78	5.29	5.17	4.71	4.48	25.68	Jan. 13,1962	1959-65
Wami, Mandera	36,450	64.26	51.45	94.55	48.62	76.33	194.59	99.94	36.56	27.24	20.49	15.65	17.42	673.38	Nov. 8,1961	1959-65
Ruvu, Kibungo	419	26.29	15.63	19.17	12.87	22.56	44.61	24.50	12.91	9.95	7.97	7.81	11.65	230.22	Nov. 7,1961	1959-65
Ruvu, Morogoro Road Bridge	15,190	105.82	80.37	100.31	43.13	73.42	209.99	119.45	45.88	28.30	20.16	15.16	18.51	1,176.85	Apr. 24,1963	1959-65
Morogoro, Morogoro	23.3	1.05	0.40	0.36	0.30	0.54	1.92	1.01	0.50	0.43	0.24	0.20	0.32	13.93	Mar. 31,1964	1959-65
Mgeta, Mgeta	101	5.04	2.50	6.56	2.18	2.89	5.74	3.13	2.00	1.49	1.29	1.15	2.10	70.01	Jan. 4,1962	1959-65
Rufiji, Stiegler's Gorge	158,200	157.58	192.97	718.12	648.38	1,252.40	2,310.45	1,736.18	748.80	414.08	296.37	221.79	183.19	5,440.38	Apr. 6,1960	1959-63
Rufiji, Pangani Rapids	158,000	304.42	487.75	1,462.71	1,694.16	2,587.56	3,666.06	1,865.02	748.79	485.70	355.73	274.37	225.17	5,711.52	Mar. 24,1964	1963-65
Great Ruaha, Mtera	67,950	8.51	77.28	255.67	298.81	377.71	343.28	187.50	77.09	38.04	23.35	15.52	9.59	907.50	Jan. 9,1962	1959-65
Kimani, Great North Road	448	0.81	5.36	15.62	17.44	25.34	13.24	4.65	2.45	1.49	1.11	0.88	0.75	91.96	Mar. 4,1960	1959-65
Mbarali, Igawa	1,600	5.53	15.41	32.86	36.39	55.18	31.16	17.36	11.87	9.09	7.44	6.18	5.55	186.54	Jan. 5,1962	1959-65
Little Ruaha, Ihimbu	2,480	6.39	11.36	23.99	33.12	47.88	39.34	24.61	17.51	14.22	11.71	9.19	7.48	131.76	Mar. 14,1962	1959-65
Lukosi, Mtandika	2,893	17.27	18.06	24.76	23.80	28.95	44.10	28.93	20.82	19.42	17.84	16.18	15.32	133.56	Apr. 19,1960	1959-65
Great Ruaha, Msembe	24,620	8.25	12.89	63.21	192.03	208.05	219.40	100.65	41.16	22.33	15.32	10.98	8.10	517.97	Mar. 20,1964	1963-65
Kilombero, Swero	33,400	157.81	238.29	498.12	665.52	785.00	1,634.73	1,249.12	495.19	311.92	249.63	203.20	166.40	3,044.24	Apr. 28,1963	1959-65
Sonjo, Sonjo	67.6	0.67	0.71	1.33	1.28	3.20	5.71	2.27	1.08	0.70	0.52	0.37	0.31	46.86	Mar. 30,1964	1962-65
Lufirio, Ipinda	1,425	12.71	25.95	49.42	49.71	78.65	96.13	63.08	34.14	21.24	14.93	12.70	12.07	181.99	Apr. 16,1965	1959-65
Simuyu, Ndagalu Primary School	6,160	2.11	60.38	9.58	21.26	13.67	54.36	11.52	0.19	0.03	0.02	0.00	0.16	591.82	Feb. 10,1964	1963-65

Mean monthly discharge, m³/s

TABLE 2-79. TANZANIA—WATER DEMAND, 1970-2000
(Source: Lwegarulila, Ministry of Water Development and Power, 1972)

Demand category	1970			1980			2000		
	Number (1000)	Per capita use l/d	Annual water demand million m³	Number (1000)	Per capita use l/d	Annual water demand million m³	Number (1000)	Per capita use l/d	Annual water demand million m³
RURAL									
Rural population	12,000	30	131	16,000	30	175	29,000	40	424
Cattle	13,000	20	95	20,000	20	145	50,000	30	547
Goats, etc.	1,300	4	2	2,000	4	3	5,000	4	7
Total			228			323			978
URBAN									
D'Salaam	330	180	22	660	180	43	2,640	180	173
Other towns	470	180	31	705	180	49	2,140	180	120
Total	800		53	1,365		92	4,780		293
AGRICULTURE			200			700			5,700
Total Tanzania			481			1,115			6,971

TABLE 2-80. TOGO—DISCHARGE OF SELECTED RIVERS
(Source: ORSTOM, 1969)

River and station	Basin area km²	Mean monthly discharge, m³/s												Max flow m³/s	Year	Period of record
		May	Jun.	Jul.	Aug.	Sep.	Oct.	Nov.	Dec.	Jan.	Feb.	Mar.	Apr.			
Oti, Sansanne Mango	35,650	3.03	25.9	75.0	294	743	489	62.0	11.5	4.52	2.77	1.53	0.920	1,750	1962	1953-66
Kara, Lama-Kara	1,560	1.90	12.3	40.4	64.3	113	32.5	6.84	1.05	0.350	0.190	0.370	1.44	1,370	1956	1954-66
Mono, Tetetou	20,500	9.04	60.6	206	259	380	272	53.0	10.2	2.82	1.29	2.35	4.40	1,490	1963	1951-66
Sio, Kpedji	1,810	5.05	18.5	21.1	15.5	16.4	22.8	6.14	2.02	1.07	0.780	1.40	2.93	132	1962	1953-66

TABLE 2-81. TUNISIA—WATER BALANCE DATA

(Source: Ben Osman, Water for Peace, 1963)

CLIMATIC ZONES		AVERAGE ANNUAL RAINFALL		HYDROLOGICAL DATA				
Zone	Area km^2	mm	Volume 10^{9}m^3	Streams mm	Rivers 10^{9}m^3	Eva- poration 10^{9}m^3	Infil- tration 10^{9}m^3	Total 10^{9}m^3
Kroumirie Mogdos Dorsal peaks	9,600	600-1,600	6.8	100-700	2.0	4.6	0.1	6.7
Medjerda plain Dorsal slopes Sahel of Sousse	32,200	300- 600	12.7	10-100	0.4	12.0	0.1	12.6
South of Dorsal Kairouan Gafsa Sfax Gabes Matmata	56,200	100- 300	10.0	3- 10	0.1	9.7	0.3	10.1
South of the Chotts	69,000	under 100	3.5	under 3	0.0	3.5	0.2	3.6
Total	167,000		33		2.5	29.8	0.7	33.0

TABLE 2-82. TUNISIA—AVAILABLE GROUND WATER RESOURCES

(Source: Ben Osman, Water for Peace, 1968)

Aquifer type	Estimated available annual yield 10^6 m^3
Water-table aquifers	133
Deep aquifers	501
Total	634

TABLE 2-83. TUNISIA—WATER USED FOR PUBLIC SUPPLIES

(Source: Ben Osman, Water for Peace, 1963)

Year	Number of connections (subscribers)	Population served	Annual water consumption m^3
1956	38,470	477,461	26,657,000
1966 1	85,770	943,179	52,440,000

1 Estimated

TABLE 2-84. TUNISIA—DAMS AND RESERVOIRS

(Source: Framji and Mahajan, 1969, Irrigation and Drainage in the World, ICID)

Dam	Storage capacity		Volume controlled annually		Type of dam	Height		Irrigated area	
	10^6 m^3	acre-ft	10^6 m^3	acre-ft		m	feet	ha	acres
Nebaana	86	69,720	27	21,889	Zoned dike	62	203.4	5,500	13,590
Lakhmess	8	6,486	6	4,864	Homogeneous dike	37	121.4	1,500	3,710
Chila	7	56,759	3.5	2,837	do	26	85.3	1,200 Supplementary irrigation	2,965
Bezirk	6	4,864	3.5	2,837	do	22	72.2	2,000 Supplementary irrigation	4,490
Masri	7	56,759	3	2,432	Zoned dike	38	124.7	do	
Kasseb	80	64,856	40	32,428	Thin concrete arch	58	190.3	Drinking water for Tunis	

TABLE 2-85. UNITED ARAB REPUBLIC—DISCHARGE OF PRINCIPAL RIVERS IN THE NILE RIVER BASIN

(Source: UNESCO, 1971)

River and station	Basin area km²	Mean monthly discharge, m³/s												Year	Period of record
		Jan.	Feb.	Mar.	Apr.	May	Jun.	Jul.	Aug.	Sep.	Oct.	Nov.	Dec.		
Victoria Nile, Namasagali	262,000	628	623	635	670	714	725	693	666	656	644	632	657	662	1939-62
Albert Nile, Pakwatch	391,000	734	696	729	726	711	731	701	691	700	772	791	845	756	1955-62
White Nile, Malakal	—	829	634	553	525	574	742	897	1,030	1,130	1,200	1,200	1,100	872	1912-62
Blue Nile, Khartoum	325,000	282	188	156	138	182	461	2,080	5,950	5,650	3,040	1,030	499	1,640	1912-62
Atbara, Kilo 3	69,000	8	2	0	0	1	35	640	2,100	1,420	340	79	25	389	1912-62
Main Nile, Kajnarty	—	1,280	975	828	765	679	755	1,900	7,210	8,430	5,480	2,770	1,680	2,730	1912-62
Main Nile, D.S. Aswan Dam	—	1,110	1,020	834	819	698	1,340	1,910	6,570	8,180	5,200	2,270	1,400	2,650	1912-62
Rosetta Br., D.S. Rosetta Br.	—	868	150	12	2	4	18	88	2,450	4,170	3,070	1,470	579	1,080	1920-62
Damietta Br., D.S. Zifta Br.	—	170	38	16	20	24	34	68	1,180	1,660	961	556	148	407	1920-62

TABLE 2-86. UNITED ARAB REPUBLIC—MUNICIPAL, INDUSTRIAL AND DOMESTIC WATER USE, 1960

(Source: Zayed, Ministry of Municipal and Rural Affairs, Cairo, 1958)

Type of system and region	Population	Daily production m³	Per capita use liters/day
Large cities			
Alexandria	1,500,000	250,000	170
Cairo	3,300,000	606,000	180
Municipal councils	7,000,000	500,000	70
Rural districts			
Artesian waterworks	5,000,000	100,000	20
Filtered waterworks	4,000,000	250,000	60
Public potable water	20,800,000	1,700,000	80
Remaining area 1)	5,300,000	100,000	20
Total UAR	26,100,000	1,800,000	70

1) Per capita use assumed to be same as for Rural Artesian Waterworks.

TABLE 2-87. UNITED ARAB REPUBLIC—WATER BALANCE OF LAKE NASSER

(Source: Wafa and Labid, Seepage Losses from Lake Nasser, in Man-made Lakes:
Their Problems and Environmental Effects, Geophysical Monograph 17, 1973.
Copyright by Am. Geophysical Union)

Period	Maximum reservoir level, meters [1]	Inflow, $10^9 m^3$	Outflow plus water accumulation in reservoir, $10^9 m^3$	Actual reservoir losses, $10^9 m^3$ [2]	Theoretically calculated losses, $10^9 m^3$		
					Absorption (bank storage)	Evaporation	Total [3]
1964 to 1965	127.60	119.430	119.082	0.348	0.091	0.942	1.033
1965 to 1966	132.70	79.550	78.497	1.053	0.295	1.552	1.847
1966 to 1967	142.45	69.120	67.760	1.360	1.190	2.679	3.869
1967 to 1968	151.25	92.370	87.290	5.080	2.427	4.485	6.912
1968 to 1969	156.55	71.102	62.740	8.362	4.034	5.773	9.807
1969 to 1970	161.30	69.539	61.263	8.276	2.719	6.975	9.694
1970 to 1971	164.88	81.790	70.373	11.417	3.593	8.177	11.770

[1] Before the High Dam the maximum reservoir level was 120 meters.
[2] These losses were obtained by subtracting outflow plus water accumulation in the reservoir (column 4) from inflow (column 3).
[3] The total losses were obtained by adding absorption (column 6) and evaporation (column 7).

FIGURE 2-8. UNITED ARAB REPUBLIC — NILE RIVER
AND AREA OF NEW VALLEY PROJECT

(Source: Ezzat, Egyptian General Desert Dev. Organization, 1964)

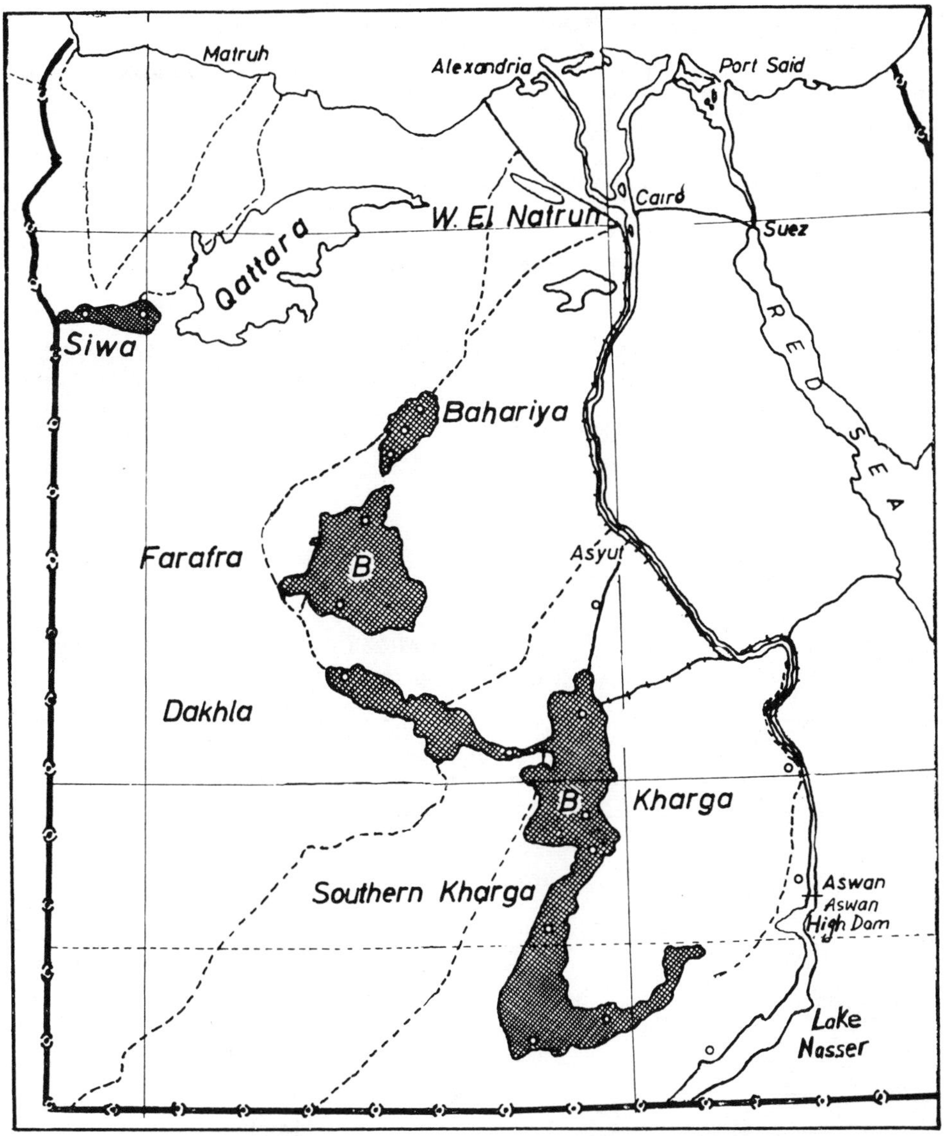

Matruh

Alexandria

Port Said

W. El Natrun

Cairo

Suez

Qattara

RED SEA

Siwa

Bahariya

Farafra

Asyut

B

Dakhla

Kharga

B

Southern Kharga

Aswan
Aswan
High Dam

Lake
Nasser

○   General location of test wells
(observation) for hydrologic
and geologic information

B   Area of New Valley Project
Dakhla and Kharga Oases
Area of detailed geologic and
hydrologic study, gravimetric
survey and detailed soil survey

TABLE 2-88. UNITED ARAB REPUBLIC—DATA ON WATER WELLS OF THE NEW VALLEY PROJECT (KHARGA OASIS)

(Source: Shata, Wasserwirtschaft in Afrika, 1963)

[Production is from "Nubian Type Sandstones"]

Name of well	Ground elevation m	Total depth m	Initial discharge m3/d	Present [1] discharge m3/d	Static head m	Temperature °C	Salinity ppm
El Mahariq No. 2	+ 55.5	652	10,750	7,300	41	38.5	410
El Mahariq No. 1	+ 55	650	10,000	8,000	21.5	39.5	430
El Kharga No. 2	+ 89	602	3,500	2,400	24	39.5	184
El Kharga N.T.W. No. 1	+ 83	637.5	13,500	9,000	23	40.0	192
El Shorafa	+ 55	650	1,600	1,360	17	36.5	364
New Ginah	+ 68	650	9,600	7,800	33.7	37.7	280
Bulaq	+ 28	500	7,600	4,200	45.2	37.0	424
Beris No. 2	+ 37.5	597	16,000	14,050	33.2	38.0	430
Beris No. 1	+ 51	500	2,400	2,720	16.8	36.5	462

[1] Discharge in August 1958.

TABLE 2-89. UNITED ARAB REPUBLIC—NATURAL DISCHARGE OF NUBIAN SANDSTONE AQUIFER

(Source: Ezzat, Egyptian General Desert Dev. Organization, 1964)

Area	Total spring discharge m3/d	Used water m3/d	Unused water (lost by evaporation and transpiration) m3/d
Kharga	500,000	400,000	100,000
Dakhla	800,000	600,000	200,000
Farafra	3,000	3,000	—
Bahariya	520,000	125,000	395,000
Siwa	190,000	190,000	—
Qattara	3,000,000	—	3,000,000
Nile	137,000	—	137,000
Total	5,150,000	1,318,000	3,832,000

TABLE 2-90. UNITED ARAB REPUBLIC—DATA ON NUBIAN SANDSTONE AQUIFER
(Source: Ezzat, General Desert Dev. Organization, 1964)

Ground water reservoir (estimated productive)

Area — 520,000 km2

Thickness — 150 m [1]

Sand thickness — 75 m

Volume — 39,000 billion m3

Porosity 0.20

Volume of fresh water in storage — 7,800 billion m3

Average effective specific yield — 0.3

Usable fresh water stored in upper 150 meters — 2,340 billion m3

Ground water recharge [2]

Inflow across Libyan border — 1.2 million m3/d

Inflow across Sudan border — 412,500 m3/d

Total inflow . . . . 1.6 million m3/d or 594 million m3/yr

[1] Average thickness of total Nubian reservoir is 750 m.
[2] Assuming a coefficient of permeability of 10 m3/d/m2.

TABLE 2-91. UPPER VOLTA–DISCHARGE OF SELECTED RIVERS
(Source: ORSTOM, 1969)

River and station	Basin area km2	Mean monthly discharge, m3/s												Max flow m3/s	Year	Period of record
		May	Jun.	Jul.	Aug.	Sep.	Oct.	Nov.	Dec.	Jan.	Feb.	Mar.	Apr.			
Goudebo, Yacouta	1,640	0.003	0.203	1.41	8.58	8.79	0.004	0.000	0.000	0.000	0.001	0.000	0.000	60.3	1964	1964-66
Black Volta, Nwokuy	15,000	9.30	11.1	17.3	43.0	87.9	112	110	50.9	18.0	12.7	10.2	9.00	157	1954	1954-66
Massili, Lumbila	2,120	0.000	0.700	0.900	5.20	14.9	0.600	0.000	0.000	0.000	0.000	0.000	0.000	200	1961	1961-66
Comoe, Karfiguela	812	3.80	5.60	7.60	19.4	17.3	9.30	6.10	5.10	4.50	4.10	3.60	3.50	168	1954	1952-66

TABLE 2-92. ZAMBIA–DISCHARGE OF SELECTED RIVERS
(Source: Framji and Mahajan, ICID, 1969)

River	Basin area km2	Mean annual discharge 106 m3	Maximum flow		Minimum flow	
			m3/s	Year	m3/s	Year
Zambezi	1,193,462	40,000	—	—	—	—
Tributaries:						
Kafau	154,000	7,400	—	—	—	—
Luangwa	141,333	—	100	1947	0	1936
			4,568	1959	162	1949

AFRICA
REFERENCES CITED

References cited in more than one country:

UNESCO, Discharge of Selected Rivers of the World; Vol. I. General and Regime Characteristics of Stations Selected (1969); Vol. II. Monthly and Annual Discharges Recorded at Various Selected Stations (1971); Vol. III. Mean Monthly and Extreme Discharges 1965-1969 (1971).Copyright by UNESCO/IASH/WMO. Reprinted by permission.

International Statistical Institute, 1970, International Statistical Yearbook of Large Towns. Vol. V., The Hague, The Netherlands.

Proceedings of the International Conference on Water for Peace, 1967, Superintendent of Documents, Washington, D.C. 20402.

Office de la Recherche Scientifique et Technique Outre-Mer (ORSTOM), 1969, Annales Hydrologiques-Annees 1964-1965, Paris; also Annuaire Hydrologique de la France d'Outre-Mer, Annees 1963-1964-1965, Paris 1968.

Framji, K.K., and Mahajan, I.K., 1969, Irrigation and Drainage in the World, Internat. Commission on Irrigation and Drainage (ICID), New Delhi.

AFRICA GENERAL

Economic Commission for Africa/World Meteorological Organization, 1966, Major Deficiencies in Hydrologic Data in Africa.

Welcomme, R.L., 1971, Preliminary List of the Inland Waters of Africa and their Characteristics, Dept. of Fisheries, Circ. 134, and The Inland Waters of Africa, CIFA Tech. Pap. No. 1, 1972, FAO, Rome.

Dekker, G., 1972, A Note on the Nile, Water Resources Research, Vol. 8, No. 4, August.

Central Office of Information Republic of the Sudan, 1966, Sudan Almanac 1965-66, Khartoum.

Afifi, A.K., 1971, Preliminary Estimation of Evaporation from Lake Victoria by the Water-Budget Method, in The Role of Hydrology and Hydrometeorology in the Economic Development of Africa, ECA/WMO Rep. 301.

Southern African Regional Commission for the Conservation and Utilization of the Soil (SARCCUS), 1970, Regional Symposium on Water for Progress (Lourenco Marques), Pretoria.

U.S. Federal Power Commission, 1968, World Power Data, Publ. P-40.

ALGERIA

Service des Etudes Scientifiques, Annuaire Hydrologique de l'Algerie, Annee 1960-61, Algiers.

Ministere de l'Industrie et de l'Energie, 1973, Surface Water Resources in Northern Algeria (Report prepared by Hydrotechnic Corp.).

BOTSWANA

Jennings, C.M.H., 1971, Note on Hydrological Research in Botswana with Special Emphasis on Research in the Hydro-geological Field, South African J. of Sci., Vol. 67, No. 1.

FAO/UNDP 1972, Surveys and Training for the Development of Water Resources and Agricultural Production-Botswana, Review of Resources, AGL:DP/BOT/67/501, Rome.

ETHIOPIA

Department of Water Resources, Hydro-Meteorologic Summary Abbay Basin (Blue Nile) 1970, Addis Ababa.

FRENCH AFRICAN DEPENDENCIES

United Nations, 1964, Water Desalination in Developing Countries, ST/ECA/82.

GHANA

Gill, H.E., 1969, A Ground-Water Reconnaissance of the Republic of Ghana with a Description of Geohydrologic Provinces, U.S. Geol. Survey Water-Supply Paper 1757-K.

KENYA

Ojany, F.F., 1973, The Status of Kenya's Water Resources and Their Development, Proc. First World Congress on Water Resources, Internat. Water Resources Assoc., Chicago.

LIBYA

Jones, J.R., 1971, Ground-Water Provinces of Libyan Arab Republic, U.S. Geol. Survey.

El Rabati, S., 1974, General Water Authority, Tripoli (Private Communication).

MALAWI

Pike, J.G., and Rimmington, G.T., 1965, Malawi—Geographical Study, Oxford Univ. Press, London.

MAURITIUS

Sentenac. R., 1962, Recherches d'Eau Souterraine a l'Ile Maurice, Mauritius Sugar Industry Research Institute.

MOROCCO

Combe, M., 1968, Ressources en Eau du Maroc et Pourcentage d'Utilisation Etat des Connaissances en 1968, Direction de l'Hydraulique.

Direction de l'Hydraulique, 1972, Les Grands Barrages du Maroc.

Monition, L., 1963, Eaux Souterraines et Alimentation en Eaux des Principales Villes du Maroc, Wasserwirtschaft in Afrika, Deutsche Afrika - Gesellschaft, Bonn.

NIGERIA

van Blommestein, W.J., 1969, Water Resources in Nigeria, FAO, Rome.

PORTUGUESE AFRICAN DEPENDENCIES

ANGOLA. Quintela Gois, Carlos, 1970, Water Balance in Angola for the Decade 1958-68, World Water Balance, Proc. Reading Symp. IASH—UNESCO—WMO, 1972.

MOZAMBIQUE. de Ataida Fonseca, C.E., 1972, La Planification des Ressources en Eau au Mozambique, Internat. Symp. on Water Resources Planning, Mexico, D.F.

RHODESIA

Fessler, W., 1972, Wasserwirtschaft in Sudafrika und Rhodesien, Osterreichische Wasserwirtschaft, Vol. 24, No. 5/6.

Haviland, P.H., 1969, Kariba Hydro-Electric Scheme July 1968 to June 1969, J. Inst. of Water Engineers, Vol. 23, No. 8, London.

SOUTH AFRICA

Commission of Enquiry into Water Matters, 1970, Report of the Commission R.P. 34

Enslin, J.F., 1964, Ground-Water Supplies in the Republic of South Africa and Their Development, Use and Control, Dept. of Water Affairs, Pretoria.

Clayton, A.J., and Pybus, P.J., 1972, Windhoek Reclaiming Sewage for Drinking Water, Civil Engineering, ASCE, Sept. 1972.

SPANISH AFRICAN DEPENDENCIES

Dingman, R.J., and Nunez, Jose, 1969, Hydrogeologic Reconnaissance of the Canary Islands, Spain, U.S. Geol. Survey Prof. Paper 650-C.

Troyano Lobaton, F. and Others, 1972, Mezcla de Aguas de Diferentes Origines en los Sistemas de Distribucion, Agua, Sept.-Oct., No. 74. (Ceuta Water System).

TANZANIA

Water Development and Irrigation Division, 1967, Hydrological Year-Book 1960-1965, Dar es Salaam.

Lwegarulila, F.K., 1972, Forecasting Water Requirements in Developing Countries with Special Reference to Tanzania, Un Ad Hoc Group of Experts on Water Requirements Forecasting, Budapest.

UNITED ARAB REPUBLIC

Wafa, T.A., and Labib, A.H., 1973, Seepage Losses from Lake Nasser, in Man-Made Lakes: Their Problems and Environmental Effects, Geophys. Monograph 17, Am. Geophysical Union, Washington, D.C.

Shata, A., 1963, The Geology of the Ground Water Supplies in the New Valley Project Area (Egypt, UAR), in Wasserwirtschaft in Afrika, Deutsche Afrika Gesellschaft, Bonn.

Ezzat, M.A., 1964, New Valley Project - Ground Water Condition, Egyptian General Desert Development Organization, Tunis.

Zayed, A.H., 1958, Review of Underground Water Sources, Availabilities and Uses, Ministry of Municipal and Rural Affairs, Cairo.

GREENLAND

Arctic Ocean

ICELAND

NORWAY

SWEDEN

FINLAND

U. S. S. R.

Lena

Yenisey

Ob

Amur

Pacific

JAPAN

Lake Baikal

KOREA

POLAND

ROMANIA

Dnepr

Volga

Don

MONGOLIA

Ocean

Black Sea

Aral Sea

Syrdarya

Caspian Sea

TURKEY

CYPRUS

LEBANON

SYRIA

ISRAEL

IRAQ

JORDAN

Amu Darya

AFGHANISTAN

Yellow

C H I N A

TAIWAN

IRAN

Yangtze

Brahmaputra

HONG KONG

PHILIPPINES

KUWAIT

PAKISTAN

Indus

NEPAL

B

SAUDI ARABIA

QATAR

U A E

OMAN

Ganges

BANGLA-DESH

BURMA

Salween

LAOS

NORTH VIETNAM

SOUTH

South China Sea

Red Sea

YEMEN

INDIA

THAILAND

Mekong

CAMB

ETHIOPIA

YEMEN D R

AFARS AND ISSAS

Arabian Sea

Bay of Bengal

BRUNEI

SOMALI REPUBLIC

MILES
0 200 400 600

0   600
KILOMETERS

SRI LANKA

M A L A Y S I A

SINGAPORE

I N D O N E S I A

Indian Ocean

np

TABLE 3-1. ASIA—BASIC DATA OF MAJOR INTERNATIONAL RIVERS IN THE ECAFE[1] REGION

(Source: UN Water Resources Series No. 29, 1966)

River	Riparian countries (A)	Total Length km	Total Drainage area sq km	Precipitation and Runoffa(B)			Dischargeb(B)					Silt flowc(B) (by weight;1m3 = 1.65 tons)				
				Mean annual rainfall mm	Mean annual runoff mm	Runoff Coefficient %	Max m3/sec	Min m3/sec	Average m3/sec	Specific flood discharge m3/sec/sq km	Average unit discharge m3/sec/sq km	Max content %	Min content %	Average content %	Average annual silt runoff 1,000 tons	Average annual silt runoff per unit area ton/sq km
Mekong	China, Burma, Laos, Thailand, Cambodia, Viet-Nam	4,350	795,000	1,380	722	52.4	67,000	1,250	14,800	0.014	0.0229	0.31	—	0.06	170,000	435
Red	China, Viet-Nam	1,200	120,000	1,500	1,090	72.7	35,000	700	3,900	0.31	0.0345	0.70	0.01	0.106	130,000	1,080
Brahmaputra	China, India, Bangladesh	2,580	580,000	2,125	1,177	55.4	72,460	2,680	19,200	0.149	0.0372	0.30	—	0.12	735,000	1,370
Meghna	India, Bangladesh	950	80,200	3,500	1,715	49.0	13,100	—	3,515	0.203	0.0543	—	—	—	—	—
Ganges	India, Bangladesh	2,200	977,500	1,250	367	29.4	61,200	1,170	11,610	0.0628	0.0119	0.30	—	0.13	480,000	492
Kosi	Nepal, India	730	86,900	1,790	643	36.0	23,000	150	1,770	0.265	0.0204	5.0	—	0.286	160,000	1,840
Gandak	Nepal, India	425	45,800	—	1,375	—	23,000	200	2,000	0.434	0.0386	—	—	0.312	196,000	4,275
Gogra	Nepal, India	1,020	132,000	960	525	54.7	14,809	320	2,200	0.112	0.0167	—	—	—	—	—
Indus	China, India, Pakistan	2,900	970,000	612	460	75.0	31,200	490	6,770	0.078	0.022	—	—	0.33	680,000	2,230
Sutlej	China, India, Pakistan	1,450	86,000	480	280	58.4	13,900	78	532	0.23	0.009	—	—	—	—	—
Kabul	Afghanistan, Pakistan	480	77,850	450	317	70.5	6,700	118	680	0.10	0.101	1.52	almost nil	0.505	20,400	805
Helmand	Afghanistan, Iran	1,050	370,000	250	32.4	13.0	17,000	—	286	0.061	0.0012	—	—	—	—	—

NOTE: (A) Those countries having within their boundaries a significant portion of the drainage of the stream under consideration.
(B) The location or portion of the drainage area commanded by the gaging stations for precipitation and runoff, discharge and silt flow given under (a),(b) and (c) for the respective rivers are shown below. Figure within brackets indicates the drainage area above the station in sq km.

Mekong
  (a) Rainfall—whole basin; Runoff—Kratie(646,000).
  (b) Kratie(646,000).
  (c) For max. silt content—Vientiane(299,000);for average silt content and annual silt runoff-Mukdahan(391,000).

Red
  (a) Whole river basin
  (b) Max. and average at Vietri(113,000), min. at Hanoi
  (c) Max. min. and av.—at Hanoi; silt runoff—whole basin.

Brahmaputra
  (a) Whole river basin
  (b) Max. and Min—Pandu(424,309); Av. Bahadurabad(536,600)
  (c) Bahadurabad(536,600)

Meghna
  (a) Rainfall—whole basin; Runoff—Bhairab Bazar(64,700)
  (b) Bhairab Bazar(64,700)

Ganges
  (a) Rainfall—whole basin; Runoff—Hardinge Bridge
  (b) Hardinge Bridge(976,200).
  (c) Hardinge Bridge(976,200).

Kosi
  (a) Whole river basin. (b) Chatra(86,900) (c) Chatra (86,900).
Gandak
  (a) Whole river basin (b) Outfall
Gogra
  (a) Whole river basin (b) Outfall
Indus
  (a) Hills catchment
  (b) Max.—Sukkur, Min.—Kalabagh(305,000), av.—Sukkur; specific discharge and average unit discharge—Attock.
  (c) Hills catchment—Kalabagh(305,000)

Sutlej
  (a) Rupar(60,500)
  (b) Rupar(60,500)

Kabul
  (a) Whole river basin
  (b) Warsak(67,500)
  (c) Sarobi(25,350)

Helmand
  (a) Whole river basin
  (b) Qal eh Bist(278,000)

— Information not available.
1 ECAFE—Economic Commission for Asia and the Far East (United Nations).

TABLE 3-2. ASIA–SUMMARY OF BASIC DATA OF MEKONG RIVER

(Source: UN Water Resources Series No. 29, 1966)

Total drainage area	795,000 sq km	307,000 sq mi	Average annual discharge		
Drainage area of the lower basin in Laos, Cambodia, Thailand and Viet-Nam	609,000 sq km	236,000 sq mi	at Chiang Saen (1961-63)	2,770 m3/sec	97,700 cfs
Drainage area at Chiang Saen (near Burma border)	189,000 sq km	73,000 sq mi	at Vientiane (1923-44 and 1948-63)	4,575 m3/sec	161,500 cfs
Drainage area at Vientiane	299,000 sq km	115,500 sq mi	at Kratie (1933-44, 1946-53 and 1960-63)	14,800 m3/sec	522,000 cfs
Drainage area at Kratie (547 km from the sea)	646,000 sq km	250,000 sq mi	Minimum discharge		
			at Chiang Saen	570 m3/sec	20,100 cfs
Length of main river	4,350 km	2,700 mi	at Vientiane	701 m3/sec	24,800 cfs
from source to Chiang Saen (upper basin)	1,955 km	1,215 mi	at Kratie	1,250 m3/sec	44,100 cfs
from Chiang Saen to the sea (lower basin)	2,395 km	1,485 mi	Specific flood discharge at Kratie	0.104 m3/sec/sq km	9.50 cfs/sq mi
			Average unit discharge at Kratie	0.0229 m3/sec/sq km	2.10 cfs/sq mi
Slope of river			Maximum annual runoff at Kratie (1939)	567 billion m3	460 million acre ft
from source to China border	1:400		Average annual runoff		
from China border to Vientiane	1:2,900		at Chiang Saen (1961-63)	87.3 billion m3	70.7 million acre ft
from Vientiane to river mouth	1:16,000		at Vientiane (1923-44 and 1948-63)	150 billion m3	121.5 million acre ft
			at Kratie (1933-44, 1946-53 and 1960-63)	467 billion m3	378 million acre ft
Average annual precipitation over river basin	1,380 mm	54.5 in	Mean annual runoff expressed in depth at Kratie (1936)	722 mm	28.4 in
			Minimum annual runoff at Kratie (1936)	391 billion m3	317 million acre ft
Maximum flood discharge			Maximum silt content at Vientiane	3,076 ppm or	0.31 %
at Chiang Saen	11,900 m3/sec	420,000 cfs	Average silt content at Mukdahan (1963)	597 ppm or	0.06 %
at Vientiane	20,800 m3/sec	735,000 cfs	Average annual silt runoff at Mukdahan (1963)	170 million metric tons	
at Kratie	67,000 m3/sec	2,360,000 cfs	Average annual silt runoff per unit area at Mukdahan (1963)	435 m tons/sq km	1,100 long tons / sq mi

TABLE 3-3. ASIA—WATER RESOURCES AND IRRIGATION WATER DEMAND, 1970-90

(Source: ECAFE, UN Water Resources Series No. 40, 1971)

Country	Mean annual runoff a) (billion m3)	1970			1990		
		Irrigated harvested area (million ha)	Irrigation requirement b) (billion m3)	(per cent of mean annual runoff)	Irrigated harvested area (million ha)	Irrigation requirement b) (billion m3)	(per cent of mean annual runoff)
Afghanistan	50	1.7	17	34.0	2.3	23	46.0
Burma	680	0.7	7	1.0	1.3	13	1.9
Cambodia	80	0.1	1	1.2	0.3	3	3.8
Ceylon	43	0.6	6	13.9	1.3	13	30.3
China (Taiwan)	53	1.0	10	18.9	1.6	16	30.2
India	1,508	42.9	430	28.5	77.1	770	51.0
Indonesia	1,940	4.5	45	2.3	8.1	81	4.2
Iran	100	3.0	30	30.0	4.0	40	40.0
Korea, Rep. of	70	1.0	10	14.3	1.6	16	22.8
Laos	270	0.0	0	0.0	0.3	3	1.1
Malaysia	650	0.3	3	0.5	0.9	9	1.4
Mongolia	25	0.0	0	0.0	0.1	1	4.0
Nepal	170	0.1	1	0.6	0.3	3	1.8
Pakistan	298	15.6	160	53.7	28.2	280	94.0
Philippines	390	1.9	19	4.9	3.9	39	10.0
Thailand	220	2.1	21	9.6	3.6	36	16.4
Viet-Nam, Rep. of	116	0.7	7	6.0	1.3	13	11.2
Total	6,663	76.2	767	11.5	136.2	1,359	20.4

a) Excluding runoff from contiguous States.
b) Assuming 1,000 mm consumptive use.

TABLE 3-4. ASIA—INDUSTRIAL WATER DEMAND, 1970-80

(Source: ECAFE, UN Water Resources Series No. 40, 1971)

Country	Increase in gross domestic product, 1970/80			Increase in industrial water use, 1970/80 (million m³/annum)				
	Total (US $ million)	Industry b) (per cent total)	Industry b) (US $ million)	Boiler feed	Consumed by products	Product treatment and cleaning	Cooling and temperature conditioning	Total including other uses
Japan a)	90,820	28	25,100	506	201	6,300	11,368	19,370
Burma	980	12	120	2	1	30	54	93
Cambodia	610	14	90	2	1	23	41	69
Ceylon	1,150	11	130	3	1	33	59	100
China (Taiwan)	4,090	27	1,110	22	9	278	501	855
India	28,800	17	4,900	99	39	1,230	2,220	3,780
Indonesia	6,960	16	1,120	23	9	281	506	862
Iran	10,900	35	3,810	78	31	956	1,725	2,940
Korea, Rep. of	8,720	22	1,920	39	16	482	870	1,480
Malaysia, West	2,990	23	690	14	55	173	312	531
Nepal	870	13	110	2	1	28	50	85
Pakistan	13,680	14	1,920	39	16	482	870	1,480
Philippines	5,520	22	1,210	24	10	304	548	933
Thailand	6,610	18	1,190	24	10	299	539	918
Viet-Nam, Rep. of	1,420	14	200	4	2	50	91	154
Other countries	93,300	20	18,520	375	201	4,649	8,386	14,280
	2,930	20	580	12	5	146	262	447
Developing region	96,230	20	19,100	387	206	4,795	8,648	14,727
Per cent				2.6	1.4	32.6	58.7	100.0

a) Gross domestic product and industrial water usage in 1966.
b) 1965/67 percentages for developing countries increased by 2 per cent.

TABLE 3-5. ASIA—HYDROELECTRIC AND THERMAL POWER GENERATING CAPACITY AND PRODUCTION, 1968

(Source: U.S. Federal Power Commission, 1971)

Country	Installed capacity (MW.)[1]			Energy production (GWh.) [2]			Population (1,000)	Kwh. per capita
	Hydro	Thermal	Total	Hydro	Thermal	Total		
Afghanistan	241	34	275	298	24	322	16,113	20
Burma	103	157	260	341	199	540	26,389	20
Cambodia	0	57	57	0	122	122	6,557	19
Ceylon	113	104	217	472	199	671	11,964	56
China (Mainland)	4,200	11,700	15,900	13,000	29,000	42,000	730,000	58
China-Taiwan	721	1,341	2,062	3,864	6,172	10,036	13,466	745
Hong Kong	0	1,054	1,054	0	3,718	3,718	3,925	947
India	5,910	8,404	14,314	20,117	26,776	46,893	523,893	90
Indonesia	283	382	665	941	742	1,683	112,825	15
Iran	309	1,642	1,951	848	3,988	4,836	27,081	179
Iraq	0	651	651	0	1,542	1,542	9,030	171
Israel	0	1,020	1,020	0	5,244	5,244	2,745	1,910
Japan	17,841	35,346	53,187	73,928	190,717	264,645	101,080	2,618
Jordan	0	60	60	0	150	150	2,103	71
Korea (North)	2,210	515	2,725	10,800	3,000	13,800	13,000	1,062
Korea (South)	327	1,126	1,453	927	5,147	6,074	30,470	199
Laos	0	13	13	0	32	32	2,825	11
Lebanon	198	176	374	756	260	1,016	2,580	394
Macau	0	18	18	0	46	46	260	177
Malaysia 3	195	507	702	796	2,163	2,959	10,305	287
Mongolia	0	185	185	0	342	342	1,210	283
Nepal	14	24	38	40	15	55	10,652	5
Pakistan	913	1,102	2,015	3,100	3,205	6,305	123,405	51
Philippines	547	1,819	1,736	1,703	5,020	6,723	35,883	187
Saudi Arabia 4	0	904	904	0	3,058	3,058	7,840	390
Singapore	0	464	464	0	1,549	1,549	1,988	779
Southern Yemen	0	83	83	0	172	172	1,195	144
Syria	16	200	216	44	700	744	5,701	131
Thailand	381	516	897	1,390	1,566	2,956	33,693	88
Turkey	733	1,240	1,973	3,172	3,657	6,829	33,539	204
USSR 5	....							
Vietnam (North)	5	182	187	10	460	470	20,700	23
Vietnam (South)	164	274	438	15	753	768	17,414	44
Total 1968	35,424	70,645	106,069	136,562	299,738	436,300	1,939,435	225

1 MW.—Megawatts—Thousand Kilowatts.
2 GWh.— Gigawatt-hours—Million Kilowatt-hours.
3 Includes Malaya, Sabah, Sarawak.
4 Includes Saudi Arabia, Bahrain, Kuwait.
5 Included in USSR-Europe.

TABLE 3-6. AFGHANISTAN—ESTIMATED SURFACE WATER RESOURCES

(Source: FAO, 1965)

	Estimated annual runoff, km³
Rivers of Amu-Darya river basin and of closed rivers of Turkmenistan	11.6 ¹)
Rivers of the Indus river basin	23.4
Rivers of the Seistan depression	13.4
Total	50.0

1) Flow of Amu-Darya (border river) excluded.

TABLE 3-7. AFGHANISTAN—SUMMARY OF BASIC DATA OF KABUL RIVER

(Source: UN Water Resources Series No. 29, 1966)

Total drainage area, source to Attock about	77,850 sq km	30,000 sq mi		
Drainage area at Sarobi	25,350 sq km	9,800 sq mi		
Drainage area at Warsak dam (Warsak is about 80 km (50 mi) above the outfall of the river at Attock)	67,500 sq km	26,000 sq mi		
Drainage areas at Afghan border	66,235 sq km	25,600 sq mi		
Drainage areas of tributaries				
Drainage area of Loghar	9,110 sq km	3,520 sq mi		
Drainage area of Panjshir	12,875 sq km	4,960 sq mi		
Drainage area of Laghman	6,890 sq km	2,660 sq mi		
Drainage area of Kunar	23,940 sq km	9,250 sq mi		
Drainage area of Swat (about)	10,350 sq km	4,000 sq mi		
Average annual precipitation over river basin, about	450 mm	18 in		
In the mountains and at Kabul	500 mm	20 in		
At Peshawar	330 mm	13 in		
Maximum recorded flood discharge at Warsak(1924)	6,700 m³/sec	236,000 cfs		
Estimated 1,000 year flood discharge	10,200 m³/sec	360,000 cfs		
Average annual discharge at Warsak	680 m³/sec	24,000 cfs		
Minimum recorded discharge at Warsak	118 m³/sec	4,180 cfs		
Specific flood discharge	0.1	m³/sec/sq km	9.1 cfs/sq mi	
Average unit discharge	0.0101	m³/sec/sq km	0.92 cfs/sq mi	
Average annual runoff at Warsak	21.4 billion m³	17.35 million acre ft		
Mean annual runoff over the basin expressed in depth	317 mm	12.5 in		
Maximum silt content at Sarobi(April 59)	15,200 ppm or	1.52 %		
Average silt content at Sarobi (1959)	5,050 ppm or	0.505 %		
Minimum silt content at Sarobi(Oct.-Dec.)	almost nil			
Length of main river	480 km	300 mi		
Average annual silt runoff at Sarobi	20.4 million metric tons			
Average annual silt runoff per unit area at Sarobi	805 m tons/sq km	2,060 long tons		
Slope of river between Afghan border and Warsak dam	1:920			

TABLE 3-8. AFGHANISTAN—SUMMARY OF BASIC DATA OF HELMAND RIVER

(Source: UN Water Resources Series No. 29, 1966)

Total drainage area	370,000 sq km	143,000 sq mi
Drainage area of Helmand proper at Qal eh Bist	278,000 sq km	107,189 sq mi
Drainage area of Arghandab system including Tarnak, Arghanstan and Dori	51,600 sq km	19,935 sq mi
Length of main river	1,050 km	655 mi
Slope of river	— — —	— — —
Average annual precipitation over river basin	250 mm	10 in
upper reaches	340 mm	13.43 in
lower reaches in desert area	125 mm	5 in
Maximum estimated flood discharge of Helmand proper at Qal eh Bist	17,000 m3/sec	600,000 cfs
Average annual discharge of Helmand proper at Qal eh Bist	286 m3/sec	10,000 cfs
Specific flood discharge	0.061 m3/sec/sq km	5.58 cfs/sq mi
Average unit discharge	0.0012 m3/sec/sq km	0.104 cfs/sq mi
Average annual runoff of whole system	11.6 billion m3	9.4 million acre ft
of Helmand proper at Qal eh Bist	9 billion m3	7.3 million acre ft
Mean annual runoff of the whole basin expressed in depth	32.4 mm	1.28 in

TABLE 3-9. AFGHANISTAN—DISCHARGE OF SELECTED RIVERS

(Source: FAO, 1965)

River	Location	Basin area in km^2	Specific discharge in $l/sec/km^2$	Average annual discharge m^3/sec	Mean annual flow km^3
Hari Rud	Assarasum	7,430	4.5	33.0	1.0
Hari Rud	Tagaw-Kaza	11,700	3.8	45.0	1.4
Hari Rud	Marwa	22,000	3.0	65.0	2.0
Adraskand	Adraskand village	1,950	2.2	4.3	0.14
Adraskand	At confluence with Rud-i-Ghaz	4,100	2.1	8.6	0.27
Adraskand	Mouth of the river	22,000	0.3	6.6	0.21
Farah Rud	Lashkagar	15,300	2.7	41.0	1.3
Farah Rud	Alikinai	18,300	2.4	43.0	1.35
Farah Rud	Bakshabad	19,400	2.3	44.0	1.4
Farah Rud	Farah (city)	26,600	1.8	48.0	1.55
Kabul	Tangi-Shaidan	1,900	3.2	6.0	0.19
Kabul	Tangi-Garu	14,370	1.6	22.6	0.71
Kabul	Naghlu	30,010	4.7	142.0	4.47
Kabul	Darunta	40,770	5.1	206.0	6.50
Kabul	At confluence with Kunar	70,740	9.5	670.0	21.1
Kabul	At boundary with Pakistan	75,400	9.9	742.0	23.4
Ghazni	Nurbuja	1,590	1.25	2.0	0.06
Ghazni	At the mouth	12,370	0.9	11.0	0.35
Nahar	At the mouth	3,710	0.8	3.0	0.09

TABLE 3-10. AFGHANISTAN—DISCHARGE OF STREAMS INTO SEISTAN SINK

(Source: UN Water Resources Series No. 29, 1966)

River basin	Area sq km	Estimated annual runoff mm	Normal runoff period	Estimated peak discharge m^3/sec	Annual runoff million m^3	Average flood discharge during the month of peak flood $m^3/sec/month$
Helmand proper at Qal eh Bist	278,000	32.5	Annual	17,000	9,000	6,100
Helmand proper above Kajakai	40,400	159.0	Annual	5,670	(6,400)	
Arghandab proper (total)	19,500	76.7	Annual	—	1,500	
Arghandab proper above dam	15,250	85.0	Annual	—	(1,300)	—
Arghandab proper below dam	4,270	25.2	January to March	—	(107)	—
Tarnak	8,600	35.8	February to May	—	321	—
Arghastan	14,950 a	24.6	February to May	1,190	370	—
Dori	8,500	25.4	February to May	—	216	—
Helmand below Qal eh Bist	16,500 b	6.35	February to April	—	105	—
Dasht-i-Margo	24,600 b	3.81	February to April	—	92.5	—
Chan Baigi	3,060	4.06	February to April	—	12.3	—
Rud-i-Khar	1,765	3.56	February to April	—	6.2	2.83
Khuspas	4,160	17.8	February to April	100	74.0	56.5
Khash	13,400	44.2	January to June	1,390	592	638
Farah	27,500	42.7	January to June	3,820	1,170	2,830
Harut	33,800	25.9	February to June	1,275	875	1,275
Musa Kala	6,680	50.8	January to June	2,010	339	—
Sanguin	7,760	25.4	February to May	—	197	—
Total for Afghan basins					14,977	

a Area of 16,600 sq km (6,437 sq mi) above Lake Ab-i-Istada is excluded.
b An estimated 114,000 sq km (44,056 sq mi) of Registan and Dasht-i-Margo do not contribute to runoff.

TABLE 3-11. BANGLADESH—DISCHARGE OF BRAHMAPUTRA RIVER AT BAHADURABAD

(Source: UN Water Resources Series No. 29, 1966)

[Period 1956-62]

Month	Maximum		Minimum		Mean	
	m^3/s	cfs	m^3/s	cfs	m^3/s	cfs
January	6,600	233,000	4,360	154,000	5,210	184,000
February	4,980	176,000	3,400	120,000	4,330	153,000
March	5,970	211,000	3,760	133,000	4,673	165,000
April	8,600	304,000	5,150	182,000	7,140	252,000
May	24,000	849,000	7,980	282,000	17,800	629,000
June	38,800	1,370,000	26,500	936,000	32,250	1,140,000
July	45,300	1,600,000	33,700	1,190,000	40,200	1,420,000
August	55,500	1,960,000	30,850	1,090,000	43,900	1,550,000
September	48,400	1,710,000	24,150	853,000	35,400	1,250,000
October	32,250	1,140,000	14,080	497,000	21,450	759,000
November	14,980	529,000	8,490	300,000	10,590	374,000
December	9,360	331,000	5,660	200,000	6,650	235,000
Annual	21,750	768,000	17,970	635,000	19,200	678,000

TABLE 3-12. BANGLADESH—DISCHARGE OF GANGES RIVER AT HARDINGE BRIDGE

(Source: UN Water Resources Series No. 29, 1966)

[Period 1934-62]

Month	Maximum		Minimum		Mean	
	m^3/s	cfs	m^3/s	cfs	m^3/s	cfs
January	3,680	130,000	2,065	73,000	3,110	110,000
February	4,750	168,000	1,895	67,000	2,740	97,000
March	3,600	127,000	1,610	57,000	2,350	83,000
April	2,970	105,000	1,273	45,000	2,040	72,000
May	3,140	111,000	1,388	49,000	2,120	75,000
June	9,680	342,000	2,350	83,000	4,360	154,000
July	29,700	1,050,000	10,750	380,000	18,070	639,000
August	52,600	1,860,000	23,600	834,000	39,400	1,390,000
September	56,000	1,980,000	25,000	884,000	36,500	1,290,000
October	42,200	1,490,000	8,350	295,000	17,700	626,000
November	16,500	584,000	4,390	155,000	7,190	254,000
December	6,740	238,000	2,860	101,000	4,190	148,000
Annual	16,350	578,000	7,810	276,000	11,650	412,000

TABLE 3-13. BANGLADESH—DISCHARGE OF MEGHNA RIVER AT BHAIRAB BAZAR

(Source: UN Water Resources Series No. 29, 1966)

[Period 1957-62]

Month	Maximum		Minimum		Mean	
	m³/s	cfs	m³/s	cfs	m³/s	cfs
January	650	23,000	510	18,000	595	21,000
February	540	19,000	370	13,000	480	17,000
March	1,105	39,000	510	18,000	625	22,000
April	1,120	40,000	735	26,000	905	32,000
May	2,460	88,000	1,370	49,000	1,905	68,000
June	5,400	190,000	3,350	119,000	4,200	148,000
July	9,100	322,000	5,700	201,000	7,280	257,000
August	9,150	323,000	6,680	236,000	7,750	274,000
September	9,510	336,000	6,430	227,000	7,730	273,000
October	8,100	286,000	5,270	186,000	6,690	236,000
November	3,315	117,000	1,765	63,000	2,770	98,000
December	1,270	45,000	795	28,000	990	35,000
Annual	3,940	139,000	3,000	106,000	3,515	124,000

TABLE 3-14. BANGLADESH–WATER BALANCE DURING LOW FLOW SEASON

(Source: FAO, 1973)

[Period: December through March]

Type of Resource	Average low-flow season 10^6 Acres	Type of Resource	Average low-flow season 10^6 Acre-ft
LAND AREAS		**WATER DEMAND IN 1978**	
Area of Bangladesh — gross	35	From locally available water resources	
net	29	Irrigation by low-lift pumps	9
Area of Bangladesh cultivated — net	22	Irrigation by tube-wells	6
Area of Bangladesh — fresh free water surface	6	Domestic water supply	7
		Livestock	5
		Industries	3
WATER POTENTIAL		Ecological constraint — water for ecosystem	70
LOCALLY AVAILABLE WATER RESOURCES	10^6 Acre-ft	Total demand by 1978	100
Precipitation — average annual (country)	235		
Precipitation (country) 10% — average annual Dec. to March	24		
Channel precipitation (local)	3	**WATER BALANCE (local) 1978: SURPLUS MAF**	37
Local runoff December to March	45		
Surface water storage between 5 to 30 ft depth	150	Conclusion: For an average year the local water resources available show a surplus of 37 MAF after allowing 50% of available local water to meet the ecological constraint. In a dry year such as 1972/73 the available local water in December can be as much as 30% less, hence:	
Groundwater storage (local) between 5 and 30 ft depth	44		
Evaporation from surface water (country)	20		
Evaporation from surface water (local)	15		
Available water (local)	137		
IMPORTED WATER RESOURCES		Available local water	96
Ganges river average December to March	18	Total demand by 1978	100
Brahmaputra average December to March	32	Meghna river average December to March	2
Total available water (imported)	52	**WATER BALANCE (local): DEFICIT**	4

TABLE 3-15. BANGLADESH—IRRIGATION WATER REQUIREMENTS FOR FIVE-YEAR PLAN 1973-78
(Source: FAO, 1973)

| | Unit | Target 1978 | |
		Without ecological constraint	With ecological constraint
FROM LOCAL WATER RESERVES:			
For low-lift pumps (temporary)			
Area to be irrigated	10^6 Acres	2.25	1.75
Water duty	Ft	4	4
Water consumption	10^6 Acre-ft	9	7
For tube-wells (permanent)			
Area to be irrigated	10^6 Acres	1.369	0.85
Water duty	Ft	4	4
Water consumption	10^6 Acre-ft	6	4
		15	11
FROM IMPORTED WATER RESOURCES:			
For pump-feeder canal system			
Area to be irrigated	10^6 Acres		1
Water duty	Ft		4
Water consumption	10^6 Acre-ft		4
Irrigation from local water	10^6 Acre-ft	15	11
Irrigation from imported water	10^6 Acre-ft	0	4
Total:	10^6 Acre-ft	15	15

TABLE 3-16. BURMA—FLOOD DATA
(Source: UN Water Resources Series No. 28, 1965)

Name of river	Drainage area at gaging station sq km	Peak discharge m3/s	Date
Mu	12,504	5,440	9 July 1928
Zawgyi	4,087	1,980	24 Sep. 1949
Paung laung	2,577	1,700	Oct. 1926
Meiktila Lake	620	4,000	4 Nov. 1935
Nyaungyan – Minhla Tank	1,200	6,000	5 May 1920
Thitson	376	510	1917
Salin	2,100	5,206	8 Oct. 1948
Mon	5,310	8,980	8 Oct. 1948
Man	1,500	897	25 Oct. 1950
Chaungmagyi	3,424	4,110	20 Oct. 1913
Yenwe	912	2,200	1937
Pegu	2,260	1,210	1926
Irrawaddy	360,000	63,700	1877

TABLE 3-17. CAMBODIA—DISCHARGE OF SELECTED RIVERS
(Source: UNESCO, 1971)

River and station	Basin area km2	Mean monthly discharge, m3/s												Year	Period of record
		Jan.	Feb.	Mar.	Apr.	May	Jun.	Jul.	Aug.	Sep.	Oct.	Nov.	Dec.		
Mekong, Kratie	646,000	2,880	2,100	1,620	1,570	2,930	9,140	25,000	37,900	41,900	25,900	12,800	6,280	14,200	1933-53; 1961-66
Se San, Ban Komphun	48,200	477	328	253	269	678	1,220	2,820	4,020	3,600	2,660	1,240	829	1,530	1961-66

TABLE 3-18. CEYLON–DISCHARGE OF SELECTED RIVERS
(Source: UNESCO, 1971)

River and station	Basin area km2	Mean monthly discharge, m3/s												Year	Period of record
		Jan.	Feb.	Mar.	Apr.	May	Jun.	Jul.	Aug.	Sep.	Oct.	Nov.	Dec.		
Mahaweli, Manampitiya	7,343	422	288	152	171	157	159	147	132	128	206	286	462	226	1941-66
Kelani, Nagalagam	2,085	108	100	107	132	258	270	185	161	160	266	228	147	177	1924-60
Malwattu Oya, Kappachchi	2,121	44.0	6.0	7.0	8.0	5.0	2.0	2.0	1.0	1.0	5.0	15.0	69.0	14	1945-66
Walawe, Moraketiya	1,580	49.0	35.0	46.0	79.0	59.0	34.0	21.0	19.0	21.0	37.0	75.0	70.0	45	1942-66

TABLE 3-19. CEYLON–FLOOD DATA
(Source: UN Water Resources Series No. 28, 1965)

Name of river	Drainage area at gaging station km2	Peak discharge m3/s	Date
Gin-ganga at Agaliya	684	1,470	18 May 1940
Kalu-ganga at Putuhaula	2,599	2,970	16 Aug. 1947
Kelani-ganga at Colombo	2,086	6,660	17 Aug. 1947
Mahawell-ganga at Teunekwena	1,417	6,300	15 Aug. 1947
Nahaweli-ganga at Waragantota	4,037	5,850	5 Aug. 1947
Nilwala-ganga at Bopagoda	412	2,640	16 Aug. 1947
Walawe-ganga at Embilipitiya	1,581	2,370	15 Aug. 1947

TABLE 3-20. CEYLON–MUNICIPAL WATER SUPPLY SYSTEM OF COLOMBO
(Source: Internat. Statistical Institute, 1971)

Year	Population served	Quantity available (1,000 m3)			Annual consumption (1,000 m3)	
		Production	Purchased from other plants	Total	Total	of which domestic
1967	551,200	31,062	14,938	46,000	46,000	39,448
1969	N.A.	33,186	16,593	49,778	49,778	43,348

N.A. – Not available.

TABLE 3-21. CHINA—PRECIPITATION AND RUNOFF IN MAJOR RIVER BASINS

(Source: Tojin Sha, 1964)

River system	Basin area km²	Mean annual precipitation mm	Mean annual runoff mm	Mean annual evaporation mm	Runoff coefficient %
Sungari	523,580	512	141	371	27.6
Liao	219,000	465	74	301	15.9
Yellow (Hwang Ho)	745,100	415	65	350	15.7
Upper Yellow (Lanchow)	216,190	427	148	279	35.0
Fen (Hochin)	38,650	471	50	321	11.0
Lo (Chuangt'oul)	62,700	436	28	408	6.5
Ching (Chiangchia Shan)	41,800	462	47	416	10.1
Wei (Hu Hsien)	63,550	573	142	431	25.0
Hwai	164,560	840	198	642	23.6
Yangtze	1,808,500	1,050	568	482	54.1
Chinsha	502,050	662	329	333	50.0
Min	133,570	1,100	722	378	65.6
Chialing	159,810	892	408	484	46.0
Wu	88,220	1,135	586	549	52.0
Tungting	261,130	1,445	852	593	59.0
Han	174,350	900	356	544	40.0
Poyang	158,680	1,670	971	699	58.2
Ch'ient'ang	49,930	1,650	940	710	57.0
Min	60,800	1,710	1,074	636	62.8
Han	29,700	1,655	982	673	59.0
Pearl (Si Kiang)	437,230	1,480	890	590	60.2
Tung	26,300	1,758	1,203	555	68.4
Pel	45,600	1,885	1,370	515	73.0
Si (Wuchow)	328,000	1,370	773	603	56.0
Upper Si (Nanning)	74,310	1,340	646	694	48.0
Total Nation	9,597,000	650	280	370	43.1

TABLE 3-22. CHINA—DISCHARGE OF PRINCIPAL RIVERS

(Source: Tojin Sha, 1964)

River and station	Basin area km^2	Length of river km	Mean annual discharge m^3/s	Total annual discharge km^3
Amur, mouth	1,844,300	4,370	11,000	346
Khabarovsk	—	—	8,600	—
Sungari, mouth	523,580	1,927	2,450	—
Pinkiang (Harbin)	—	—	2,830	—
Yellow (Hwang Ho), mouth	745,000	4,845	1,500	47.6
Yangtze, mouth	1,808,500	5,800	34,000	1,063
Hankow	—	—	23,800	—
Pearl, mouth	425,700	2,100	12,500	380(est.)

TABLE 3-23. CHINA—PRECIPITATION AND RUNOFF IN AMUR RIVER BASIN

(Source: Tojin Sha, 1964)

Basin area	Mean annual precipitation mm	Runoff mm	Evaporation mm	Runoff Coefficient %
Middle reaches of Amur River	450	149	301	33.1
Sungari River	520	169	351	32.5
Ussuri River	640	337	303	52.7
Lower reaches of Amur River	620	346	274	55.8
Entire Basin	506	188	318	37.2

TABLE 3-24. CHINA—DISCHARGE OF RIVERS IN AMUR RIVER BASIN

(Source: Tojin Sha, 1964)

River	Length of river km	Basin area km²	Mean annual discharge m³/s	Total annual discharge km³	Percent of total annual basin discharge
Halkhain Gol	1,600	200,900	440	13.9	4.0
Halkhain Gol	1,520	169,700	400	12.6	3.6
Tse-ya River	1,210	233,000	1,800	56.7	16.4
Bureya	950	69,700	950	29.9	8.6
Sungari River	1,927	523,200	2,450	77.2	22.3
Ussuri River	890	187,000	2,000	63.0	18.2
Tungusic	—	38,100	385	12.1	3.5
Amgun	665	54,900	600	18.9	5.5
Others	—	363,800	1,975	62.2	18.0
Nunkiang (Nonni)	1,490	242,900	900	—	—
Total		1,843,000	11,000	346.5	100

TABLE 3-25. CHINA—DISCHARGE OF RIVERS IN YANGTZE RIVER BASIN

(Source: Tojin Sha, 1964)

Station or tributary	Distance from mouth	Basin area km²	Percent of total river basin	Mean annual discharge m³/s	Total annual discharge km³	Percent of total river basin flow
Yangtze System	5,800	1,808,500	100	33,700	1,062.9	100
Pat'ang	4,550	187,500	10.0	1,050	33.1	3.1
Yalung River	3,561	144,280	7.0	2,000	63.1	5.9
Ipin, An-pien	2,944	501,750	27.7	5,500	173.3	16.3
Min River	2,885	133,570	—	3,040	95.9	9.0
Ipin, Lichuang	—	640,530	35.4	8,560	270.0	25.4
Chialing River	2,511	159,810	8.8	2,000	63.1	5.9
Chungking Tsun-tan	—	867,860	48.0	11,700	369.0	34.7
Wu River	2,389	88,220	4.9	1,750	55.2	5.2
Ichang	1,850	1,007,530	—	14,300	451.0	42.4
Tungting System	1,162	261,130	14.4	7,200	227.1	21.4
Han River	1,125	174,350	9.6	1,690	53.3	5.0
Hankow	1,125	1,490,070	82.4	23,800	750.7	70.6
Po Yang System	835	158,680	8.8	5,740	181.0	17.0
Tatung	—	1,704,260	94.2	81,060	979.6	92.2
Hwai River	319	187,000	—	1,033	32.6	3.1

TABLE 3-26. CHINA—DISCHARGE OF RIVERS IN YELLOW RIVER BASIN

(Source: Tojin Sha, 1964)

River	Length of river km	Basin area km^2	Mean annual discharge km^3
T'ao	555	32,000	4.40
Huang (Shui)	327	33,960	4.80
Shan-shui (Ch'ing-Shui)	200	14,000	1.30
Wu-ting	369	23,533	1.73
Fen	635	39,360	1.57
Pei-lo	660	28,440	.79
Ching	451	45,335	1.66
Wei	787	60,608	6.94

TABLE 3-27. CHINA—DISCHARGE OF RIVERS IN TARIM RIVER BASIN

(Source: Tojin Sha, 1964)

River and station	Mean annual discharge m^3/s	Total annual discharge hm^3
Manass, Hungshan Tsui	45.2	1,425
Tasi, Hungshan Wan	5.72	180
Chinkou, Charkus	11.1	350
Payinkou, Anchi Bridge	6.31	199
Kueitun (or Wusu), Kueitun Bridge	14.9	470
Ssukoshu	9.65	304
Total Basin	—	2,978

TABLE 3-28. CHINA—DISCHARGE AND SILT LOAD OF RIVERS IN PEARL RIVER (SI KIANG) BASIN

(Source: Tojin Sha, 1964)

River	Station	Basin area km2	Mean annual flow m3/s	Mean depth mm	Mean silt content grams/m3	Total annual silt load (Million Tons)
Tung River Basin						
Tung	Hui-yang	25,100	957.2	1,205	140	2.45
Hsin-feng	Hui-lung	5,870	233.5	1,252	—	—
Pei River Basin						
Pei	Shih-chiao	38,200	1,496.8	1,237	140 (Shen-yuan)	5.14
Lien	Han-Kuang	7,990	334.2	1,313	—	—
Weng	Huang-Kang	4,700	187.9	1,251	—	—
Sui	Ssu-hui	7,030	320.5	1,437	60 (Liu-chi)	—
Si River Basin						
Si	Wuchow	32,800	8,049.0	775	410	86.00
Nan-pan	Kai-yuan	16,400	120.0	230	—	—
Pei-pai	Pan-chiang-chia	14,350	214.3	474	—	—
Hung-shui	Chien-chiang	126,000	2,120.0	535	760 (Tung Lan)	42.90
Ling	Wu-hsuan	196,000	4,443.8	714	—	—
Hsun	Kweiping	289,000	6,467.3	705	—	—
Liu	Liuchow	45,600	1,356.0	914	220	9.93
Lung	Ishan	13,600	347.7	757	—	—
Lo-ching	Lo-yung	6,720	225.4	1,054	—	—
Yu	Pai-se	18,050	347.9	608	—	—
Tso	Lungchow	17,710	329.0	583	—	—
Yu	Kwei-hsien	87,280	1,781.2	647	230 (Nanning)	8.01
Kwei	Chao-ping	14,700	570.0	1,225	30 (Kweilin)	—
Ho	Kai-chien	7,703	326.5	1,335	—	—

TABLE 3-29. CHINA—DATA ON PRINCIPAL LAKES

(Source: Tojin Sha, 1964)

Lake Name	Location (Province)	Area km2	Depth m	Elevation m	Volume km3
Hsingkai (Kokaiko)	Heilungkiang	4,400	10	69	
Chinghai (Koko Nor)	Chinghai	4,200 4,297 *	38	3,205	105
Tungting	Hunan-Hupei	3,915 4,350 *	—	—	35.4
Poyang	Kiangsi	3,780 5,100 *	—	—	36.3
Lo-pu-po	Sinkiang	3,000	—	—	
Hungtze	Kiangsu	2,700	—	—	
Lake Tai	Kiangsu	2,213	—	12	
Namu (Teng-ri-Nor)	Tibet	1,900	—	4,627	
Tsilin	Tibet	1,825	—	4,495	
Po-ssu-teng	Sinkiang	1,500	—	1,030	
Hu-lun-chih (Furunchi)	Inner Mongolia	1,100	—	534	
Ai-pi (Ebiko)	Sinkiang	1,000	—	190	
Lake Chao	Anhwei	1,000	—	—	
Pu-mu-wei	Tibet	881	—	4,936	
Yang-cho-yung	Tibet	880	—	4,419	
Kaoyu	Kiangsu	700	—	—	
Buluntokhoi	Sinkiang	700	—	468	
Lake Moncalm	Tibet	700	—	4,960	
Oling	Chinghai	648	—	4,139	
Buyer Nor	Inner Mongolia	600	—	830	
Hwang Lake	Kiangsu	600	—	—	
Ziling	Chinghai	570	—	4,155	
Weishan	Shantung	500	—	—	
I-chi-chi	Tibet	500	—	4,787	
Liangtze	Hupei	500	—	—	
Paishui	Hupei	400	—	—	
Taerh	Inner Mongolia	350	—	1,228	
Ma-na-sa-lo-tun	Tibet	343	—	4,557	
Chen-chih	Yunnan	340	5-8	1,886	
Erh-hai	Yunnan	300	15	1,980	
Tien-chi	Kirin	30	320	1,960	

* High water level

TABLE 3-30. CHINA—WATER USE IN YELLOW AND YANGTZE RIVER BASINS

(Source: Tojin Sha, 1964)

[Unit: 100 Million m³]

Water use	Yellow River	Yangtze River	Total
Non-Productive [1]	22.4	15.0	37.4
Industrial and Urban	12.8	—	12.8
Navigation	13.9	7.0	20.9
Irrigation	233.0	238.0	471.0
Fishing	11.8	—	11.8
Totals	293.9	260	553.9

[1] Non-productive water uses include:
1. 740 million m³ of Tung-ping Lake water losses.
2. One billion m³ of water losses from the Yellow River.
3. 500 million m³ of water losses from irrigation in all the provinces.
4. One billion 500 million m³ of water losses from diverting the courses of the Han, Chi and Yellow Rivers.

TABLE 3-31. CHINA—WITHDRAWAL OF IRRIGATION WATER BY PROVINCE

(Source: Tojin Sha, 1964)

[Water withdrawn for irrigation, 100 million m³]

Province	Yellow River	Han River	Total
Hopei	67.0	58.5	125.5
Honan................................	68.0	74.6	142.6
Shantung	83.0	69.7	152.7
Anhwei	4.8	27.3	32.1
Kiangsu	10.2	7.9	18.1
Total	233.0	238.0	471.0

TABLE 3-32. CHINA—WATER USE DATA BY PROVINCE

(Source: Tojin Sha, 1964)

[Water use in 100 million m³]

Water use	Hopei Province	Shantung Province	Honan Province	Kiangsu Province	Total
Area under irrigation (unit 10,000 mou) [1]	47,650	80,000	80,400	8,400	216,450
Irrigation	5,400	4,930	8,310	540	19,180
Urban industry	1,976	1,185	2,490	280	5,931
Navigation	1,300	2,850	1,500	—	5,650
Fishing	—	500	—	—	500

[1] 1 mou is approximately 1/15 hectare.

TABLE 3-33. CYPRUS—DISCHARGE OF KOURIS RIVER
(Source: UNESCO, 1971)

River and station	Basin area km²	\- Mean monthly discharge, m³/s \- Jan.	Feb.	Mar.	Apr.	May	Jun.	Jul.	Aug.	Sep.	Oct.	Nov.	Dec.	Year	Period of record
Kouris, Erimi	351	2.0	1.8	1.9	0.7	0.4	0.05	0.02	0.00	0.08	0.1	0.2	1.3	0.7	1954-66

TABLE 3-34. CYPRUS—WATER BALANCE
(Source: Konteatis, Dept. of Water Development, 1967)

Consumption of Water and Extent of Land

Utilization	Surface Runoff Excluding River Bed Infiltration — Million gallons	%	Extent of land thousand donums¹	%	Ground Water — Pumped and Subsurface — Million gallons	%	Extent of land thousand donums	%	Ground Water — Springs — Million gallons	%	Extent of land thousand donums	%	Evapotranspiration plus Consumption by Direct Rainfall — Million gallons	%	Extent of land thousand donums	%	Totals — Million gallons	%	Extent of land thousand donums	%
Irrigated crops including irrigated cereals	34,000	3.4	435	6.3	51,000	5.1	225	3.3	6,000	0.6	30	0.4	30,000	3	690	10	121,000	12.1	690	10
Dry farming crops (cereals, carobs, olives, tobacco, vines)													200,000	20	2,000	29	200,000	20	2,000	29
Forests, shrubs													200,000	20	1,400	20	200,000	20	1,400	20
Pastures, grass													100,000	10	1,200	17.5	100,000	10.0	1,200	17.5
Evaporation from bare land, fallow land, built up areas, water surfaces															1,610	23.5	344,500	34.5	1,610	23.5
Domestic, animal husbandry and industrial water					4,400	0.5			1,300	0.2							5,700	0.6		
Ground water over extraction					−26,700	−2.7											−26,700	−2.7		
Subsurface losses					15,000	1.5											15,000	1.5		
Surface losses to the sea	40,500	4															40,500	4.0		
Totals	74,500	7.4	435	6.3	43,700	4.4	225	3.3	7,300	0.8	30	0.4	530,000	53	6,900	100	1,000,000	100	6,900	100

1) 1 donum = 14,400 sq ft.

TABLE 3-35. CYPRUS—ESTIMATED SAFE YIELD AND WITHDRAWAL FROM MAIN AQUIFERS

(Source: Konteatis, Dept. of Water Development, 1967)

[As of 1965]

Aquifer	Safe yield of aquifer (Permissible withdrawal) million gallons per year	Number of Boreholes and Wells	Number of Springs	Actual withdrawal million gallons in 1965	Balance million gallons in 1965	Remarks
Western Messaoria Alluvial-Pleistocene-Pliocene	11,000	1,052	41	18,600	− 7,600	
Eastern Messaoria	4,800	3,750	1	9,800	− 5,000	Alluvial, Plio-Pleistocene
Akrotiri-Phassouri	2,500	784	−	5,100	− 2,600	
Kyrenia Limestone	3,600	20	10	200	+ 1,200	Jurassic fissured and karstic limestones
Other Areas	14,100	(?)	(?)	2,200	− 12,700	
Subsurface Flow	15,000			26,800	+ 15,000	
Totals	51,000			62,700	− 11,700	

TABLE 3-36. CYPRUS—RURAL DOMESTIC WATER SUPPLIES, 1966

(Source: Konteatis, Dept. of Water Development, 1967)

District	Satisfactory Piped Supply — Villages with House-to-House Connection			Villages with Public Fountains				Unsatisfactory Piped Supply — Villages with House-to-House Connection			Villages with Public Fountains				No Piped Supply — Villages				Total No. of Villages	Total Population
	No.	%	Pop.	No.	%	Pop.	%	No.	%	Pop.	No.	%	Pop.	%	No.	%	Pop.	%		
Nicosia	86	48.32	14,928	63	35.39	25,885	16.56	−	−	−	27	15.17	14,988	9.60	2	1.12	478	0.30	178	156,279
Kyrenia	20	42.55	14,661	14	29.80	5,145	18.70	1	2.12	3,496	12	25.53	4,215	15.32	−	−	−	−	47	27,517
Famagusta	55	56.12	56,340	22	22.45	7,271	9.13	3	3.06	2,864	18	18.37	13,140	16.50	−	−	−	−	98	79,615
Limassol	69	60.52	46,569	34	29.82	10,168	15.96	−	−	−	10	8.78	6,606	10.37	1	0.88	370	0.58	114	63,713
Paphos	48	36.36	25,025	83	62.88	23,801	48.50	−	−	−	1	0.76	250	0.51	−	−	−	−	132	49,076
Larnaca	23	38.98	22,885	21	35.59	8,247	21.23	3	5.09	2,983	11	18.65	4,551	11.72	1	1.69	170	0.44	59	38,836
Total	301	47.93	280,408	237	37.74	80,517	19.41	7	1.12	9,343	79	12.57	43,750	10.54	4	0.64	1,018	0.24	628	415,036

Note: Schemes with less than 10 gallons per day per capita are classified as unsatisfactory.

TABLE 3-37. CYPRUS—MUNICIPAL WATER SUPPLY SYSTEM OF NICOSIA

(Source: Internat. Statistical Institute, 1971)

Year	Population served	Quantity available (m3)			Annual consumption (m3)	
		Production	Purchased from other plants	Total	Total	of which domestic
1967	109,000	5,399,172	1,598,218	6,997,390	6,582,763	5,924,487
1969	114,000	5,811,702	1,871,929	7,683,631	7,176,557	6,458,901

TABLE 3-38. INDIA—DISCHARGE OF SELECTED RIVERS

(Source: UNESCO, 1971)

River and station	Basin area km2	Mean monthly discharge, m3/s												Year	Period of record
		Jan.	Feb.	Mar.	Apr.	May	Jun.	Jul.	Aug.	Sep.	Oct.	Nov.	Dec.		
Godavari, Dowlaishwaram	299,320	243	195	140	113	73.0	915	7,960	11,900	10,600	4,260	1,110	404	3,180	1901-60
Krishna, Vijayawada	251,355	111	67	44	38	194	598	5,490	6,550	4,340	2,740	1,010	254	1,730	1901-60
Mahanadi, Kaimundi	132,090	142	106	100	103	76	224	4,140	7,710	5,700	1,600	591	187	1,710	1947-65
Narbada, Garudeshwar	89,345	156	118	80	59	37	194	2,150	4,690	5,540	1,500	436	207	1,260	1949-62
Tapti, Kathore	61,575	–	–	–	–	–	250	1,160	1,670	1,950	680	221	–	–	1940-61
Pennar, Nellore	53,290	25	13	15	9	44	15	47	93	171	235	277	190	95	1934-47
Damodar, Rhondia	19,920	36	36	28	22	35	214	777	1,190	992	439	96	50	329	1934-61

TABLE 3-39. INDIA—DISCHARGE OF BRAHMAPUTRA RIVER AT PANDU

(Source: UN Water Resources Series No. 29, 1966)

[Period 1955-62]

Month	Maximum		Minimum		Mean	
	m³/s	cfs	m³/s	cfs	m³/s	cfs
Januray	3,995	141,000	3,200	113,000	3,740	132,000
February	3,795	134,000	3,030	107,000	3,510	124,000
March	5,690	201,000	3,540	125,000	4,360	154,000
April	8,920	315,000	6,060	214,000	7,105	251,000
May	26,870	949,000	10,450	369,000	19,030	672,000
June	34,520	1,219,000	21,690	766,000	29,780	1,050,000
July	44,145	1,559,000	35,850	1,266,000	35,850	1,266,000
August	46,835	1,654,000	22,765	804,000	36,275	1,281,000
September	45,955	1,623,000	20,501	724,000	28,485	1,006,000
October	22,255	786,000	14,215	502,000	16,960	599,000
November	8,865	313,000	6,625	234,000	7,845	277,000
December	5,155	182,000	4,330	153,000	4,755	168,000
Annual	17,360	613,000	15,345	542,000	16,225	573,000

TABLE 3-40. INDIA—AVAILABLE SURFACE WATER RESOURCES

(Source: Ministry of Irrigation and Power, 1972)

Region	Basin area (square miles)	Annual Normal rainfall (inches)	Annual Mean temperature (°F)	Annual Loss (inches)	Annual Runoff (inches)	Annual Runoff (million acre feet)
Rivers flowing into Arabian Sea (excluding the Indus System)	189,790	47.95	77.9	23.11	24.84	251.46
Indus Basin in India	136,673	21.86	54.7	13.02	8.84	64.43
Rivers flowing into Bay of Bengal, other than Ganga and Brahmaputra systems	467,309	42.77	79.0	29.37	13.40	334.03
Ganga system	376,818	43.76	62.2	24.00	19.76	397.09
Brahmaputra System	195,460	48.11	46.8	18.47	29.64	308.95
Rajputana	64,887	11.48	79.1	11.48	—	—
Total	1,430,937	41.03	—	23.26	17.77	1,355.96

TABLE 3-41. INDIA—AVAILABLE GROUND WATER RESOURCES

(Source: Ministry of Irrigation and Power, 1972)

[All units in million acre feet unless otherwise stated]

State	Amount of contribution of rainfall to ground-water recharge	Possible recharge due to canal infiltration	Total	Evapo-transpiration and sub-surface runoff losses	Net ground-water-recharge	Annual draft by the end of 1967-68	Net ground water recharge available for further groundwater development	Area irrigated by groundwater at present (million acres)
Andhra Pradesh	20.0	4.6	24.6	30	17.2	3.57	13.6	1.4
Assam Region (including Nagaland, NEFA etc.)	40.7	1.2	41.9	60	16.7	0.003	16.7	—
Bihar	26.9	4.4	31.3	30	21.9	2.35	19.5	1.2
Delhi	0.5	—	0.5	30	0.3	N.A.	—	—
Gujarat	14.1	0.4	14.5	30	10.2	4.13	6.1	1.75
Haryana	3.0	1.6	4.6	25	3.5	0.75	2.7	0.75
Himachal Pradesh	2.3	—	2.3	60	0.9	N.A.	—	0.003
Jammu and Kashmir	9.6	0.4	10.0	60	4.0	0.001	4.03	0.03
Kerala	9.5	1.2	10.7	50	5.4	0.004	5.4	0.016
Madhya Pradesh	43.1	1.4	44.5	40	26.7	4.22	22.5	1.00
Madras and Pondicherry	12.9	3.5	16.4	30	11.5	3.47	8.00	2.30
Maharashtra	19.8	1.2	21.0	40	12.6	3.41	9.2	2.00
Mysore	14.8	1.8	16.6	40	10.0	1.03	9.0	0.75
Punjab	5.1	4.0	9.1	25	6.9	3.3	3.6	3.5
Orissa	19.4	3.4	22.8	30	16.0	0.15	15.8	0.20
Rajasthan	4.0	1.8	5.8	40	3.4	2.07	1.4	3.00
Uttar Pradesh	34.5	10.0	44.5	20	35.5	17.92	17.6	9.00
West Bengal	19.6	3.4	23.0	30	16.1	0.36	15.7	0.10
Total India	**299.8**	**44.3**	**344.1**	**30**	**218.8**	**46.76**	**170.8**	**27.00**

N.A. — Not available.

FIGURE 3-1. INDIA — RIVER BASINS AND PROPOSED GANGA-CAUVERY LINK

(Source: Chaturvedi, Indian Inst. of Technology, 1973)

[For data on river basins see Table 3-42]

TABLE 3-42. INDIA–WATER, LAND AND POPULATION DATA BY RIVER BASIN

(Source: Chaturvedi, Indian Inst. of Technology, 1973)

No.	River Basin (see map on page 203)	Water Resources (Million Cubic Meters)					Population			Land			Per Unit Resources Availability			Resources Developed	
		Potential		Utilized													
		Surface runoff	Ground water	Surface runoff	Ground water	Total	Total (millions)	Density)km2	% Rural Population	Culturable (1,000 Ha)	Net irrigated (1,000 Ha)	Cropping intensity	Culturable land per capita (9/6) (Ha)	Water resource per capita (1/6) (m3)	Water resource/culturable land (13/12) (m)	% Gross water resource (5/1)	% Culturable irrigated (10/9)
		1	2	3	4	5	6	7	8	9	10	11	12	13	14	15	16
1	Indus	76,907	11,109	46,048	8,515	54,563	24.63	77	68.7	9,638	3,302	1.436	0.3913	3,123	0.792	71.0	34.3
2	Ganga	509,760	82,698	132,421	40,955	173,376	221.19	457	91.8	60,161	11,356	1.160	0.2720	2,305	0.848	34.0	18.9
3	Brahmaputra	499,914	20,803	6,058	220	6,278	17.65	94	91.1	12,146	691	1.000	0.6880	28,324	4.132	1.3	5.7
4	Luni and River of Saurashtra and Kutch	12,278	4,937	3,812	8,389	12,201	21.82	68	74.5	23,447	1,093	1.112	1.0749	563	0.052	99.4	4.7
5	Sabarmati	3,663	2,467	1,470	1,826	3,296	4.77	220	81.3	1,548	216	1.041	0.3245	768	0.237	90.0	14.0
6	Mahi	7,702	2,467	3,830	1,083	4,913	5.03	144	86.1	2,210	135	1.100	0.4413	1,531	0.450	63.8	6.1
7	Narmada	44,331	4,937	2,786	1,372	4,158	10.60	107	81.0	5,901	202	1.060	0.5095	4,182	0.821	9.4	3.4
8	Tapi	17,982	5,184	5,383	1,545	6,928	8.75	134	78.3	4,417	178	1.320	0.5037	2,055	0.408	37.7	4.0
9	East flowing North of Godavari	32,024	5,184	11,756	299	12,055	15.37	203	85.8	4,603	840	1.196	0.2994	2,084	0.692	38.8	18.2
10	Brahmani Baitarni	27,363	5,184	4,320	136	4,456	7.77	150	88.4	3,201	310	1.442	0.4120	3,058	0.744	18.8	9.7
11	Mahanadi	66,644	12,343	21,287	277	21,564	17.80	126	90.0	7,994	1,044	1.228	0.4474	3,744	0.837	32.4	13.1
12	Godavari	117,997	14,194	32,732	5,389	38,121	35.46	113	85.5	18,931	1,618	1.240	0.5336	3,328	0.624	32.3	8.5
13	Krishna	62,784	9,628	52,437	6,513	58,950	38.50	149	80.9	20,299	1,819	1.160	0.5275	1,631	0.312	93.9	9.0
14	Pennar	6,858	2,374	5,031	1,390	6,421	6.78	123	78.0	3,551	358	1.327	0.5238	1,011	0.194	93.65	10.08
15	Cauveri	18,601	5,801	17,930	2,557	20,487	21.60	246	78.0	5,797	1,255	1.247	0.2684	861	0.321	110.1	21.7
16	West flowing	217,894	13,577	13,467	N.A.	13,467	40.55	362	76.6	6,279	661	1.325	0.1548	5,373	3.460	6.2	10.5

Note: Groundwater data have not been accounted in total resource availability.
N.A. – Not available.

TABLE 3-43. INDIA—MAJOR STORAGE RESERVOIRS
(Source: Ministry of Irrigation and Power, 1972)

River Systems/Storage Site	Actual Storage	
	10^6 m^3	10^6 Acre-ft
Storage Reservoirs of more than 2,500 million m^3		
1. Bhakra	7,450	6.04
2. Pong	6,970	5.65
3. Rihand	8,980	7.28
4. Grandhisagar	6,900	5.60
5. Hirakud	5,830	4.73
6. Nagarjunasagar	7,730	6.27
7. Pochampad	3,170	2.57
8. Ukai	7,100	5.76
9. Srisailam	5,090	4.13
10. Sharavathi	6,540	5.30
11. Koyna	2,690	2.18
12. Tungabhadra	3,710	3.01
13. Mettur	2,660	2.16
14. Balimela	2,840	2.30
Storage Reservoirs of more than 1,250 million m^3, but less than 2,500 million m^3		
1. Bhadra	1,790	1.45
2. Kadana	1,220	0.99
3. Rana Pratapsagar	1,590	1.29
4. Mahi Bajajsagar	2,010	1.63
5. Hidkal (Ghataprabha)	1,420	1.15
6. Krishnarajasagar	1,250	1.01
7. Jayakwadi	2,070	1.68
8. Bhima	1,700	1.38
9. Tawa	2,100	1.70
10. Iddiki	1,470	1.19
11. Maithon	1,360	1.10
12. Panchet	1,330	1.08
13. Ramganga	2,210	1.79

TABLE 3-44. INDIA—FLOOD DAMAGE

(Source: Kumra, Water for Peace, 1967)

[Average annual flood damage, period 1955-67]

	Brahmaputra River Region	Ganga River Region	North-West River Region	Central India and Deccan River Region	Total for the country
Total area (in million acres)	79	237.0	120.0	373.0	809
Total cropped area (in million acres)	14.00	113.00	33.00	174.0	334.0
Area affected by floods (million acres)	3.0	8.0	2.00	3.00	16.00
Crop area affected (million acres)	0.50	3.0	1.00	1.0	5.5
Value of crops damaged (million Rs.)	63.00	240.00	78.00	49.00	430.00
Damage to property and public utilities (million Rs.)	18.00	55.00	27.0	40.00	140.00
Total direct damage (million Rs.)	82.00	297.00	107.00	84.00	570.00

TABLE 3-45. INDIA—MUNICIPAL WATER SUPPLY SYSTEMS

(Source: Internat. Statistical Institute, 1971)

City	Year	Population served	Quantity available (1,000 m^3) Production	Quantity available (1,000 m^3) Purchased from other plants	Quantity available (1,000 m^3) Total	Annual consumption(1,000m^3) Total	Annual consumption(1,000m^3) of which domestic
Ahmedabad (metrop.)	1967	1,574,743	110,843	0	110,843	110,843	N.A.
	1969	1,672,507	106,667	0	106,667	106,667	N.A.
Bangalore	1967	1,441,000	46,720	N.A.	46,720	33,580	16,790
	1969	1,529,000	46,720	N.A.	46,720	33,580	16,790
Bombay	1967	4,903,000	343,129	0	343,129	343,129	186,374
	1969	5,534,000	333,673	0	333,673	333,673	202,896
Kanpur	1967	N.A.	71,313	0	71,313	71,313	42,909
	1969	1,144,000	74,377	0	74,377	74,377	43,774
Madras	1967	2,000,000	61,522	0	61,522	61,522	50,338
	1969	2,010,000	61,018	0	61,018	61,018	50,855
New Delhi	1966/67	314,377	519,149	0	519,149	519,149	454,596
	1968/69	N.A.	610,977	0	610,977	610,977	535,059
Poona	1967	671,000	1,253	30,397	31,650	31,650	N.A.
	1969	718,000	102	32,161	32,263	32,263	N.A.

N.A. — Not available.

TABLE 3-46. INDONESIA—DISCHARGE OF PRINCIPAL RIVERS

(Source: United Nations, 1968)

River	Province	Basin area km^2	Length km	Discharge m^3/s
Tjimanuk	West Java	3,650.2	150	1,300
Tjitarum	"	5,969.0	225	2,400
Tjitanduj	"	3,561.1	120	1,595
Tjiudjung	"	2,045.4	90	1,000
Bekasi	"	1,451.9	70	—
Tjipunagara	"	1,492.9	75	1,073
Pemali	Central Java	1,391.5	65	240
Serang	"	4,826.5	125	—
Djuana	"	1,328.0	80	—
Seraju	"	3,737.4	110	166
Progo	"	2,476.5	100	345
Opak	"	1,679.5	50	—
Bengawan Sala	East Java	15,440.0	350	7,596
Brantas	"	2,176.2	275	810
Bondojudo	"	1,820.1	35	—
Sampejan	"	1,359.4	95	—
W. Sekampung	Lampung	4,939.0	140	—
W. Seputih	"	7,061.6	160	—
Asahan	West Sumatra	7,477.3	100	—
Kapuas	West Kalimantan	—	1,010	—
Barito	South Kalimantan	—	650	—
Luwu	Sulawesi	—	—	—
Djeneberang	"	—	60	—
Tukad Badung	Bali	—	50	—
Tukad Buleleng	"	—	—	—

— Not available.

TABLE 3-47. INDONESIA—PEAK DISCHARGE OF SELECTED RIVERS

(Source: UN Water Resources Series No. 28, 1966)

River	Drainage area at gaging station km^2	Peak discharge m^3/s
Tjisadane — Masing	129	335
Tjiliwung — Rawajati Djatinegara	318	100
Tjitarum — Palumbon	4,150	2,930
Tjimanuk — Tjidjeundjing	1,608	910
Kali Tuntang — Padasmalang	291	149
Kali Tanggulangin	4,968	181
Tjilaki - Tjiheulang	163	188
Tjibuni — Tanggeung	289	640
Kali Brantas — Sengguruh	1,618	920
Asahan	4,000	237
Musi	55,584	1,400
Sakampaeng	4,839	460
Brantas	8,900	1,390
Marangin	3,900	3,130

TABLE 3-48. INDONESIA—RAINFALL AND RIVER DISCHARGE ON JAVA

(Source: Tjan, Water for Peace, 1967)

[Mean monthly values in millimeters]

River: Province: Drainage area:	Tjiudjung West Java 1,418 km²		Pemali Central Java 863 km²		Tjomal Central Java 541 km²		Brantas East Java 897 km²		Sampean East Java 1,196 km²	
	rain- fall	dis- charge	rain- fall	dis- charge	rain- fall	dis- charge	rain- fall	dis- charge	rain- fall	dis- charge
January	410	345	474	329	620	406	462	89	372	100
February	351	255	493	381	685	606	380	97	363	126
March	287	219	467	349	646	563	264	92	443	145
April	373	244	319	195	383	331	127	60	198	92
May	202	165	166	107	340	273	66	47	131	83
June	176	108	187	89	344	251	70	50	48	70
July	159	66	122	92	214	178	37	53	22	65
August	146	58	59	44	200	150	3	36	20	51
September	188	79	65	30	151	112	11	30	40	47
October	307	129	150	42	320	179	87	30	92	47
November	278	141	250	83	423	250	233	33	182	53
December	389	204	437	205	508	285	233	30	327	74
Year	3,266	2,002	3,179	1,946	4,834	3,584	1,973	647	2,238	958
Water losses evapotrans- piration	1,264		1,233		1,250		1,326		1,280	

TABLE 3-49. INDONESIA—PRINCIPAL LAKES

(Source: UN Water Resources Series No. 38, 1968)

Lake	Province	Surface area, km²
Toba	North Sumatra	1,148.0
Manindjau	West Sumatra	98.0
Singkarak	West Sumatra	110.0
Kerintji	West Sumatra	121.9
Ranau	Lampung	43.8
Belida	West Kalimantan	117.5
Kanahan Djempung	East Kalimantan	225.0
Prion	East Kalimantan	548.5
Poso	Central Sulawesi	281.3
Tempe	South Sulawesi	46.9
Towuti	South Sulawesi	578.1
Batur	Bali	—

TABLE 3-50. INDONESIA—MAJOR DAMS AND RESERVOIRS

(Source: UN Water Resources Series No. 38, 1968)

Dam	Province	Volume of dam million m³	Reservoir capacity million m³	Status
Djatiluhur	West Java	10.00	3,000	Completed 1965
Riam Kanan	South Kalimantan	0.65	1,200	Under construction
Karangkates	East Java	6.00	343	''
Seloredjo	East Java	1.30	62	''
Sempor	Central Java	0.50	36	''

TABLE 3-51. INDONESIA—HYDRO-ELECTRIC POWER PLANTS
(Source: UN Water Resources Series No. 38, 1968)

Location	River, lake or reservoir project	Province	Installed capacity kW
EXISTING			
Tarutung	River	North Sumatra	120
Tes	"	South Sumatra	1,320
Sungai Penuh	"	West Sumatra	70
Tonsea Lama	Lake	North Sulawesi	4,440
Mendalan	River	East Java	23,000
Siman	"	"	10,000
Sengguruh	"	"	2,970
Giringan	"	"	3,200
Golang	"	"	2,700
Klontjing	"	"	52
Djolok	Lake	Central Java	20,480
Susukan	"	"	2,400
Timo	"	"	12,000
Ketenger	River	"	7,040
Wonosobo	"	"	124
Bandjarnegara	"	"	256
Parakankendang	"	West Java	10,000
Plengan	Lake	"	5,150
Lamadjan	"	"	19,200
Tjikalong	"	"	19,200
Bengkok/Dago	River	"	3,850
Tjidjedil	"	"	552
Ubrug	"	"	17,100
Kratjak	"	"	16,575
Djatiluhur	Reservoir	"	100,000
Total			282,599
UNDER CONSTRUCTION			
Asahan	Lake	North Sumatra	160,000
Batang Agam	River	West Sumatra	10,000
Riam Kanan	Reservoir	South Kalimantan	30,000
Sawito	River	South Sulawesi	550
Garung	"	Central Java	10,000
Ngebel	Lake	East Java	2,250
Total			212,800

TABLE 3-52. INDONESIA—MUNICIPAL WATER SUPPLY SYSTEMS
(Source: UN Water Resources Series No. 38, 1968)

[Cities with public water supply systems of 100 l/s capacity and over]

City	Province	System capacity l/s	Remarks
Medan	North Sumatra	500	In operation
Padang	West Sumatra	250	Under construction
Palembang	South Sumatra	300	In operation
Djakarta	Djawa	2,574	In operation
		3,000	Under construction
Bandung	West Java	865	In operation
Tjirebon	West Java	105	In operation
Semarang	Central Java	735	In operation
Jogjakarta	Central Java	200	In operation
Surabaja	East Java	540	In operation
		1,000	Under construction
Pontianak	West Kalimantan	110	In operation
Menado	North Sulawesi	30	In operation
		125	Under construction
Makasar	South Sulawesi	150	In operation

Note: In 1970, 198 cities and towns were served by piped water. Total capacity of public water supply systems was 10,425 l/s; only 20 percent of urban population had access to piped supplies.

TABLE 3-53. IRAN—DISCHARGE OF PRINCIPAL RIVERS
(Source: UN Water Resources Series No. 38, 1968)
[Period of record 1956-61]

Name of rivers	Name of stations	Mean discharge m³/s	Name of rivers	Name of stations	Mean discharge m³/s
CASPIAN SEA BASIN			**LAKE REZAYEH BASIN**		
Gorgan	Gonbad Gabus	3.726	Tajiyarsarab	Asbgharan	1.276
Gorgan	Pahlavi Dej	7.336	Aji Chai	Vaniar	1.089
Tajan	Solaiman Tangeh	6.637	Lighvan	Lighvan	0.658
Tajan	Righcheshmeh	13.050	Zarinehrud	Sarighamish	40.060
Talar	Kasilian	4.807	Siminehrud	Dashband	10.610
Talar	Kia Kola	10.950	Baranduzchai	Babarud	5.510
Talar	Shirgah	8.850	Shahrchai	Band Rezayeh	6.310
Babol	Garantalar	8.460	Nazluchai	Tapik	10.420
Babol	Bahol	20.476			
Lar	Pulur	10.670			
Heraz	Karesang	33.970			
Chalus	Polizoghal	12.646			
Samush	Heratbar	1.970			
Gharangu	Miyaneh	10.598			
Ghizil Uzan	Polidokhtar	35.796			
Ai Doghmush	Miyaneh	2.880			
Sefid Rud	Rudbar	108.456			
Shahrud	Lushan	30.130			
PERSIAN GULF BASIN			**CLOSED BASIN OF CENTRAL IRAN**		
Malayer	H.A. Moradian	3.124	Vafragan	B. Shahabbasi	10.970
Nahavand	Gusheh Vaghas	3.528	Ghom	Abbasabad	5.900
Tuesargan	Firuzabad	1.365	Kordan	Dehsomaeh	3.370
Khuramrud	Aran	4.112	Karaj	Polikhab	10.870
Dinavar	Bisetun	8.380	Karaj	Bilegan	14.058
Gamasiab	Polichehr	31.806	Jajerud	Hajiabad	5.550
Gharasu	Doab	5.018	Jajerud	Latian	7.526
Razavar	Polichubi	8.272	Hablehrud	Bonehkuh	7.228
Gharasu	Polikohneh	16.378	Zayandehrud	P. Zamankhan	28.810
Gharasu	Giharbagestan	16.766	Zayandehrud	P. Qalleh	27.670
Khurramabad	Chamanjir	9.380	Zayandehrud	P. Mazraeh	26.570
Kashgan	Afrineh	35.440	Zayandehrud	P. Marnan	12.468
Chulhul	Afrineh	2.180	Zayandehrud	P. Khaju	12.680
Kashgan	Polidokhtar	4.190	Zayandehrud	P. Varzaneh	3.390
Karkheh	Paypol	155.967	Kor	Durudzan	23.310
Karkheh	Hamidieh	133.302	Bar	Arieh	0.345
Karun	Gotwand	325.866			
Marboreh	Dorud	8.460			
Abe Tireh	Dorud	14.660			
Darrehtakht	Darrehtakht	1.208			
Abe Sabzeh	Chamchit	6.100			
Abe Sesar	Sefiddasht	44.216	**GHARA GHUM RIVER BASIN**		
Abe Zaz	Sefiddasht	14.570			
Abebakhtiari	Tangeh Panj	119.688			
Abe Dez	Talehzang	232.200	Turugh	Kertian	0.380
Dez	Dezful	211.610	Kashafrud	Aghdarband	2.410
Marun	Behbehan	32.650	Durungar	Sangesurakh	0.858
Jarrahi	Khalafabad	41.600			
Zohreh	Gachsaran	37.317			

FIGURE 3-2. IRAN — CROSS SECTION AND PLAN OF GHANAT WATER SYSTEM

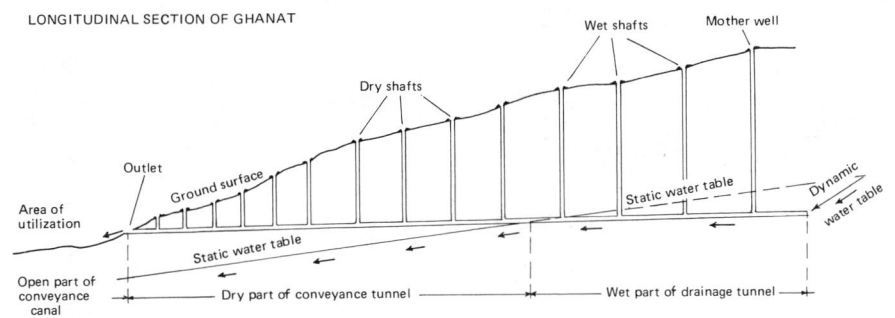

LONGITUDINAL SECTION OF GHANAT

CROSS SECTION OF DRY TUNNEL

CROSS SECTION OF WET TUNNEL

PLAN OF GHANAT

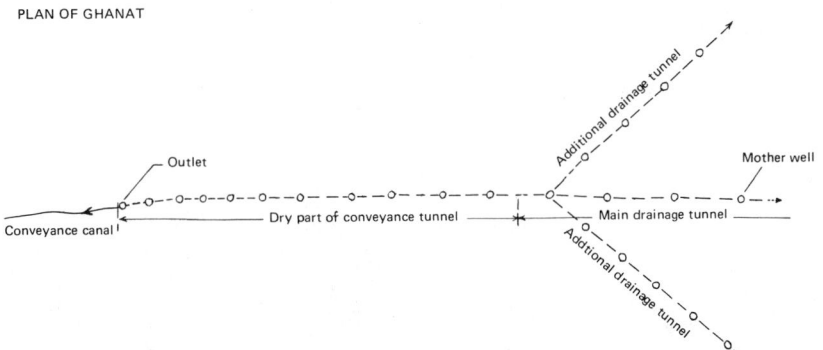

TABLE 3-54. IRAN—SUMMARY OF GROUND WATER PRODUCTION
(Source: UN Water Resources Series No. 38, 1968)

Method	Total number (approximative)	Total discharge in million m³/year	Cost of 1 m³ Rial (US Cent)	Discharge per unit l/s	Executed in 1961 - 1965
Ghanat	25,000	12,000	< 0.5 Rials (0.67)	10 to 300	few
Deep Well	3,000	1,800	0.5 to 1 R. (0.67 to 1.34)	20 to 200	>1,200
Shallow Well	>10,000	2,000	0.25 to 0.5 (0.67 to 0.34)	5 to 20	5,000

Total ground water discharge in Iran: 16,000 million m³/year or 533 m³/s.

TABLE 3-55. IRAN–MAJOR DAMS AND RESERVOIRS

(Source: Vahidi, Ministry of Water and Power, 1973)

Name of Dam	River	Location	Type	Length of dam crest m	Height above foundation m	Total reservoir capacity 106m3	Annual volume of regulated water 106m3	Area of land to be irrigated ha	Total generating capacity KW	Year of completion
EXISTING										
Shah Esmail	Golpayegan	Akhtakhan	Rockfill	360	56	44.4	65	5,000	–	1957
Amir Kabir	Karadj	Karadj	Concrete thin arch	390	180	205	400	21,000	2x45,000	1961
Mohamad Reza Shah Pahlavi	Dez	Dezfull	Concrete thin arch	212	203	3,340	6,940	96,300	8x65,000	1962
Shahbanou Farah	Sefid Rud	Mangil	Concrete Butress	425	106	1,800	2,000	240,000	5x17,600	1962
Shahnaz Pahlavi	Ab Shineh	Hamadan	Concrete Butress	286	53	8	17	200	–	1963
Farahnaz Pahlavi	Jaj Rud	Latiyan	Concrete Butress	450	107	95	245	30,000	2x22,500	1967
Kurosh Kabir	Zarineh Rud	Bokan	Earthfill	720	50	650	535	95,000	2x5,000	1970
Shahpoor Aval	Mahabad	Mahabad	Rockfill	700	46	230	333	20,000	2x2,880	1970
Shah Abbass Kabir	Zayandeh Rud	Isfahan	Concrete arch	450	100	1,250	900	95,000	3x18,400	1970
Aras	Aras	Ghezel Gheshlagh	Earthfill	945	38	1,350	1,400	90,000	2x11,000	1970
Voshmgir	Gorgan	Gorgan	Earthfill	430	19	79	225	20,000	–	1970
Daryush Kabir	Kur	Dorudzan	Earthfill	700	60	993	526	7,700	–	1972
UNDER CONSTRUCTION										
Reza Shah Kabir	Karun	Masjed Solaiman	Concrete thin arch	380	200	2,900	9,260	38,000	4x250,000	1974
Minab	Minab	Minab	Concrete thin arch	450	59	344	236	14,000	–	1975
Jiroft	Halil Rud	Tang Narab	Concrete thin arch	250	130	440	230	10,500	2x15,000	1975
Lar	Lar	Pollur	Earthfill	1,500	105	960	300	65,000	140,000	1978
PROJECTED										
Nader Shah	Marun	Tang Takab	Rockfill	320	172	1,620	1,150	55,000	–	1978
Taleghan	Shah Rud	Sangban	Earthfill	1,000	70	208	450	56,000	76,000	1978
Saveh	Vafraghan	Saveh	Concrete thin arch	262	88	290	328	20,500	2,000	1978
Chah Nime	Hirmand	Chah Nime	Earthfill		17	648	–	45,000	–	1977

TABLE 5-56. IRAN—SOURCES OF MUNICIPAL WATER SUPPLIES

(Source: Amouzegar, Water for Peace, 1967)

Name of Ostan or Province	No. of Cities	Wells			River	Ghanat	Spring
		Deep	Semi-deep	Dug shallow well			
Ostans							
Markazy	29	18	11				3
Gilan	21	16	6				1
Mazandaran	26	22	28				3
Azabaijane Shargi	24	28		6		1	2
Azabaijane Gharbi	13	8		2			4
Kermanshahan	12	3	1			2	4
Khuzestan	12	3			9		
Fars and Kohkiloye	18	10	23				3
Kerman	18	25			1	2	
Khorasan	36	40	11			3	3
Esfahan and Yazd	25	21	2	7			1
Sistan and Balochestan	4	7					
Kordestan	5	2		2		1	
Banader and Jazayer Jabob		3	12		1		
Provinces							
Lorestan	16	3	6				1
Ilam	12			5	4		
Hamedan	10	5		1			
Total	281	214	100	23	15		25

TABLE 3-57. IRAQ–DISCHARGE OF TIGRIS AND EUPHRATES RIVERS

(Source: UNESCO, 1971)

River and station	Basin area km2	Mean monthly discharge, m3/s												Year	Period of record
		Jan.	Feb.	Mar.	Apr.	May	Jun.	Jul.	Aug.	Sep.	Oct.	Nov.	Dec.		
Tigris, Mosul	54,900	549	780	1,100	1,610	1,500	722	314	180	145	176	266	377	643	1931-66
Tigris, Fatha	107,600	1,080	1,640	2,290	3,210	2,940	1,540	737	443	356	363	539	727	1,320	1931-66
Tigris, Baghdad	134,000	952	1,400	2,020	2,720	2,680	1,480	700	384	286	293	430	629	1,160	1931-66
Euphrates, Hit D.S.	264,100	677	812	1,140	2,140	2,380	1,250	545	322	271	328	453	570	906	1932-66
Euphrates, D.S. Hindiya B.	274,100	437	505	747	1,400	1,750	934	360	205	160	176	209	312	600	1931-66

TABLE 3-58. IRAQ–AVAILABLE SURFACE WATER RESOURCES

(Source: Ubell, K., 1971, The Institute for Applied Research on Natural Resources, Iraq, UNESCO, Paris)

River and station	Multi-annual average discharge (m3/s)				Potential water resources
	1940-49	1950-59	1960-69	1940-69	
Euphrates River					
at Hit	1,004	871	1,151	1,009	1,009
at Hindiya	730	548	632	637	
Tigris River					
at Mosul	697	659	853	736	736
at Fatha	1,498	1,297	1,521	1,439	
at Baghdad	1,367	1,138	1,076	1,194	
at Kut	1,279	979	810	1,023	
Khazir River					
at Manquba	30	28	30	30	30
Greater Zab River					
at Eski Kelek	461	414	500	458	458
Lesser Zab River					
at Altun Kupri	225	239	270	245	245
Adhaim River					
at Injana	23	23	23	23	23
Diyala River					
at discharge site	180	192	162	178	178
Total Iraq					2,679

TABLE 3-59. IRAQ—SURFACE WATER USE, 1960-69

(Source: Ubell, K., 1971, The Institute for Applied Research on Natural Resources, Iraq, UNESCO, Paris)

River	Mean annual water use Million m^3/year
Euphrates between Hit and Hindiya	17,213
Tigris and tributaries between Mosul and Fatha	4,190
Tigris between Fatha and Baghdad	14,052
Diyala	5,139
Tigris between Baghdad and Kut	8,614
Total	49,208

TABLE 3-60. IRAQ—MUNICIPAL WATER SUPPLY SYSTEM OF BAGHDAD

(Source: Internat. Statistical Institute, 1971)

Year	Population served	Annual water consumption 1,000 m^3
1967	1,884,151	97,070
1969	2,016,438	126,590

TABLE 3-61. ISRAEL—DISCHARGE OF PRINCIPAL RIVERS

(Source: Israel Hydrological Service, 1971)

River and station	Basin area km²	Mean monthly discharge, 1,000 m³												Max flow m³/s	Date	Period of record
		Oct.	Nov.	Dec.	Jan.	Feb.	Mar.	Apr.	May	Jun.	Jul.	Aug.	Sep.			
Mediterranean Drainage																
N. Keziv at Gesher ha'Ziv	131	0	14	92	400	406	81	1	0	0	0	0	0	37	1-26-63	1944-67
N. Bet ha'Emeq at Shavei Tsiyon	72	0	20	281	342	342	130	68	0	0	0	0	0	19	12-18-51	1944-59
N. Qishon at Quarry	685	29	695	2,514	4,153	3,705	2,378	969	189	46	20	14	8	200	1-29-62	1953-67
N. Tsippori at Tel'Alil	220	25	39	104	475	1,325	834	431	253	127	64	39	27	17	12-22-61	1944-51
N. Daliya at Bat Shelomo	42	0	176	1,193	1,812	1,743	822	104	4	0	0	0	0	28	12-13-61	1955-67
N. Daliya at Tel Aviv-Haifa Rd.	70	0	146	1,053	2,191	1,763	835	78	0	0	0	0	0	23	1-13-50	1950-67
N. Tanninim on Zikhron Ya'aqov-Giv at Ada Rd.	65	0	116	1,403	2,247	2,307	1,341	426	157	73	36	18	0	46	12-14-61	1955-67
N. Tanninim at Northern Roman Bridge	99	4,640	4,700	5,382	6,195	6,468	6,718	5,701	5,323	4,796	4,635	4,490	4,318	850	1959/60	1955-67
N. 'Ada on Binyamina-Pardes Hanna Road	60	0	17	2,047	2,634	1,978	1,121	513	193	45	5	6	3	24	12-13-61	1955-62
N. 'Ada at Southern Roman Bridge	94	364	512	935	1,393	1,756	1,421	874	615	467	351	318	291	—	—	—
N. Hadera at Gan Shemu'el	576	17	2,196	2,740	1,807	1,698	889	459	0	0	0	0	0	145	12-22-61	1960-67
N. Alexander at Ma'abarot	518	117	614	2,992	2,962	889	440	9	0	0	0	0	0	260	1-30-58	1938-67
N. Yarqon at Railway Bridge	1)	18,537	17,961	18,603	18,600	16,923	18,868	18,733	18,631	17,679	18,121	17,904	17,420	8.7	2-18-47	1941-55
N. Yarqon at Herzliya Rd. Bridge	953	865	1,906	5,305	4,861	2,485	1,117	28	0	0	0	0	0	508	11- 9-55	1940-67
N. Qana at Yarhiv	244	0	11	210	85	210	80	12	0	0	0	0	0	78	2-17-46	1960-67
N. Shillo at Nahshonim	357	10	46	930	1,320	699	576	46	0	0	0	0	0	104	1-29-67	1960-67
N. Natuf near Lod	251	0	36	585	184	148	30	1	0	0	0	0	0	340	12-21-51	1953-67
N. Ayalon at Bet Dagan-Yehud Road	802	566	1,945	4,851	3,372	2,588	1,407	28	0	0	0	0	0	319	12-24-51	1938-55

Station														Max	Date	Period
N. Soreq at Hartuv	202	0	0	6	81	244	347	6	0	0	0	0	0	127	12-25-41	1955-60
N. Soreq at Railway Bridge	401	0	80	384	470	576	329	14	0	0	0	0	0	103	12-25-41	1955-66
N. Soreq near Gedera	492	0	197	2,147	1,605	845	371	8	0	0	0	0	0	63	1-16-50	1955-67
N. Soreq at Yavne	515	0	369	1,078	561	263	493	31	0	0	0	0	0	32	12- 7-55	1953-58
N. Gamli'el on Coastal Road	73	0	126	2,491	2,021	1,369	231	6	0	0	0	0	0	93	12-11-64	1960-67
N. Lakhish near Ein Tsurim	281	0	161	1,027	766	276	105	1	0	0	0	0	0	42	1-11-52	1949-60
N. Lakhish on Yavne-Ashqelon Road	992	1	65	3,733	3,826	992	339	15	0	0	0	0	0	180	12-19-66	1955-67
N. Guvrin near Shafir	204	2	6	240	375	118	46	2	2	0	0	0	0	36	12-19-66	1949-67
N. ha'Ela at Kefar Ahim	328	0	0	386	307	140	72	4	1	0	0	0	0	25	1-11-52	1950-63
N. Adoraiym at Railway Bridge	211	4	48	596	1,547	377	85	25	0	0	0	0	0	130	12- 2-63	1954-67
N. Shigma at Beror Hayil	382	0	73	639	397	717	157	6	0	0	0	0	0	110	2- 8-61	1954-67
N. Be'er Sheva at Be'er Sheva	1,090	0	848	3,944	1,757	2,298	534	650	185	0	0	0	0	560	12- 2-63	1943-64
Eastern Drainage																
N. Senir at Ma'ayan at Barukh	629	3,663	3,000	7,051	19,522	27,908	25,824	16,880	10,904	6,705	5,038	4,180	3,787	250	1-29-40	1939-67
N. Hermon at She'ar Yashuv	141	4,440	4,788	8,269	16,446	17,996	18,455	14,094	10,162	7,383	5,773	5,042	4,436	80	2- 4-52	1939-67
N. Dan at Foot Bridge	24 [2]	18,619	17,446	17,994	19,160	19,455	22,101	21,724	22,423	20,638	20,822	19,968	18,256	15	2-12-63	1939-63
Jordan at Benot Ya'aqov Bridge	1,530	25,884	30,624	37,849	67,321	92,429	85,044	61,393	45,354	32,947	26,849	23,731	22,360	96	2- 3-45	1935-55
Upper Jordan Southern Station	1,467	21,284	26,623	40,136	52,720	75,861	72,341	56,588	40,427	25,300	19,379	18,174	19,850	125	1-26-63	1959-67
N. 'Iyon at Metulla	51	28	169	788	1,576	2,336	1,991	990	285	27	0	0	0	14	12-18-51	1949-67

1) Flow consists of base flow due to springs.
2) Bulk of flow is base flow due to springs.

FIGURE 3-3. ISRAEL – WATER SUPPLY NETWORK AND SCHEMATIC DIAGRAM OF NATIONAL WATER RESOURCES SYSTEM

(Source: Jacobs, M. and Litwin, Y., 1973)

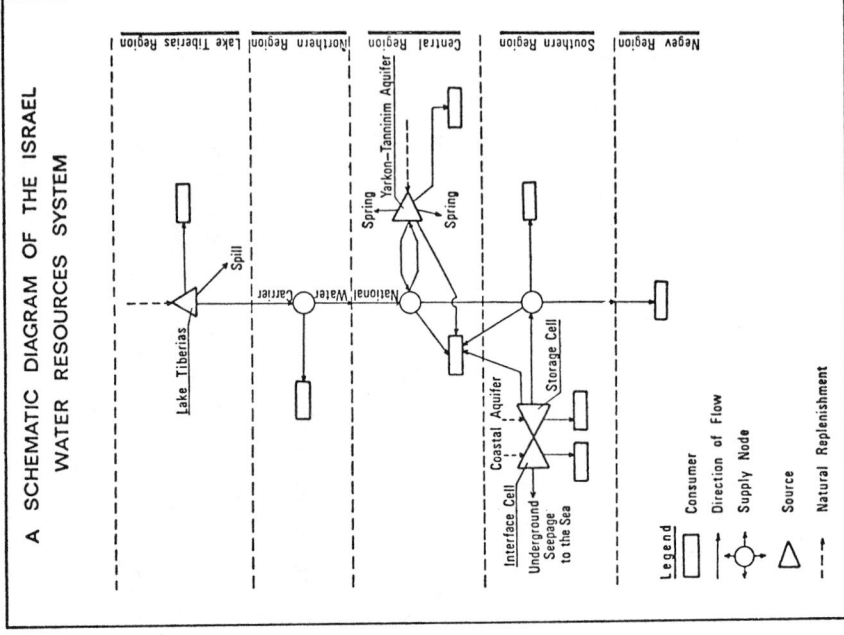

A SCHEMATIC DIAGRAM OF THE ISRAEL WATER RESOURCES SYSTEM

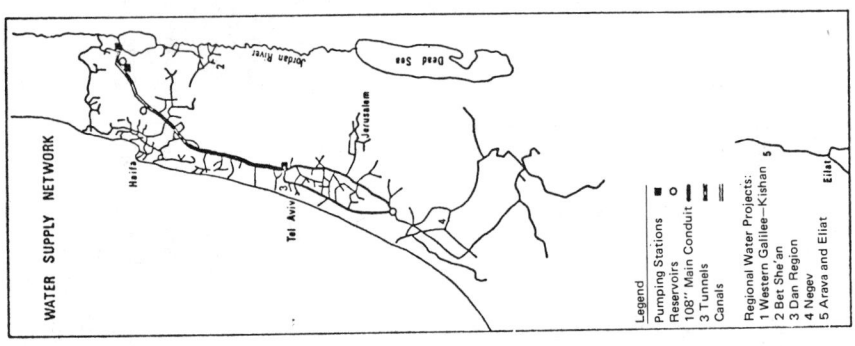

WATER SUPPLY NETWORK

TABLE 3-62. ISRAEL—PRESENT AND FUTURE WATER SOURCES

(Source: Shelef, Israel Inst. of Technology, 1973)

Water Source	Annual Supply				
	1972			1990 (est.)	
	Million cubic meters		Percent	Million cubic meters	Percent
Ground Water		830	56.1	840	44.5
Sand Aquifer	240			230	
Limestone Aquifer	590			610	
Surface Water		610	41.2	650	34.4
Jordan Watershed	570			570	
Flood & Storm Waters	40			80	
Renovated Wastewaters		39	2.6	360	19.0
Domestic	35			300	
Industrial	4			60	
Desalinated Waters [1]		1	0.1	40	2.1
Total		1,480	100.0	1,890	100.0

[1] Including seawater and brackish water desalination.

TABLE 3-63. ISRAEL—ANNUAL WATER USE BY QUALITY AND USE CATEGORY, 1962-69
(Source: Israel Hydrological Service, 1970)

[In thousand cubic meters]

Year	Total Use			Agricultural			Domestic			Industrial		
	Total water	Fresh water	Saline water	Total water	Fresh water	Saline water	Total water	Fresh water	Saline water	Total water	Fresh water	Saline water
1962	1,373,101	1,271,535	101,566	1,144,223	1,058,723	85,500	173,757	173,507	250	55,121	39,305	15,816
1963-64	1,288,369	1,182,005	106,364	1,038,569	953,569	85,000	192,585	191,385	1,200	57,215	37,051	20,164
1964-65	1,328,849	1,228,239	100,610	1,075,367	992,867	82,500	199,053	198,520	533	54,429	36,852	17,577
1965-66	1,418,453	1,302,940	115,513	1,152,917	1,058,612	94,305	206,388	206,202	186	59,148	38,126	21,022
1966-67	1,474,459	1,350,552	123,907	1,202,981	1,103,443	99,538	210,706	210,514	192	60,772	36,595	24,177
1967-68	1,410,613	1,291,658	118,955	1,133,254	1,040,234	93,020	211,380	211,121	259	65,979	40,303	25,676
1968-69	1,536,804	1,422,109	114,695	1,235,359	1,145,076	90,283	231,215	230,994	221	70,230	46,039	24,191

TABLE 3-64. ISRAEL—PRESENT AND FUTURE FRESH WATER DEMAND
(Source: Jacobs, M. and Litwin, Y., 1973)

Year	Water demand, million m³			
	1971	1975	1980	1985
Agriculture	1,170	1,210	1,250	1,250
Domestic	250	300	350	430
Industry	55	100	140	220
Losses	50	50	50	60
Total	1,525	1,660	1,790	1,960

TABLE 3-65. ISRAEL—PRECIPITATION AND GROUND-WATER WITHDRAWAL

(Source: Israel Hydrological Service, 1971)

[Period 1957-67]

| Precipitation, mm | | | | | Pumping 106m3 | | | | | | |
| Mountain area | Coastal Plain | | | Hydro-logical year | Yarqon-Tanninim basin 1) | | | Coastal Plain | | | Year 2) |
Jerusalem	Nitsanim	Tel-Aviv	Hadera		Net pumpage	Artificial recharge	pumpage	Net pumpage	Artificial recharge	pumpage	
342	501	460	629	1957-58	296	—	296	402	—	402	1957
413	405	465	421	1958-59	333	—	333	493	—	493	1958
206	240	316	449	1959-60	314	—	314	431	—	431	1959
476	444	531	488	1960-61	312	—	312	450	9	459	1960
447	271	648	657	1961-62	292	—	292	424	7	431	1961
227	276	422	428	1962-63	302	—	302	478	5	483	1962
689	625	619	622	1963-64	269	—	269	418	13	431	1963-64
639	624	808	666	1964-65	281	11	292	347	33	380	1964-65
359	226	326	436	1965-66	272	9	281	343	38	381	1965-66
753	590	623	684	1966-67	311	15	326	326	86	412	1966-67

Average precipitation

486	453	564	542	1931-1960

1) Pumpage from the Yarqon-Tanninim basin includes the spring flow from Rosh-Ha'ayin till 1962.
2) Pumpage between 1957-1962 refers to the calendar year. From 1963-64 the pumpage refers to the fiscal year (starting April). Precipitation is according to the hydrological year (starting October).

TABLE 3-66. ISRAEL—ARTIFICIAL RECHARGE OF GROUND WATER

(Source: Harpaz, Tahal, 1973)

[Years 1967-68 and 1970-71]

Aquifer	Injection into wells — No. of wells		Injection into wells — Quantity million m3		Surface recharge — Type	Surface recharge — Quantity million m3		Total million m3	
	1967-68	1970-71	1967-68	1970-71		1967-68	1970-71	1967-68	1970-71
Coastal Plain 1) Sandstone aquifer	115	94	68	47	Spreading grounds	26	38	94	85
Judean Hills Limestone-dolomite aquifer	19	17	35	30	Reservoirs	2	3	37	33
Yavne'el Valley Basalt aquifer	1	–	1	–		–	–	1	–
Total	135	111	104	77		28	41	132	118

1) Includes Dan Region Sewage Disposal and Reclamation Project.

TABLE 3-67. ISRAEL—QUALITY OF INJECTED WATER AND OF GROUND WATER

(Source: Harpaz, Tahal, 1973)

[In mg/l, except where otherwise stated]

Constituent	Lake Kinneret after treatment	Source of recharge water — Southern coastal aquifer1)	Source of recharge water — Northern limestone aquifer1)	Source of recharge water — Southern limestone aquifer1)	Source of recharge water — Mixed lake and ground waters	Source of recharge water — Treated Dalya stream water	Source of recharge water — Yavne'el basalt aquifer2)
Cl	355	180	53	184	225	50	60
SO4	57	41	14	49	47	32	14
NO3	1	27	12	1	7	12	32
HCO3	137	335	357	320	181	290	238
Ca	57	51	76	74	57	103	30
Mg	35	42	37	32	30	8	19
K	10	3	–	5	8	1	3
Na	173	134	24	117	112	25	72
Fe	0.05	0.04	0.10	0.16	0.4	0.02	–
TDS	778	685	417	645	643	560	400
Dissolved O2	10.3	6.9	6.0	0.2	9.0	6.7	7.9
CO2	22	29	37	30	9	15	12
pH	7.9	7.2	7.0	7.2	7.4	7.6	7.4
Temperature, in degrees centigrade	17	24	23	25	21	15	24
Turbidity (JU)	2	0	–	0	0	5-10	0
Permanganate demand, mg/l of O2	1.8		–	–	1.3	3.0	0
Coli confirmed (MPN per 100 ml)	less than 2	less than 2	–	less than 2	0	240	–
Algae (ASU)	150-800	0	–	0	486	–	–

1) Also a replenished aquifer.
2) A replenished aquifer only.

TABLE 3-68. ISRAEL—WATER QUALITY OF DEAD SEA

(Source: Israel Hydrological Service, 1971)

[Average ion content and total salinity, in grams/liter]

	Volume km^3	Depth m	Ca^{++}	Mg^{++}	Na$^+$	K$^+$	Cl$^-$	Br$^-$	SO$_4^{--}$	HCO$_3^-$	Salinity (Sum of Ion Concentration)
Average for entire water body	136	0-400	16.86	40.65	39.15	7.26	212.14	5.12	0.47	0.22	322.13

TABLE 3-69. ISRAEL—MUNICIPAL WATER SUPPLY SYSTEMS

(Source: Internat. Statistical Institute, 1971)

City	Year	Population served	Quantity available (1,000 m^3)			Total annual consumption (1,000 m^3)
			Production	Purchased from other plants	Total	
Jerusalem	1967	197,000	11,846	0	11,846	N.A.
	1969	285,000	17,281	0	17,281	N.A.
Tel Aviv-Yafo	1967	388,000	895	51,446	52,341	39,014
	1969	382,900	2,178	46,695	48,873	42,789

N.A. — Not available

TABLE 3-70. JAPAN—DISCHARGE CHARACTERISTICS OF PRINCIPAL RIVERS

(Source: ECAFE—UN Water Resources Series No. 38, 1968)

River	Total drainage area km2	Length of river km	Gaging station	Drainage area at gaging station km2	Discharge, m3/sec							Annual runoff m3×106	Year of record
					Maximum	95 days	185 days	275 days	355 days	Minimum	Average		
Ishikari	14,300	262	Ishikari-Ohashi	12,697	4,566.9	507.5	347.8	249.0	196.5	22.3	491.1	15,496.6	1954-65
Kitakami	10,200	247	Kozenji	7,060	4,073.3	324.6	224.7	174.0	123.1	45.8	296.9	9,369.69	1952-65
Tone	16,840	298	Kurihashi	8,588	10,692.0	286.6	184.2	128.3	88.3	6.1	270.2	8,275.18	1938-65
Shinano	12,050	367	Oijya	9,843	5,996.0	587.7	390.8	297.6	194.3	54.4	503.4	15,885.62	1951-65
Kiso	9,100	193	Unuma	4,684	11,145.0	336.9	204.1	135.7	92.1	49.0	326.9	10,366.5	1951-65
Yodo	8,258	75	Hirakata	7,281	7,800.0	312.1	221.6	165.3	117.4	74.0	306.8	9,681.37	1952-65
Ota	1,690	110	Kumura	1,481	4,267.8	76.1	49.6	34.0	20.3	3.7	73.8	2,326.98	1953-65
Yoshino	3,650	194	Iwatsu	2,768	15,000.0	154.0	81.4	46.4	28.6	3.8	169.8	5,354.73	1953-65
Chikugo	2,860	123	Senoshita	2,315	6,070.0	91.0	57.0	41.9	25.3	0.7	116.1	3,663.60	1950-65

TABLE 3-71. JAPAN—MONTHLY DISCHARGE OF PRINCIPAL RIVERS

(Source: UNESCO, 1971)

River and station	Basin area km2	Mean monthly discharge, m3/s												Year	Period of record
		Jan.	Feb.	Mar.	Apr.	May	Jun.	Jul.	Aug.	Sep.	Oct.	Nov.	Dec.		
Yodo, Hirakata	7,280	230	261	224	250	274	443	577	334	386	296	181	201	305	1952-64
Shinano, Oijya	9,845	300	280	526	964	700	493	688	381	497	470	398	400	504	1951-64
Tone, Kurihashi	8,590	140	154	151	283	263	286	396	436	406	368	213	150	271	1938-64
Chikugo, Senoshita	2,315	51.9	58.0	65.0	101	126	211	302	135	152	85.6	57.0	41.1	116	1950-64
Ishikari, Ishikari-Ohashi	12,700	232	232	313	1,060	746	430	394	503	480	414	348	343	421	1954-64

TABLE 3-72. JAPAN—RAINFALL, RUNOFF AND GROUND-WATER USE

(Source: ECAFE, UN Water Resources Series No. 38, 1968)

Rainfall and runoff

Annual rainfall	1,600 mm
Average annual runoff	400,000,000,000 m3
Equivalent depth	1,100 mm
Average runoff per capita	4,100 m3
Population (1965 Census)	98,300,000

Use of groundwater in 1965

Domestic	....	1,500,000,000 m3
Industrial	....	5,200,000,000 m3
Irrigation	....	1,800,000,000 m3
Total	....	8,500,000,000 m3

TABLE 3-73. JAPAN—SOURCES OF WATER FOR PUBLIC SUPPLY AND IRRIGATION

(Source: ECAFE, UN Water Resources Series No. 3, 1968)

[Year 1966]

	Public Supply		Irrigation	
Source	**Facilities No.**	**Intake million m3**	**Source**	**Percentage**
Surface water	676	4,636 1)	Rivers and lakes	70.6
Riverbed water	499	936	Ponds and reservoirs	16.8
Groundwater	634	114	Groundwater	4.0
Other sources	150	738	Mountain streams	3.7
			Rainwater	4.9
Total	**1,949**	**7,425**	**Total**	**100.0**

1) Including 933 million m3 from reservoirs.

TABLE 3-74. JAPAN—WATER DEMAND, 1965-75
(Source: Nakazawa, Japan Water Resources Div., 1972)

[In million m3/year]

Region		1965				1975			
		Public supply	Indus-trial	Agricul-tural	Total	Public supply	Indus-trial	Agricul-tural	Total
SURFACE WATER									
Hokkaido		310	690	2,630	3,630	1,170	2,710	3,430	7,310
Tohoku		530	860	9,480	10,870	1,330	3,520	11,070	15,920
Kanto	Inland	170	260	3,840	4,270	970	2,100	4,490	7,560
	Coastal	1,900	770	2,350	5,020	6,090	4,210	2,480	12,780
	Total	2,070	1,030	6,190	9,280	7,060	6,310	6,970	20,340
Hokuriku		70	300	2,120	2,490	290	1,140	2,240	3,940
Tokai		410	890	3,280	4,580	1,890	5,110	4,370	11,370
Kinki	Inland	170	240	1,560	1,970	700	800	1,560	3,060
	Coastal	1,030	1,000	1,770	3,800	2,800	3,950	1,840	8,590
	Total	1,200	1,240	3,330	5,770	3,500	4,750	3,400	11,650
Chugoku	Sanin	60	100	990	1,150	120	370	1,110	1,600
	Sanyo	250	960	2,670	3,880	900	2,630	2,820	6,350
	Total	310	1,060	3,660	5,030	1,020	3,000	3,930	7,950
Shikoku		150	530	1,730	2,400	460	1,990	2,130	4,580
Kyushu	Northern Kyushu	260	300	2,840	3,400	1,050	2,260	3,530	6,840
	Southern Kyushu	40	220	2,240	2,500	360	1,490	4,330	6,180
	Total	300	520	5,080	5,900	1,410	3,750	7,860	13,020
Grand Total		5,350	7,110	37,500	49,960	18,130	32,550	45,400	96,080
GROUND WATER		1,480	5,580	12,500	19,600	1,980	6,820	12,930	21,780
Gross national water use		6,830	12,690	50,000	69,520	20,110	39,370	58,380	117,860

TABLE 3-75. JAPAN—INDUSTRIAL WATER USE, 1965-71
(Source: ECAFE, UN Water Resources Series No. 44, 1973)

[in 1,000 m3/day]

Use category	1965	1970 E	1971 E
Boiler feed	1,343 (2.7)	1,920 (2.3)	2,100 (2.2)
Water for materials	703 (1.4)	610 (0.7)	650 (0.7)
Product treatment and cleaning	16,740 (34.0)	20,500 (24.1)	21,500 (22.4)
Cooling	24,497 (49.8)	50,000 (58.8)	58,000 (60.4)
Air conditioning	2,900 (5.9)	4,500 (5.3)	4,900 (5.1)
Other	2,997 (6.2)	7,511 (8.8)	8,750 (8.2)
Total....................	49,162 (100.0)	85,041 (100.0)	95,900 (100.0)

Note: Percentage of total use is given in brackets.
E — Estimated.

TABLE 3-76. JAPAN—INDUSTRIAL WATER USE BY TYPE OF INDUSTRY
(Source: Min. of Health and Welfare, Water for Peace, 1967)

[1,000 m3/day]

Year	Industry					Total Water Use
	Foods	Steel	Chemical	Paper & Pulp	Others	
1956	2,535 (10.9)	2,071 (8.9)	6,728 (28.9)	6,216 (26.7)	5,692 (24.6)	23,243 (100)
1962	2,535 (7.1)	4,313 (12.0)	10,317 (28.7)	9,747 (27.1)	9,020 (25.1)	35,931 (100)
1964	2,936 (6.1)	6,181 (12.9)	16,250 (34.0)	11,575 (24.2)	10,813 (22.8)	47,755 (100)

Note: Figures in parentheses represent percentage of total use.

TABLE 3-77. JAPAN—MAJOR HYDROELECTRIC POWER PLANTS

(Source: ECAFE, UN Water Resources Series No. 32, 1967)

Plants	Type	Capability MW	Max. discharge m3/s	Head m	Dam Type	Dam Height m	Dam Storage capacity 106m3	Annual output 109kWh	Turbine Type	Turbine MW-No. of units	Owner
EXISTING											
Tagokura	S	380	420	105	Grav.	150	370	580	VF	102-4	PDC
Okutadami	S	360	249	170	Grav.	157	458	898	VF	127-3	PDC
Sakuma	S	350	306	133.5	Grav.	155	205	1,387	VF	96-4	PDC
Kurobe-4	S	258	54	560.2	Arch	186	164	1,037	VP	96-2 (95-1)	KPG
Shiroyama	S	250	192	153	R.F.	73	4	526	VFPT	65-4	DPC
Miboro	S	215	130	192.1	R.F.	131	330	733	VF	117-2	DPC
Hitotsuse	S	180	137	152.4	Arch	130	155	428	VF	94-2	KYP
Sinano	R	165	171	109.8	—	—	—	1,271	VF	39-5	TEP
Hatanagi-1	S	137	160	107.7	Hollow	125	80	298	VF	47-1 (44-2)	CEP
Kinugawa	R	127	45	330.4	—	—	—	142	VF	69-2	TEP
Maruyama	S	125	186	80.8	Grav.	88	18	622	VF	70-2	KEP
Wadagawa-2	S	122	32.2	457.4	Grav.	140	200	364	VF	69-2	HEP
Senju	R	120	250	55.8	—	—	—	716	VF	36-5	NRC
Sendaigawa-1	S	120	60	93.1	Grav.	117	77	447	VF	70-2	PDC
UNDER CONSTRUCTION								Target year of completion			
Azumi	S	623	540	135.7	Arch	155	94	1971	VF / VFPT	109-2 / 107-4	TEP
Ikehara	S	340	342	116.5	Arch	116	220	1966	VF / VFPT	80-2 / 105-2	PDC
Takene-1	S	340	300	136	Arch	130	40	1969	VFPT	90-4	CEP
Naruha	S	303	424	84	Arch	104	80	1969	VF / VFPT	78-1 / 77-3	CHP
Yagisawa	S	240	300	93.5	Arch	131	176	1967	VFPT	87-3	TEP
Nagano	S	220	266	96.5	RF	125	223	1968	VFPT	113-2	PDC
Midono	S	245	360	79.7	Arch	95.5	4	1970	VF / VFPT	64-2 / 63-2	TEP
Kiso	S	116	60	225.9	Grav.	36	2	1967	VF	126-1	KEP

VF: Vertical Francis
S: Storage type plant
R: Run-of-river-type plant
KYP: Kyushu Electric Power Co.
NRC: National Railroad Corp.

VP: Vertical Pelton
PDC: Electric Power Development Co.
TEP: Tokyo Electric Power Co.
KPG: Kanagawa Prefectural Government

P.T.: Pumped storage type plant
KEP: Kansai Electric Power Co.
CEP: Chubu Electric Power Co.
R.F.: Rockfill
VFPT: Vertical Francis Pump Turbine

TABLE 3-78. JAPAN—MUNICIPAL WATER SUPPLY SYSTEMS

(Source: Internat. Statistical Institute, 1971)

City	Year	Population served	Quantity available (1,000m^3)			Annual consumption (1,000m^3)	
			Production	Purchased from other plants	Total	Total	of which domestic
Kawasaki	1967	857,906	137,983	0	137,983	115,151	58,878
	1969	905,631	155,069	0	155,069	133,534	72,556
Kyoto	1967	1,336,500	170,196	0	170,196	120,967	N.A.
	1969	1,369,900	186,206	0	186,206	135,021	N.A.
Kitakiushu	1967	978,815	108,789	0	108,789	75,775	75,775
Kobe	1967	1,180,417	50,512	101,663	152,175	107,697	68,623
	1969	1,212,643	44,328	114,703	159,031	122,002	90,567
Nagoja[1]	1967	1,931,956	258,048	2,003	260,051	186,001	109,764
	1969	1,993,103	296,782	2,389	299,171	221,831	130,542
Tokyo (metrop.)	1967	11,048,467	N.A.	N.A.	1,433,839	1,027,784	878,422
	1969	11,347,189	N.A.	N.A.	N.A.	1,216,508	1,034,564
Yokohama	1967	2,007,476	300,191	0	300,191	208,878	93,826
	1969	2,098,040	317,909	0	317,909	226,396	106,635

[1] Including some surrounding districts or suburbs.

N.A.—Not available.

TABLE 3-79. JORDAN—QUALITY OF WATER IN JORDAN VALLEY

(Source: Wilson and Wozab, 1954)

Name	Date of Collection	Temperature °F	pH	Spec. cond. (Micromhos at 25°C)	Total dissolved solids	Sod-Potassium (as Na)	Calcium	Magnesium	Chloride	Carbonate	Bicarbonate	Sulfate	Nitrate	Boron	Percent Sodium	Aquifer
Surface Waters																
Jordan River	Mar. 3, 1954	68	7.3	950	698	128	58	30	240	—	188	54	12	Nil	50.8	
Yarmouk River	Mar. 5, 1954	63	7.7	480	395	44	42	19	55	—	195	32	8	Nil	34.2	
Wadi Arab	Feb. 18,1954	66	7.4	620	584	40	67	35	52	—	305	60	25	Nil	21.8	
Wadi Yabis	Mar. 4, 1954	64	7.5	400	369	18	51	20	30	—	224	6	20		15.6	
Wadi Ziglab	Mar. 26,1954	69	8.2		547	15	75	44	60	26	312	15			8.0	
Wadi Kufrinja	Mar. 4, 1954	62	7.5		321	11	48	18	26	—	195	8	15		11.0	
Wadi Rajib	Mar. 4, 1954	62	7.4		341	10	51	19	24	—	210	12	15		9.6	
Wadi Zerqa	Feb. 18,1954	60	7.4	740	572	55	70	28	98	—	237		40	Nil	29.1	
Wadi Shu'eib	Mar. 22,1954	70	8.3		335	11	53	22	48	12	173	16			9.3	
Wadi Malih	Mar. 23,1954	78	8.0		3,868	814	330	163	1,780	19	317	445			54.2	
Wadi Auja	Mar. 23,1954	64	8.1		325	10	47	23	28	17	195	5			9.0	
Springs																
Ein Maqla	Jan. 12,1954	104			1,139	159	132	55	320	—	350	130	Nil	Nil	38.6	Nubian sandstone facies (?)
Ein Zahar	Feb. 3, 1954	78	7.9	610	577	17	86	32	36	—	371	Tr.	35	Nil	11.0	Belqa series
Ein Tabiqa	Feb. 3, 1954	74	8.0	680	611	36	109	26	64	—	395	36	5	Nil	17.1	Ajlun series
Ein Abu Zayid	Feb. 3, 1954	78	7.9	750	800	73	115	33	60	—	439	68	12		20.0	Belqa series
Zerqa Hot Springs	Feb. 3, 1954	90	7.5	7,500	6,452	1,210	700	82	1,930	—	1,035	1,490	Nil	Nil	55.8	Zerqa Group (?)
Ein Khanezir	Feb. 3, 1954		7.9	3,000	2,630	351	464	84	580	pres	275	1,072	10		33.6	Kurnub sandstone
Ein Bassat el Faras	Feb. 3, 1954	80	7.9		2,792	466	240	142	688	—	356	880	20		46.1	Zerqa group (?)
Hamma el Malih	Mar. 25,1954	99	7.3	3,000	2,303	453	238	79	990	—	268	275			51.8	Albian-Aptian group
Ein el Hamma	Mar. 25,1954	88	7.5	960	845	124	78	43	236	—	314	50			42.1	Albian-Aptian group (?)
Ein el Faria	Jan. 23,1954			580	418	23	77	12	58	4.5	229	16			17.1	Senonian
Ein Makhna	Jan. 23,1954	66		330	329	13	64	7	23	—	202	19			12.8	Senonian
Ein el Fashura	Jan. 23,1954			310	343	25	60	5	26		195	31			23.9	Lower Eocene
Beit el Ma'an	Jan. 23,1954	66		445	385	18	76	9	52		236	5			14.6	Recent (?)
Ras el Ein	Jan. 23,1954	64		320	324	8	68	6	23		213	5			7.8	Lower Eocene
Ein Qudeira	Jan. 23,1954	65		400	356	8	77	8	36	7.5	210	9.1			7.4	Senonian
Ein el Beida	Mar. 25,1954	78	7.5		561	32	75	33	72	—	332	17			17.5	Ghor marls & gravel

Name	Date		pH											Geology
Makhrug	Mar. 25,1954	78	7.3	3,000	300	152	132	880	—	381	22		41.5	Ghor marls & gravel
Ein Fasayil	Mar. 15,1951	71		460	Tr.	36	27	36	—	140	15		Tr.	Cenomanian
Ein Sultan	Mar. 24,1954	70	7.5	400	11	60	19	25	—	244	15		9.8	Senonian (?)
Ein Fawnar	Mar. 25,1954	69	7.5		10	61	16	26	—	234	8	3	9.0	Cenomanian (?)
Ein Qinya	Jan. 23,1954				18	62	17	24	—	268	10		14.4	Albian
Upper Ein Arik	Jan. 23,1954	67		350	14	55	18	27	—	238	10		12.6	Albian
Ein Deir Hajla	Mar. 25,1954	75	7.4	880	137	78	54	160	—	366	75	5	31.8	Ghor marls & gravel
Ein Fashka	Jan. 23,1954	78		4,500	460	209	242	1,660	17	244	54		38.2	Lower Cenomanian
Water wells														
Ghor el Khibid 1	Dec. 13,1950				2,022	404	98	2,855	—	1,129	827		75.6	Ghor marls
Kerameh A	Jan. 1,1951				181	94	79	466	—	274	67		41.2	Ghor marls
Ghor Nimrin 7	Apr. 8, 1951				320	113	101	689	15	305	97		49.7	Ghor marls
Sweima Rd. Bore	Mar. 17,1954	80	8.2	2,350	289	113	119	620	—	292	244	5	45.0	Ghor marls
Marj Naja 2	Mar. 25,1954	84	7.5	1,400	211	81	58	420	—	288	65	5	51.0	Ghor marls
PWD 13	Mar. 15,1951	84			232	94	114	489	—	500	100		41.7	Ghor marls
Hejaz Farm	Mar. 24,1954		7.5		36	66	49	80	—	381	17	2	17.8	Ghor marls
Bassa 2	Mar. 26,1954				151	42	49	128	—	513	30		52.0	Ghor marls

TABLE 3-80. JORDAN—MUNICIPAL WATER SUPPLY SYSTEM OF AMMAN

(Source: Internat. Statistical Institute, 1971)

Year	Population served	Annual water consumption 1,000 m3
1967	320,220	9,300
1969	500,000	11,700

TABLE 3-81. NORTH KOREA—DISCHARGE CHARACTERISTICS OF PRINCIPAL RIVERS
(Source: Marynov, 1960)

River	Station	Discharge, m³/s				Ratio of maximum discharge to minimum discharge
		in normal years		in dry years		
		Maximum discharge	K*	Minimum discharge	K*	
Amnokan	At the Supkhun hydropower plant	814	0.26	242	0.40	3.4
Chanchingan	At 1st dam	33.9	0.28	10.3	0.31	3.3
	At 2nd dam	5.8	0.28	1.8	0.31	3.2
Puchongan	At 1st dam	14.8	0.30	4.8	0.28	3.1
	At 2nd dam	0.9	0.30	0.3	0.28	3.0
Khochongan	At Rentupkhen reservoir	16.6	0.41	5.7	0.36	2.9
	At Hwansuwon reservoir	7.6	0.41	2.6	0.36	2.9
Toknogan	At Toknogan hydropower plant	106.5	0.24	31.1	0.40	3.4

K* — deviation from long-term average.

TABLE 3-82. SOUTH KOREA—RAINFALL, RUNOFF AND WATER USE
(Source: Bo Young Choi, Water for Peace, 1967)

Mean annual precipitation . . . . (two-thirds falls in 30 days in summer)	1,159 mm equivalent to 110,000 million m³
Evapotranspiration and infiltration (36%)	40,000 million m³
Surface water runoff (64%)	70,000 million m³
Flood flow (67%)	47,000 million m³
Ordinary flow (37%)	23,000 million m³
Total water use (1964)	5,100 million m³

TABLE 3-83. SOUTH KOREA—WATER RESOURCES OF THE HAN, NAKTONG, GEUM AND YONGSAN RIVER BASINS
(Source: ECAFE, UN Water Resources Series No. 40, 1971)

River basin	Han	Naktong	Geum	Yongsan
Drainage area (km²)	26,219	23,656	9,886	2,789
Length of river (km)	488	512	401	115
Cultivated area (ha)	368,269	485,700	156,951	100,709
Population (millions)	8.70	5.36	2.93	1.33
Annual rainfall (mm)	1,200	1,106	1,220	1,285
Annual rainfall (million m³)	310	255	120.6	35.9
Annual losses (million m³)	137	107	64.1	18.2
Annual runoff (million m³)	173	148	56.5	17.7
Annual flood runoff (million m³)	117	105	29.2	13.9
Annual sub-surface runoff (million m³)	56	43	27.3	3.8

TABLE 3-84. SOUTH KOREA—DISCHARGE OF PRINCIPAL RIVERS
(Source: UNESCO, 1971)

River and station	Basin area km2	Mean monthly discharge, m3/s												Year	Period of record
		Jan.	Feb.	Mar.	Apr.	May	Jun.	Jul.	Aug.	Sep.	Oct.	Nov.	Dec.		
Han, Indogyo	25,046	51	79	168	355	279	300	2,340	1,150	623	194	135	94	485	1917-64
Nagdong, Sam Nang Jin	22,916	56	83	163	335	265	393	1,110	907	771	178	115	94	375	1924-64

TABLE 3-85. SOUTH KOREA—DAMS AND RESERVOIRS
(Source: ECAFE, UN Water Resources Series No. 32, 1967)

[Dams and reservoirs constructed or planned according to ten-year integrated development program]

Name of river	Name of dam	Type of dam	Function	Height m	Length m	Volume of dam 1,000 m3	Total storage capacity 1,000 m3	Regulation of flood control m3/sec	Irrigation ha	Increase of low flow m3/sec	Max. generating capacity kW	Annual power production kWh	Begun	Completed
Han	Soyang Gang	P.F.I.W.	Rockfill dam	88	420.5	3,200	1,010,000	3,000	33,500	32	90,000	287,600,000	1967	1971
Han	Chung Ju	P.F.I.W.	Concrete gravity dam	76	343	580	1,800,000	5,000	65,000	50	150,000	520,000,000	1967	1971
Han	Tan Yang	P.F.I.W.	Concrete gravity dam	64	435	440	450,000	3,000	13,500	30	86,400	225,000,000	1972	1975
Han	Han Gang	N.I.W.P.	Concrete gravity low submerged dam	24.6	1,360	330	145,500		35,300	34	20,000	113,004,000	1971	1974
Naktong	Nam Gang	F.I.W.P.	Earth dam	22.8	969	1,120	86,000	6,000	9,800	26	10,500	33,510,000	1962	1968
Naktong	An Dong	F.P.I.W.	Rockfill type dam	54.12	540	11,325	473,000	2,000	18,500	21.8	29,700	75,802,000	1967	1970
Naktong	Hyupchon	F.P.I.W.	Concrete gravity dam	71	380	487.5	570,000	1,500	20,000	11.3	37,900	118,526,000	1971	1974
Naktong	Im Ha	F.P.I.W.	Concrete gravity dam	54	265	185	200,000	1,500	11,800	7.8	19,750	52,639,000	1972	1975
Keum	Yong Dam	P.I.W.F.	Concrete gravity dam	54	344	312	520,000	3,000	15,000	17	54,000	255,000,000	1968	1971

F = Flood control
P = Power generation
I = Irrigation
W= Municipal and Industrial Water Supply
N = Inland navigation

TABLE 3-86. SOUTH KOREA—WATER DEMAND IN HAN AND NAKTONG RIVER BASIN, 1968-2001

(Source: ECAFE, UN Water Resources Series No. 40, 1971)

[In million m3]

Year	Han River Basin					Naktong River Basin				
	1968	1971	1981	1991	2001	1968	1971	1976	1981	1986
Municipal water.........	360	505	1,061	1,767	2,776					
Industrial water.........	114	282	1,834	5,108	13,106	Municipal and industrial water........... 90	330	620	870	1,100
Irrigation water.........	845	1,050	1,772	2,043	2,443	Irrigation 1,120	1,220	1,280	1,460	1,550
Total demand	1,319	1,837	4,667	8,918	18,125	Miscellaneous uses...... 1,050	1,050	1,050	1,050	1,050
Surface water.........	988	1,377	3,500	6,688	13,594	Total 2,260	2,600	2,950	3,380	3,700
Ground water.........	331	460	1,167	2,230	4,531					

TABLE 3-87. SOUTH KOREA—GROUND WATER DEVELOPMENT FOR IRRIGATION, 1966-69

(Source: ECAFE, UN Water Resources Series No. 40, 1971)

Year	Wells excavated			Wells drilled			Infiltration galleries			Total	
	Number	Irrigated area (ha)	Cost (million won) [1]	Number	Irrigated area (ha)	Cost (million won)	Number	Irrigated area (ha)	Cost (million won)	Irrigated area (ha)	Cost (million won)
1966	—	—	—	443	1,779	36	4	493	56	2,272	92
1967	—	—	—	228	1,601	32	16	819	86	2,420	118
1968	605	1,700	104	635	2,844	63	33	1,658	144	6,202	311
1969	32,103	95,915	4,146	4,267	18,229	721	3,293	51,850	4,692	165,994	9,559
Total	32,708	97,615	4,250	5,633	24,452	852	3,346	54,820	4,978	176,888	10,080

1) US $1.00 = 300 won.

TABLE 3-88. SOUTH KOREA—INDUSTRIAL WATER SUPPLY, 1969

(Source: ECAFE, UN Water Resources Series No. 40, 1971)

Classification	Installed capacity (m^3/day)	Industrial areas supplied (No.)
Coastal industrial complexes	230,000	4
Inland industrial complexes	80,000	2
Industrial water supply in cities	321,000	—
Private industrial water supply	368,000	—
Total	999,000	

Note: The demand for water for industry is expected to increase rapidly from 1.00 million m^3/day in 1969 to 2.41 million m^3/day in 1971, then to 10.58 million m^3/day in 1981.

TABLE 3-89. SOUTH KOREA—PUBLIC WATER DEMAND, 1967-81

(Source: ECAFE, UN Water Resources Series No. 40, 1971)

Year	1967	1968	1971	1976	1981
Total population (million)	29.8	30.8	32.4	35.6	40.7
Population supplied (million)	7.4	8.2	11.6	17.7	22.3
Population supplied (per cent)	25.0	26.5	36.0	50.0	55.0
Supply capacity (million m^3/day)	1.13	1.61	2.70	4.50	8.70
Maximum demand (million m^3/day)	0.88	1.40	2.42	4.03	7.80
Average demand (million m^3/day)	0.81	1.04	2.16	3.50	6.00
Average demand (liters/capita/day)	110	128	187	200	270
Municipalities supplied	86	94	132	162	186

TABLE 3-90. SOUTH KOREA—MUNICIPAL WATER SUPPLY SYSTEMS

(Source: Internat. Statistical Institute, 1971)

City	Year	Population served	Total Quantity available 1,000 m^3	Annual consumption (1,000 m^3) Total	Annual consumption (1,000 m^3) of which domestic
Pusan	1967	987,744	49,275	22,039	N.A.
	1969	1,198,365	74,825	32,304	N.A.
Seoul	1967	2,972,378	184,804	101,368	56,303
	1969	3,967,400	251,086	137,351	76,141
Tagu	1967	616,580	29,200	N.A.	N.A.
	1969	765,000	65,700	N.A.	N.A.

N.A. — Not available.

TABLE 3-91. KUWAIT–DESALTING PLANTS

(Source: Al Adsani, Kuwait Water Res. Dev. Centre, 1973)

Plant	Date commissioned	No. of units	Total output gallons per day	TDS in distillate ppm	Type of plant
M.E.W.[1] Shuwaikh C	1958	2	1,000,000	80	Flash
" D	1958	2	1,000,000	80	Flash
" E	1960	2	2,000,000	80	MSF [2]
" F	1965	2	2,000,000	30	MSF
" G	1968	2	4,000,000	30	MSF
" B	1968	2	4,000,000	30	MSF
" A	1970	1	4,000,000	30	MSF
M.E.W. Shuaiba A	1966	3	3,000,000	30	MSF
" B	1968	1	2,000,000	30	MSF
" C	1968	2	4,000,000	30	MSF
" D	1971	1	5,000,000	30	MSF
M.E.W.Shuaiba South A.	1971	4	20,000,000	30	MSF
Total Dual Purpose Plants			52,000,000		
M.E.W. Electrodialysis		1	200,000 (discontinued)		
Total M.E.W.			52,200,000		
Kuwait Oil Co.		3	1,500,000		MSF
Aminoil Co.		1	380,000		LTMSF [3]
Grand Total			54,080,000		

1) Ministry of Electricity and Water.
2) MSF – Multi-Stage Flash Distillation.
3) LTMSF – Not explained.

TABLE 3-92. KUWAIT–DESALTING COST AND CAPITAL INVESTMENT

(Source: Al Adsani, Kuwait Water Res. Dev. Centre, 1973)

[Year 1971-1972]

Name	No. of Distillation Units	Station		"Actual" Cost of Distillate produced U.S. cents per 1,000 gallons	Capital Investment	
		Combined rated capacity million 1) gallons per day	Average capacity per unit million 1) gallons per day		Total U.S. Dollars	Average Capital Cost of one million gallons per day capacity U.S.Dollars
Shuwaikh	13	21.6	1.662	101.09	26,404,568	1,222,434
Shuaiba North	7	16.8	2.40	116.29	18,803,612	1,119,262
Shuaiba South	4	24.0	6.0	not available	22,606,068	941,919

1) Capacity in U.S. gallons.

TABLE 3-93. KUWAIT–PRODUCTION AND CONSUMPTION OF WATER, 1953-70

(Source: El-Shamy, et al., J. Am. Water Works Assoc., Dec. 1971. Copyright by AWWA. Reprinted by permission)

Year	Desalted water produced			Ground water pumpage			Drinking Water Consumption million gallons
	Shuwaikh desalting plant	Shuaiba desalting plant	Electro-dialysis	Rawdatain Field	Umm Al-Aish Field	Sulaibiya Field	
1953	142					138	—
1954	285					281	306
1955	423					314	434
1956	608					625	625
1957	878					638	778
1958	1,096					971	1,124
1959	1,397					1,419	1,423
1960	1,615					1,731	1,675
1961	1,870					2,151	1,909
1962	1,969			396		3,563	2,158
1963	1,595			950		4,011	2,515
1964	1,872			1,003		4,094	2,716
1965	1,900		34	899	396	5,141	3,070
1966	2,366	543	57	918	158	5,569	3,990
1967	2,443	433	46	1,113	418	5,011	5,141
1968	2,816	1,809	64	866	297	5,332	6,010
1969	3,378	2,506	58	643	258	6,704	7,031
1970	4,147	2,560	52	658	240	6,906	7,966

TABLE 3-94. KUWAIT—QUALITY OF WATER

(Source: El-Shamy, et al., J. Am. Water Works Assoc., Dec. 1971. Copyright by AWWA. Reprinted by permission)

	Distillate Analyses, mg/l		Ground Water Analyses, mg/l	
	Shuwaikh	Shuaiba	Rawdatain	Sulaibiya
TDS	45.0	71.0	1,100.0	4,757.0
pH	7.2	8.3	7.9	7.7
Chloride (Cl)	14.0	16.0	178.0	1,048.0
Sulfate (SO₄)	5.0	10.0	389.0	1,682.0
Sodium (Na)	7.6	9.0	205.0	569.0
Potassium (K)	0.43	0.32	5.0	23.0
Calcium (Ca)	2.1	1.3	104.0	550.0
Magnesium (Mg)	0.94	0.98	11.0	192.0
Iron (Fe)	0.05	0.08	0.05	0.18
Manganese (Mn)	0.0099	0.008	0.000	0.005
Copper (Cu)	0.139	0.15	0.025	0.02
Zinc (Zn)	0.1	0.097	0.01	0.016
Total alkalinity as CaCO₃	4.2	4.2	124.0	129.0
Total hardness as CaCO₃	7.9	8.8	292.0	2,200.0

TABLE 3-95. LAOS–DISCHARGE OF MEKONG RIVER

(Source: UNESCO, 1971)

River and station	Basin area km^2	Mean monthly discharge, m^3/s												Year	Period of record
		Jan.	Feb.	Mar.	Apr.	May	Jun.	Jul.	Aug.	Sep.	Oct.	Nov.	Dec.		
Mekong, Vientiane	299,000	1,750	1,420	1,170	1,140	1,500	3,420	7,050	11,580	11,100	7,090	5,020	2,800	4,590	1960-65
Mekong, Pakse	545,000	2,630	2,030	1,660	1,520	2,280	10,400	16,900	26,200	27,900	18,700	9,250	4,300	10,300	1960-65

TABLE 3-96. LAOS–PEAK DISCHARGE OF PRINCIPAL RIVERS

(Source: ECAFE, UN Water Resources Series No. 28, 1965)

Name of river	Basin area at gaging station km^2	Peak discharge m^3/s	Date
Mekong at Luang Prabang	269,000	18,500	18 Aug. 1960
Mekong at Vientiane	299,000	25,300	25 Aug. 1929
Mekong at Thakkek	373,000	33,800	31 Aug. 1939
Mekong at Pakse	545,000	46,200	31 Aug. 1939
Mekong at B. Chan Noi	549,000	45,600	28 Sep. 1961
Nam Khan	6,100	5,280	25 Jul. 1963
Nam Ngum	16,500	3,330	13 Aug. 1963
Seban Fai	8,560	3,320	25 Aug. 1961
Seban Hieng	19,400	6,400	28 Sep. 1961
Sedone	6,170	2,420	22 Aug. 1961
Nam Ou	25,800	3,800	Aug. 1960
N. Suong	5,780	1,490	Aug. 1960
N. Khan	6,100	2,180	Aug. 1960

TABLE 3-97. LEBANON—RAINFALL, RUNOFF AND WATER USE

(Source: Fawaz, Water for Peace, 1967)

	Million m³
Precipitation	8,000
Water losses (evapotranspiration)	4,500
Surface water runoff and infiltration	3,500
Water available for exploitation	1,500
Water use (1966)	
Public water supply	94
Irrigation	400

TABLE 3-98. LEBANON—PRESENT AND PROJECTED DEMAND FOR PUBLIC WATER SUPPLIES

(Source: Fawaz, Water for Peace, 1967)

District	Water demand in 1967 m³/day	Available water supply in 1967 m³/day	Available water supply in 1970 m³/day	Water demand in 1985 m³/day	Water demand in 2005 m³/day
Northern Lebanon	73,000	45,000	110,000	107,000	144,000
Mount Lebanon	80,000	90,000	160,000	118,000	158,000
Beirut	90,000	75,000	125,000	132,000	177,000
Southern Lebanon	50,000	23,000	103,000	73,000	98,000
Bekaa	42,000	24,000	52,000	62,000	83,000
Total	335,000	257,000	550,000	492,000	660,000

TABLE 3-99. MALAYSIA–DISCHARGE OF PRINCIPAL RIVERS IN WEST MALAYSIA

(Source: Drainage & Irrigation Dept. West Malaysia, 1961)

River and station	Basin area Mi²	Monthly discharge in cubic feet per second per square mile of basin area												Max. flow	Date	Min. flow	Date	Year or Period of record
		Jul.	Aug.	Sep.	Oct.	Nov.	Dec.	Jan.	Feb.	Mar.	Apr.	May	Jun.					
Discharge to Straits of Malacca																		
Sungei Pandang Terap at Lengkuas	492	1.42	1.85	2.77	5.15	5.92	2.37	0.73	0.38	0.35	0.97	2.02	1.32	17.68	Nov.'55	0.06	Mar.'55	1946-60
Sungei Muda at Batu Pekaka	1,290	2.51	2.48	3.60	5.76	6.21	3.45	1.69	1.25	1.35	2.22	2.86	2.51	15.16	Sep.'47	0.05	Feb.'49	1947-60
Sungei Perak at Iskandar Bridge	3,000	1.93	1.83	2.29	3.91	4.81	4.17	3.80	2.57	2.11	2.40	2.96	2.18	133.00	Dec.'26	0.27	Aug.'23	1948-60
Sungei Sungkai at 8th Mile Telok Anson-Kampar	189	5.66	5.72	6.25	7.98	9.40	8.75	6.96	5.16	7.02	8.19	8.56	5.21	17.09	Dec.'55	1.43	Jul.'58	1954-60
Sungei Bernam at Lima Blas Estate	422	3.43	2.92	4.76	5.62	7.11	5.66	5.59	4.87	5.75	5.62	6.51	4.31	24.17	Jan.'51	0.73	Mar.'40	1949-55
Sungei Selangor at Rantau Panjang	560	2.28	1.85	2.76	4.80	7.55	5.83	4.72	3.22	3.89	6.42	5.82	3.27	25.15	Jan.'27	0.26	Jul.'38	1950-60
Sungei Klang at Puchong	275	2.12	2.18	2.60	4.49	4.75	4.20	3.20	2.96	3.38	4.99	4.25	2.75	20.90	Oct.'57	0.15	Sep.'49	1948-60
Sungei Langat at Dengkil	478	2.02	1.70	2.38	3.69	5.64	5.75	3.47	3.12	3.35	5.07	4.60	2.86	14.45	Jun.'54	0.32	Aug.'51	1949-60
Sungei Muar at Buloh Kasap	1,210	0.58	0.44	0.53	1.02	2.26	2.49	1.98	1.43	0.88	1.44	1.57	0.80	15.00	Dec.'54	0.05	Mar.'59	1952-60
Discharge to China Sea																		
Sungei Trengganu at Kampong Tanggol	1,290	2.10	2.13	3.10	4.44	—	—	—	—	—	2.73	2.94	2.25	> 40	—	0.91	Jul.'58	1948-60
Sungei Kelantan at Guillemard Bridge	4,600	2.75	3.01	3.78	5.66	6.81	8.49	7.45	4.32	3.07	2.70	3.67	3.25	55.09	Jan.'51	0.54	Apr.'59	1949-60

TABLE 3-100. MALAYSIA—RUNOFF AND WATER USE IN WEST MALAYSIA

(Source: Framji and Mahajan, ICID, 1969)

	Million m3
Surface water runoff	190,920
Estimated water use (1970)	
Irrigation	7,398
Industrial and domestic	678

TABLE 3-101. MALAYSIA—MUNICIPAL WATER SUPPLY SYSTEM OF KUALA LUMPUR

(Source: Internat. Statistical Institute, 1971)

Year	Population served	Quantity of water available	Annual consumption(1,000 m3)	
			Total	of which domestic (Est.)
1967	435,000	60,871	43,183	30,000
1969	576,300	62,792	47,828	33,000

TABLE 3-102. NEPAL—DISCHARGE OF PRINCIPAL RIVERS

(Source: ECAFE, UN Water Resources Series No. 38, 1968)

River	Drainage area km^2	Maximum discharge m^3/sec	Minimum discharge m^3/sec
Kosi at Chatra	59,500	23,000	300
Gandaki at Tribeni	38,800	20,000	230
Karnali at Chisapani	42,890	10,190	214

TABLE 3-103. NEPAL—MUNICIPAL WATER SUPPLY SYSTEMS

(Source: ECAFE, UN Water Resources Series No. 38, 1968)

[As of 1968]

Number of public water supply systems	39
Source of water	streams and lakes
Total water consumption.........................	4,045 million gallons per year
Population served	600,000
Percent of total population served	6.5

TABLE 3-104. PAKISTAN—SUMMARY OF BASIC DATA OF INDUS RIVER

(Source: ECAFE, UN Water Resources Series No. 29, 1966)

Total drainage area	970,000 sq. km	375,000 sq. mi
Drainage area of hill catchment of the river and its tributaries	450,500 sq. km	174,190 sq. mi
Drainage area at Kalabagh	305,000 sq. km	118,000 sq. mi
Length of main river	2,900 km	1,800 mi
Slope of river		
from source to Attock	1:300	
from Attock to Mithankot	1:1,250	
from Mithankot to the sea	1:12,000	
Average annual precipitation over the hill catchment	612 mm 24.1 in	
over the foothill areas	550 mm 21.6 in	
over the plain less than	250 mm 9.8 in	
Maximum recorded flood discharge		
at Attock (1929)	23,200 m3/sec	820,000 cfs
at Sukkur (1958)	31,200 m3/sec	1,100,000 cfs
Average annual discharge at Sukkur	6,700 m3/sec	240,000 cfs
Minimum discharge at Kalabagh	490 m3/sec	17,300 cfs
Maximum recorded annual runoff of the hill catchment (1941-42)	264 billion m3	213.6 million acre ft
Average annual runoff of the hill catchment of the plains....	207 billion m3	168.4 million acre ft
Minimum recorded annual runoff of the hill catchment (1926-27)	171 billion m3	138.9 million acre ft
Specific discharge at Attock	0.078 m3/sec/sq km	8.35 cfs/sq mi
Average unit discharge at Attock	0.022 m3/sec/sq km	2 cfs/sq mi
Average silt content	3,300 ppm or	0.33 %
Annual silt runoff at Kalabagh	680 million tons	
Average annual silt runoff per unit area above Kalabagh.....	2,230 m tons/sq km	5,700 long tons/sq mi

FIGURE 3-4. PAKISTAN — INDUS RIVER SYSTEM AND SALINITY CONTROL
AND RECLAMATION PROJECTS

(Source: West Pakistan Water and Power Development Authority,
ECAFE UN Water Resources Series No. 23, 1963 — Amended)

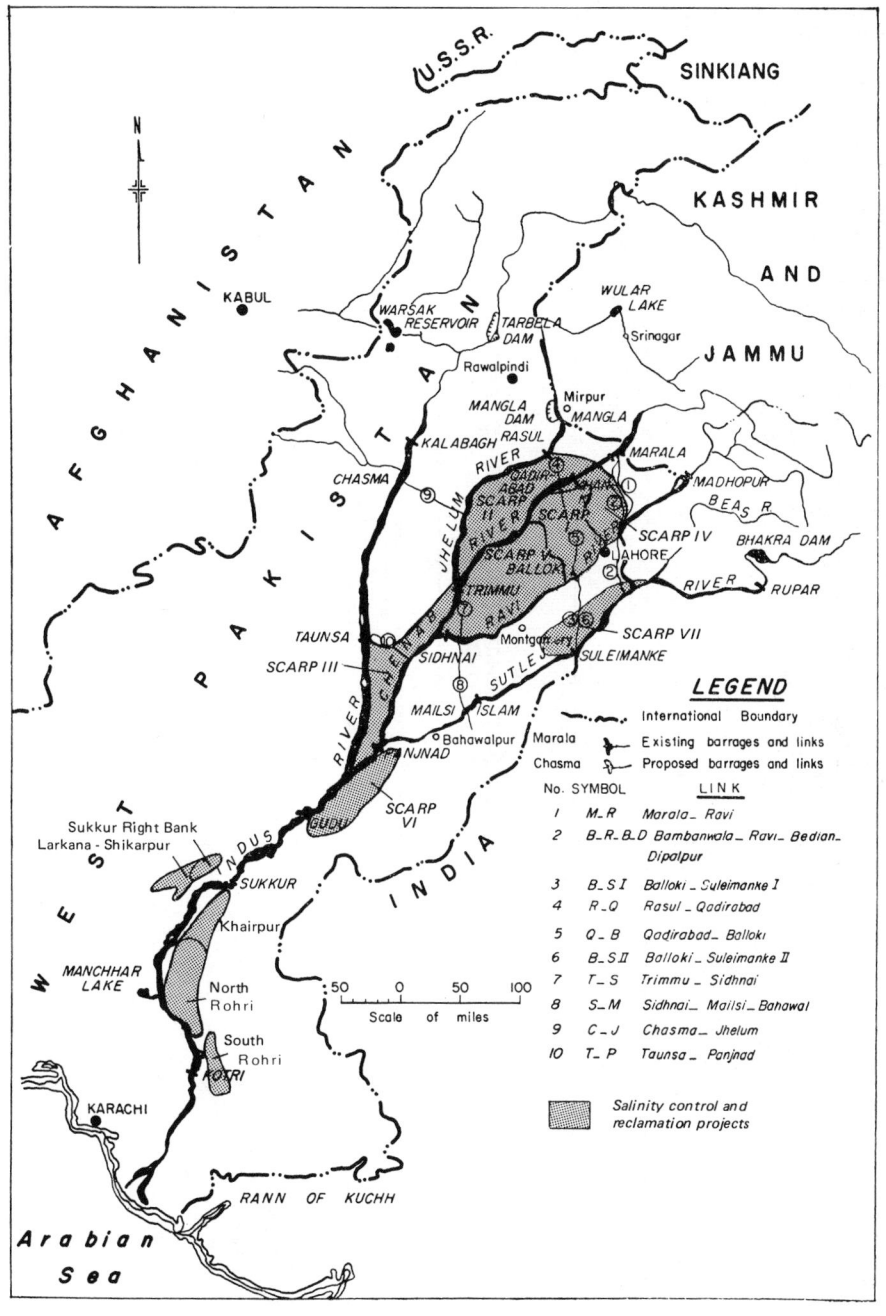

LEGEND

International Boundary

Marala	Existing barrages and links
Chasma	Proposed barrages and links

No.	SYMBOL	LINK	
I	M_R	Marala_ Ravi	
2	B_R_B_D	Bambanwala_ Ravi_ Bedian_ Dipalpur	
3	B_S I	Balloki _ Suleimanke I	
4	R_Q	Rasul _ Qadirabad	
5	Q _ B	Qadirabad_ Balloki	
6	B_S II	Balloki_ Suleimanke II	
7	T _ S	Trimmu _ Sidhnai	
8	S_ M	Sidhnai_ Mailsi_ Bahawal	
9	C _ J	Chasma_ Jhelum	
10	T_ P	Taunsa_ Panjnad	

Salinity control and
reclamation projects

TABLE 3-105. PAKISTAN—RUNOFF AND DISCHARGE OF INDUS RIVER AND MAJOR TRIBUTARIES AT RIM STATIONS

(Source: ECAFE, UN Water Resources Series No. 29, 1966)

River	Gaging station	Drainage area at gaging station		Precipitation		Runoff		Runoff coeffi-cient in per cent	Mean annual volume of flow		Maximum recorded annual volume of flow		Minimum recorded annual volume of flow	
		sq km	sq mi	mm	inch	mm	inch		106 m3	106ac.ft.	106 m3	106ac.ft.	106 m3	106ac.ft.
Indus	Kalabagh	305,000	118,000	430	17	355	14	82	110,300	89.5	142,000	115.0	85,000	68.8
Jhelum	Mangla	32,400	12,445	1,120	44	890	35	80	27,850	22.6	35,650	28.9	18,500	15.0
Chenab	Merala	29,500	11,399	1,240	49	960	38	78	29,000	23.5	36,400	29.5	23,500	19.0
Ravi	Madhopur	9,200	3,562	1,340	53	810	32	60	7,900	6.4	11,500	9.3	5,200	4.2
Beas	Mandi-Plain	13,900	5,384	1,420	56	1,140	45	80	15,650	12.8	24,950	20.2	8,200	6.6
Sutlej	Rupar	60,500	23,400	480	19	280	11	58	16,800	13.6	21,500	17.4	12,600	10.2
	Total	450,500	174,190	–	–	460	18	–	207,500	168.4	272,000	220.3	153,000	123.8

TABLE 3-106. PAKISTAN—MAXIMUM AND MINIMUM DISCHARGE OF INDUS RIVER AND MAJOR TRIBUTARIES

(Source: ECAFE, UN Water Resources Series No. 29, 1966)

River	Gaging station		Maximum discharge		Average maximum discharge		Minimum discharge		Average minimum discharge	
			m3/s	cfs	m3/s	cfs	m3/s	cfs	m3/s	cfs
Indus	Attock	1929a)	23,200	820,000	15,000	529,400	490	17,300	–	–
	Sukkur	1958	31,200	1,100,000	–	–	–	–	–	–
Jhelum	Mangla	1929	29,600	1,043,000	–	–	112	3,940	254	8,940
Chenab	Merala	1959	24,700	871,000	5,900	207,700	102	3,580	129	4,560
Ravi	Shahdara	1955	15,400	542,000	–	–	34	1,220	–	–
Beas	Mandi-Plainb)	1942	9,800	346,200	–	–	60	2,130	–	–
Sutlej	Suleimanki	1955	12,000	422,000	3,850	135,900	78	2,770	112	3,980

a) The years indicated apply only to maximum discharge.
b) Mandi-Plain gaging station is at few kilometers above Harike town.

TABLE 3-107. PAKISTAN—DATA ON DRAINAGE AND IRRIGATION PROJECTS

(Source: Khan, Ministry of Fuel, Power and Natural Resources, 1973)

Project (for location see p. 245)	Irrigable area 10⁶ acres	Number of Tubewells		Length of surface drains miles	Number of pumping stations
		Irrigation	Drainage		
I Punjab North/Upper Indus Plains					
1. *SCARP-I Central Rechna	1.14	1,796	—	—	—
2. SCARP-II Chaj Doab	2.10	2,437	460	450	—
3. SCARP-III Lower Thal Doab	1.05	1,635	100	150	—
4. SCARP-IV Upper Rechna	1.24	1,942	—	—	—
5. SCARP-V Lower Rechna	2.10	2,292	—	—	—
6. SCARP-VI Panjnad Abbasia	1.47	1,782	30	650	—
7. SCARP-VII Depalpur & CBDC	1.00	1,170	—	—	—
Sub-Total Punjab	10.10	13,054	590	1,250	—
II Sind (South/Lower Indus Plains)					
1. Khairpur	0.32	165	377	352	5
2. Larkana-Shikarpur Stages I & II	0.51	—	—	541	6
3. North Rohri (Fresh ground water)	0.69	1,218	—	—	—
4. South Rohri (Fresh ground water)	0.48	1,692	—	—	—
5. Ghotki	0.40	1,388	—	—	—
6. Sukkur Right Bank	0.13	352	—	—	—
7. Left Bank Outfall Drain	6.00	—	—	273	—
8. Khairpur East Tile Drainage	0.04	—	—	670	—
Sub-Total Sind	8.57	4,815	377	1,836	11
III NWFP (Upland Area)					
1. Pabbi Pilot Project	0.04	39	—	—	—
2. Peshawar City Anti-Water-Logging Project	—	20	—	2	—
Sub-Total NWFP	0.04	59	—	2	—

*SCARP—Salinity Control and Reclamation Project.

TABLE 3-108. PHILIPPINES—DISCHARGE OF PRINCIPAL RIVERS
(Source: UNESCO, 1971)

River and station	Basin area km²	Mean monthly discharge, m³/s												Year	Period of record
		Jan.	Feb.	Mar.	Apr.	May	Jun.	Jul.	Aug.	Sep.	Oct.	Nov.	Dec.		
Luson Island															
Chico River at Abbot, Pinukpuk, Mt. Province	3,349	119	109	82.8	70.1	95.5	168	275	329	419	367	341	250	208	1955-64
Cacayan River at Pangal, Echague, Isabela	4,244	271	222	158	83.1	70.0	102	230	297	273	373	599	469	263	1958-65
Agno River at Carmen, Rosales, Pangasinan	2,209	41.0	26.4	19.8	21.3	33.6	88.7	193	324	360	216	123	72.8	127	1945-65
Pampanga River at San Augustin, Arayat, Pampanga	6,487	86.6	52.2	36.7	32.3	34.6	130	352	569	584	398	271	177	228	1944-65
Panay Island															
Jalaur River at Caluan, Pototan, Iloilo	1,499	55.1	33.1	16.8	21.5	19.1	43.9	103	103	106	112	128	62.9	64.0	1956-65
Negros Island															
Ilog River at Pandan, Orong, Kabankalan	1,453	24.5	14.9	11.5	14.6	27.7	72.0	140	151	181	125	75.5	54.8	59.3	1956-65
Mindanao Island															
Agusan River at Poblacion, Talacogon, Agusan	7,390	1,030	1,280	783	449	397	427	399	448	459	489	480	923	605	1955-63
Polangui River at Inug-ug, Pikit, Cotabato	12,999	318	351	285	251	357	484	462	463	472	451	401	342	350	1955-65

TABLE 3-109. PHILIPPINES—HYDROELECTRIC POWER PLANTS

(Source: ECAFE, UN Water Resources Series No. 38, 1968)

[Data as of 1968]

Name of project	Type	Installed capacity kW	Cost (thousand pesos)
Major plants:			
Ambuklao	Reservoir	75,000	127,488
Angat	Reservoir	212,000	238,750
Binga	Reservoir	100,000	74,517
Caliraya	Reservoir	36,000	8,296
Maria Cristina	Reservoir	100,000	38,593
Minor plants:			
Agusan River	Pondage-run-of-river	1,600	1,743
Amburayan Chute	Run-of-river	200	257
Amlan Falls	Run-of-river	800	2,031
Balongbong River	Run-of-river	200	286
Cawayan River	Run-of-river	400	671
Digos River	Run-of-river	200	502
Loboc River	Run-of-river	1,200	2,590
Lake Buhi-Barit River	Reservoir	1,800	2,830
Penaranda Chute	Run-of-river	300	295
Talomo River No. 2	Run-of-river	600	82
Talomo River No. 2A	Run-of-river	400	460
Talomo River No. 2B	Run-of-river	300	458
Talomo River No. 3	Pondage-run-of-river	1,600	1,920
Total		532,600	501,769

TABLE 3-110. PHILIPPINES—STATUS OF PUBLIC WATER SUPPLY
(Source: Lesaca, National Water & Air Pollution Control Commission, 1971 and ECAFE, 1968)

WATER WORKS SYSTEM

NWSA [1] - operated	1,142 systems
Provincial, City, Municipal and small community water systems locally operated	1,287 systems
Population served at Manila & suburbs by NWSA	3,340,000
Population served by provincial systems	7,350,000
Total population served	10,690,000

ARTESIAN WELLS

Number of wells	20,240
Population served	2,500,000

DEVELOPED SPRINGS

Number of improved springs	2,062
Population served	1,312,000

IMPROVED DUG WELLS

Number of dug wells	42,250
Population served	2,113,000

POPULATION SERVED

Total population (1970)		37,000,000
Total population served:		
Water works system	10,690,000	
Artesian wells	2,500,000	
Developed springs	1,312,000	
Improved dug wells	2,113,000	
	16,615,000	or 45%

WATER USE (1968), m³/year

Production of waterworks:	
Surface water	544,800,000
Ground water	120,000,000
Total	664,800,000

PER CAPITA CONSUMPTION, m³/year

Major cities	138
Major towns	69
Villages	14

[1] National Waterworks and Sewerage Authority .

TABLE 3-111. SAUDI ARABIA—GEOLOGIC FORMATIONS AND AQUIFER CHARACTERISTICS

(Source: G. Otkun, 1969)

Age	Formation	Lithologic Description	Thickness m	Aquifer Characteristics
Quaternary and Tertiary	Surficial deposits and basalt	Gravel, sand, silt and basalt	28	Produce variable quality and quantity of water depending upon recharge by rainfall. Basalt yields little water in western Saudi Arabia.
Tertiary	Kharj	Limestone, lacustrine limestone, gypsum	95	Generally called Neogene aquifer. Irregular occurences of water. Artesian and non-artesian conditions. Prolific aquifer in the areas of Hofuf, Wadi Miyah and some others in Eastern Province.
	Hofuf	Sandy marl and sandy limestone	91	
	Dam	Marl, shale, subordinate sandstone	84	
	Hadrukh	Calcareous, silty sandstone	33	
	Dammam	Limestone, dolomite	56	Produces moderate amount of water under artesian and non-artesian conditions.
	Rus	Marl, chalky limestone	243	
	Umm er Radhuma	Limestone, dolomitic limestone	142	One of the most prolific aquifers of the Kingdom with high transmissibility varying between 500,000 and 3 million gpd/ft.
Cretaceous	Aruma	Limestone	42	Yields little water of low quality.
	Wasia (Sakaka sandstone northwest Arabian)	Sandstone, subordinate shale	425	Low productive or even dry near outcrop, very highly productive under artesian and non-artesian conditions in Eastern Province. Hydraulically connected with Biyadh near outcrops.
	Riyadh	Sandstone, subordinate shale	180	Moderately productive sandstone aquifer, hydraulically interconnected with Wasia near outcrop.
	Buwaib	Biogenic calcarenite and calcarenite limestone	46	
	Yamama	Biogenic calcarenite	170	
	Sulaiy	Chalky aphanitic limestone		

TABLE 3-111. SAUDI ARABIA–GEOLOGIC FORMATIONS AND AQUIFER CHARACTERISTICS (continued)

Age	Formation	Lithologic Description	Thickness m	Aquifer Characteristics
Jurassic	Hith	Anhydrite	90	Always yields mineralized water.
	Arab	Calcarenite, calcarenite and aphanitic limestone	124	Yields small amounts of water, mostly mineralized. Irregular occurence of water.
	Jubaila	Aphanitic limestone	118	Similar to Arab Formation above.
	Hanifa	Aphanitic limestone	113	———————
	Tuwaiq Mountain	Aphanitic limestone	203	———————
	Dhruma	Aphanitic limestone sandstone south of 22° N, and north of 26° N.	375	Produces moderate amounts of water north of 26° N and south of 22° N where it is generally represented by sandstone. South of 22° hydraulically connected with Minjur.
	Marrat	Shale and aphanitic limestone	103	Yields little water of fair to poor quality.
Triassic	Minjur	Sandstone, shale	315	Generally highly productive sandstone aquifer under flowing and non-flowing artesian conditions.
	Jilh	Aphanitic limestone, sandstone and shale	326	Mostly hydraulically interconnected with Minjur, produces low-quality water.
	Sudair	Red and green shale	116	———————
Permian	Khuff	Limestone and shale sandstone south of 21° N.	171	Moderately productive limestone aquifer, yields mostly mineralized water.
?	Wajid	Sandstone	950	Highly productive sandstone aquifer, under flowing and non-flowing artesian conditions.
Devonian	Jauf	Limestone, shale and sandstone	299	Productive, generally in al Jauf area.
Ordovician and Silurian	Tabuk	Sandstone and shale	1,072	Productive sandstone aquifer under flowing and non-flowing artesian conditions.
Cambrian	Saq Umm Sahm Ram Quweira Siq	Sandstone	600	One of the most productive sandstone aquifers of Saudi Arabia, under flowing and non-flowing artesian conditions.
Precambrian Basement Complex				

FIGURE 3-5. SAUDI ARABIA – KHOBAR DESALTING PLANT AND WATER – SUPPLY SYSTEM

(Source: Shaheen, Water & Sewage Works, May, 1973; reprinted with permission)

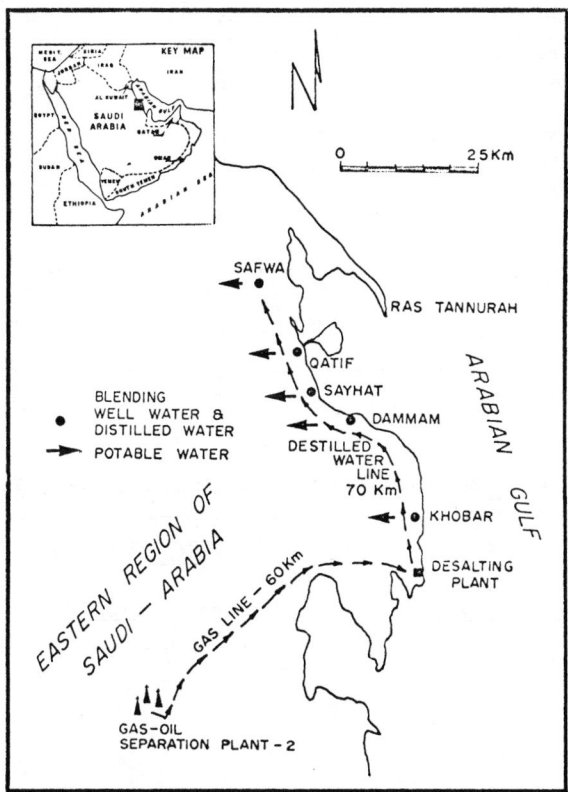

TABLE 3-112. SAUDI ARABIA – QUALITY OF WATER FOR KHOBAR WATER SUPPLY SYSTEM

(Source: Shaheen, Water & Sewage Works, May, 1973; reprinted with permission)
[Typical quality; samples collected 3/8/70]

Test	Well No. 4	Well No. 3
Color	Clear	Clear
Odor	No smell	No smell
pH	7.1	7.1
Total D.S. mg/l	2,855	2,754
Chloride, Cl^- mg/l	1,110	1,050
Nitrite, NO_2^- mg/l	0.0	0.0
Nitrate, NO_3^- mg/l	6	6.5
Sulfate, $SO_4^=$ mg/l	409	414
Free Ammonia, NH_4^+ mg/l	Trace	Trace
Dissolved oxygen, mg/l	2.15	2.25
Coliform group	Neg.	Neg.

TABLE 3-113. SAUDI ARABIA—QUALITY OF GROUND WATER IN DAMMAM

(Source: Shaheen, Water & Sewage Works, May, 1973. Reprinted by permission)

[Typical average quality; results expressed in mg/l]

Test	12/4/71	8/3/71	12/5/71	5/10/71	6/1/72
Free ammonia	0.54	0.74	1.55	0.81	0.80
Calcium	206	205	228	196	204
Iron	—	—	Trace	Trace	Trace
pH	7.2	7.1	7.1	7.3	7.1
Chloride	925	900	895	738	950
Sulfate	535	386	427	536	127
Total D.S.	2,266	2,190	2,017	2,139	2,499
Alkalinity	180	158	180	164	192
MPN *	Neg.	Neg.	Neg.	Neg.	Neg.
Temp.	26° C	27° C	28° C	34° C	29° C
Color	Normal	Normal	Normal	Normal	Normal

* MPN is an indication of coliform group.

TABLE 3-114. SAUDI ARABIA—QUALITY OF WATER AT RIYADH WATER PLANT

(Source: Shaheen, Water & Sewage Works, May, 1973. Reprinted by permission)

[Average quality during June 1971 at various plants]

	Before Coolers	After Pumps
Malez plant		
Temperature, °F	122°	86°
pH	7.2	8.5
Chlorides, mg/l	—	277
Total hardness as CaCO₃, mg/l	604	280
Free chlorine, mg/l	—	0.5
Shemessy plant		
Temperature, °F	113°	84°
pH	7.2	8.7
Chlorides, mg/l	—	202
Total hardness as CaCO₃ , mg/l	590	265
Free chlorine, mg/l	—	0.5
Manfouha plant		
Temperature, °F	100°	82°
pH	7.2	8.0
Chlorides, mg/l	177	110
Total hardness as CaCO₃ , mg/l	495	220
Free chlorine, mg/l	—	0.5
Hayir pretreatment plant		
Temperature, °F	124°	86°
pH	7.0	7.9
Chlorides, mg/l	230	181
Total hardness as CaCO₃ , mg/l	585	425

TABLE 3-115. SINGAPORE—WATER USE, 1971

(Source: Chen, Water & Sewage Works, 1972. Reprinted by permission)

Use	Water Consumption	
	mgd	%
Domestic	50	45
Commercial and industrial	24	22
Other (shipping, government, public standpipes)	36	33
	110	100

Population (1971) 2,100,000
Per capita consumption .... 50 gpd
Projected total water demand (1975).... 150 mgd
Source of water: rivers in Johore (Malaysia)
Capacity of impounding reservoirs in Singapore:
    MacRitchie Reservoir........... 815 million gallons
    Peirce Reservoir.............. 757 million gallons
    Seletar...................... 5,300 million gallons

TABLE 3-116. SYRIA—DISCHARGE OF EUPHRATES RIVER

(Source: UNESCO, 1971)

River and station	Basin area km2	Mean monthly discharge, m^3/s												Year	Period of record
		Jan.	Feb.	Mar.	Apr.	May	Jun.	Jul.	Aug.	Sep.	Oct.	Nov.	Dec.		
Euphrates, Youssef Pacha	97,000	435	839	1,580	2,710	1,670	597	335	237	216	215	315	781	830	1950-66

TABLE 3-117. SYRIA—GROUND WATER QUALITY WEST OF EUPHRATES RIVER

(Source: Burdon and Mazloum, 1956)

Aquifer	No. of samples	Total soluble salts, mg/l			Dominant anion			
		Min.	Average	Max.	HCO$_3$	SO$_4$	Cl	Mixed
Quaternary	5	175	514	877	3	1	0	1
Limestone	28	235	617	2,142	21	0	3	4
Basalt	8	154	710	2,215	6	1	0	1
Terrestrial	4	245	750	1,450	2	0	1	1
Rmah Chert	5	436	1,652	2,926	1	3	0	1
Fars Formation	8	426	3,434	5,571	0	4	3	1
Altered marl	12	809	3,983	13,536	1	0	9	2
Totals	70	—	—	—	34	9	16	11

TABLE 3-118. TAIWAN—AVAILABLE WATER RESOURCES
(Source: ECAFE, UN Water Resources Series No. 40, 1971)

Category	Quantity of water
Average annual precipitation equivalent to..............	2,430 mm 84.1 billion m3
River runoff (68%).............	57.2 billion m3
Evapotranspiration (28%).............	23.5 billion m3
Ground water recharge (4%).............	3.4 billion m3

TABLE 3-119. TAIWAN—MONTHLY DISCHARGE OF CHO—SHUI RIVER
(Source: UNESCO, 1971)

River and station	Basin area km2	Mean monthly discharge, m3/s												Year	Period of record
		Jan.	Feb.	Mar.	Apr.	May	Jun.	Jul.	Aug.	Sep.	Oct.	Nov.	Dec.		
Cho-shui, Chi Chi	2,311	51.4	43.3	56.2	89.6	138	289	218	268	260	111	79.6	60.6	139	1941-66

TABLE 3-120. TAIWAN–CHARACTERISTICS OF PRINCIPAL RIVER BASINS

(Source: Wang, ECAFE, 1963)

River basin	Drainage area km2	Length km	Head elev. m	Slope	Discharge, m3/s Maximum	Discharge, m3/s Minimum	Mean precipitation mm
Tan Shui Ho	2,726	158.7	3,529	1:45	13,000	29.3	3,010
Tou Chien Chi	566	63.0	2,233	1:28	4,700	1.0	2,259
Hou Lung Chi	537	58.0	2,580	1:22	3,400	0.4	2,142
Ta An Chi	758	95.8	3,296	1:29	6,100	2.7	2,876
Ta Chia Chi	1,236	140.2	3,639	1:39	10,600	14.2	2,572
Wu Chi	2,026	116.8	2,596	1:45	13,900	5.0	2,210
Cho Shui Chi	3,155	186.4	3,416	1:55	22,000	23.9	2,466
Pei Kang Chi	645	81.9	516	1:159	2,200	0.1	1,944
Pu Tzu Chi	400	75.7	1,421	1:53	1,100	0.2	2,144
Pa Chang Chi	475	80.9	1,940	1:42	1,800	0.4	2,477
Chi Shui Chi	379	65.1	550	1:118	1,300	0.2	2,442
Tseng Wen Chi	1,177	138.5	2,440	1:57	5,500	0.2	2,749
Erh Tseng Hang Chi	350	65.2	460	1:142	1,500	0.1	2,032
Hsia Tan Shui Chi	3,257	170.9	3,997	1:43	22,000	1.0	3,385
Lin Pien Chi	344	42.2	2,880	1:15	2,500	0.0	4,462
Ilan Cho Shui Chi	979	73.1	3,535	1:21	7,800	13.7	3,101
Hua Lien Chi	1,507	57.3	2,260	1:25	9,000	14.5	2,748
Hsiu Ku Luan Chi	1,790	81.2	2,360	1:34	10,500	20.9	2,174
Pei Nan Ta Chi	1,603	84.4	3,666	1:23	9,000	2.0	2,255

TABLE 3-121. TAIWAN–WATER USE, 1964-70

(Source: ECAFE, UN Water Resources Series No. 40, 1971)

[million m3]

Year	Public Supply Piped supply[a]	Public Supply Scattered piped supplies	Public Supply Sub-total	Public Supply Per cent	Industrial Use	Industrial Per cent	Agriculture[b] Use	Agriculture[b] Per cent	Total
1964	–	–	430	4.1	150	1.4	9,800	94.5	10,380
1970	590	60	650	5.6	540	4.7	10,300	89.7	11,490

a Urban supply plus rural supply.
b Effective rainfall plus irrigation water excluding fishery and large-scale husbandry consumption.

TABLE 3-122. TAIWAN—PUBLIC WATER SUPPLY, 1968

(Source: ECAFE, UN Water Resources Series No. 40, 1971)

Production capacity

Production (m³/day)	Plants (No,)	Serviced area (ha)
Above 50,000	5	26
30,001 to 50,000	4	35
10,001 to 30,000	9	43
3,001 to 10,000	27	46
1,001 to 3,000	33	79
Below 1,000	55	103
Total	133	332

Usage

Users (No.)	Plants (No.)	Serviced area (ha)
Above 50,000	18	115
30,001 to 50,000	13	27
10,001 to 30,000	38	81
3,001 to 10,000	40	72
1,001 to 3,000	18	30
Below 1,000	4	5
Total	131 [a]	330 [a]

Total production (m³/day)	1,265,924
Water users (No.)	5,580,127
Usage (%)	40.88
Consumption (lit/capita)	227

[a] Excluding industrial usage.

TABLE 3-123. TAIWAN—MUNICIPAL WATER SUPPLY SYSTEM OF TAIPEI

(Source: Internat. Statistical Institute, 1971)

Year	Population served	Total quantity available 1,000 m³	Annual consumption (1,000 m³)	
			Total	of which domestic
1967	1,224,642	117,884	74,352	56,051
1969	1,689,723	156,060	101,499	77,950

TABLE 3-124. THAILAND–DATA ON MAIN RIVER SYSTEMS
(Source: ECAFE, UN Water Resources Series No. 38, 1968)

River	Gaging station	Drainage area km2	Mean annual peak discharge m3/sec	Average minimum flow m3/sec	Ratio of dry-season flow to mean peak discharge	Mean annual runoff million m3
Chao Phraya	Nakhon Sawan	110,371	3,363	32	1:105	22,880
Ping	Yanhee	26,396	1,675	10	1:167	5,410
Wang	Wang Krai	10,407	445	1	1:445	1,370
Yom	Sukhothai	12,658	1,415	1	1:1415	2,620
Nan	Uttaradit	16,775	2,100	12	1:175	6,420
Chi	Yasothon	47,406	1,253	5	1:250	7,300
Mun	Ubon	106,673	3,179	6	1:530	18,300
Mae Klong	Tha Muang	27,220	2,400	48	1:50	12,860
Pa Sak	Kaeng Khoi	14,522	704	5	1:141	2,620
Petch	Petch barrage	4,060	472	2	1:235	1,170

TABLE 3-125. THAILAND–DISCHARGE OF MEKONG AND CHAO PHYA RIVERS
(Source: UNESCO, 1971)

River and station	Basin area km2	Mean monthly discharge, m3/s												Year	Period of record
		Jan.	Feb.	Mar.	Apr.	May	Jun.	Jul.	Aug.	Sep.	Oct.	Nov.	Dec.		
Mekong, Mukdahan	391,000	2,480	1,950	1,600	1,560	2,440	8,040	14,880	22,430	22,400	12,400	6,240	3,570	8,330	1936; 1938-40 1942-44;1948-66
Chao Phya, Nakohn-Sawan	111,435	231	156	112	96	180	480	662	1,310	2,190	2,950	1,700	529	883	1905-66

TABLE 3-126. QUALITY OF RIVER WATER IN NORTHEAST THAILAND
(Source: USAID/UN, 1968)

[Average of monthly samples from July 1956 to June 1957 in ppm]

River	Sampling point	Ca	Mg	Na	HCO3	SO4	Cl	SiO2	PO4	N	Total dissolved solids
Pao	Kalasin	9.0	2.1	23.6	45.9	0.3	35.5	15.7	0.00	0.31	119.3
Chi	Chaiyaphum	22.0	2.9	14.1	76.7	6.3	16.8	11.3	0.06	0.42	120.5
Chi	Khon Kaen	21.1	4.0	57.1	69.7	5.2	94.5	10.1	0.00	0.20	244.7
Mun	Surin	10.6	1.2	28.7	21.6	1.3	56.6	12.8	0.01	0.89	148.5
Mun	Ubon	10.9	2.3	40.0	42.4	2.0	61.6	10.8	0.00	0.28	165.2

TABLE 3-127. THAILAND—SUMMARY OF WATER RESOURCES DEVELOPMENT
(Source: Binson, 1972)

Description	Units	Completed up to 1971	Under Construction in 1972
Storage of water	million m3	16,500	12,300
Irrigation			
Drainage			
Water conservation	hectares	2,129,600	762,500
Flood control			
Reclamation			
Ditches and Dikes	hectares	1,125,000	624,900
Inland waterways	kilometers	322	352
Hydro-power	kilowatts	475,300	40,000

TABLE 3-128. THAILAND—BANGKOK WATER SUPPLY SYSTEM
(Source: ECAFE, UN Water Resources Series No. 38, 1968)

Item	Year		
	1966	1967	1968
Population in the distribution area (No.)	1,936,549	1,995,688	2,042,453
Connections (No.)	133,980	141,318	150,943
Length of pipe larger than 4 in diameter (m) [a]	59,276	46,556	17,501
Average consumption liters per head per day [a]	300	330	355
Water produced annually by treatment plants (m3)	177,998,780	183,619,220	186,399,490
Water produced annually by deep wells (m3)	33,933,330	57,266,857	77,789,460
Capacity of raw water intake canal (m3/day)	1,000,000	1,000,000	1,000,000

[a] Annual increments.

FIGURE 3-6. TURKEY–HYDROLOGIC CYCLE AND WATER USE

(Source: Noyan, et al., Water for Peace, 1967)

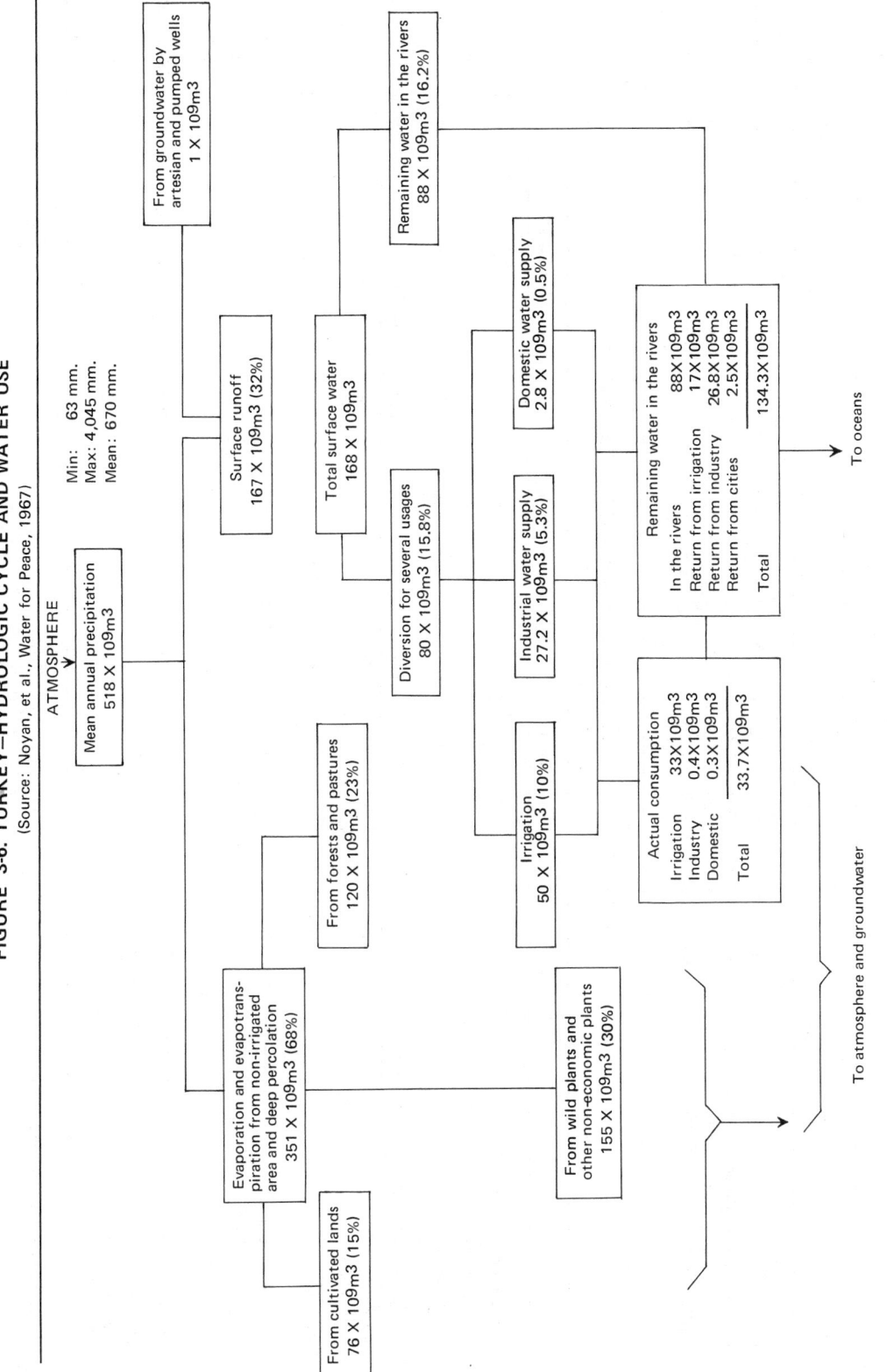

TABLE 3-129. TURKEY—DISCHARGE OF SELECTED RIVERS

(Source: UNESCO, 1971)

River and station	Basin area km^2	Mean monthly discharge, m^3/s Jan.	Feb.	Mar.	Apr.	May	Jun.	Jul.	Aug.	Sep.	Oct.	Nov.	Dec.	Year	Period of record
Sakarya, Botbasi	55,320	242	328	350	354	273	162	101	73.8	92.1	115	123	213	202	1961-66
Kizilirmak, Inozu	75,120	174	234	308	260	190	172	89.5	62.0	87.1	98.9	97.7	132	159	1962-66
Euphrates (Firat), Keban	63,835	308	380	723	1,990	1,780	792	361	246	217	248	300	304	639	1937-66
Buyuk Menderes, Soke	23,889	167	190	168	124	83.7	53.7	28.1	23.8	31.6	38.9	67.7	103	90.0	1951-66
Ceyhan, Ceyhan	19,767	265	324	421	446	326	153	66.9	47.6	48.3	61.2	70.9	265	199	1954-66

TABLE 3-130. TURKEY—DATA ON EUPHRATES RIVER SYSTEM

(Source: Garbrecht, Wasser und Boden, 1968)

	Turkey		Syria		Iraq		Saudi Arabia		Total
Drainage area (1,000 km^2)	125		76		177		66		444
% of Total area	28		17		40		15		100
% of Country	16		41		39		5		
Annual discharge ($10^9 m^3$) *	INFLOW	OUTFLOW	INFLOW	OUTFLOW	INFLOW	OUTFLOW	INFLOW	OUTFLOW	
Minimum	0	12.6	12.6	14.0	14.0	14.0	0	0	14
Mean	0	28.4	28.4	32.4	32.4	32.4	0	0	32.4
Maximum	0	42.0	42.0	45.0	45.0	45.0	0	0	45.0
% of Mean discharge	88		12		0		0		100
Length (km)	1,230		710		1,060		0		3,000
% of Length	41		24		35		0		100
Elevation (m)	2,500-3,000	326	326	166	166	0	—	—	2,500-3,000
Fall (m)	2,674-2,174		160		166		—		2,500-3,000
% of Total Fall	88		6		6		—		100

* Based on period 1937-1963.

TABLE 3-131. TURKEY–RIVER DISCHARGE INTO THE BLACK SEA
(Source: Ozturgut, CENTO Symp. on Hydrology and Water Resources, 1966)

[In m3/s]

River	Jan.	Feb.	Mar.	Apr.	May	Jun.	Jul.	Aug.	Sep.	Oct.	Nov.	Dec.	Total km3/year
Sakarya	514	704	582	444	440	321	140	81	101	93	87	256	9.9
Kizilirmak	195	328	275	310	352	319	149	67	77	101	60	141	6.2
Yesilirmak	172	259	259	478	422	313	106	45	48	25	31	70	5.9
Miscellaneous	353	394	464	593	615	244	218	138	117	126	111	274	10.4
Danube													206.0
Dnieper	711	804	1,506	3,816	5,225	2,120	1,045	844	735	735	865	784	50.5
Don	247	296	1,051	3,544	1,754	375	255	202	181	186	195	190	22.2
Rion	305	373	427	624	647	624	513	274	279	252	274	358	13.0
Dniester	167	216	491	591	422	374	398	326	271	245	248	223	10.3
Kuban	233	209	333	430	650	743	677	435	273	231	225	260	12.3
Others													22.6
Total													370

TABLE 3-132. TURKEY—WATER

(Source: Cecen, Wasser

No.	River basin	Area km^2	Mean precipi- tation mm	Available surface water resources				Available ground water resources 10^{6}m^3	Estimated irrigable area km^2	Irrigated area km^2
				10^{9}m^3	mm	m^3/s	l/s/km^2			
1	Meric-Ergene	14,560	500	1.06	72.5	33.5	2.3	—	11,700	1,125
2	Marmara	24,100	800	5.85	242.6	185.6	7.7	—	246.9	730
3	Susurluk	22,399	600	4.16	185.9	132.0	5.9	—	1,950	1,650
4	North Aegean	10,003	650	1.74	173.2	55.2	5.5	27	205.0	670
5	Gediz	18,000	600	2.27	126.0	72.0	4.0	21	2,579	941
6	K. Menderes	6,907	650	1.04	151.2	33.15	4.8	—	1,970	100
7	B. Menderes	24,976	600	3.15	126.0	100.0	4.0	250	3,860	2,000
8	W. Mediterranean	20,953	900	6.86	327.6	217.0	10.4	45	1,837	1,400
9	Antalya	19,577	1,100	14.33	491.4	454.9	23.2	80	1,800	880
10	Burdur Lakes	6,374	600	0.9	141.8	28.7	4.5	61	2,362	170
11	Akarcay	7,605	500	0.91	119.7	28.9	3.8	56	2,016	377
12	Sakarya	58,160	450	4.1	69.3	130.0	2.2	133	17,541	2,471
13	W. Black Sea	29,598	800	9.32	315.0	296.0	10.0	17	1,080	900
14	Yesilirmak	36,144	500	4.57	126.0	145.0	4.0	185	2,423	733
15	Kizilirmak	78,180	400	5.66	72.5	179.8	2.3	178	18,622	8,800
16	Int. basin Konya	53,850	400	3.91	72.5	124.0	2.3	319	16,171	3,450
17	E. Mediterranean	22,048	900	9.03	409.5	286.6	13.0	84	637	260
18	Seyhan	20,450	880	5.92	289.8	187.9	9.2	—	1,674	1,674
19	Asi	7,796	900	1.47	189.0	46.8	6.0	95	1,673	1,400
20	Ceyhan	21,982	950	7.25	330.8	230.0	10.5	66	4,200	4,153
21	Euphrates	125,540	700	28.35	220.5	900.0	7.0	427	13,000	12,970
22	E. Black Sea	24,077	1,400	11.34	469.4	360.0	14.9	—	190	150
23	Coruh	19,872	600	4.88	245.7	155.0	7.8	—	463	270
24	Aras	27,548	500	3.71	207.9	181.38	6.6	60	4,999	2,000
25	Int. basin Van	19,405	700	3.54	182.7	113.0	5.8	16	1,419	690
26	Tigris	57,614	850	19.49	337.0	618.8	10.7	14	2,500	2,030
	Total	77,688	670	166.81	214.0	5,300.0	6.8	2,134	120,185	51,995

RESOURCES DATA BY RIVER BASIN
und Boden, 1968)

Flood protected area km2	Drained area km2	Hydropower		Drinking water use 106m3	Volume in storage 109m3	No. of dams	No. of hydroelectric installations
		Installed capacity MW	Generated energy 106kWh				
347	35	—	—	—	0.650	13	—
200	—	20.00	70	660	0.800	13	—
30	24	6.50	19	33	1.090	14	1
20	—	20.40	51	—	0.688	13	3
50	50	60.00	193	—	2.200	5	1
—	—	—	—	—	—	—	—
500	—	111.40	621	10	2.184	23	6
99	8	777.00	3,119	—	5.581	23	19
100	4	213.00	864	—	3.460	12	7
—	2	—	—	—	0.076	8	—
200	5	—	—	—	0.100	4	—
147	136	420.00	1,462	—	4.659	41	8
250	—	380.86	1,740	718	—	21	33
300	—	800.00	3,013	—	4.374	5	11
300	500	1,183.00	6,000	—	15.000	96	26
50	—	10.00	27	—	0.434	16	1
120	47	335.00	1,700	15	4.800	5	11
—	—	68.30	419	40	—	2	—
110	20	61.00	227	—	0.700	3	1
332	—	751.00	2,621	—	9.095	13	9
—	—	4,800.00	30,000	—	96.600	67	68
—	—	260.00	1,033	—	—	1	—
—	—	1,250.00	5,000	—	—	3	—
150	—	160.00	800	—	1.238	15	6
—	—	28.69	135	—	0.681	11	6
—	—	1,200.00	4,309	—	—	—	—
3,155	831	12,585.29	61,891	1,476	154.40	427	217

FIGURE 3-7. U.K. ASIAN DEPENDENCIES: HONG KONG — EXISTING AND POTENTIAL STORAGE RESERVOIRS

(Source: Robertson, A.S., J. Institution of Water Engineers, 1969)

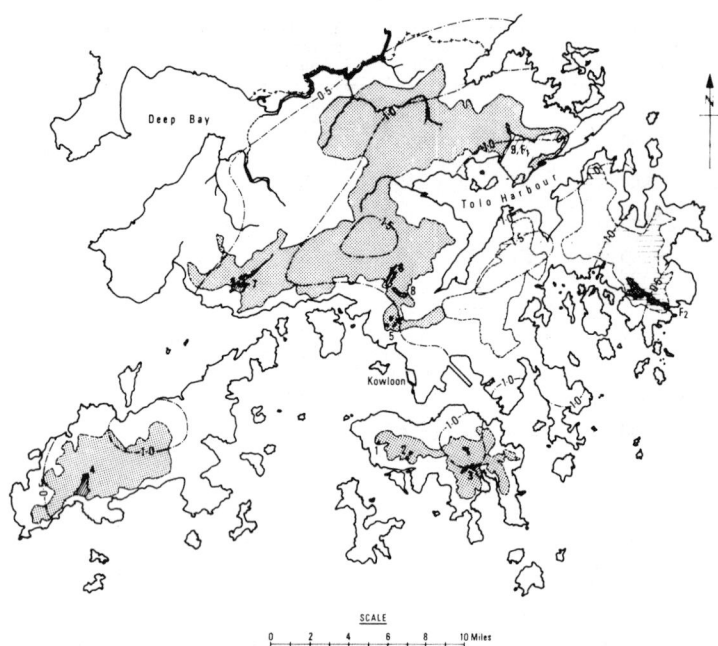

Key		Catchment	Shading used	Area, acres	Average gross yield, m.g. per year
▨	Existing reservoir				
■	Possible future reservoir	Existing High Island Scheme (projected)	dots	70,600	76,000
—1·5—	Approximate line of equal average gross yield: m.g. per acre per year		horizontal lines	13,500	13,000
+-+-+	Hong Kong–China border			84,100	89,000

TABLE 3-133. U.K. ASIAN DEPENDENCIES: HONG KONG—STORAGE RESERVOIRS

(Source: Robertson, A.S., J. Institution of Water Engineers, 1969)

[Location of reservoirs shown on adjacent map]

Existing reservoirs			Possible future reservoirs		
Ref. No.	Reservoir	Capacity m.g.	Ref. No.	Reservoir	Rough potential capacity m.g.
1	Pokfulam	57	FI	Plover Cove (raising)	14,500
2	Aberdeen, upper and lower	280	F2	High Island	70,000
3	Tai Tam Group	2,016			
4	Shek Pik	5,381			
5	Kowloon Group	682			
6	Jubilee	2,921			
7	Tai Lam Chung	4,507			
8	Lower Shing Mun	948			
9	Plover Cove	37,380			
	Total existing storage	54,172		Total projected storage	84,500

TABLE 3-134. U.K. ASIAN DEPENDENCIES: HONG KONG—WATER USE
(Source: Kwei, Hong Kong Water Authority, 1972)

Population served by public water supply	About 3.9 million
Percentage of total population served	About 99 %
Per capita water use	46.5 gpd (1971)
Public water use (overall total includes industrial water)	181.1 mgd (1971)
Industrial water use	Approx. 46.6 mgd (1970)
Annual water importation from China .... (pumped from East River in Kwangtung Province)	15,000 million gallons (taken over a period of 9 months each year)

TABLE 3-135. NORTH VIETNAM—FLOW CHARACTERISTICS OF RED RIVER
(Source: ECAFE, UN Water Resources Series No. 29, 1966)

Total drainage area	120,000 sq km		46,000 sq mi
Area of delta....................	7,000 sq km		2,700 sq mi
Length of river	1,200 km		750 mi
Average annual precipitation over river basin	1,500 mm		59 in
Average annual precipitation over the part of river basin in Viet-Nam	1,800 mm		71 in
Maximum recorded discharge at Vietri(1945)	35,000 m³/sec		1,250,000 cfs
Average discharge at Vietri	3,900 m³/sec		138,000 cfs
Minimum discharge at Hanoi	700 m³/sec		25,000 cfs
Specific flood discharge	0.31 m³/sec/sq km		28.4 cfs/sq mi
Average unit discharge	0.0345 m³/sec/sq km		3.15 cfs/sq mi
Average annual runoff at Vietri	123 billion m³		100 million acre ft
Average annual runoff expressed in depth	1,090 mm		43 in
Average silt content at Hanoi	1,060 ppm	or	0.106 %
Maximum silt content	7,000 ppm	or	0.7 %
Low water average silt content	100 ppm	or	0.01 %
Average annual silt runoff	130 million tons		
Average annual silt runoff per unit area	667 m³/sq km		1.40 acre ft/sq mi
or	1,080 tons/sq km		2,840 tons/sq mi

TABLE 3-136. SOUTH VIETNAM—DATA ON PRINCIPAL RIVERS

(Source: Ton-That Ngo, Water for Peace, 1967)

Name of the river	Province	Basin area km2	Streamflow Flood flow m3/s	Streamflow Base flow m3/s	Streamflow Annual flow 106m3	Water quality Average pH	Water quality Max. salt content mg/liter	Observations
Thach-Han	Quang-Tri	1,415	552	19	2,506	5.5	130	
O-Lau	Thua Thien	572	251	8	1,080	5.5	110	
Bo	"	798	605	11.25	4,298	5.5	170	
An-Nong	"	1,532	1,990	28	3,671	5.5	140	
Thu-Bon	Quang-Nam	3,340	10,000 (estimated)	39	–	5.5	140	Very irregular flow
Tam-Ky	Quang-Tin	246	1,500 (estimated)	39	–	5.5	140	
Tra-Bong	QuangNgai	380	800	1.56	–	5.5	140	
Tra-Khuc	QuangNgai	2,800	5,200	36	–	5.5	140	
Ve	"	815	1,150	10	2,730.8	5.25	140	
Tra-Cau	"	209.5	228	1	–	5.25	210	
Lo-Bo	"	70	53.5	0.6	–	5.25	180	
Kone	Binh-dinh	1,424	1,070	10	1,770.7	5.25	180	
Tam-Giang	Phu-Yen	1,840	5,520	4	–	5.25	180	
Da-Rang	"	12,000	9,757	30	–	5.25	180	
Cay	Khanh-Hoa	291	180	2.4	–	5.25	180	
Cai	"	1,749	3,900	40	4,781.9	5.25	180	
Cai	NinhThuan	2,049	45	21	–	5.25	180	
La-Nga	Binh-Tuy	1,883	342	5	1,441	5.25	180	
Dong-Nai		22,000	12,000	240	–	5.25	180	At Bien-Hoa
Srepok	Darlac	8,813	970	20.5	5,210	–	–	
Mekong		795,000	61,300 (27.8.61)	1,250 (12.4.60)	–	–	–	At Stung-Treng
Bassac	Chau-Doc	–	7,140	–	–	–	–	

TABLE 3-137. SOUTH VIETNAM–WATER BALANCE ELEMENTS

(Source: Djang, FAO, 1961)

[Estimated valves]

Element	Quantity
Total land area...............................	178,028 km^2
Mean annual precipitation.......................	2,000 mm
equivalent to...............................	356 km^3
Evapotranspiration losses and infiltration	237 km^3
Surface water runoff	119 km^3
Inflow of Mekong River	346 km^3
(almost totally lost to sea)	
Total surface water runoff lost to sea	465 km^3

TABLE 3-138. SOUTH VIETNAM–MUNICIPAL WATER SUPPLY SYSTEM OF SAIGON

(Source: Internat. Statistical Institute, 1971)

Year	Population served	Annual consumption (1,000 m^3)	
		Total	of which domestic
1967	1,736,880	72,971	63,343
1969	1,640,000	105,490	94,087

TABLE 3-139. SOUTH VIETNAM—HYDROELECTRIC POWER PLANTS
(Source: ECAFE, 1971)

Name of plant	Installed capacity (kW)	Net head (m)	Firm power (kW)	Average annual energy output (106 kWh)	Year of completion
Ankroet	3,000	–	–	15.3	1961
Drayling, Darlac	480	13	–	–	1933
Ea-Nao, Darlac	280	20	–	2.2	1939
Danhim No. 1a	160,000	741	97,000	1,000.0	1964
Total	163,760				

a Operation suspended owing to war damage.

ASIA
REFERENCES CITED

References cited in more than one country:

UNESCO, Discharge of Selected Rivers of the World; Vol. I. General and Regime Characteristics of Stations Selected (1969); Vol. II. Monthly and Annual Discharges Recorded at Various Selected Stations (1971); Vol. III. Mean Monthly and Extreme Discharges 1965-1969 (1971).Copyright by UNESCO/IASH/WMO. Reprinted by permission.

International Statistical Institute, 1970, International Statistical Yearbook of Large Towns. Vol. V., The Hague, The Netherlands.

Proceedings of the International Conference on Water for Peace, 1967, Superintendent of Documents, Washington, D.C. 20402.

ECAFE, 1966, A Compendium of Major International Rivers in the ECAFE Region, UN Water Resources Series No. 29.

ECAFE, 1965, Proceedings, Sixth Regional Conference on Water Resources Development in Asia and the Far East, UN Water Resources Series No. 28.

ECAFE, 1968, Proceedings, Eighth Regional Conference on Water Resources Development in Asia and the Far East, UN Water Resources Series No. 38.

ECAFE, 1971, Proceedings, Ninth Session of the Regional Conference on Water Resources Development in Asia and the Far East, UN Water Resources Series No. 40.

ASIA GENERAL

U.S. Federal Power Commission, 1968, World Power Data, Publ. P-40.

AFGHANISTAN

FAO, 1965, Report on Survey of Land and Water Resources, Afghanistan, Vol. III, Hydrology.

BANGLADESH

FAO, 1973, Report of the Agricultural Team for Bangladesh, Vol. II, Appendices, FAO/DEN/TF96.

CHINA

Tojin Sha (Publisher), 1964, Water Resources of China; Translated by U.S. Government, CCM Information Corp., N.Y.

CYPRUS

Konteatis, C.A.C., 1967, The Water Resources of Cyprus, Their Conservation and Development, Dept. of Water Development, Cyprus.

INDIA

Ministry of Irrigation and Power, 1972, Report of the Irrigation Commission, New Delhi, (in Four Volumes).
Chaturvedi, M.C., 1973, Indian National Water Plan and Grid, First World Congress on Water Resources, International Water Resources Association, Chicago.

IRAN

Vahidi, Iraj, 1973, Water Resources Development in Iran, First World Congress on Water Resources, International Water Resources Association, Chicago.

IRAQ

Ubell, K., 1971, The Institute for Applied Research on Natural Resources, Iraq: Iraq's Water Resources, in Nature and Resources, Vol. VII, No. 2, UNESCO, Paris. Copyright by UNESCO. Reprinted by permisssion.

ISRAEL

Israel Hydrological Service, 1971, Hydrological Year-Book of Israel—Summary of Records Prior to October 1967, Jerusalem.
Shelef, G., 1973, Wastewater Reclamation and Reuse in Israel, First World Congress on Water Resources, International Water Resources Association, Chicago.
Israel Water Allocation Department, 1970, Water in Israel, Consumption and Extraction 1962-1968, Water Commissioners Office, Tel Aviv.
Jacobs, M. and Litwin, Y., 1973, A Survey of Water Resources Development, Utilization and Management in Israel, First World Congress on Water Resources, International Water Resources Association, Chicago.
Harpaz, Yoav, 1973, Artificial Groundwater Recharge in Israel, in Water in Israel Part A—Selected Articles, Ministry of Agriculture, Water Commission, Tel Aviv.

JAPAN

Nakazawa, Kazuto, 1972, Water Demand Forecasting in Japan, International Symposium on Water Resources Planning, Mexico, D.F.

ECAFE, 1967, Proceedings of the Seventh Regional Conference on Water Resources Development in Asia and the Far East, UN Water Resources Series No. 32.

ECAFE, 1973, Proceedings of the Tenth Session of the Regional Conference on Water Resources Development in Asia and the Far East, UN Water Resources Series No. 44.

JORDAN

Wilson, G.R., and Wozab, D.H., Chemical Quality of Waters Occurring in the Jordan Valley Area, Publ. No. 37, International Association of Scientific Hydrology, Rome, 1954.

NORTH KOREA

Marynov, V.V., 1961, The Water Resources of the People's Democratic Republic of Korea and Their Utilization, National Technical Information Service, JPRS 6565, Springfield, Va.

KUWAIT

Al Adsani, A.M.S., 1973, The Progress of Desalting Technology in Kuwait, First Congress of Water Resources, International Water Resources Association, Chicago.

El-Shamy, H.K., and others, 1971, Drinking Water in Kuwait, J. Am. Water Work Association, Vol. 63, Dec. Reprinted by permission. AWWA Inc. 6666 West Quincy Ave. Denver, Co. 80235.

MALAYSIA

Drainage & Irrigation Department West Malaysia, Hydrological Data-Streamflow Records 1941-1960, Kuala Lumpur.

Framji, K.K., and Mahajan, I.K., 1969, Irrigation and Drainage in the World, International Commission on Irrigation and Drainage, New Delhi.

PAKISTAN

M. Khan, S.K., 1973, Report on Water Resources Activity in Pakistan, First World Congress on Water Resources, International Water Resources Association, Chicago.

PHILIPPINES

Lesaca, R.M., 1971, National Water and Air Pollution Control Commission, Manila, Status of Water Supply, (Private Communication).

SAUDI ARABIA

Otkun, Galip, 1969, Outline of Ground Water Resources of Saudi Arabia. Paper Presented at International Conference on Arid Lands in a Changing World, Tucson, June.

Shaheen, E.J., 1973, Water Supply and Treatment in Two Regions of Saudi Arabia, Water and Sewage Works, May. 434 S. Wabash, Chicago, Illinois 60605.

SINGAPORE

Chen, Willie, 1972, Water is Precious in the Republic of Singapore, Water & Sewage Works, Sept. 434 S. Wabash, Chicago, Illinois 60605.

SYRIA

Burdon, D.J., and Mazloum, S., 1956, Some Chemical Types of Ground Water from Syria, Salinity Problems in the Arid Zones, UNESCO.

TAIWAN

Wang, K.T., 1963, Ground Water Development in Taiwan, in The Development of Groundwater Resources with Special Reference to Deltaic Areas, ECAFE, UN Water Resources Series No. 24.

THAILAND

USAID, UN, 1968, Atlas of Physical, Economic and Social Resources of the Lower Mekong Basin.

Binson, Boonrod, 1972, Water Resources Development in Thailand, International Symposium on Water Resources Planning, Mexico City.

TURKEY

Cecen, K., 1968, Landeskunde der Turkei, Wasser und Boden, V. 20, No. 2.

Garbrecht, G., 1968, Gediz und Euphrat als Beispiele groszraumiger Wasserwirtschaftsplanungen, Wasser und Boden, V. 20, No. 2.

Ozturgut, Erdogan, 1966, Water Balance of the Black Sea and Flow through the Bosporus, CENTO Symposium on Hydrology and Water Resources Development, Ankara.

U.K. ASIAN DEPENDENCIES
HONG KONG

Robertson, A.S., and La Touche, M.C.D., 1969, Assessing the Yield of Hong Kong's Reservoirs, J. Institution of Water Engineers, Vol. 23, No. 8.

Kwei, S.K., 1972, Office of the Water Authority, Hong Kong, (Private Communication).

SOUTH VIETNAM

Djang, Y.H., 1961, Vietnam's Water Resources and Their Development, FAO.

ECAFE, 1971, Development of Water Resources in the Republic of Vietnam, ECAFE Water Resources Journal, March.

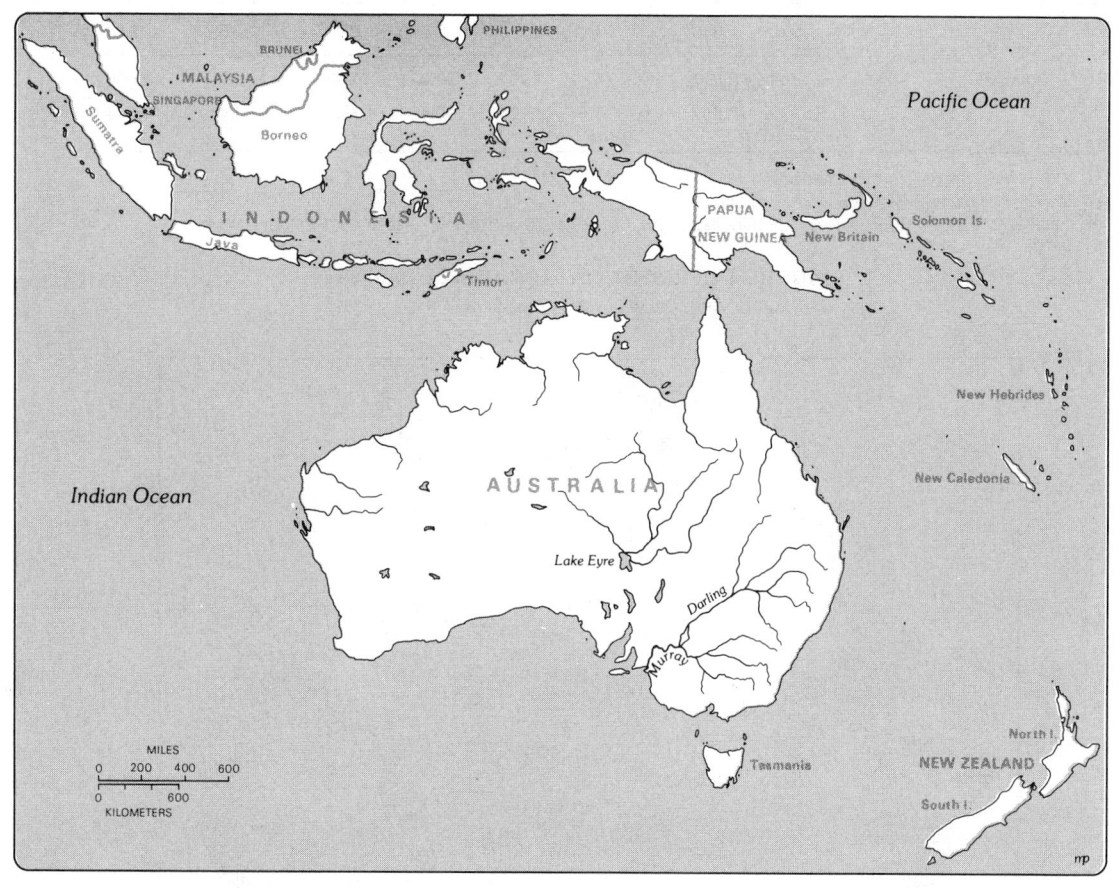

PHILIPPINES

MALAYSIA
BRUNEI
SINGAPORE
Borneo
Sumatra

I·N·D·O·N·E·S·I·A
Java
Timor

Pacific Ocean

PAPUA
NEW GUINEA
New Britain
Solomon Is.

New Hebrides

Indian Ocean

AUSTRALIA

Lake Eyre

Darling

Murray

New Caledonia

MILES
0   200   400   600

0   600
KILOMETERS

Tasmania

NEW ZEALAND

North I.

South I.

mp

TABLE 4-1. OCEANIA—HYDROELECTRIC AND THERMAL POWER GENERATING CAPACITY AND PRODUCTION, 1967-68

(Source: U.S. Federal Power Commission, 1971)

Geographical divisions	Year	Installed capacity (MW.) [1]			Energy production (GWh.) [2]			Population (1,000)	Kwh. per capita
		Hydro	Thermal	Total	Hydro	Thermal	Total		
Australia	1967	2,980	8,515	11,495	7,546	32,172	39,718	11,810	3,363
	1968	3,147	9,319	12,466	7,531	35,074	42,605	12,031	3,541
New Zealand	1967	2,273	702	2,975	9,908	1,607	11,515	2,726	4,224
	1968	2,439	698	3,137	10,257	1,764	12,021	2,751	4,370
Islands [3]	1967	97	610	707	500	2,292	2,792	3,189	876
	1968	97	636	733	345	2,630	2,975	3,259	913
Sub-total	1967	5,350	9,827	15,177	17,954	36,071	54,025	17,725	3,048
	1968	5,683	10,653	16,336	18,133	39,468	57,601	18,041	3,193

[1] MW. — Megawatts=Thousand Kilowatts.
[2] GWh. — Gigawatt-hours=Million Kilowatt-hours.
[3] Includes Brunei, Fiji Islands, French Polynesia, Guam, Mauritius, New Caledonia, Reunion, Ryukyu Island, Samoa (U.S. and Western).

FIGURE 4-1. AUSTRALIA — DRAINAGE BASINS

(Source: ECAFE, UN Water Resources Series No. 32, 1967)

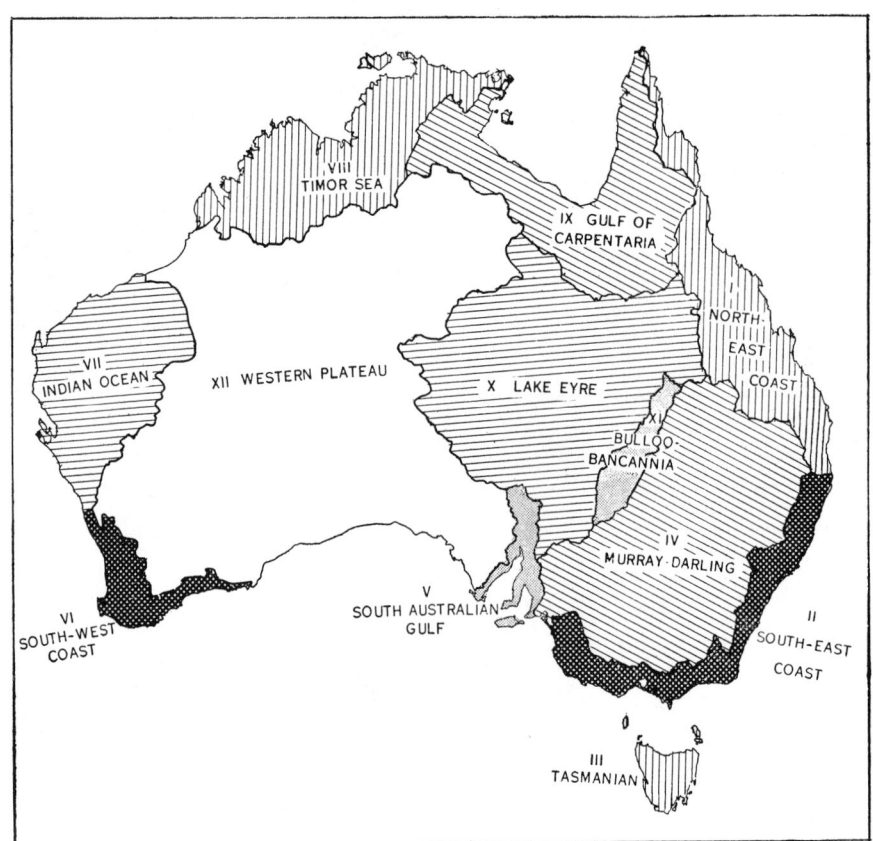

TABLE 4-2. AUSTRALIA—AREA AND ESTIMATED RUNOFF OF DRAINAGE DIVISIONS
(Source: ECAFE, UN Water Resources Series No. 32, 1967)

[Drainage divisions shown on adjacent map]

Division	Area km^2 (sq mi)	Average annual runoff	
		Million m^3 (1,000 ac ft)	mm (in)
I Northeast Coast	454,027 (175,300)	82,836 (67,155)	182.4 (7.18)
II Southeast Coast	268,324 (103,600)	36,425 (29,530)	135.6 (5.34)
III Tasmanian	68,376 (26,400)	47,152 (38,226)	689.6 (27.15)
IV Murray-Darling	1,056,720 (408,000)	23,654 (19,176)	22.4 (0.88)
V South Australian Gulf	75,369 (29,100)	532 (431)	7.1 (0.28)
VI South-West Coast	140,119 (54,100)	7,220 (5,853)	51.6 (2.03)
VII Indian Ocean	519,813 (200,700)	6,158 (4,992)	11.9 (0.47)
VIII Timor Sea	539,497 (208,300)	74,257 (60,200)	137.7 (5.42)
IX Gulf of Carpentaria	640,766 (247,400)	63,121 (51,172)	98.6 (3.88)
X Lake Eyre	1,143,744 (441,600)	4,459 (3,615)	3.8 (0.15)
XI Bulloo-Bancannia	100,751 (38,900)	407 (330)	4.1 (0.16)
Total divisions I-XI	5,007,506 (1,933,400)	346,219 (280,680)	69.1 (27.2)
XII Western Plateau	2,679,355 (1,034,500)	0 (0)	0 (0)
Australia	7,686,861 (2,967,900)	346,219 (280,680)	45.0 (1.77)

Note: Mean precipitation of Australia is 420 mm (16.5 in).

TABLE 4-3. AUSTRALIA—DISCHARGE OF SELECTED RIVERS
(Source: Australian Water Resources Council, 1967)

Basin	River	Station	Latitude	Longitude	Basin area sq mi	Annual discharge (thousands acre ft.)					Based on period
						Mean	Max	(year)	Min.	(year)	
I—North-East Coast											
	Burdekin	Home Hill	19°40'	147°20'	50,075	7,036	21,200	(1954-55)	127	(1920-21)	1920-55
	Fitzroy	Riverslea	23°35'	149°54'	51,020	4,601	23,126	(1953-54)	144	(1930-31)	1922-62
	Burnett	Walla	25°08'	151°59'	12,440	1,312	7,876	(1955-56)	24	(1919-20)	1910-64
II—South-East Coast											
	Clarence	Lilydale	29°31'	152°42'	6,440	3,017	12,103	(1950)	502	(1957)	1922-64
	Hunter	Singleton	32°34'	151°10'	6,350	720	4,062	(1950)	75	(1919)	1898-1964
	Snowy	Jarrahmond	37°40'	148°22'	5,100	1,682	3,254	(—)	776	(—)	42 years
	Glenelg	Dartmoor	37°56'	141°17'	4,200	526	1,308	(1956)	66	(1950)	16 years
III—Tasmania											
	South Esk	Launceston	41°25'	147°10'	3,380	1,775	4,230	(1956)	525	(1914)	1902-65
IV—Murray-Darling											
	Murray	Mildura Weir	34°10'	142°10'	92,000	9,177	25,939	(1870)	4,466	(1902-03)	39 years
	Murrumbidgee	Balranald	34°39'	143°34'	63,750	2,056	9,527	(1956)	323	(1944)	1887-1964
	Darling	Bourke	30°05'	145°57'	149,000	3,582	18,199	(1956)	377	(1946)	1944-63
	Darling	Menindee	32°23'	142°26'	220,000	2,805	23,644	(1890)	1	(1902)	1881-1961
VI—South-West Coast											
	Warren	Pemberton	34°31'	115°59'	1,430	267	570	(1955)	72	(1959)	1940-62
	Murray	Hughes Bridge	32°46'	116°05'	2,720	240	720	(1955)	28	(1940)	1940-62
VIII—Timor Sea											
	Ord	Coolibah Pocket	16°08'	128°44'	17,800	3,200 E	13,300 E	(1959)	<100 E	(1952)	1930-65
X—Lake Eyre											
	Diamantina	Birdsville	25°55'	139°22'	44,480	747	3,085	(1949-50)	13	(1951-52)	1948-64

E—Estimated.

TABLE 4-4. AUSTRALIA—LARGE DAMS AND RESERVOIRS
(Source: ECAFE, UN Water Resources Series No. 32, 1967)

[Over 1,000 million m³ capacity]

Name	Location	Capacity: million m³ (1,000acre-ft.)	Height a) of dam: meters (feet)	Completed (year)	Remarks
EXISTING					
Eucumbene	Eucumbene river, New South Wales	4,760 (3,860)	116 (381)	1958	Part of the Snowy Mountains hydro-electric and irrigation scheme.
Eildon	Upper Goulburn river, Victoria	3,392 (2,750)	79 (260)	1955	Storage for irrigation and for the generation of hydro-electric power.
Hume	Murray river near Albury, New South Wales	3,084 (2,500)	43 (142)	1961	Storage for domestic, stock and irrigation purposes; hydro-electric power also developed.
Menindee Lake Storage	Darling river near Menindee, New South Wales	2,467 (2,000)	15 (50)	1960	Irrigation and water supply.
Warragamba	Warragamba river, New South Wales	2,060 (1,670)	113 (369)	1960	Sydney water supply; also a small hydro-electric power station.
Miena	Great Lake, Tasmania	2,107 (1,709)	18.6 (61)	1967	Increases storage in Great Lake.
Burrinjuck	Murrumbidgee river, New South Wales	1,032 (837)	80 (264)	1956	Storage for irrigation and production of hydro-electric power.
Burrendong	Macquarie river near Wellington, New South Wales	1,678 (1,361)	76 (250)	1967	Rural water supplies, flood mitigation and possible hydro-electric power generation.
Blowering	Tumut river, New South Wales	1,627 (1,320)	100.6 (330)	1968	Regulates discharges from stations of Snowy Mountains scheme, primarily for irrigation but also for power generation.
UNDER CONSTRUCTION					
Wyangala	Lachlan river, New South Wales	1,233 (1,000)	85 (280)		Strengthening and enlarging of existing dam for increased water supply and hydro-electric power generation.
Fairbairn	Nogoa river, Central Queensland	1,440 (1,180)	45 (148)		Irrigation near Emerald.
Ord	Ord river, Western Australia	5,672 (4,600)	67 (220)		Irrigation and hydro-power (later).
Main Gordon Storage	Gordon river, Tasmania	12,330 (10,000)	137 (450)		Gordon river power development, stage 1, middle Gordon scheme.
PROPOSED					
Burdekin Falls	Burdekin river, North Queensland	8,118 (6,584)	46 (150)		Hydro-electric power, irrigation and flood mitigation.
Chowilla	Murray river in South Australia, near Victorian border	6,250 (5,070)	21 (68)		Regulation of lower Murray river.

a) Height of dam above river bed.
Note: Total reservoir capacity of major dams in 1965 was 31×10^9 m³ (25×10^6 acre-ft.).

FIGURE 4-2. AUSTRALIA — SNOWY MOUNTAINS SCHEME

(Source: Warrell, Water for Peace, 1967)

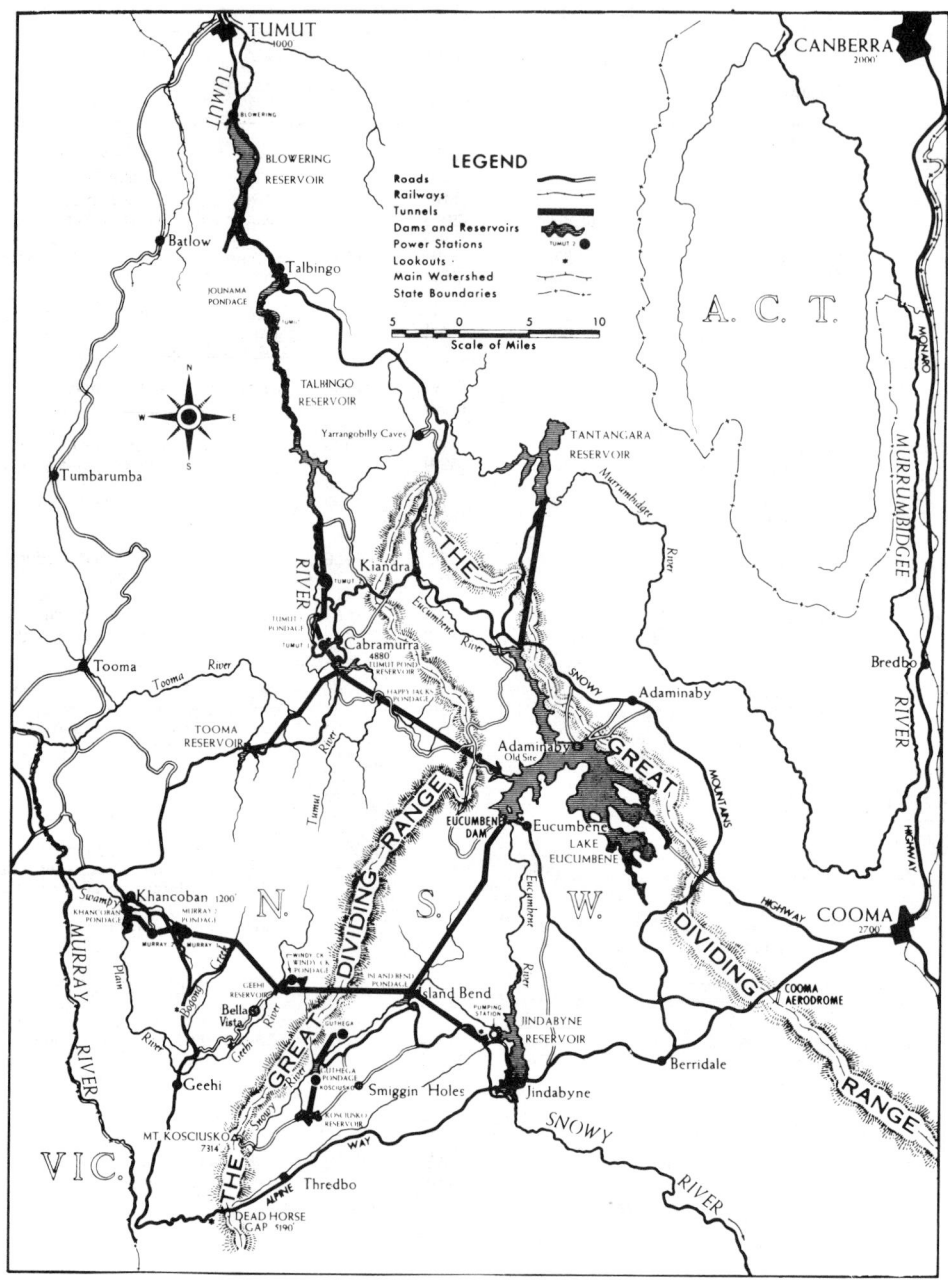

TABLE 4-5. AUSTRALIA—MAJOR WORKS OF SNOWY MOUNTAINS SCHEME

(Source: Warrell, Water for Peace, 1967)

Name	Type [1]	Height (feet)	Crest length (feet)	Dam volume (cubic yards)	Reservoir gross capacity (acre feet)	Year of completion
DAMS						
Talbingo	R	530	2,300	18,500,000	747,000	1971
Eucumbene	ER	381	1,900	8,800,000	3,890,000	1958
Blowering	R	368	2,650	11,300,000	1,320,000	1968
Geehi	R	300	870	1,830,000	17,100	1966
Jindabyne	R	235	1,100	1,150,000	558,000	1967
Tooma	ER	220	1,000	1,440,000	22,800	1961
Jounama	R	144	1,700	807,000	35,300	1968
Khancoban	ER	60	3,500	740,000	21,600	1965
Tumut Pond	CA	283	713	185,000	42,800	1958
Island Bend	CG	160	480	79,000	2,450	1965
Tumut 2	CG	152	390	62,700	2,170	1962
Tantangara	CG	148	710	97,000	206,000	1960
Murray 2	CA	140	430	23,500	1,900	1968
Guthega	CG	110	456	57,000	1,480	1955

Notes: 1. R=Rockfill (at least 50% rock), ER=Earth and Rockfill (less than 50% rock),
CA=Concrete Arch, CG=Concrete Gravity.

Name	Length (miles)	Excavated diameter (feet)	Percentage lined	Lined diameter (feet)	Year of completion
TUNNELS					
Guthega	2.9	19	12	17	1955
Eucumbene-Tumut	13.9	24	28	21	1959
Tumut 1 Pressure	1.5	24	100	21	1959
Tooma-Tumut	9.0	13	21	11	1961
Murrumbidgee-Eucumbene	10.3	12	18	10	1961
Tumut 2 Pressure and Tailwater	7.0	23	100	21	1962
Eucumbene-Snowy	15.2	22	20	20	1965
Snowy-Geehi	9.1	22	13	20	1965
Murray 1 Pressure	7.5	25	100	23	1966
Jindabyne-Island Bend	6.1	14	10	12½	1968
Murray 2 Pressure	1.6	26	100	24½	1967

Name	Number of units	Total rated capacity	Rated head (ft)	Year of completion
POWER STATIONS				
Guthega	2	60,000 MW	810	1955
Tumut 1	4	320,000 MW	960	1959
Tumut 2	4	280,000 MW	860	1961
Tumut 3	6	1,500,000 MW	495	1972
Blowering	1	80,000 MW	284	1969
Murray 1	10	950,000 MW	1,510	1966
Murray 2	4	550,000 MW	867	1969
Jindabyne Pumping Station	2	900 cusecs	760	1968
Tumut 3 Pump Units	3	10,500 cusecs	509	1973

TABLE 4-6. AUSTRALIA—PRINCIPAL GROUND-WATER BASINS
(Source: ECAFE, UN Water Resources Series No. 32, 1967)

Basin	State	Area km²	Area sq mi
Great Artesian	Queensland, New South Wales, South Australia and Northern Territory	1,751,487	676,250
Canning	Western Australia	388,500	150,000
Murray	Victoria, New South Wales and South Australia	282,310	109,000
Georgina	Northern Territory and Queensland	279,720	108,000
Eucla	Western Australia	191,660	74,000

TABLE 4-7. AUSTRALIA—OCCURRENCE AND QUALITY OF GROUND WATER
(Source: Morse, Water for Peace, 1967)

State or Territory	Area of underground water occurrence. Sq miles and percentage of State or Territory a)	Quality of water and area in which available				Area in which occurrence of water or water quality is unknown. Sq miles percentage of total area
		Less than 1,000 ppm. Sq miles and percentage of total area	1,000-3,000 ppm. Sq miles and percentage of total area	3,000-7,000 ppm. Sq miles and percentage of total area	7,000-14,000 ppm. Sq miles and percentage of total area	Greater than 14,000 ppm. Sq miles and percentage of total area
New South Wales	422,300 (135b)	95,200 (23)	119,700 (28)	33,700 (8)	42,700 (10)	94,500 (22)	36,500 (9)
Victoria	122,500 (139c)	28,200 (23)	20,900 (17)	15,200 (12)	3,900 (3)	15,000 (12)	0
Queensland	1,117,000 (167d)	359,800 (32)	163,800 (15)	19,000 (2)	0	0	318,600 (29)
South Australia	389,200 (102g)	64,000 (16)	71,300 (18)	49,500 (13)	29,400 (8)	51,500 (13)	123,500 (32)
Western Australia	1,129,500 (115e)	3,000 (6)	37,500 (3)	51,500 (5)	0	46,600 (4)	983,200 (87)
Tasmania	26,100 (100)	10,900 (42)	500 (2)	0	0	0	14,000 (54)
Northern Territory	575,000 (110f)	84,100 (15)	58,600 (10)	11,500 (2)	1,400 (0)	700 (3)	274,600 (50)
Australian Capital Territory	910 (100)	460 (50)	440 (50)	0	0	8,100 (1)	10 (1)
AUSTRALIA	3,782,500 (130)	645,700 (17)	472,700 (12)	180,400 (5)	77,400 (3)	216,400 (6)	1,750,400 (46)

a) Aquifers often overlap so that the summation of their areas can give a total greater than the area of the State concerned. These areas indicate potential water availability over wide areas, rather than known occurrences.
b) Excludes 65,600 sq miles of porous rock areas containing water of poor quality.
c) Includes 39,300 sq miles of area containing water that ranges widely in quality beyond the categories listed above.
d) Includes 255,800 sq miles of areas containing water that ranges widely in quality beyond the categories listed above.
e) Includes 7,700 sq miles of areas containing water that ranges widely in quality beyond the categories listed above.
f) Includes 136,700 sq miles of areas containing water that ranges widely in quality beyond the categories listed above.
g) Excludes 18,500 sq miles of porous rock areas containing water of poor quality.

TABLE 4-8. AUSTRALIA—WATER DEMAND FOR MINERAL DEVELOPMENT
(Source: Dept. of National Development, Canberra. ECAFE, UN Water Resources Series No. 44, 1973)

Region and major minerals		Established annual water demand current and projected		Sources of supply; remarks
		million m³	acre-feet	
Queensland:	Cape York Peninsula bauxite	141	114,000	
	Leichhardt and Cloncurry River Basins copper	44	36,000	
	phosphate	37	30,000	
	Isaac River Basin coal	197	16,000	
Northern Territory:	Gove Peninsula bauxite			
	Groote Island manganese	1970: 14	11,400	Existing supply ground water but ample surface water for future.
	Tennant Creek copper	1980: 25	20,300	
	Arnhem Land uranium			
	McArthur River silver-lead-zinc			
Western Australia:	Pilbera Region iron	1970: 8.3	6,700	Current demand being met by ground water; combined use likely in future.
		1990: 83	677,000	
East Murchison:	Windarra, Agnew Mt. Keith nickel			Large-scale ground water investigation in progress financed by mining companies.
	Yeelirrie uranium	58	47,000	
	Barrambie vanadium			

TABLE 4-9. AUSTRALIA—MUNICIPAL WATER SUPPLY SYSTEMS
(Source: ECAFE, UN Water Resources Series No. 32, 1967)

City	Population served	Total storage capacity (million m³)	Per capita storage (m³)	Annual supply (million m³)	Per capita annual supply (m³)
Sydney	2,760,000	2,600	942	450	163
Melbourne	2,260,000	300	133	320	142
Brisbane	725,000	340	469	115	158
Adelaide	700,000	105 a)	136	130	169
Perth	525,000	280	533	100	190
Hobart	130,000	4 b)	28	30	213
Canberra	110,000	90	818	30	270
Newcastle	330,000	195 c)	591	55	166
Geelong	113,000	61	540	22	195

a) A major part of the supply comes from the Murray river which is regulated by upstream storage.
b) The source of supply, the Derwent river, is regulated for hydro-electric power generation. The capacity is used as a holding storage.
c) Part of the supply comes from groundwater.

TABLE 4-10. AUSTRALIA—TERRITORY OF PAPUA & NEW GUINEA.
DISCHARGE OF PRINCIPAL RIVERS

(Source: Jones, Water for Peace, 1967)

[Data to January 1, 1965; drainage divisions shown on adjacent map]

River and station	Basin area (sq miles)	Period of record (years)	Gaged flows in cusecs		
			Highest	Mean	Lowest
South East Coast Division (13,630 sq miles)					
Laloki	120	13	15,800	490	70
Tauri	1,085	9	41,000	4,800	810
Angabunga	613	12	19,800	3,900	750
Oreba	390	11	14,200	1,700	280
Vanapa	750	1½	44,800	4,900	970
Eilogo	6.1	3	640	40	4
Laloki	154	3	10,300	690	70
Brown	828	10	17,600	2,800	500
Eworogo	30	10	3,600	160	21
Laloki	62	9	13,300	220	12
Eilogo	12	7	1,100	60	14
Goldie	31	3	6,300	280	70
Musgrave	34	3	14,500	300	50
Warama	0.35	2	10	—	4
Gulf Division (29,195 sq miles)					
Goroka	15	7½	2,100	90	21
Asaro	86	5	7,700	510	190
Dunantina	108	6	29,000	410	120
Aibe	6.5	1½	1,200	—	10
Bena	125	5½	15,200	500	120
Omahaga	13.4	4½	2,200	80	22
Purari	11,100	3	362,000	87,000	27,000
Tuma	15.1	1½	1,800	—	13
Huon Division (10,230 sq miles)					
Gabensis	21	3	1,300	100	15
Oomsis	13.5	4	800	20	0
Gabensis	31	7	680	75	6
Snake	165	7	4,800	420	130
Wanton	50	5½	37,200	310	60
North Central Division (9,790 sq miles)					
Ramu	343	6½	16,000	1,140	200
North East Coast Division (10,165 sq miles)					
Musa	1,625	6	38,000	9,300	2,100
Waria	580	2½	12,800	3,300	1,300
Mambare	820	3	132,000	7,600	2,500

FIGURE 4-3. AUSTRALIA — TERRITORY OF PAPUA AND NEW GUINEA
DRAINAGE DIVISIONS

(Source: Jones, Water for Peace, 1967)

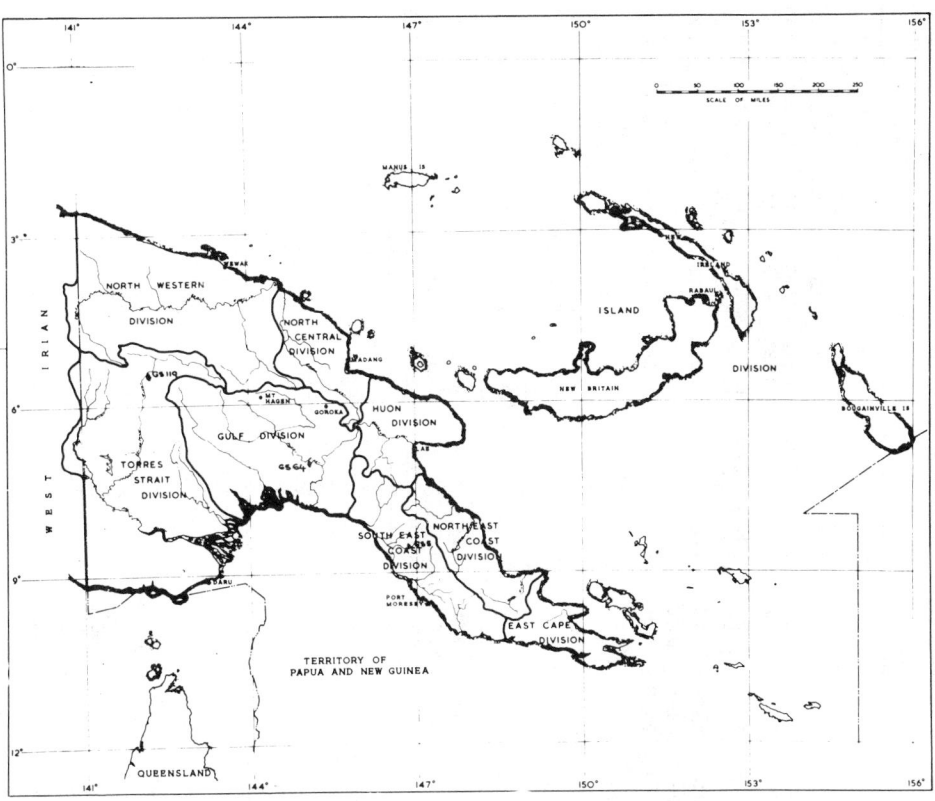

TABLE 4-11. FRENCH PACIFIC DEPENDENCIES-NEW CALEDONIA-DISCHARGE OF SELECTED RIVERS

(Source: ORSTOM, 1968)

River and station	Basin area km²	Jan.	Feb.	Mar.	Apr.	May	Jun.	Jul.	Aug.	Sep.	Oct.	Nov.	Dec.	Year	Period of record
Plaine des Lacs, Goulets	61	8.1	9.7	6.7	9.1	4.8	4.4	2.5	3.2	2.1	1.1	2.7	2.8	4.7	1956-65
Hienghene, Hienghene	114	19.2	12.8	9.46	8.19	5.31	4.89	3.57	1.75	2.24	0.67	0.62	4.34	5.21	1954-65
Tipindje, Ouen-Cout	247	12.7	14.9	13.1	11.6	7.19	8.40	4.45	3.05	3.30	1.21	1.18	4.03	7.04	1955-65
Quinne, Mouth	143	5.75	26.0	26.7	34.9	15.7	11.6	3.33	6.27	4.52	2.38	15.5	5.83	13.1	1963-65
Dumbea Nord, Station	32.2	0.43	3.08	2.69	5.20	1.62	1.61	0.64	1.01	0.79	0.36	1.38	0.38	1.58	1963-65
Dumbea Est, Dam	56.2	1.40	6.57	5.12	9.64	3.23	2.87	0.99	1.97	1.22	0.46	2.52	0.54	3.00	1963-65

TABLE 4-12. NEW ZEALAND—DISCHARGE OF PRINCIPAL RIVERS
(Source: UNESCO, 1971)

River and station	Basin area km²	Mean monthly discharge, m³/s												Year	Period of record
		Jan.	Feb.	Mar.	Apr.	May	Jun.	Jul.	Aug.	Sep.	Oct.	Nov.	Dec.		
Buller, Te Kuha	6,350	874	176	388	294	555	244	572	657	617	589	518	524	501	1964
Manganui, Tariki Road	81	2.77	3.93	4.64	3.11	4.56	8.89	7.30	11.1	3.54	4.25	14.2	6.11	6.20	1965;
		5.63	4.78	5.66	4.07	6.37	9.79	8.74	10.29	7.78	7.92	7.98	4.98	6.99	1962-65
Selwyn, Whitecliffs	151	4.98	5.60	3.45	4.47	1.87	2.83	3.88	3.59	2.52	2.97	3.79	1.53	3.45	1965;
		2.38	2.41	2.01	4.19	3.85	4.19	7.90	4.44	3.88	2.49	3.42	1.94	3.59	1962-65
Cleddau, Milford	155	33.7 *	25.4 *	35.9 *	21.0 *	24.8 *	21.5 *	8.12 *	4.50 *	22.0 *	20.9 *	41.3 *	35.1 *	24.5 *	1965;
		36.2 *	20.1 *	30.6 *	18.5 *	24.8 *	12.5 *	13.4 *	16.2 *	20.2 *	21.3 *	41.9 *	29.7 *	23.8 *	1964-65
Omakere, Fordale	54	0.001	0.0002	3.03	0.314	0.236	1.73	1.47	6.71	1.06	0.303	0.154	0.179	1.27	1965
Moutere, Old House Road	62	0.079	0.020	0.012	0.050	0.058	0.631	1.18	1.45	0.277	0.120	0.144	0.050	0.340	1965;
		0.074	0.174	0.196	0.183	0.719	1.29	1.44	1.49	0.931	1.48	0.532	0.161	0.724	1962-65
Clutha, Clyde	12,020	439	365	390	413	504	526	492	450	359	430	450	719	462	1964;
		436	397	384	386	377	394	416	465	445	512	561	509	440	1960-64
Waiau, Lake Manapouri Outlet	4,620	504	470	303	487	354	291	291	205	173	226	269	461	334	1966;
		402	390	368	393	405	374	340	308	348	441	467	444	391	1932-66
Waitaki, Kurow	9,750	174	314	323	317	273	289	340	286	180	149	144	42	235	1964
Waikato, Taupo Outlet	3,290	147	138	120	99.6	191	165	163	182	164	120	87.2	89.4	139	1966;
		133	127	121	112	113	117	130	138	140	142	140	133	129	1906-63
Jollie, Mount Cook Station	145	13.3	10.4	7.13	5.43	5.29	4.98	3.37	3.68	5.15	8.46	14.0	15.6	8.07	1965
Hutt, Kaitoke	85	2.38	3.34	16.7	3.62	4.44	11.8	8.52	17.9	7.33	5.97	23.5	5.12	9.22	1965;
		5.09	3.54	7.05	5.06	6.85	10.7	11.9	12.5	12.2	6.82	13.1	4.41	8.26	1960-65

* Incomplete record.

TABLE 4-13. NEW ZEALAND—DATA ON MAJOR LAKES
(Source: ECAFE, UN Water Resources Series No. 38, 1968)

	Lake	Surface area sq miles	Drainage area sq miles	Mean discharge cfs	Height above sea level feet
North Island	Taupo[a]	243	1,270	4,500	1,170
	Rotorua	31	203	—	920
	Waikaremoana[a]	21	165	620	2,000
	Wairarapa	31	150	—	2
South Island	Tekapo[a]	37	550	3,000	2,340
	Pukaki[a]	32	523	4,500	1,620
	Ohau	23	460	2,300	1,725
	Hawea[a]	46	567	2,200	1,100
	Wanaka	74	982	7,200	900
	Wakatipu[a]	113	1,150	6,200	1,010
	Te Anau	133	1,275	9,700	680
	Manapouri	55	1,785	13,900	600
	Hauroko	27	225	1,100	510
	Ellesmere[a]	70	745	—	0
	Benmore[a]	30	3,000	11,500	1,180

[a] Lake level controlled for power generation.

TABLE 4-14. NEW ZEALAND—NON-CONSUMPTIVE USE OF WATER BY HYDROELECTRIC POWER PLANTS, 1965-66

(Source: ECAFE, UN Water Resources Series No. 36, 1968)

Power-station	Static head H feet	Generation 1965/66 GWh	Water used [a] million acre-feet	Net water used million acre-feet	Remarks
Karapiro	100	525	6.04		
Arapuni	175	824	5.42		
Waipapa	53	255	5.52		
Maraetai	200	851	4.90		
Whakamaru	124	489	4.54	6.04	Successive re-use
Atiamuri	81	276	3.92		
Ohakuri	115	382	3.82		
Aratiatia	110	284	2.97		
Kaitawa	443	96	0.25		
Tuai	676	205	0.35	0.42	Successive re-use
Piripaua	370	135	0.42		
Mangahao	896	71	0.09	0.09	
Cobb	1,950	170	0.10	0.10	
Arnold	42	27	0.74	0.74	
Highbank	330	112	0.39	0.39	
Coleridge	490	135	0.32	0.32	
Tekapo	100	82	0.94		
Benmore	305	1,635	6.15	6.45	Successive re-use
Waitaki	70	393	6.45		
Roxburgh	150	1,283	9.84	9.84	
Monowai	154	36	0.36	0.36	
Total government generation		8,266	63.43	24.75	
Non-government generation		333	2.55 [b]	2.55	
Total national generation		8,599	65.98	27.30	

[a] 1.15 kW/H.
[b] Pro rata water use assumed.

TABLE 4-15. NEW ZEALAND—AGRICULTURAL WATER USE, 1970

(Source: Burton and Heiler, New Zealand Water Conference, 1970)

Unit	Quantity	Estimated Unit Consumption	Total annual consumption (million of gals.)
Sheep	60.029 X 10⁶ head	1 gphd	22,000
Dairy Cattle	3.505 X 10⁶ head	30 gphd	38,400
Beef Cattle	4.241 X 10⁶ head	10 gphd	15,400
Pigs	0.602 X 10⁶ head	2½ gphd	550
Irrigated land	204,597 acres	2.16 ft/acre/yr	120,000

TABLE 4-16. NEW ZEALAND—INDUSTRIAL WATER DEMAND, 1967-79

(Source: Ferrier, New Zealand Water Conference, 1970)

Industry	Estimated annual Water-use 1967-68 million gallons	Predicted increase to 1978-79 percent	Forecast annual Water-use 1978-79 million gallons
Aerated soft drinks	33	27	42
Aluminium smelting	170	—	260
Brewing	395	27	500
Cement	430	90	820
Coal Mining	360	54	555
Dairy	9,740	82	17,700
Milk supply	2,800	27	3,550
Food processing	2,180	27	2,770
Gas			
A. Manufactured	180	Static	180
B. Natural	480	50	720
Meat	1,290	33	1,720
Oil refining	72	77	128
Pulp and paper	23,700	92	45,500
Steel	8.7	—	24
Wool textile	2,900	33	3,860
	45,000		78,310
Deduct 20% allowance ex public supplies (say)	8,000		17,310
Sub-totals	37,000		61,000
Add allowance for other industries ex public supplies	18,500	30 (say)	24,000
Totals	55,500	65	85,000

TABLE 4-17. NEW ZEALAND–SOURCES OF WATER FOR PUBLIC SUPPLIES

(Source: Kingsford, New Zealand Water Conference, 1970)

Source of water	Population served	% of Total population	No. of water supply systems
Surface water...............	1,150,817	54.9	181
Ground water...............	580,858	27.7	112
Mixed supply	365,730	17.4	12
Total Population	2,097,405	100	

TABLE 4-18. NEW ZEALAND–MUNICIPAL WATER SUPPLY SYSTEMS

(Source: ECAFE, UN Water Resources Series No. 38, 1968)

Local authority	Population served thousands	Type of supply	Average annual mgd	Demand			Price cents/ 1,000 gallons
				Daily	Industrial	Domestic	
					gallons per capita		
Auckland Regional Authority	500	Upland reservoirs	12,000	72	22	50	30
Christchurch City	150	Artesian wells	4,500	84	–	–	10
Wellington, Porirua, Tawa and Upper Hutt	170	Upland reservoirs	5,000	90	28	62	25
Lower Hutt and Eastbourne	55	Artesian wells	1,600	78	6	72	15
Dunedin City	90	Upland reservoirs and groundwater	3,000	89	20	69	15
Hamilton	55	Waikato river	1,700	85	–	–	15
Palmerston North	45	Tiritea stream	2,000	83	20	63	20
Invercargill and Bluff	40	Oreti river	1,000	60	13	47	20
Wanganui	35	Upland reservoirs	1,200	98	–	–	13
New Plymouth	32	L. Mangamahoe	1,300	111	–	–	23
Timaru	25	Pareora river	900	94	–	–	15
Napier	25	Artesian wells	1,200	123	22	101	13
Hastings	25	Artesian wells	1,000	112	–	–	11
Nelson and Richmond	30	Upland reservoirs	–	–	–	–	–
Gisborne	25	Upland reservoirs	650	81	12	69	25
Rotorua	20	Ututina spring	600	85	5	80	10
Tauranga	20	Waiorohi stream	600	80	10	70	18
Whangarei	20	Upland reservoirs	700	92	–	–	28
Forty towns of 4,000 to 16,000 population	Over 250		Over 10,000	50-120	–	–	10-25

TABLE 4-19. UNITED STATES PACIFIC DEPENDENCIES—GUAM
DISCHARGE OF SELECTED STREAMS
(Source: Ward, et al., U.S. Geological Survey, 1962)

Name	Period of record	Drainage area sq mi	Average discharge		Instantaneous maximum discharge cfs	Maximum daily discharge cfs	Minimum daily discharge cfs
			cfs	Inches per year			
Umatac River	1952-62	2.04	7.99	53.2	7,460	500	0.25
Inarajan River	1952-62	4.49	15.6	47.2	—	1,580	1.16
Ugum River	1952-62	6.96	27.5	53.7	5,900	1,790	3.40
Ylig River	1952-62	6.58	23.5	48.5	4,040	2,050	.18
Pago River	1951-62	6.18	22.7	49.9	5,310	2,540	0

TABLE 4-20. WESTERN SAMOA—MUNICIPAL WATER USE
(Source: ECAFE, 1968)

Total population	137,260 (1966)
Population served by public water supply systems	80,000
Per capita consumption	15-25 gallons/day
Annual water consumption	600 million gallons
Source of water	Rivers and streams

TABLE 4-21. WESTERN SAMOA—DATA ON HYDROELECTRIC POWER PLANTS
(Source: ECAFE, 1968)

Station	Alaoa	Fuluasou	Magiagi
Height of station above sea level (ft)	394	108	N.A.
Installed capacity (kVA)	1,250	287	80
Static head (ft)	455	206	185
Length of pipeline (ft)	1,550	8,245	N.A.
Size of pipeline (diameter/inches)	48, 33	33, 32, 31	22, 20, 18
Reservoir capacity (cu ft)	400,000	1,400,000	N.A.
Water consumption at full load (cu ft/min)	3,200	1,600	N.A.
Minimum stream flow (cu ft/min)	N.A.	960	N.A.
Average annual generation (million kWh)	3.5	1.3	0.6

N.A. — Not available.

FIGURE 4-4. WESTERN SAMOA – WATER SUPPLY SYSTEMS

(Source: ECAFE, UN Water Resources Series No. 36, 1968)

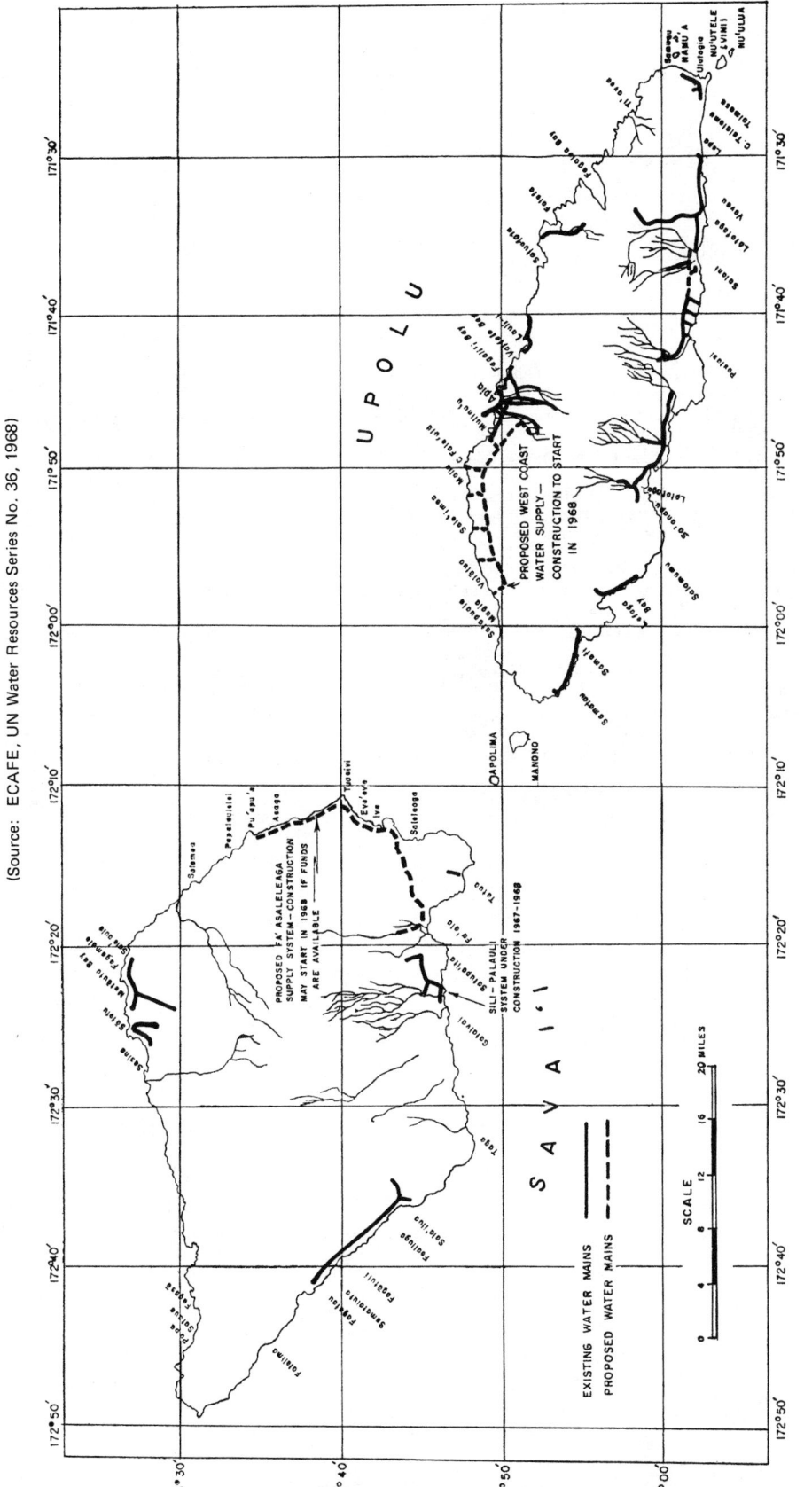

OCEANIA
REFERENCES CITED

References cited in more than one country:

Proceedings of the International Conference on Water for Peace, 1967, Superintendent of Documents, Washington, D.C. 20402.

ECAFE, 1968, Multiple-Purpose River Basin Development-Part 2E-Water Resources Development in Australia, New Zealand and Western Samoa, UN Water Resources Series No. 36.

OCEANIA GENERAL

Federal Power Commission, 1968, World Power Data Publ. P-40.

AUSTRALIA

ECAFE, 1967, Proceedings of the Seventh Regional Conference on Water Resources Development in Asia and the Far East, UN Water Resources Series No. 32.

ECAFE, 1973, Proceedings of the Tenth Session of the Regional Conference on Water Resources Development in Asia and the Far East, UN Water Resources Series No. 44.

Australian Water Resources Council, 1967, Stream Gaging Information, June 1965, Canberra.

FRENCH PACIFIC DEPENDENCIES—NEW CALEDONIA

Office de la Recherche Scientifique et Technique Outre-Mer (ORSTOM), Annuaire Hydrologique de la France d'Outre-Mer, Annees 1963-1964-1965, Paris, 1968.

NEW ZEALAND

Kingsford, Michael, and others, 1970, The Chemistry of New Zealand Drinking Water Supplies, Proc. New Zealand Water Conference, Auckland.

Burton, J.R., and Heiler, T.D., 1970, Efficient Use of Water in Agriculture, Proc. New Zealand Water Conference, Auckland.

UNESCO, 1969, Discharge of Selected Rivers of the World, Vol. I. General and Regime Characteristics of Stations Selected.

ECAFE, 1968, Proceedings of the Eighth Session of the Regional Conference on Water Resources Development in Asia and the Far East, UN Water Resources Series No. 38.

Ferrier, D.A., 1970, Water Use by Industry, Proc. New Zealand Water Conference, Auckland.

UNITED STATES PACIFIC DEPENDENCIES—GUAM

Ward, P.E., and others, 1962, Geology and Hydrology of Guam, Mariana Islands, U.S. Geological Survey.

Arctic Ocean

U.S.S.R.

ICELAND

GREENLAND

Yukon

Mackenzie

Great Bear Lake

Great Slave Lake

Hudson Bay

C A N A D A

Nelson

St. Lawrence

Columbia

Great Lakes

Missouri

Pacific Ocean

Atlantic Ocean

U N I T E D   S T A T E S

Ohio

Colorado

Arkansas

BERMUDA

Mississippi

Rio Grande

M E X I C O

Gulf of Mexico

BAHAMAS

PUERTO RICO

CUBA

DOMINICAN REPUBLIC

HAITI

JAMAICA

BELIZE

Caribbean Sea

HONDURAS

VENEZUELA

GUATEMALA
EL SALVADOR

NICARAGUA

PANAMA

COSTA RICA

COLOMBIA

MILES
0   200   400   600

0   600
KILOMETERS

np

TABLE 5-1. NORTH AMERICA—AVAILABLE WATER RESOURCES IN CENTRAL AMERICA

(Source: CEPAL, 1972)

[By country and drainage basin]

Country	Country Total					Atlantic Slope					Pacific Slope				
	Available water resources					Available water resources					Available water resources				
	Mean flow m3/s	Low flow m3/s	Ground water m3/s	l/s/km2 1)	1,000m3/ capita/yr 1)	Mean flow m3/s	Low flow m3/s	Ground water m3/s	l/s/km2 1)	1,000m3/ capita/yr 1)	Mean flow m3/s	Low flow m3/s	Ground water m3/s	l/s/km2 1)	1,000m3/ capita/yr 1)
Costa Rica	3,019	496	334	59.6	54.2	1,579	270	195	67.0	145.3	1,440	226	139	53.0	32.1
El Salvador	601 2)	90 3)	83	30.0	5.6	—	—	—	—	—	601 2)	90 3)	83	30.0	5.6
Guatemala	3,697	510	194 4)	28.2	22.3	2,744	298	—	25.3	30.2	953	212	194	40.8	12.7
Honduras	3,229	434	288	28.1	39.8	2,947	394	254	30.9	59.8	282	40	34	14.1	8.9
Nicaragua	5,520	564	527	42.5	90.7	5,302	552	488	45.4	111.8	218	12	39	16.4	16.2
Panama	4,038	592	105	—	—	1,588	314	14	—	—	2,450	278	91	—	—

1) Calculated from mean flow.
2) Includes 154 m3/s originating in Honduras and Guatemala.
3) Includes 13 m3/s originating in Honduras and Guatemala.
4) Partial estimate.

TABLE 5-2. NORTH AMERICA—CHARACTERISTICS OF IMPORTANT RIVER BASINS IN CENTRAL AMERICA

(Source: CEPAL, 1972)

Country and river basin	Area 1,000 km2	Available water resources m3/s	Population (1,000)		Total Water Use m3/s	
			Total	Urban	1970	1980
Costa Rica	20.8	959	1,424.9	566.3	133.8	215.0
Rio Tempisque and adjacent rivers	11.8	368	282.6	59.2	22.0	35.0
Rio Grande de Tarcoles	2.1	101	905.3	451.6	58.3	58.3
Rio Reventazon, Pacuare and others	6.9	490	237.0	55.5	58.5	72.3
Rio San Juan a)						49.4
El Salvador	14.0	519	2,885.3	1,307.4	186.8	432.2
Rio Paz a)	1.0	25	163.2	57.2	0.2	7.9
Rio Lempa a)	9.0	409	2,340.3	1,115.0	160.2	367.3
Rio Sonsonate and Banderas	1.6	28	199.9	68.2	14.6	17.8
Rio Grande de San Miguel	2.4	57	181.9	67.0	11.8	39.2
Guatemala	92.4	2,238	4,268.9	1,641.4	89.3	240.5
Rio Samala, Nahualate, etc.	7.7	417	858.3	254.6	13.0	84.6
Rio Achiguate, Maria Linda	4.0	180	620.3	377.3	23.2	78.2
Rio Los Esclavos	3.2	58	167.0	37.9	12.7	27.4
Rio Lempa and Paz a)	4.1	53	279.7	72.2	1.1	8.8
Rio Motagua a)	14.0	189	1,336.5	751.3	19.2	21.1
Rio Selegua, Usumacinta and Hondo a)	59.4	1,341	1,007.1	148.1	20.1	20.4
Honduras	83.2	2,282	1,816.1	563.9	116.4	337.6
Rio Choluteca	7.9	71	466.1	259.9	15.2	30.9
Rio Chamelecon and Ulua	27.3	670	1,000.6	253.4	64.5	222.4
Rio Patuca, Aguan and others	48.0	1,541	349.4	50.6	36.7	84.3
Nicaragua	57.7	1,743	1,524.3	743.8	89.0	514.0
Rio Estero Real and Brito	11.4	192	371.8	180.7	44.0	71.3
Rio San Juan a)	26.6	789	836.9	506.6	24.8	370.4
Rio Grande de Matagalpa	19.7	762	315.6	56.5	20.2	72.3
Total Central America	268.1	7,741	11,919.5	4,822.8	615.3	1,739.3

a) International river basin.

TABLE 5-3. NORTH AMERICA—WATER USE IN CENTRAL AMERICA, 1970-80
(Source: CEPAL, 1972)

Country and year	Total use, m3/s			Irrigation			Drinking water and industrial supply		Hydroelectric power generation		River navigation m3/s
	Water withdrawn	Net use	Water consumed	Area under irrigation 1,000 ha	Total water demand m3/s	Water consumed m3/s	Total water demand m3/s	Water consumed m3/s	Installed capacity MW	Water use m3/s	
Costa Rica											
1970	217.1	43.0	20.5	45.7	39.6	19.4	3.4	1.1	166	69.6	104.5
1980	285.2	61.7	29.3	55.2	56.2	27.5	5.5	1.8	346	119.0	104.5
El Salvador											
1970	187.9	27.4	12.7	23.6	23.6	11.6	3.8	1.1	102	141.5	19.0
1980	398.8	91.1	43.2	81.6	84.0	41.2	7.1	2.0	305	288.7	19.0
Guatemala											
1970	106.5	23.2	10.8	19.1	19.1	9.4	4.1	1.4	99	38.9	44.4
1980	262.8	152.0	72.8	144.3	144.3	70.7	7.7	2.1	289	66.4	44.4
Honduras											
1970	118.8	42.6	20.4	49.8	40.8	20.0	1.8	0.4	32	19.2	57.0
1980	360.3	169.9	82.8	190.9	167.0	82.0	2.9	0.8	412	133.4	57.0
Nicaragua											
1970	168.3	51.1	24.8	43.3	49.4	24.3	1.7	0.5	50	12.7	104.5
1980	541.0	123.0	59.6	115.4	119.1	58.5	3.9	1.1	100	23.0	395.0
Panama											
1970	433.8	27.8	151.1	23.7	23.8	11.5	4.0	3.6	62	115.4	290.6
1980	705.1	75.1	199.6	56.1	68.7	33.8	6.4	5.8	167	315.4	314.6
Total Central America											
1970	1,232.4	215.1	240.3	205.2	196.3	96.2	18.8	8.1	511	397.3	620.0
1980	2,553.2	672.8	487.3	643.5	639.3	313.7	33.5	13.6	1,619	945.9	934.5

TABLE 5-4. NORTH AMERICA—HYDROELECTRIC AND THERMAL POWER GENERATING CAPACITY AND PRODUCTION, 1968

(Source: U.S. Federal Power Commission, 1971)

Country	Installed capacity (MW)[1]			Energy production (GWh.)[2]			Population (1,000)	Kwh per capita
	Hydro	Thermal	Total	Hydro	Thermal	Total		
NORTH AMERICA								
Canada	24,957	10,951	35,908	134,973	41,405	176,378	20,772	8,491
Greenland	0	32	32	0	61	61	45	1,356
Mexico	2,941	3,440	6,381	12,621	10,110	22,731	47,267	473
United States	51,874	258,307	310,181	225,874	1,210,155	1,436,029	201,177	7,138
Subtotal	79,772	272,730	352,502	374,368	1,261,731	1,635,199	269,261	6,073
CENTRAL AMERICA								
British Honduras	0	5	5	0	16	16	116	138
Canal Zone.	47	68	115	274	318	592	56	10,571
Costa Rica	179	58	237	758	65	823	1,634	504
El Salvador.	109	63	172	412	158	570	3,266	175
Guatemala	43	107	150	156	434	590	4,864	121
Honduras	32	59	91	162	112	274	2,413	114
Nicaragua	57	100	157	288	186	474	1,842	257
Panama	16	119	135	54	586	640	1,372	466
Subtotal	483	577	1,060	2,104	1,875	3,979	15,563	256
WEST INDIES								
Bahamas	0	46	46	0	209	209	177	1,181
Bermuda	0	52	52	0	182	182	50	3,640
Cuba	45	1,290	1,335	119	4,427	4,546	8,074	563
Dominican Republic. . . .	15	188	203	54	611	665	4,029	165
Haiti	0	35	35	0	111	111	4,671	24
Jamaica	21	258	279	118	950	1,068	1,913	558
Lesser Antilles [3]	1	145	146	1	417	418	952	439
Netherlands Antilles. . . .	0	259	259	0	1,167	1,167	215	5,428
Puerto Rico	105	1,117	1,222	195	5,369	5,564	2,723	2,044
Trinidad and Tobago . . .	0	253	253	0	1,064	1,064	1,021	1,042
Subtotal	178	3,643	3,830	487	14,507	14,994	23,825	629

[1] MW. — Megawatts=Thousand Kilowatts.
[2] GWh. — Gigawatt-hours=Million Kilowatt-hours.
[3] Includes Barbados, Guadeloupe, Martinique and Virgin Islands.

TABLE 5-5. BARBADOS—AVAILABLE WATER RESOURCES AND WATER USE IN 1962
(Source: United Nations, 1964)

Water Resources:

Land area	430 sq km
Annual precipitation	1,016-1,524 mm
Estimated available fresh water resources (ground water and surface water)	53.3 million m3/yr

Water Use:	Millions of cubic meters
Households	9.4
Industry	7.0
Sugar mills	(4.0)
Irrigation	4.4
Losses and waste	6.2
Total	27.0

TABLE 5-6. BARBADOS—DATA ON PUBLIC WATER SUPPLY SYSTEM
(Source: Barbados Waterworks Dept., 1972)

Population (1970)	238,000
Source of water	wells 96%
	springs 4%
Capacity	35 mgd
Consumption	22 mgd
Storage capacity...........................	13.1 million gallons
Water quality	
Total dissolved solids	300 ppm
Total hardness	172 ppm

TABLE 5-7. BELIZE—MUNICIPAL WATER SUPPLY SYSTEMS [1]
(Source: National Water and Sewerage Authority, 1972)

Municipality	Source of water	Description of system
Belize City (Pop. 40,000)	Ground water	Water pumped from 3 shallow wells 10 miles from city to 400,000 gallon cap. storage tank in Belize City.
Belmopan	Surface water	Water from Belize River is treated and distributed.
Benque Viejo del Carmen	Ground water	Spring water pumped to storage reservoir.
Corozal Town	Ground water	Water from 2 wells 1 mile west of town stored in 50,000 gallon cap. storage tank.
San Ignacio	Surface water	Water from Mocal River pumped into 2 reservoirs with total cap. of 200,000 gallons.
Stann Creek Town	N.A.	Capacity of system 150,000 gallons.
Punta Gorda	Rain water and ground water	Rain water augmented by well; total storage cap. 360,000 gallons.

[1] Total population in urban areas served by public water systems: 33,000 (about 50% of urban population).
N.A. — Not available.

FIGURE 5-1. CANADA. DRAINAGE BASINS
(Source: Canadian National Committee IHD, 1972)
[Basin names and discharge data shown on Table 5-8]

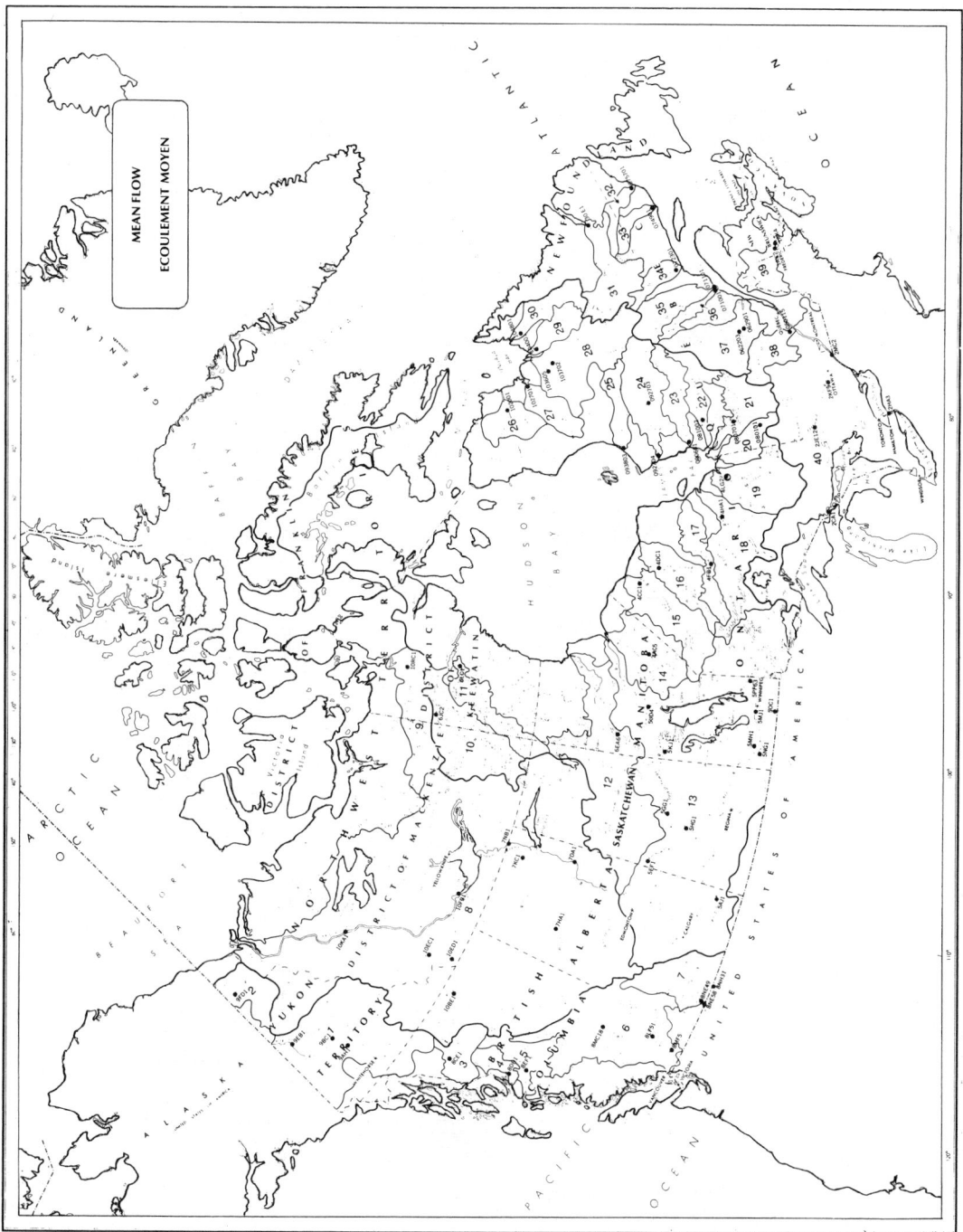

TABLE 5-8. CANADA—DRAINAGE BASINS AND MEAN ESTIMATED RIVER FLOW

(Source: Canadian National Committee IHD, 1972)

[Flow at mouths or points where rivers cross Canadian-USA border]

	Area No.	River basin	Drainage Area		Mean Discharge	
			mi^2	km^2	cfs	m^3/s
Pacific Ocean	1	Yukon	115,000 [1]	298,000	82,000	2,320
	2	Porcupine	21,500 [2]	55,700	16,000	456
	3	Stikine	19,000	49,200	39,100	1,100
	4	Nass	8,000	20,700	33,800	957
	5	Skeena	21,200	54,900	61,100	1,730
	6	Fraser	84,800	220,000	125,000	3,540
	7	Columbia	59,700 [3]	154,600	98,700	2,800
Arctic Ocean	8	Mackenzie	697,000	1,810,000	343,000	9,700
	9	Back	41,400	107,226	18,500	524
Hudson Bay	10	Thelon	55,000	142,000	29,700	841
	11	Kazan	27,600	71,500	19,200	544
	12	Churchill	109,000	281,000	42,400	1,200
	13	Nelson	414,005 [4]	723,000	83,600	2,370
	14	Hayes	41,700	108,000	20,800	589
	15	Severn	38,900	101,000	17,000	482
	16	Winisk	26,000	67,300	14,900	422
	17	Attawapiskat	19,400	50,200	15,000	425
	18	Albany	51,700	134,000	49,300	1,400
	19	Moose	41,900	108,000	48,600	1,380
	20	Harricana	11,300	29,300	20,000	566
	21	Nottaway	25,400	65,800	41,900	1,190
	22	Rupert	16,700	43,200	31,700	897
	23	Eastmain	17,900	46,400	32,800	928
	24	La Grande Riviere	37,700	97,700	59,700	1,690
	25	Grande Riviere de la Baleine	16,500	42,700	24,000	679
	26	Arnaud	19,100	49,400	23,700	671
	27	Aux Feuilles	16,400	42,500	20,800	589
	28	Koksoak	51,500	133,000	89,900	2,550
	29	Riviere a la Baleine	12,300	31,800	20,400	577
	30	George	16,100	41,700	33,100	937
Atlantic Ocean	31	Churchill	30,800	79,800	55,700	1,580
	32	Petit Mecatina	7,560	19,600	18,200	515
	33	Natashquan	6,220	16,100	14,400	407
	34	Moisie	7,410	19,200	17,300	490
	35	Manicouagan	17,700	45,800	36,100	1,020
	36	Aux Outardes	7,360	19,100	14,100	399
	37	Saguenay	34,000	88,100	61,700	1,750
	38	St. Maurice	16,700	43,200	25,900	733
	39	Saint John	21,400 [5]	55,400	39,800	1,130
	40	St. Lawrence	396,000 [6]	1,026,000	348,000	9,860
SUMMARY		Pacific Drainage	417,000 [7]	1,080,000	750,000	21,200
		Arctic Drainage	1,380,000	3,580,000	547,000	15,500
		Hudson Bay	1,560,000	4,040,000	1,040,000	29,400
		Atlantic Drainage	790,000 [8]	2,040,000	1,190,000	33,700
		Gulf of Mexico Drainage....	10,600	27,500	900	25
		TOTAL	4,160,800	10,776,000 [9]	3,530,000	99,800

[1] Including 9,000 sq mi USA.
[2] Including 700 sq mi USA.
[3] Including 20,000 sq mi USA.
[4] Including 57,300 sq mi USA.
[5] Including 7,600 sq mi USA.
[6] Including 195,000 sq mi USA.
[7] Including 29,700 sq mi USA.
[8] Including 203,000 sq mi USA.
[9] Including 290,000 sq mi USA.

TABLE 5-9. CANADA—MONTHLY DISCHARGE OF PRINCIPAL RIVERS

(Source: Canadian National Committee IHD, 1972)

River and station	Basin area km2	Mean monthly discharge, m3/s												Year	Period of record
		Jan.	Feb.	Mar.	Apr.	May	Jun.	Jul.	Aug.	Sep.	Oct.	Nov.	Dec.		
Yukon, Dawson	275,000	558	463	421	516	3,370	5,880	4,740	3,680	2,860	1,970	1,020	690	2,210	1945-70(19-24)[1]
Nass, Shumal Creek	18,000	173	158	131	336	1,250	2,090	1,700	1,230	872	832	491	253	810	1929-70(17-36)[1]
Skeena, Usk	42,200	202	179	147	370	1,870	2,810	1,720	953	736	816	564	317	916	1928-70(28-39)[1]
Fraser, Hope	203,000	924	856	796	1,640	4,830	7,050	5,580	3,570	2,420	1,990	1,580	1,140	2,710	1912-70
Columbia, Birchbank	88,000	641	610	644	1,020	3,210	6,020	4,840	2,720	1,610	1,210	1,330	1,320	1,630	1937-70
Kootenay, Copeland	34,700	130	139	162	495	1,240	1,410	693	292	204	195	178	154	444	1929-70
Mackenzie, Norman Wells	1,570,000	4,400	3,340	3,060	3,130	9,490	15,300	15,200	12,400	10,700	9,280	6,780	4,980	7,340	1943-70(3-21)[1]
Slave, Fitzgerald	606,000	1,750	1,430	1,180	1,390	4,430	6,530	6,230	5,150	4,390	3,910	2,390	1,770	3,440	1921-70(11-20)[1]
Peace, Peace River	186,500	353	340	347	1,140	3,710	5,780	3,570	1,700	1,210	1,200	756	455	1,730	1915-32-1957-70
Saskatchewan, The Pas	324,000	168	161	171	626	1,130	1,220	1,450	1,110	768	540	352	196	676	1913-70
North Saskatchewan, Prince Albert	119,500	40.8	40.4	44.3	242	336	530	610	451	317	174	84.4	51.1	244	1910-70
South Saskatchewan, Saskatoon	139,500	87.5	83.7	119	376	413	757	589	280	222	170	124	94.7	275	1911-70
South Saskatchewan, Medicine Hat	58,400	70.1	73.5	112	198	386	675	387	170	131	119	101	74.6	210	1911-70
Red, Emerson	104,000	19.1	18.0	43.9	330	219	143	100	44.9	38.4	35.1	32.1	23.8	87.8	1912-70
Assiniboine, Headingley	162,000	7.95	7.20	11.5	121	150	91.2	66.3	35.4	21.8	20.4	16.1	10.9	46.2	1913-70
Assiniboine, Brandon	92,000	4.57	4.06	7.46	81.9	105	60.2	47.7	22.9	15.9	16.1	12.8	6.95	31.2	1906-70
Souris, Wawanesa	63,400	0.562	0.512	2.12	28.9	28.6	18.4	10.7	5.22	2.60	2.29	1.87	0.776	8.54	1912-70(16-58)[1]
Winnipeg, Slave Falls	126,000	721	716	692	744	937	1,040	1,060	883	781	774	785	755	835	1907-70
Aux Outardes, Power Plant	18,900	149	114	104	171	948	900	497	405	378	400	311	216	382	1922-70
Saguenay, Isle-Maligne	73,000	917	911	917	1,090	2,760	2,550	1,630	1,410	1,380	1,530	1,390	1,030	1,470	1913-70
Saint-Maurice, Grand-Mere	42,000	565	564	568	880	1,460	821	644	584	570	622	636	565	708	1922-70
Saint John, Pokiok	38,800	308	223	332	1,800	2,410	809	458	334	328	473	762	535	734	1918-67
St. Lawrence, Cornwall	766,000	6,040	6,160	6,110	6,290	6,440	6,710	6,760	6,730	6,570	6,400	6,360	6,400	6,430	1958-70
Niagara, Queenston	660,000	5,490	5,340	5,520	5,780	6,060	6,110	6,000	5,890	5,770	5,670	5,680	5,680	5,750	1860-1970
Ottawa, Chats Falls	89,600	823	784	887	1,910	2,430	1,690	1,110	788	705	824	1,010	928	1,170	1914-70
St. Marys, Sault Ste. Marie	209,500	1,920	1,860	1,830	1,890	2,060	2,170	2,290	2,350	2,370	2,340	2,270	2,070	2,120	1860-1970

1 Incomplete record; number of months or years used to calculate mean flows in parenthesis.

TABLE 5-10. CANADA—FLOW CHARACTERISTICS OF PRINCIPAL RIVERS

(Source: Canadian National Committee IHD, 1972)

River and station	Mean annual flow m³/s	Maximum daily flow		Minimum daily flow m³/s
		m³/s	Date	
Yukon, Dawson	2,210	14,900	June 11-'64	180
Nass, Shumal Creek	810	4,360	June 18-'31	24.4
Skeena, Usk	916	6,540	June 14-'50	51.8
Fraser, Hope	2,710	15,200	May 31-'48	340
Columbia, Birchbank	1,630	10,600	June 9-'61	365
Kootenay, Copeland	444	2,830	May 23-'54	38.2
Mackenzie, Norman Wells	7,340	22,500	June 4-'67	2,110
Slave, Fitzgerald	3,440	8,750	July 16-'65	544
Peace, Peace River	1,730	14,600	July 11-'65	149
Saskatchewan, The Pas	676	3,000	June 11-'48	50.7
North Saskatchewan, Prince Albert	244	5,300	July 2-'15	11.2
South Saskatchewan, Saskatoon	275	3,940	June 15-'53	14.2
South Saskatchewan, Medicine Hat	210	4,080	June 11-'53	10.2
Red, Emerson	87.8	2,670	May 13-'50	0.025
Assiniboine, Headingley	46.2	615	April 27-'16	0.566
Assiniboine, Brandon	31.2	651	May 7-'23	0.198
Souris, Wawanesa	8.54	323	April 22-'69	0.000
Winnipeg, Slave Falls	835	2,800	June 30-'54	261
Aux Outardes, Plant	382	2,830	May 29-'43	10.5
Saguenay, Isle Maligne	1,470	9,260	May 31-'28	51.0
Saint-Maurice, Grand-Mere	708	5,100	May 24-'47	110
Saint John, Pokiok	734	8,160	May 2-'23	27.5
St. Lawrence, Cornwall	6,430	8,690	June 28-'60	4,500
Niagara, Queenston	5,750	8,470	Nov. 17-'55	2,440
Ottawa, Chats Falls	1,170	4,110	May 9-'67	323
St. Marys, Sault. Ste. Marie	2,120	N.A.		N.A.

N.A. — Not available.

TABLE 5-11. CANADA—MAJOR LAKES, BY PROVINCE

(Source: Dominion Bureau of Statistics, 1970)

Province and lake	Elevation ft.	Area sq miles
NEWFOUNDLAND		
Melville	sea level	1,133
Michikamau	1,521	566
QUEBEC		
Eau Claire (a l')	790	535
Mistassini	1,229	840
ONTARIO		
Erie (total 9,889) part	572	4,912
Huron, including Georgian Bay (total 23,860) part	580	15,353
Nipigon	855	1,870
Ontario (total 7,313) part	245	3,849
Seul (reservoir)	1,172	539
Superior (total 32,483) part	602	11,524
Woods, Lake of the (total 1,695) part (reservoir)	1,060	953
MANITOBA		
Cedar	830	517
Island	744	550
Manitoba	814	1,817
Moose	838	525
Southern Indian	835	1,060
Winnipeg	713	9,465
Winnipegosis	833	2,103
SASKATCHEWAN		
Athabasca (total 3,120) part	699	2,180
Lac la Ronge	1,198	552
Reindeer (total 2,467) part	1,150	2,096
Wollaston	1,300	796
ALBERTA		
Athabasca (total 3,120) part	699	940
Claire	699	545
NORTHWEST TERRITORIES		
Baker	30	975
Dubawnt	764	1,600
Great Bear	511	12,275
Great Slave	513	10,980
La Martre	870	685
Maguse	—	540
Nueltin (total 850) part	875	580
Yathkyed	461	860

Note: Areas given are for mean water levels. All elevations are in feet above mean sea level. "Total" refers to the area of the whole lake; "part" refers to the area within the designated province or territory.

TABLE 5-12. CANADA—HYDROPOWER RESOURCES, BY PROVINCE

(Source: Dominion Bureau of Statistics, 1970)

[As of January 1, 1969]

Province or Territory	Undeveloped Water Power			Developed Water Power
	Available continuous power at 88% efficiency			Installed generating capacity kw
	at Q 95[1] kw	at Q 50[2] kw	at Q m[3] kw	
Newfoundland	1,195,000	3,450,000	4,641,000	820,000
Prince Edward Island	—	1,000	2,000	—
Nova Scotia	21,000	112,000	161,000	163,000
New Brunswick..............	29,000	106,000	276,000	563,000
Quebec....................	7,791,000	27,657,000	86,276,000	11,049,000
Ontario....................	462,000	1,088,000	1,635,000	6,412,000
Manitoba	2,964,000	5,501,000	5,853,000	1,184,000
Saskatchewan	650,000	1,171,000	1,434,000	584,000
Alberta....................	895,000	3,244,000	4,866,000	616,000
British Columbia.............	4,697,000	15,954,000	23,984,000	3,538,000
Yukon Territory	664,000	3,237,000	5,689,000	18,000
Northwest Territories	864,000	2,232,000	3,322,000	35,000
Canada	20,232,000	63,753,000	88,189,000	24,982,000

[1] Power equivalent of flow available 95% of the time.

[2] Power equivalent of flow available 50% of the time.

[3] Power equivalent of arithmetical mean flow.

TABLE 5-13. CANADA—WATER PUMPED FROM SELECTED MINES

(Source: Brown, Geological Survey of Canada, 1967)

[Approximate figures only; water introduced for mining purposes not separated]

Mine	Location	Ore mined	Depth ft.	Water pumped (gpd)
Nova Scotia				
MacBean	Thorburn	coal	1,275	600,000
Greener mine	Sydney Mines	coal	280	575,000
Walton mine	Walton	barite	970	2,700,000
New Brunswick				
Wedge	Newcastle	copper-zinc	1,150	150,000
Heath Steele	Newcastle	copper-zinc	400	720,000
Quebec				
Gaspe	Murdochville	copper	700	1,100,000 to 6,000,000
East Malartic	Norrie	gold	4,950	500,000
Johnsons	Thetford Mines	asbestos	1,200	115,000
Kilmar	Kilmar	magnesite	850	1,300,000
Newfoundland				
Rothermere	Buchans	lead-zinc	2,500	20,000
Tilt Cove	Tilt Cove	copper	2,700	600,000
Ontario				
Langis Silver	New Liskeard	silver	400	200,000
No. 2 mine	Caledonia	gypsum	75	75,000
Madsen	Madsen	gold	4,075	109,000
Errington	Atikokan	iron	1,480	2,100,000
Frood	Sudbury	nickel-copper	3,600	300,000
Lake Shore	Kirkland Lake	gold	8,075	200,000
Manitoba				
San Antonio	Bissett	gold	4,930	7,500
Amaranth	Amaranth	gypsum	130	1,000
Flin Flon	Flin Flon	copper-zinc	4,000	616,000
Saskatchewan				
Beaverlodge	Uranium City	uranium	4,000	280,000
Alberta				
Atlas	East Coulee	coal	400	86,000
British Columbia				
Bluebell	Riondel	silver-lead-zinc	1,110	5,900,000
Britannia	Britannia Beach	copper-zinc	4,000	up to 6,000,000
Northwest Territories				
Con-Rycon	Yellowknife	gold	3,500	470,000
Yukon Territory				
Hector and Calumet	Elsa	silver-lead-zinc	1,300	50,000

TABLE 5-14. CANADA–POPULATION SERVED WITH FLUORIDATED WATER

(Source: Connor, Water for Peace, 1967)

[As of January 1, 1966]

Province	Living in communities with controlled fluoride		Percent of total population in communities with controlled fluoridation
	Population	Percent	
Newfoundland	6,000	0.1%	1.2%
Prince Edward Island	–	0.0	0.0
Nova Scotia	191,200	4.0	25.1
New Brunswick	–	0.0	0.0
Quebec	410,900	8.5	7.3
Ontario	3,166,100	66.0	47.0
Manitoba	573,300	11.9	59.6
Saskatchewan	279,600	5.8	29.4
Alberta	76,200	1.6	5.3
British Columbia	86,900	1.8	4.9
Yukon	5,100	0.1	34.0
Northwest Territories	6,500	0.2	26.0
Canada	4,801,800	100.0	24.5

TABLE 5-15. CANADA–WATER SUPPLY OF DWELLINGS NOT SUPPLIED BY MUNICIPAL WATER SUPPLY SYSTEMS

(Source: Brown, Geol. Survey of Canada, 1967)

[Sources of water unknown but probably largely from wells; data as of 1962]

Type of supply	Number of dwellings	Average occupancy	Estimated water supply gpd per dwelling	Estimated water supply 1,000's of gal. per day
Piped into dwelling	759,349	3	200	151,870
Not piped into dwelling	377,401	3	100	37,740
Total Canada	1,136,750			189,610

TABLE 5-16. CANADA–WATER USED BY MUNICIPAL WATER SUPPLY SYSTEMS [1]

(Source: Brown, Geol. Survey of Canada, 1967)

[Data as of 1963]

Province	No. of systems	No. supplying groundwater entirely or partly	Water supplied in 1,000's of gal. per day		Population served in 1,000's	Per capita supply gal. per day
			Groundwater	Total		
Alberta	64	26	3,755	90,537	810	112
British Columbia	83	20	9,664	160,711	1,278	126
Manitoba	42	10	1,333	85,819	1,048	82
New Brunswick	23	16	10,185	57,019	250	228
Newfoundland	21	1	300	23,217	155	150
Nova Scotia	34	14	4,660	38,408	341	113
Ontario	294	120	97,824	669,638	6,443	104
Prince Edward Island	3	3	2,908	2,908	29	100
Quebec	272	81	28,156	573,660	4,286	134
Saskatchewan	47	33	10,691	51,348	254	202
Totals	883	324	169,476	1,753,265	14,894	135 Average

Groundwater supplied % of total 9.7

[1] Serving populations of 1,000 or more.

TABLE 5-17. COSTA RICA—MONTHLY DISCHARGE OF SELECTED RIVERS

(Source: CEPAL, 1971)

River	Station	Basin area km2	Mean monthly discharge, m^3/s												Year
			Jan.	Feb.	Mar.	Apr.	May	Jun.	Jul.	Aug.	Sep.	Oct.	Nov.	Dec.	
San Carlos	Jabillos	545	65.7	27.4	28.4	17.9	23.5	53.5	77.1	66.0	79.6	73.2	72.5	93.1	56.5 1)
Reventazon	Pascua	1,685	147.1	74.7	72.3	51.4	77.0	139.6	175.5	163.4	196.9	189.0	180.0	121.4	132.4 2)
Pacuare	Siquirres	763	74.0	28.5	30.4	16.3	27.0	48.8	66.7	60.6	76.7	92.6	90.7	50.9	55.3 3)
Banano	Asuncion	91	18.4	12.5	10.0	14.9	16.2	15.8	17.9	15.3	13.4	15.5	22.4	25.8	16.5 4)
Tempisque	Guardia	955	19.7	14.0	11.0	9.6	15.6	34.5	23.4	24.5	38.3	57.2	38.3	25.8	26.0 5)
Corobici	Corobici	332	7.7	5.5	4.1	3.3	4.2	7.8	9.5	9.5	17.1	26.9	21.3	10.4	10.6 6)
Barranca	Nagatac	191	5.0	3.9	3.2	2.9	5.2	11.2	14.5	13.9	31.1	43.6	17.2	7.3	13.2 7)
Grande de Tarcoles	Balsa	1,627	47.3	35.2	30.4	30.6	40.3	72.3	86.0	80.2	149.2	175.2	124.6	69.9	78.4 8)
Virilia	San Miguel	817	20.7	17.1	14.8	14.2	19.9	33.9	34.4	30.4	58.7	76.2	50.5	30.7	33.5 9)
Grande de Candelaria	El Rey	666	9.8	6.9	5.4	5.0	6.8	25.9	42.0	31.1	68.7	82.6	60.7	17.7	30.3 10)
Grande de Terraba	Palmar	4,843	114.5	84.1	67.3	75.4	215.9	367.2	390.8	387.8	510.8	651.6	608.9	202.6	306.4 11)

Maximum flows and month:
1) 1,210.0 (Dec.)
2) 1,436.0 (Dec.)
3) 724.3 (Dec.)
4) 789.0 (Dec.)
5) 1,518.3 (Oct.)
6) 451.4 (Oct.)
7) 712.1 (Oct.)
8) 1,425.0 (Sep.)
9) 1,176.0 (Nov.)
10) 589.6 (Sep.)
11) 3,500.0 (Nov.)

TABLE 5-18. COSTA RICA—FLOW CHARACTERISTICS OF SELECTED RIVERS
(Source: CEPAL, 1971)

River	Location	Basin area km2	Discharge, m3/s			Percentage runoff from June to November	Period of record
			Mean	Max.	Min.		
Atlantic Slope							
Frio	Cote	15	2.63	18.95	0.48	50	1965-67
San Carlos	Jabillos	545	56.49	1,210.00	15.36	62	1963-65
Arenal	San Gregado	328	54.47	286.00	18.10	52	1965-67
Arenal	Arenal	199	12.71	44.38	10.10	65	1960-65
Reventazon	Pascua	1,685	132.35	1,436.00	33.20	66	1963-65
Reventazon	Angostura	1,367	101.88	1,285.00	22.88	64	1953-67
Reventazon	El Congo	885	60.97	1,206.00	15.10	64	1962-65
Reventazon	Cachi	692	48.61	1,030.00	10.30	67	1956-66
Reventazon	Cordoncillal	254	20.48	436.40	5.20	66	1958-66
Reventazon	Tapanti	190	13.53	309.00	3.08	68	1962-66
Pejibaye	Oriente	226	35.18	940.00	5.20	65	1962-67
Pejibaye	El Humo	130	25.37	615.00	3.75	64	1953-66
Macho	Montecristo	65	4.90	219.50	1.00	68	1955-66
Macho	Belen	46	3.14	184.00	0.44	72	1960-66
Pacuare	Siquirres	763	55.27	724.30	11.00	66	1963-65
Pacuare	Pacuare	459	33.68	670.00	5.20	64	1958-67
Banano	Asuncion	91	16.50	789.000	2.17	51	1957-67
Pacific Slope							
Tempisque	Guardia	955	25.98	1,518.25	6.70	69	1951-65
Colorado	Colorado	130	4.15	223.68	0.74	63	1951-65
Corobici	Corobici	332	10.61	451.38	1.24	73	1954-65
Barranca	Nagatac	191	13.24	712.10	1.14	83	1955-65
Grande de Tarcoles	Balsa	1,627	78.42	1,425.00	19.23	73	1959-67
Grande de Tarcoles	La Presa	644	13.02	576.80	0.00	93	1959-67
Grande de Tarcoles	El Desarenador	644	18.10	22.60	11.53	53	1959-65
Poas	Tacares	183	12.33	560.80	3.20	68	1953-67
Virilla	San Miguel	817	33.45	1,176.00	8.30	71	1956-67
Grande de Candelaria	El Rey	666	30.31	589.60	3.50	53	1963-65
Grande de Terraba	Palmar	4,843	306.39	3,500.00	42.50	80	1962-65
Grande de Terraba	Cristo Rey	830	70.84	1,137.20	14.30	82	1963-65

TABLE 5-19. COSTA RICA—AVAILABLE WATER RESOURCES
(Source: CEPAL, 1971)

River basin	Basin area km2	Precipitation million m3		Surface water runoff m3/s			Ground water; estimated dependable yield b) m3/s
		Normal year	Dry year a)	Normal year	Dry year a)	95% flow	
Atlantic Slope							
San Juan c)	12,325	37,416	28,062	830	623	208	150
Tortuguero, Reventazon, Pacuare, Matina	6,869	20,601	15,452	490	367	49	44
Banano, Estrella and others	1,710	4,841	3,631	100	75	5	—
Changuinola, Sixaola c)	2,616	7,169	5,950	159	132	8	1
Total	23,520	70,027		1,579		270	195
Pacific Slope							
Tempisque, Bebedero and others	11,836	23,185	16,428	368	260	55	37
Grande de Tarcoles	2,133	4,553	3,415	101	76	35	29
Grande de Candelaria, Naranjo, Savegre	4,228	10,831	8,665	240	192	24	4
Grande de Terraba	4,871	16,343	13,565	363	301	57	38
Golfo Dulce	4,112	16,584	13,765	368	306	55	31
Total	27,180	71,496		1,440		226	139
Total Costa Rica.......	50,700	141,523		3,019		496	334

a) Not totalled as droughts usually occur in limited areas.
b) Large withdrawal of ground water will reduce 95% flow of rivers.
c) International river basin; figures represent Costa Rica portion only.

TABLE 5-20. COSTA RICA—AVAILABLE GROUND-WATER RESOURCES

(Source: CEPAL, 1971)

[Estimated replenishable resource; ground water in storage excluded; in million m3]

River basin	Total infiltration	Discharge to ocean		Evapotranspiration		Base Flow		
		Total	Recoverable[a]	Total	Recoverable[b]	Total	Recoverable	Dependable yield[e]
Atlantic Slope								
San Juan	9,900	2,010	14	—	—	7,890	4,720 d)	4,735
Chirripo, Matina	304	68	5	68	20	168	79 c)	105
Pacuare	221	45	3	68	21	108	51 c)	75
Reventazon	1,180	102	5	226	68	852	506 d)	580
Tortuguero	1,840	540	15	523	157	777	451 d)	625
Banano, Estrella	247	247	—	—	—	—	—	—
Changuinola	14	14	—	—	—	—	—	—
Sixaola	194	146	—	—	—	48	24 c)	25
Total	13,900	3,172	42	885	266	9,843	5,831	6,145
Pacific Slope								
P. Nicoya	265	140	—	125	38	—	—	40
Tempisque	1,880	91	7	205	62	15,841	785 c)	855
Bebedero and others	958	316	14	205	61	437	206 c)	280
G. Tarcoles	1,560	9	2	—	—	1,551	922 d)	925
G. Candelaria, Naranjo, Savegre	550	296	14	—	—	254	113 c)	130
G. Terraba	2,040	23	3	68	20	1,949	1,166 d)	1,190
Golfo Dulce	2,370	248	18	432	130	1,690	827 c)	975
Total	9,623	1,123	58	1,035	311	7,465	3,419	4,395
Total Costa Rica	23,523	4,320	100	1,920	577	17,308	9,250	10,540

a) Based on aquifer characteristics.
b) Estimated at 30% of total.
c) Estimated at 50% of base flow less recoverable discharge to sea.
d) Estimated at 60% of base flow less recoverable discharge to sea.
e) Total of recoverable ground water.

TABLE 5-21. COSTA RICA–HYDROELECTRIC POWER PLANTS
(Source: CEPAL, 1970)

[Data as of 1969-70]

River	Plant	Capacity MW	Energy GWh	Head m	Water use, m³/s	
					Total	Non-repetitive
Atlantic Slope						
Reventazon	Cachi	64.0	562.0	244	33.0	33.0
Macho	Rio Macho	30.0	146.0	464	4.5	
Birris	Birris - 1	1.5	6.7 *	160	0.6	
Birris	Birris - 2	2.4	10.5 *	215	0.7	
Pacuare	Pacuare	0.5	2.2 *	105	0.3	
Asuncion	Asuncion	0.7	3.2 *	68	0.1	0.1
Total		99.1	730.6		39.2	33.1
Pacific Slope						
G. Tarcoles	Garita	30.0	171.0	157	15.6	15.6
Tiribi	Avance	0.2	1.1 *	148	0.1	
Tiribi	P. Escondido	0.2	0.8 *	80	0.2	
Tiribi	Los Lotes	0.4	1.7 *	96	0.2	
Anonos	Anonos	0.6	2.6 *	28	1.3	
Virilla	Belen	5.0	21.9 *	96	3.3	
Virilla	Brazil	2.8	12.3 *	54	3.2	
Virilla	Ventanas	10.0	43.8 *	84	7.4	7.4
Virilla	Electriona	2.7	11.9 *	81	2.1	
Circuelas	Nuestro Amo	7.5	32.8 *	173	2.7	
Poas	Carrillos	2.0	8.8 *	142	0.9	
Poas	Cacao	0.7	3.0 *	42	1.0	
Poas	Tacares	5.5	24.3 *	97	3.6	
El Gallito	La Joya	0.3	1.5 *	153	0.2	
Segundo	Rio Segundo	0.3	1.1 *	57	0.3	
Total		68.2	338.6		42.1	23.0
Total Costa Rica		137.3	1,069.2		81.3	56.1

* Estimated at 4.38 X MW.

TABLE 5-22. CUBA–DISCHARGE OF PRINCIPAL RIVERS

(Source: Comite Nacional Cubano IHD, 1969)

River and station	Basin area km2	Year	Mean monthly discharge, m3/s												Year	Max. flow m3/s			Mean daily min. flow m3/s
			Jan.	Feb.	Mar.	Apr.	May	Jun.	Jul.	Aug.	Sep.	Oct.	Nov.	Dec.		Mean daily	Peak flow	Date	
San Diego, Los Gavilanes	157	1966	0.2	0.3	0.6	1.7	10.0	7.4	7.2	6.2	6.5	5.3	1.1	0.6	4.0	74	130	May 8	0.14
		1967	0.4	0.5	0.2	0.2	0.1	11.0	3.0	2.4	3.7	3.5	5.7	1.2	2.7	130	290	Jun. 18	0.03
		1968	0.6	0.4	0.2	0.1	0.7	9.0	3.2	2.4	3.1	6.5	1.7	0.7	2.4	51	120	Jul. 21	0.07
Cuyaguateje, Portales II	502	1966	1.0	0.8	4.2	3.5	29.0	19.0	26.0	14.0	6.3	35.0	3.9	1.5	12.0	250	430	May 9	0.50
		1967	1.1	1.1	0.6	0.5	1.4	49.0	12.0	13.0	30.0	20.0	9.5	3.9	12.0	280	790	Jun. 16	0.26
		1968	1.6	0.9	0.7	0.4	11.0	27.0	13.0	11.0	13.0	24.0	6.3	2.1	9.3	145	190	Jun. 2	0.35
Santa Cruz, Santa Ana	30	1966	0.5	0.8	0.7	0.8	4.0	4.8	4.1	2.0	1.0	4.1	0.7	0.5	2.0	36	135	May 30	0.26
		1967	0.3	0.2	0.2	0.2	0.2	4.6	0.8	0.4	2.4	3.4	2.4	0.8	1.3	33	130	Jun. 17	0.13
		1968	0.4	0.2	0.1	0.1	2.9	5.7	2.5	0.9	2.5	5.2	0.7	0.2	1.8	38	115	Oct. 16	0.10
San Cristobal, La Campana	100	1966	2.0	1.9	2.0	2.0	5.3	10.5	7.8	4.3	2.8	6.6	1.6	0.9	4.0	115	350	Jun. 8	0.50
		1967	0.7	0.6	0.4	0.2	0.1	6.3	0.9	0.9	5.2	7.0	4.9	1.3	2.4	79	350	Jun. 17	0.09
		1968	1.7	0.6	0.4	0.3	6.3	12.0	5.0	1.4	3.4	13.0	1.8	0.7	3.9	125	400	Oct. 16	0.20
Zaza, Paso Ventura	848	1966	1.2	4.4	3.0	2.0	37.0	66.0	31.0	13.0	7.0	7.6	3.2	1.7	15.0	630	1,350	Jun. 8	0.76
		1967	0.9	0.6	0.4	0.7	1.0	21.0	5.1	15.0	13.0	14.0	5.3	2.3	6.5	71	280	Jun. 11	0.10
		1968	1.4	0.8	0.3	0.1	7.4	78.5	17.0	12.5	25.0	53.5	18.0	4.2	18.0	468	964	Oct. 16	0.01
Gibara, El Jobo	80	1966	0.2	0.1	0.1	0.1	0.7	4.6	1.5	0.2	0.4	2.0	3.1	0.9	1.2	82	680	Jun. 26	0.01
		1967	0.7	0.4	0.2	0.1	0.1	1.0	0.1	0.3	0.3	0.1	0.7	0.2	0.3	9.7	61	Nov. 12	0.01
		1968	1.4	0.3	0.2	0.1	2.6	3.9	0.5	0.3	1.4	1.5	1.7	0.5	1.2	32	290	Jun. 9	0.01
Toa, El Toro II	326	1966	8.2	3.6	8.2	6.2	9.1	12.0	5.9	2.2	4.4	11.1	17.5	6.9	7.9	63	240	Nov. 16	1.00
		1967	5.8	4.5	2.1	7.4	1.9	7.9	2.2	2.0	3.5	4.4	12.0	2.1	4.6	67	125	Apr. 2	0.77
		1968	2.1	1.4	1.8	0.9	8.4	8.7	2.3	2.5	3.5	3.2	5.7	2.1	3.5	34	88	May 28	0.43

TABLE 5-23. CUBA—AVAILABLE WATER RESOURCES

(Source: Keshishev, Soviet Geography: Review and Translation, 1967)

[By Provinces]

	Pinar del Rio	Havana	Matanzas	Las Villas	Camaguey	Oriente	Cuba
Province area (1,000 km2)	13.5	8.22	8.44	21.41	26.35	36.6	114.5
Percent of Cuba's area	11.8	7.15	7.35	18.7	23.0	32.0	100
Precipitation layer (mm.)	1,390	1,370	1,510	1,470	1,280	1,180	1,380
Precipitation volume (km3)	18.76	11.27	12.74	31.47	33.73	43.21	151.18
Runoff coefficient	0.21	0.173	0.162	0.25	0.22	0.26	0.22
Runoff layer (mm.)	293	235	245	368	282	307	301
Runoff volume (km3)	3.94	1.95	2.06	7.87	7.43	11.23	34.48
Runoff module (liters/sec./km2)	9.3	7.4	7.8	11.6	8.9	10.0	9.5
Province population (1965; 1,000)	518.5	2,093.6	455.8	1,240.4	759.8	2,286.1	7,390.2
Runoff per capita (m3)	7,600	930	4,530	6,350	9,800	4,950	4,650

FIGURE 5-2. DOMINICAL REPUBLIC — DRAINAGE BASINS

(Source: Organization of American States, 1969)

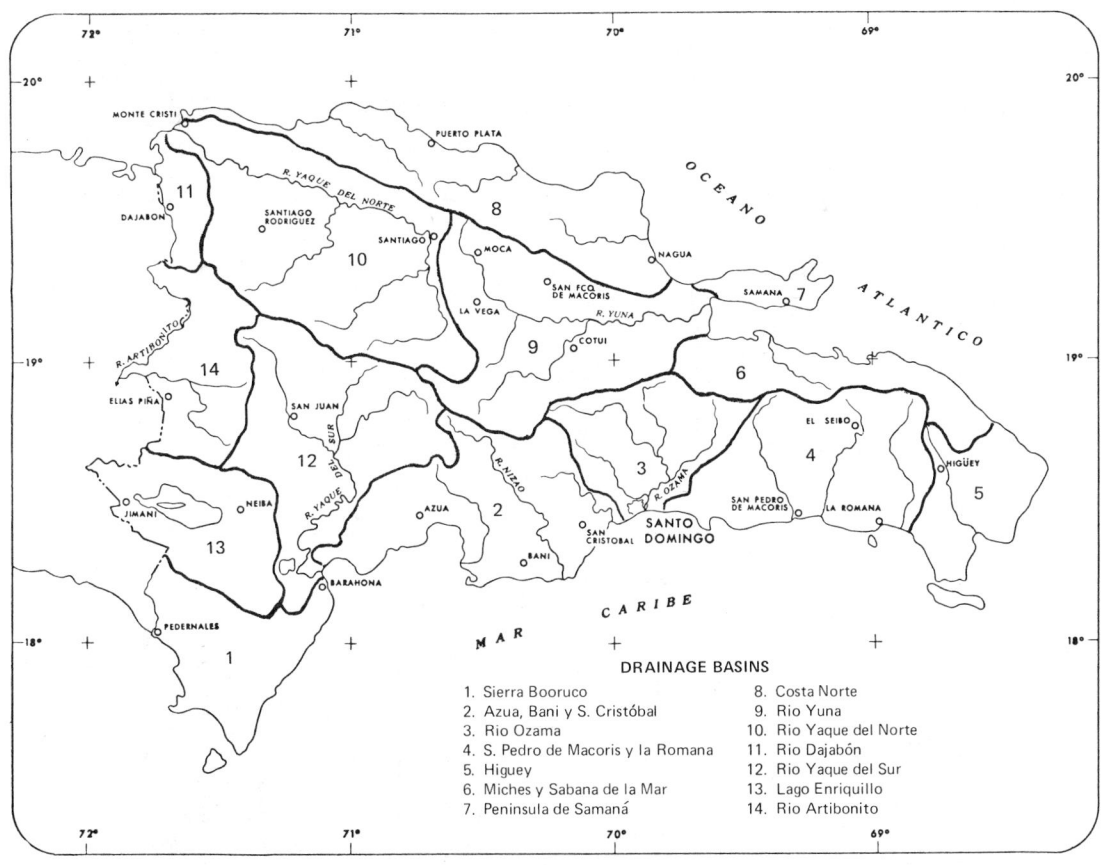

DRAINAGE BASINS

1. Sierra Booruco
2. Azua, Bani y S. Cristóbal
3. Rio Ozama
4. S. Pedro de Macoris y la Romana
5. Higuey
6. Miches y Sabana de la Mar
7. Peninsula de Samaná
8. Costa Norte
9. Rio Yuna
10. Rio Yaque del Norte
11. Rio Dajabón
12. Rio Yaque del Sur
13. Lago Enriquillo
14. Rio Artibonito

TABLE 5-24. DOMINICAN REPUBLIC–DISCHARGE OF SELECTED RIVERS

(Source: Organization of American States, 1969)

Drainage Basin River and station	Jan.	Feb.	Mar.	Apr.	May	Jun.	Jul.	Aug.	Sep.	Oct.	Nov.	Dec.
						Mean monthly discharge, m³/s						
Sierra del Bahoruco Zone												
Pedernales River	1.1	0.5	0.8	1.1	0.9	0.8	1.0	1.3	1.0	0.9	0.7	0.9
Azua, Bani and San Cristobal Zone												
Jaina River at Caobal a)	8.5	7.6	4.0	11.0	18.5	14.6	12.0	11.5	5.5	9.5	6.6	5.4
Nizao River at Higuana a)	19.5	22.0	15.0	9.6	11.4	22.0	12.7	15.0	10.0	13.0	18.0	21.0
Ocoa River at Mendez a)	1.5	1.0	1.0	1.5	2.0	2.0	1.8	2.0	2.8	3.0	2.5	1.8
Ozama River Basin												
Ozama River	Not available — average estimated flow about 50 m3/s											
San Pedro de Macoris and La Romana Zone												
Chavon River	3.1	3.4	3.0	5.1	15.1	13.0	9.0	8.1	13.4	10.7	7.2	3.3
Dulce River	0.7	1.0	0.5	1.1	3.1	3.2	2.2	1.2	3.3	1.2	0.9	1.0
Cumayasa River	1.2	1.3	0.9	2.0	5.1	5.0	4.8	4.1	4.2	4.0	3.9	1.8
Soco River	5.5	7.0	5.0	10.9	21.0	21.0	10.2	15.2	19.1	18.1	12.0	7.3
Macoris River	4.0	7.3	3.8	9.2	21.3	22.8	18.7	21.0	23.0	22.7	15.1	7.9
Higuey Zone												
Yuma River b)	0.9	0.7	0.5	1.6	2.1	1.2	1.2	1.0	1.4	1.3	1.2	1.1
Yuna River Basin												
Yuna River at Villa Riva c)	57.0	62.5	64.0	96.0	162.0	130.0	52.0	86.0	59.5	96.0	114.0	109.4
Yuna River at Hatillo c)	27.1	28.0	20.0	41.3	49.1	42.4	27.0	37.0	28.0	36.0	50.2	36.3
Jima River at Rincon c)	9.0	6.5	4.1	8.5	15.0	13.0	10.1	7.1	6.5	9.2	9.3	7.2
Camu River 10 km. below La Vega c)	8.1	8.5	5.2	12.7	23.3	12.2	10.3	8.5	11.6	15.2	15.2	9.1
Yaque del Norte River Basin												
Jimenoa River at Hato Viejo	6.3	6.5	6.2	7.3	8.4	7.5	6.4	5.4	6.0	6.4	7.7	8.3
Yaque del Norte River at Baitoa	15.5	14.3	18.4	22.4	26.5	25.2	18.1	15.0	16.4	25.6	25.5	23.3
Bao River at Bao (Janico)	7.4	7.5	8.1	9.4	16.0	22.2	9.4	8.3	11.3	15.6	14.4	11.2
Yaque del Norte River at Santiago	31.4	36.4	35.0	39.7	55.1	53.7	29.7	34.1	29.6	44.3	42.6	36.6

Amina River at Potrero	3.2	3.7	5.4	6.6	10.2	10.3	5.6	4.1	4.4	10.0	9.0	6.4
Mao River at Chorrera	11.2	8.2	8.7	10.1	18.0	33.3	18.1	18.0	21.1	24.3	20.2	11.0
Yaque del Norte River at Puente San Rafael	21.2	15.0	14.2	48.1	80.2	74.1	31.2	25.2	46.2	85.2	86.2	59.0
Guayubin River at La Antona	3.1	3.3	9.1	10.2	10.0	11.5	6.2	6.5	7.5	10.3	7.0	6.1
Dajabon River Basin												
Dajabon River c)	8.0	9.1	6.0	18.0	19.0	14.0	6.0	9.0	18.0	15.0	10.0	10.0
Yaque del Sur River Basin												
San Juan River at Guazumal	6.2	4.2	4.0	5.0	4.8	12.2	10.3	15.2	17.1	20.1	15.0	8.3
Mijo River at Cacheo	3.0	2.4	2.4	3.5	6.5	9.6	7.6	7.7	6.3	6.6	6.2	3.7
Yaque del Sur at Boca de los Rios	5.2	4.6	4.7	6.6	9.2	12.0	10.0	8.7	11.2	16.0	12.4	7.7
Las Cuevas at Sabana Yaque	3.3	2.8	1.5	2.4	4.3	5.4	4.4	3.3	6.2	6.3	8.4	5.2
Del Medio River at Boca de los Rios	5.2	4.0	3.8	3.3	7.3	13.3	7.2	6.2	11.1	11.2	8.4	5.1
Yaque del Sur at Villarpando	25.0	17.3	18.3	32.0	53.4	73.1	61.1	65.4	69.1	82.0	77.2	50.2
Yaque del Sur at Los Carosos	15.1	12.7	11.2	14.6	26.3	37.5	24.2	21.6	25.1	39.7	32.6	21.0
Artibonito River Basin												
Artibonito River c)	8.0	24.0	28.0	25.0	95.0	85.0	65.0	62.0	25.0	86.0	34.0	11.0
Macasia River c) (tributary of Artibonito River)	2.5	1.0	1.5	1.5	4.0	3.5	2.6	2.3	3.5	5.0	7.0	5.0

a) Computed from existing hydrometric data and verified empirically.
b) Based on calculations of runoff of 120 square kilometers of upper watershed and waters located upstream of irrigable lands.
c) Calculated empirically.

TABLE 5-25. EL SALVADOR—MONTHLY DISCHARGE OF SELECTED RIVERS
(Source: CEPAL, 1971)

River and discharge	Station	Basin area km²	Mean monthly discharge, m³/s												Year	Period of record
			Jan.	Feb.	Mar.	Apr.	May	Jun.	Jul.	Aug.	Sep.	Oct.	Nov.	Dec.		
Paz	La Hachadura	795	12.5	12.1	11.7	13.5	16.8	50.2	48.7	43.3	47.2	35.8	19.4	15.0	27.2	1962-66[1]
Sonsonate	Sensunapan	219	2.0	2.0	2.0	3.7	5.8	9.3	8.9	10.3	14.2	8.7	4.3	2.1	6.1	1959-66[2]
Banderas	Carretera Litoral	433	1.6	1.2	1.4	2.0	5.1	14.1	16.4	13.0	18.5	11.7	7.2	2.9	13.3	1961-66[3]
Jiboa	Carretera Litoral	229	4.2	3.7	3.3	3.2	3.9	6.9	10.3	13.5	14.8	12.5	8.1	6.3	7.6	1961-66[4]
Lempa	San Marcos	18,000	69.2	59.8	65.5	60.3	131.0	545.0	943.0	616.0	1,104.0	691.0	171.0	75.1	377.5	1961-66[5]
Torola	Osicala	760	7.8	7.2	6.0	7.5	24.9	80.1	51.7	50.8	159.0	81.7	16.1	8.6	40.1	1962-66[6]
Aceihuate	Junta Lempa	709	5.6	5.4	4.3	5.1	9.9	15.2	25.1	18.0	23.3	18.7	3.7	3.4	11.5	1962-66[7]
Sucio	Mouth	843	4.4	4.1	4.2	4.1	7.1	16.8	37.5	26.5	41.4	20.2	9.2	6.3	15.2	1961-66[8]
Suquiapa	Tecachico	308	5.2	5.1	5.0	5.1	5.9	8.7	11.1	9.2	11.9	8.3	6.7	5.7	7.3	1961-66[9]
Grande San Miguel	Vado Marin	2,027	7.6	7.6	8.0	8.5	11.7	30.4	44.7	26.8	77.7	77.5	24.9	8.5	27.8	1959-66[10]
Sirama	Sirama	329	0.3	0.2	0.2	0.2	2.8	15.1	9.2	5.3	20.2	8.6	2.7	0.4	5.4	1961-66[11]
Goascoran	Goascoran	1,750	3.7	2.5	1.9	2.2	28.5	64.8	66.0	36.3	93.6	79.4	19.6	6.1	33.7	1962-66[12]

Maximum flows and month:
1) 389.1 (Jun.)
2) 253.0 (Sep.)
3) 285.0 (Sep.)
4) 99.3 (Sep.)
5) 3,516.0 (Sep.)
6) 992.0 (Sep.)
7) 187.2 (Sep.)
8) 271.0 (Sep.)
9) 133.6 (Jul.)
10) 284.0 (Jul.)
11) 720.6 (Jun.)
12) 670.0 (Sep.)

TABLE 5-26. EL SALVADOR—WATER BALANCE ELEMENTS
(Source: CEPAL, 1971)

Element	Annual mm	Annual volume of water million m^3
INFLOW		
Precipitation	182	36,367
River inflow from other countries (Rio Lempa Basin)	24	4,841
Total	206	41,208
OUTFLOW		
Total runoff...............................	95	18,953
Direct runoff	(75)	(15,088)
Base flow	(19)	(3,865)
Ground water discharge to sea	12	2,313
Total evapotranspiration	86	17,200
of ground water	(3)	(696)
of precipitation...........................	(83)	(16,504)
Total	193	38,466

TABLE 5-27. EL SALVADOR—AVAILABLE WATER RESOURCES
(Source: Lemus, Servicio Hidrologico, 1972)

River basin	Basin area km^2	Mean annual discharge, m^3/s			Ground water, m^3/s	
		Normal	Wet year	Dry year	Total	Dependable yield
Lempa	10,000 18,500 [1]	470.0	813.0	128.0	48.8 [2]	29.9
Paz	925	31.1	48.7	13.5	4.5	4.5
Between Paz and Sonsonate	674	11.2	19.5	2.8	1.0	1.0
Sonsonate	875	20.2	33.3	7.1	3.0	3.0
Between Sonsonate and Jiboa	1,399	23.2	40.5	5.9	1.0	1.0
Jiboa	608	5.3	8.2	2.5	1.3	1.3
Between Jiboa and Lempa	956	9.5	16.4	2.6	4.0	4.0
Between Lempa and Grande de S. Miguel	968	16.3	26.4	6.1	4.0	4.0
Grande de San Miguel	2,356	39.6	64.4	14.7	15.2	15.2
Goascoran and others	2,241	27.7	51.6	3.7	1.0	1.0
Total	21,000				83.8	64.9

[1] Total estimated basin area including Honduras and Guatemala.
[2] Within El Salvador only.

TABLE 5-28. EL SALVADOR—HYDROELECTRIC POWER PLANTS, 1970
(Source: CEPAL, 1971)

River	Plant	Capacity MW	Power generated Gwh	Head m	Water use, m3/s Total	Water use, m3/s Non-repetitive
EXISTING						
El Molino	Atechuecia	0.1	0.4	60	0.1	0.1
Sonsonate	Sonsonate	0.1	0.4	16	0.3	
Sonsonate	Bululu	0.4	1.6	22	1.1	
Sonsonate	Cucumacayan	1.4	6.0	86	1.0	
Sonsonate	La Calera	0.3	1.2	36	0.5	
Sonsonate	Santa Emilia	0.2	1.0	12	1.2	1.2
Guija	Guajoyo	15.0	51.0	48	15.2	
Suquiapa	Cutumay	0.2	0.7	22	0.5	
Suquiapa	San Luis - 1	0.4	1.6	26	0.9	
Suquiapa	San Luis - 2	0.5	2.4	26	1.3	
Sucio	Quezaltepeque	1.5	6.7	28	3.4	
Lempa	5 de Noviembre	82.0	434.0	54	115.7	115.7
San Esteban	Santa Julia	0.2	0.7	32	0.3	0.3
	Total El Salvador	102.3	507.7		141.5	117.3
PROJECTED						
Lempa	Silencio	189.0	562.0	48	158.6	158.6
Lempa	Paso del Oso	140.0	299.0	70	58.0	—
Lempa	El Tigre	400.0	1,500.0	77	279.0	279.0

TABLE 5-29. EL SALVADOR—PRESENT AND PROJECTED WATER DEMAND
(Source: Lemus, Servicio Hidrologico, 1972)

[In m3/s]

River basin	1970 Irrigation	1970 Hydropower	1970 Public water supply and industry	1970 Total water demand	1980 Irrigation	1980 Hydropower	1980 Public water supply and industry	1980 Total water demand	1990 Irrigation	1990 Hydropower	1990 Public water supply and industry	1990 Total water demand
Lempa and Jiboa to Grande de San Miguel	10.07	115.07	3.02	129.42	33.02	158.06	5.72	197.52	176.00	279.00	8.84	463.84
Paz	—	0.01	0.13	0.23	7.06	0.1	0.22	7.92	23.06	0.01	0.34	24.04
Between Paz and Sonsonate	—	—	0.03	0.03	—	—	0.05	0.05	7.03	—	0.08	7.38
Sonsonate	10.04	1.02	0.14	11.74	13.04	1.02	0.27	14.87	15.03	1.02	0.41	16.91
Between Sonsonate and Jiboa	0.06	—	0.21	0.81	0.06	—	0.37	0.97	24.05	—	0.55	22.05
Grande de San Miguel	1.08	0.03	0.15	2.25	29.01	0.03	0.26	29.66	59.00	0.03	0.39	59.69
Goascoran and others	0.01	—	0.12	0.22	0.01	—	0.22	0.32	8.02	—	0.33	8.53
Total	23.60	117.3	3.80	144.70	84.0	160.20	7.11	251.31	313.09	280.60	10.94	605.44

TABLE 5-30. FRENCH AMERICAN DEPENDENCIES—GUADELOUPE—DISCHARGE OF SELECTED RIVERS

(Source: ORSTOM, 1968)

River and station	Basin area km²	Mean monthly discharge, m³/s												Year	Period of record
		Jan.	Feb.	Mar.	Apr.	May	Jun.	Jul.	Aug.	Sep.	Oct.	Nov.	Dec.		
Grand Goyave, Prise d'Eau	54.8	3.78	2.55	1.28	3.32	4.41	2.31	12.5	12.7	12.1	13.4	10.8	6.50	7.18	1965
Grand Carbet, Prise Marquisat	11.8	1.67	1.13	0.88	1.96	2.16	1.97	1.93	2.04	2.32	2.35	2.37	2.00	1.90	1953-65

TABLE 5-31. FRENCH AMERICAN DEPENDENCIES—MARTINIQUE—DISCHARGE OF SELECTED RIVERS

(Source: ORSTOM, 1968)

River and station	Basin area km²	Mean monthly discharge, m³/s												Year	Period of record
		Jan.	Feb.	Mar.	Apr.	May	Jun.	Jul.	Aug.	Sep.	Oct.	Nov.	Dec.		
La Lezarde, Soudon	56	2.90	2.18	1.25	1.31	1.41	1.50	3.30	3.35	4.35	4.30	4.35	2.99	2.75	1965
R. du Lorrain, Piroque	26.1	2.71	1.55	0.80	1.11	1.00	1.17	2.75	2.47	3.15	2.94	2.71	2.31	2.06	1965

TABLE 5-32. GUATEMALA–DISCHARGE OF SELECTED RIVERS
(Source: CEPAL, 1972)

River	Station	Basin area km²	Jan.	Feb.	Mar.	Apr.	May	Jun.	Jul.	Aug.	Sep.	Oct.	Nov.	Dec.	Year
ATLANTIC SLOPE															
Chixoy	Chixoy	6,020	24.3	19.3	16.4	18.3	18.6	55.4	74.5	60.4	108.9	98.1	48.8	32.1	47.9[1]
Cahabon	Chajcar	380	18.9	13.2	11.3	9.1	5.2	17.3	37.8	25.8	36.7	44.3	31.7	24.0	23.6[2]
Pixcaya	El Tesoro	155	0.9	0.8	0.8	0.8	0.7	1.1	1.4	1.2	2.3	2.2	1.1	0.9	1.2[3]
PACIFIC SLOPE															
Camala	Candelaria	849	8.1	7.3	6.7	6.8	7.4	11.6	11.7	9.8	12.9	14.8	11.1	8.9	9.8[4]
Nahualate	Sta. Catarina	145	1.2	1.0	0.9	0.9	0.9	2.9	2.9	1.8	5.1	4.6	2.3	1.4	2.2[5]
Madre Vieja	Palmira	344	9.1	6.6	6.4	6.6	7.2	11.3	10.9	9.8	14.1	15.2	10.9	8.2	9.8[6]
Madre Vieja	Panibaj	165	1.0	0.9	0.9	0.8	0.8	2.9	1.9	1.4	2.8	1.6	1.0	0.9	1.4[7]
Xaya	La Sierra	86	0.3	0.3	0.3	0.3	0.3	0.6	0.8	0.9	1.5	1.1	0.5	0.4	0.6[8]
Aguacapa	Agua Cailente	351	4.8	4.8	4.7	4.8	5.0	10.1	16.4	15.2	31.1	19.1	9.3	5.7	10.9[9]
Los Esclavos	La Sonrisa	1,553	3.3	3.2	2.8	3.5	4.3	17.8	19.2	11.0	27.6	25.5	6.7	3.9	10.6[10]

Maximum flows and month:
1) 443.0 (Oct.)   2) 100.0 (Oct.)   3) 8.9(Sep.)   4)46.0 (Sep.)   5) 50.0 (Sep.)   6) 52.0 (Oct.)   7) 45.0 (Jun.)   8)3.4(Oct.)
9) 115.1 (Sep.)   10) 39.8 (Sep.)

TABLE 5-33. GUATEMALA–WATER BALANCE ELEMENTS
(Source: CEPAL, 1972)

Elements	Annual value, mm		
	Atlantic Slope	Pacific Slope	Total Guatemala
Precipitation	216	224	218
Total runoff	80	129	89
Direct runoff	—	(82)	—
Base flow	—	(47)	—
Ground water discharge to sea	—	15	—
Evapotranspiration	127	95	121
of ground water	—	(4)	—
of precipitation	—	(91)	—

TABLE 5-34. GUATEMALA–AVAILABLE WATER RESOURCES

(Source: CEPAL, 1972)

River basin	Basin area km²	Precipitation million m³		Surface water runoff m³/s			Ground water estimated dependable yield [2] m³/s
		Normal year	Dry year [1]	Normal year	Dry year [1]	95% flow	
Atlantic Slope							
Selegua, Usumacinta, Hondo [3]	59,400	140,985	119,837	1,341	1,140	134	N.A.
Nuevo, Belice and others	21,487	37,817	32,144	480	408	48	N.A.
Moho, Dulce, Puerto Barrios zone	13,583	38,572	32,400	734	616	88	N.A.
Motagua [3]	13,950	17,019	14,296	189	159	28	N.A.
Total	108,420	234,393		2,744		298	N.A.
Pacific Slope							
Suchiate [3]	1,270	3,556	2,845	68	54	17	13
Between Suchiate and Samala	3,276	8,583	6,609	177	136	44	35
Samala, Nahualate, etc.	7,714	20,211	15,562	417	321	80	76
Achiguate, Maria Linda	3,950	10,349	7,969	180	139	54	52
Los Esclavos	3,200	5,248	4,093	58	45	6	11
Paz [3]	1,310	2,214	1,771	25	22	8	5
Lempa [3]	2,660	2,318	1,851	28	23	3	1
Total	23,380	52,475		953		212	193
Total Guatemala	131,800	286,868		3,697		510	N.A.

1) Not totalled as drought normally affects small areas only.
2) Large-scale withdrawals of ground water will decrease river flow.
3) International river basin; figures apply to Guatemala only.
N.A. – Not available.

TABLE 5-35. GUATEMALA–DATA ON HYDROELECTRIC POWER PLANTS

(Source: CEPAL, 1972)

[Data as of 1970]

River	Plant	Capacity MW	Power generated GWh	Head m	Water use, m3/s	
					Total	Non-repetitive
Atlantic Slope						
Cahabon	Chichaic	0.6	2.6	11	3.4	3.4
Pixcaya	E. Selle	0.9	3.9	67	0.8	
El Zapote	Zapote	0.4	1.7	28	0.9	2.2
Colorado	Rio Hondo	2.4	10.5	300	0.5	
Pacific Slope						
Samala	Sta. Maria	5.9	25.8	112	3.3	3.3
Samala	Zunil	1.0	4.4	31	2.0	
Samala	Cantel 2	0.8	3.5	47	1.1	
Samala	Cantel 1	0.7	3.1	21	2.1	
Guacalate	Modelo	0.7	3.1	18	2.5	2.5
Michatoya	Palin	1.6	7.0	84	1.2	
Michatoya	San Luis	5.0	21.9	76	4.1	
Michatoya	El Salto	5.5	24.0	91	3.8	
Michatoya	Jurun	60.0	196.0	660	4.3	4.3
Los Esclavos	Los Esclavos	13.0	66	106	8.9	8.9
Total		98.5	373.5		38.9	24.6

TABLE 5-36. HAITI—DISCHARGE OF PRINCIPAL RIVERS
(Source: OAS, 1972)

River	Station	Basin area km²	Mean annual flow m³/s
Artibonite	Pont Sonde	6,862	101 1)
Trois Rivieres	Paulin Lacorne	897	13
Estere	Pont de l'Estere	834	19
Grande Riviere du Nord	Pont Parois	699	8
Quinte	—	690	– 2)
Grand'Anse	Passe Ranja	556	27
Grande Riviere de Jacmel	Jacmel	535	5
Grande Riviere de Nippes	—	459	3
Riviere de Cavaillon	Pont de Cavaillon	380	9
Riviere du Limbe	Pont Christophe	312	7

1) Prior to construction of Peligre dam and reservoir.
2) Almost all its water is used for irrigation.

TABLE 5-37. HONDURAS—MONTHLY DISCHARGE OF SELECTED RIVERS
(Source: CEPAL, 1973)

River	Station	Basin area km²	Jan.	Feb.	Mar.	Apr.	May	Jun.	Jul.	Aug.	Sep.	Oct.	Nov.	Dec.	Year
Atlantic Slope															
Chamelecon	Chamelecon Bridge	1,780	50.9	26.4	14.9	10.3	7.4	38.8	35.8	29.0	50.1	57.9	63.9	50.7	43.6
Ulua	Pimienta Bridge	8,916	128.1	65.0	55.3	40.1	66.1	239.2	235.2	220.7	452.6	465.6	233.4	145.3	195.5
Jicatuyo	Quecoa	3,455	36.6	18.9	14.3	33.1	24.6	111.1	93.2	85.5	172.9	163.0	86.8	49.2	74.1
Humuya	La Encantada	1,960	6.5	3.7	2.1	2.2	8.7	36.4	22.8	16.9	42.4	48.6	11.5	6.4	16.6
Patuca	Cayetano	4,740	41.5	57.4	39.1	25.0	85.4	317.7	675.3	273.7	187.5	332.3	133.5	93.3	209.4
Guayape	Guayabillas	1,010	10.4	8.3	6.4	5.1	11.4	35.9	38.3	32.2	23.8	45.0	16.9	10.9	20.5
Pacific Slope															
Choluteca	Los Encuentros	6,267	5.4	3.5	2.6	2.2	33.5	72.8	62.2	27.3	57.7	96.0	22.1	9.6	33.7

TABLE 5-38. HONDURAS—FLOW CHARACTERISTICS OF SELECTED RIVERS

(Source: CEPAL, 1973)

River	Station	Basin area km²	Discharge, m³/s			Percentage runoff from June to November	Period of record
			Mean	Max	Min		
Atlantic Slope							
Chamelecon	Chamelecon Bridge	1,780	43.6	487.0	5.3	62	1956-59; 1965
Ulua	Pimienta Bridge	8,916	195.5	3,020.0	15.7	78	1956-58; 1965
Ulua	Chinda	7,950	180.4	2,508.0	17.0	79	1956-59; 1965
Ulua	Suspension Bridge	4,295	86.4	955.0	13.0	76	1956-58
G. Otoro	La Gloria	745	15.3	270.0	1.4	81	1955-58; 1965
Jicatuyo	Quecoa	3,455	74.1	1,198.0	6.0	80	1956-59
Humuya	La Encantada	1,960	16.6	726.0	0.3	85	1956-58; 1964-65
Humuya	Las Higueras	880	7.2	472.1	0.0	83	1956-59; 1964-65
Patuca	Cayetano	4,740	209.4	3,850.0	13.0	84	1956-58
Talgua	Talgua	–	4.7	53.2	0.4	64	1955-59
Telica	Telica	–	9.0	254.0	0.5	65	1956-59; 1965
Guayape	Guayabillas	1,010	20.5	728.0	2.2	78	1956-59; 1964-65
Jalan	El Delirio	775	24.8	344.0	3.4	64	1956-59
Pacific Slope							
Choluteca	Los Encuentros	6,267	33.7	890.8	0.9	85	1958-59; 1964-65
Choluteca	Paso La Ceiba	2,175	16.4	730.1	0.4	83	1956-59; 1964-65
Choluteca	Hernando Lopez	1,510	11.9	962.2	0.3	81	1954-59; 1964-65

TABLE 5-39. HONDURAS–AVAILABLE WATER RESOURCES
(Source: CEPAL, 1973)

River basin	Basin area km²	Precipitation million m³		Surface water runoff m³/s			Ground water estimated dependable yield 2) m³/s
		Normal year	Dry year 1)	Normal year	Dry year 1)	95% flow	
Atlantic Slope							
Motagua 3)	2,651	5,673	4,710	63	52	9	7
Chamelecon 4)	4,676	8,229	6,830	144	119	22	15
Ulua	22,562	33,166	27,200	526	431	79	40
Cangrejal 4)	5,417	12,188	9,510	251	196	38	25
Aguan, Sico, Platano 4)	22,448	41,079	31,220	716	544	72	41
Patuca	25,646	43,341	32,510	825	618	82	40
Guarunta and Cruta	5,057	11,681	8,760	148	111	37	55
Coco 3)	6,681	14,430	10,825	274	206	55	31
Total	95,138	169,787		2,947		394	254
Pacific Slope							
Lempa 3)	5,779	9,188	7,535	125	103	10	7
Goascoran 3)	1,243	2,398	1,920	19	15	4	3
Nacaome 4)	3,607	4,941	3,855	39	31	8	7
Choluteca 3)	7,876	8,899	6,940	71	55	14	14
Negro 3)	1,562	2,249	1,665	28	21	4	3
Total	20,067	27,675		282		40	34
Total Honduras	115,205	197,462		3,229		434	288

1) Not totalled as droughts affect small areas only.
2) Large-scale exploitation of ground water will diminish base flow of rivers.
3) International river basin; figures apply to Honduras only.
4) Some smaller river basins included.

TABLE 5-40. HONDURAS–HYDROELECTRIC POWER PLANTS

(Source: CEPAL, 1973)

[Data as of 1970]

River	Plant	Capacity MW	Energy GWh	Head m	Water use, m3/s	
					Total	Non-repetitive
Atlantic Slope						
Humuya	Comayagua *	0.06	0.3	—		
Humuya	La Paz *	0.13	0.7	—		
Humuya	Gracias *	0.10	0.4	—		
Humuya	Santa Rosa *	0.06	0.3	—		
Lago Yojoa	Canaveral	30.00	192.6	145.0	19.0	19.0
Aguan basin	Yoro *	0.02	0.1	—		
Aguan basin	Minas de Oro *	0.05	0.2	—		
Patuca basin	Catacamas *	0.08	0.3	—		
Patuca basin	Paraiso *	0.07	0.3	—		
	Total	30.35	194.3	—	19.0	19.0
Pacific Slope						
—	La Esperanza *	0.07	0.3	—		
—	Marcala *	0.04	0.2	—		
Choluteca basin	Yuscaran *	0.06	0.3	—		
Choluteca basin	Rosario *	0.13	0.7	—		
Choluteca basin	La Leona *	1.60	3.2	296.0		
Choluteca basin	Valle de Angeles *	0.08	0.3	—	0.2	0.2
	Total	1.98	5.0	—	0.2	0.2
	Total Honduras	32.55	200.2	—	19.2	19.2

* Information incomplete. Energy calculation based on 4,380 hrs. of operation.

TABLE 5-41. JAMAICA—WATER BALANCE ELEMENTS
(Source: Williams, Geol. Survey Dept., 1969)

Element		
Total area Jamaica	4,411	mi2
Mean annual precipitation	77.12	in.
equivalent to	13,500	mgd
Evapotranspiration losses.	7,500	mgd
Available water resources	6,000	mgd
Total annual water use (1969)	450	mgd

TABLE 5-42. JAMAICA—DISCHARGE OF SELECTED RIVERS
(Source: Williams, Geol. Survey Dept., 1969)

River	Mean minimum flow cfs	Mean low flow cfs	Mean maximum flow cfs	Peak flow		Mean annual flow 10^9 gal.	Period of record years
				cfs	Year		
Martha Brae	101	153	1,267	—	—	121	12
Hope	4.6	6.2	662	1,380	1956	—	7

TABLE 5-43. JAMAICA—MUNICIPAL WATER SUPPLY SYSTEM OF KINGSTON
(Source: Williams, Geol. Survey Dept., 1969)

Population	Water requirement, mgd			Source of water
	1969	1975	2000	
556,100 (1969)	30	45	140	Hermitage Reservoir (Capacity 450 mg) and Mona Reservoir (825 mg); 12 deep wells total capacity 15 mgd.

FIGURE 5-3. MEXICO — WATER RESOURCES REGIONS AND SUBDIVISIONS
(Source: del Arenal, Water Res. Bull., 1969)

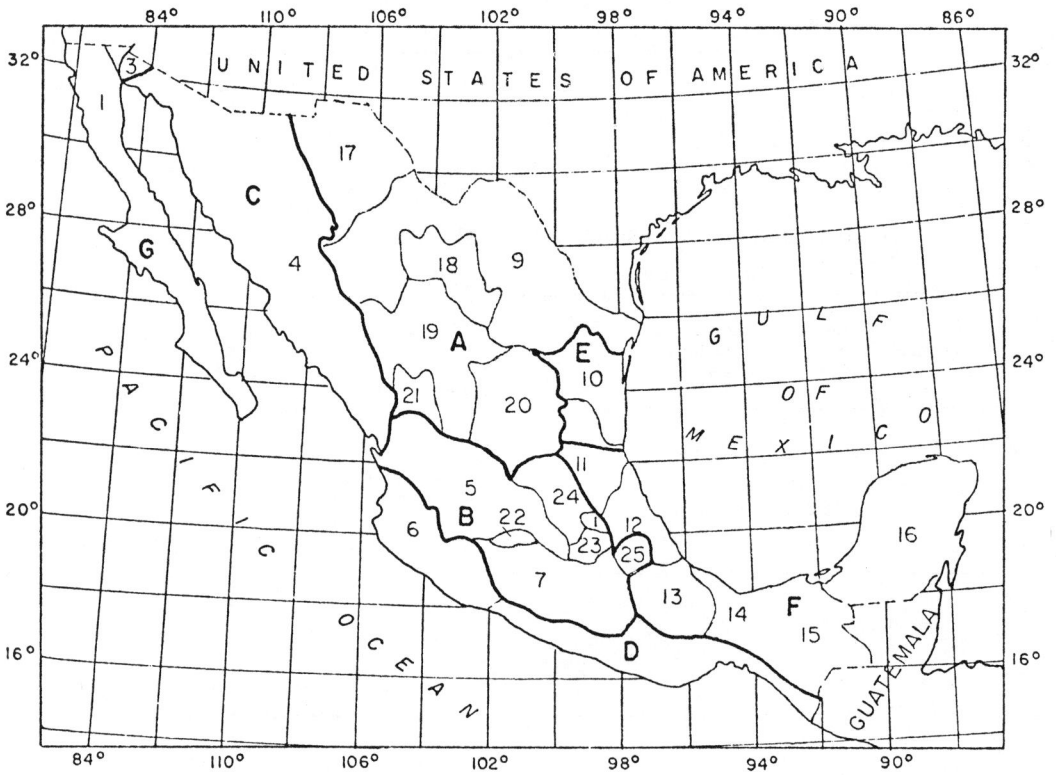

A Northern Highlands
B Central and Southern Plateaus
C North Pacific Basin
D South Pacific Basin
E North Gulf Basin
F South Gulf Basin
G Peninsula of Baja California

TABLE 5-44. MEXICO–AVAILABLE WATER RESOURCES

(Source: del Arenal, Water Res. Bull, March 1969)

[Regions shown on adjacent map]

No.	Region	Total runoff		Population			Water availability (m³/capita/year)		
		millions m³	m³/km²	1965	1970	1980	1965	1970	1980
	MEXICAN REPUBLIC	357,375	181,669	42,529,314	50,550,477	70,932,989	8,403	7,070	5,038
1	Baja California Norte	312	4,763 2)	353,110	487,832	851,592	884	640	366
2	Baja California Sur	262	3,556 2)	99,951	122,995	190,429	2,621	2,130	1,376
3	Rio Colorado	1,850	35,083	480,865	665,357	1,162,455	3,847	2,780	1,591
4	Noroeste	28,858	83,746	2,347,723	2,887,122	4,381,232	12,292	9,995	6,587
5	Cuenca Lerma-Chapala-Santiago	11,457	89,732	7,226,818	8,845,057	13,312,510	1,585	1,295	861
6	Pacifico Centro	11,333	184,067	1,127,079	1,361,300	2,024,954	10,055	8,325	5,597
7	Cuenca del Rio Balsas	12,186	114,048	4,200,521	4,781,931	6,139,724	2,901	2,548	1,985
8	Pacifico Sur	50,116	534,058	2,242,013	2,555,493	3,249,766	22,353	19,611	15,421
9	Cuenca del Rio Bravo	4,780	20,032 2)	3,606,130	4,460,873	6,706,978	1,326	1,072	713
10	Golfo Norte	3,473	69,502	276,616	312,594	407,597	12,555	11,110	8,521
11	Cuenca del Rio Panuco	18,860	229,776	2,392,407	2,661,899	3,308,507	7,883	7,085	5,700
12	Golfo Centro	28,098	763,740	2,353,720	2,659,973	3,333,600	11,938	10,563	8,429
13	Cuenca del Rio Papaloapan	41,135	892,299	1,713,276	2,018,426	2,847,380	24,010	20,380	14,447
14	Golfo Sur	25,083	979,805	408,458	553,232	1,007,346	61,409	45,339	24,900
15	Sistema Grijalva Usumacinta	108,749	1,188,513	1,645,040	1,942,654	2,721,224	66,107	55,980	39,963
16	Peninsula de Yucatan	5,811	41,386	1,000,220	1,126,626	1,420,210	5,810	5,158	4,092
17	Cuencas Cerradas del Norte (zona norte)	856	10,160 2)	250,632	283,063	345,868	3,415	3,024	2,475
18	Bolson de Mapimi	(1)	(2)	8,576	8,576	8,576			
19	Cuencas Cerradas del Norte (zona sur)	1,592	14,840 2)	926,865	889,000	569,038	1,718	1,791	2,798
20	El Salado	(1)	(2)	973,211	1,024,081	1,074,848			
21	Durango	711	41,289	364,425	401,852	472,847	1,951	1,769	1,504
22	Cuitzeo-Patzcuaro	413	71,951	555,425	655,035	922,577	744	631	448
23	Valle de Mexico	480	49,844	7,569,539	9,411,395	14,003,670	63	51	34
24	Valle de Metztitlan	740	258,741	174,588	187,615	211,969	4,239	3,944	3,491
25	Valles de Oriental, Libres y El Seco	220	39,136 2)	232,105	246,496	258,092	948	893	852

1) Practically no runoff.

2) Water deficit area.

TABLE 5-45. MEXICO–DISCHARGE OF SELECTED RIVERS
(Source: UNESCO, 1967)

River and station	Basin area km²	Mean monthly discharge, m³/s												Year	Period of record
		Jan.	Feb.	Mar.	Apr.	May	Jun.	Jul.	Aug.	Sep.	Oct.	Nov.	Dec.		
Santiago, El Capomal	128,943	63	52	42	43	38	158	563	717	663	263	95	63	234	1940-66
Usumacinta, Boca del Cerro	50,743	1,160	889	706	619	6,330	592	2,540	2,080	2,850	3,660	2,330	1,560	1,700	1948-66
Panuco, Las Adjuntas	58,115	154	132	110	125	139	424	777	490	714	716	365	234	335	1956-66

TABLE 5-46. MEXICO—MAJOR DAMS AND RESERVOIRS
(Source: del Arenal, Water Res. Bull, March 1969)

[Reservoir capacity 100 million m^3 and over]

No.	State	Name of dam	Total capacity of reservoir million m^3	Year completed	Purpose[1]	Built-by[2]
1	Chiapas	Netzahualcoyotl	13,000	1965	CA-G-R-N	SRH
2	Michoacan	Morclos (Infiernillo)	12,500	1964	F. M.	CFE
3	Oaxaca	Presidente Aleman	8,000	1955	CA-G-R-N	SRH
4	Coahuila	Internacional La Amistad	7,000	UC	R-CA-G	SRH
5	Tamaulipas	Internacional Falcon	5,038	1953	R-G-CA	SRH
6	Sonora	Plutarco Elias Calles	3,500	1961	F. M.	CFE
7	Sinaloa	Miguel Hidalgo	3,355	1956	R-G-CA	SRH
8	Sinaloa	Pte. Adolfo Lopez Matcos	3,200	1964	R-G-CA	SRH
9	Chihuahua	La Boquilla	3,150	1916	R	CARC
10	Sonora	Alvaro Obregon	3,000	1952	R-G-CA	SRH
11	Durango	Lazaro Cardenas	3,000	1947	R-CA	SRH
12	Coahuila	Venustiano Carranza	1,385	1932	R-CA	SRH
13	Tamaulipas	Marte R. Gomez	1,241	1946	R	SRH
14	Sonora	Pte. Adolfo Ruiz Cortines	1,015	1955	R-C-CA	SRH
15	Oaxaca	Pte. Benito Juarez	942	1961	R	SRH
16	Sonora	La Angostura	865	1942	R-G	SRH
17	Chihuahua	El Granero	850	UC	R-G-CA	SRH
18	Sinaloa	Sanaiona	845	1948	R-G	SRH
19	Guanajuato	Solis	800	1949	R-G	SRH
20	Michoacan	La Villita	700	UC	F.M.-R	SRH
21	Sinaloa	Josefa Ortiz de Dominguez	500	UC	R-CA	SRH
22	Mexico	Valle de Bravo	458	1947	F. M.	CFE
23	Chihuahua	Francisco I. Madero	425	1949	R	SRH
24	Jalisco	Santa Rosa	420	1963	F. M.	CFE
25	Puebla	Manuel Avila Camacho	405	1946	R	SRH
26	Michoacan	Tepuxtepec	371	1931	F. M.	CMLFM
27	Aguascalientes	Calles	340	1931	R	SRH
28	Mexico	Villa Victoria	254	1944	F. M.	CFE
29	Guerrero	Palos Altos	250	UC	R	SRH
30	Sonora	Abelardo Rodriguez	250	1948	R	SRH
31	Michoacan	El Bosque	248	1954	F. M.	CFE
32	Hidalgo	Endo	182	1951	R	SRH
33	Guanajuato	Ignacio Allende	150	UC	AA-CA	SRH
34	Jalisco	Tacotan	149	1958	R-CA-G	SRH
35	Baja California	Rodriguez	137	1937	CA-AA	SRH
36	Chihuahua	El Tintero	130	1950	R-CA	SRH
37	Mexico	Huapango	129	1939	R	SRH

[1] Abbreviations: R—Irrigation; G—Generation; CA—Flood control; AA—Water supply; N—Navigation; F.M.—Not given.

[2] Abbreviations: SRH—Hydraulic Resources Secretariat; CFE—Federal Commission on Electricity; CMLFM—Mexican Light and Power Company; CARC—Agricultural Company of Conchos River; UC—Under construction.

TABLE 5-47. MEXICO—WATER BALANCE ELEMENTS
(Source: Oriver Alba, 1969)

Element		
Area of country.....................................	1,967,200	km2
Mean annual precipitation	755	mm
equivalent to	1,483,000	106m3
Surface water runoff	350,000	106m3
Evapotranspiration losses	1,100,000	106m3
Total water use (1960)	33,000	106m3

TABLE 5-48. MEXICO—WATER USE, 1960
(Source: Mexico Ministry of Water Resources, 1963)

Category	Water use million m3	Percent
Irrigation	30,000	90
Industrial	2,240	7.5
Domestic	760	2.5
Total	33,000	100

TABLE 5-49. MEXICO—INDUSTRIAL WATER DEMAND
(Source: Lamadrid and Camhaji, Mexico, 1973)

Industry	Water demand		Percentage
	Volume, m3	l/s	
Food products (excl. beverages)	941,001,122	29,839	54
Chemical products	208,610,570	6,615	12
Metal industry	197,662,013	6,268	11
Paper and paper products	117,991,944	3,742	7
Oil refineries	57,376,225	1,819	3
Non-ferrous products	51,167,583	1,623	3
Thermopower plants	51,052,487	1,619	3
Beverages	36,389,440	1,154	2
Textile	17,524,914	556	1

TABLE 5-50. MEXICO—MUNICIPAL WATER SUPPLY SYSTEM OF MEXICO CITY
(Source: Internat. Statistical Institute, 1970)

Year	Population served	Annual consumption 1,000 m3	
		Total	of which domestic
1967	6,815,000	750,503	525,353
1969	7,425,000	877,328	614,130

TABLE 5-51. NETHERLANDS AMERICAN DEPENDENCIES—ESTIMATED ANNUAL RUNOFF ON ARUBA, BONAIRE AND CURAÇAO

(Source: Pijpers, Govt. Neth. Antilles, 1972)

Island	Area km2	Number of basin areas studied	Precipitation, mm		Estimated annual runoff, m3/ha			
			3-yrs. cycle	6-yrs. cycle	Minimum	Maximum	Mean	Median
Aruba	190	8	530	700	550	1,080	730	700
Bonaire	281	3	690	870	600	750	675	675
Curaçao	472	6	690	870	385	740	565	575
Total		17	—	—	385	1,080	660	655

TABLE 5-52. NETHERLANDS AMERICAN DEPENDENCIES—AVAILABLE WATER RESOURCES ON ARUBA, BONAIRE AND CURAÇAO

(Source: Pijpers, Govt. Neth. Antilles, 1972)

[Annually, in m3]

	Surface water		Ground water	Purified sewage	Total
	New reservoirs	Existing reservoirs			
Aruba	1,400,000	220,000	30,000	360,000	2,010,000
Bonaire	640,000	215,000	380,000	—	1,235,000
Curaçao	1,100,000	1,100,000	745,000	2,650,000	5,595,000
Grand Total	3,140,000	1,535,000	1,155,000	3,010,000	8,840,000

TABLE 5-53. NETHERLANDS AMERICAN DEPENDENCIES—GROUND WATER USE ON ARUBA, BONAIRE AND CURAÇAO, 1972

(Source: Pijpers, Govt. Neth. Antilles, 1972)

Island and well field	Estimated annual supply m3	Approx. use m3	U.S.D.A. salinity hazard classification
Aruba	All groundwater supplies are saline or brackish		
Bonaire			
Dos Pos	35,000	20,000	High
Santa Barbara	130,000 [1]	40,000	Very high
Tera Cora	190,000	50,000	Very high
Total	355,000	110,000	
Curaçao			
Santa Cruz	120,000	18,000	Very high
Barber	80,000	35,000	Very high
Klein St. Marta	30,000	5,000	High
Gr. St. Michiel [2]	100,000	6,000	Medium
Gr. Piscadera [2]	100,000	8,000	Low
Zapate [2]	100,000	0	Low
Cas Grandi [2]	100,000	26,000	Low
Gr. St. Joris [2]	50,000	6,000	Very high
Klein Kwartier	10,000	—	High
Total	690,000	104,000	

[1] Including 36,000 m3 utilized on Hato plantation.
[2] Shell property.

TABLE 5-54. NETHERLAND AMERICAN DEPENDENCIES—MUNICIPAL WATER SUPPLY SYSTEMS OF ARUBA AND CURAÇAO

(**Source**: Pijpers, Govt. Neth. Antilles, 1972)

[Data as of 1971]

ARUBA

Maximum daily desalting capacity	25,000 m³
Average daily production	13,500 m³
Average daily water use by industry	7,200 m³
Average daily water use by population, hotels, shipping	6,300 m³
Storage capacity at plant site	75,000 m³
Storage capacity on island (exclusive of industries)	28,600 m³

CURAÇAO

Average daily water production at desalination plant	18,000 m³
Storage capacity at plant	60,000 m³
Population served (1971)	149,091
Number of house connections	29,804
Water consumption per connection per month	11.7 m³
Length of distribution network	440 km
Storage capacity within distribution system	43,000 m³

TABLE 5-55. NICARAGUA—MONTHLY DISCHARGE OF SELECTED RIVERS

(Source: CEPAL, 1972)

River	Station	Basin area km²	Jan.	Feb.	Mar.	Apr.	May	Jun.	Jul.	Aug.	Sep.	Oct.	Nov.	Dec.	Year
								Mean monthly discharge, m³/s							
Coco	Guanas	5,803	19.9	15.2	10.9	9.9	16.6	77.0	81.7	76.9	84.9	131.6	76.2	28.4	52.2
Grande	Dario	771	0.1	0.0	0.0	0.0	1.2	6.8	3.1	2.3	6.5	15.6	2.9	0.2	3.2
Tuma	El Dorado	579	7.1	5.4	3.2	2.0	3.7	15.9	22.7	16.9	21.2	31.6	13.1	7.6	12.7
Tuma	Yacica	838	8.5	6.5	4.5	3.3	4.5	26.0	44.0	32.3	39.9	50.7	20.2	10.6	20.8
San Juan	Pilares [1]	29,632	660.2	545.3	474.3	410.2	381.7	469.5	516.9	553.3	609.2	694.3	737.3	710.8	563.6
Tipitapa	Tipitapa [1]	—	48.2	29.6	17.6	10.1	6.4	21.4	23.2	22.0	49.8	120.4	120.6	86.9	46.3
Rio Viejo	Santa Barbara	1,185	0.5	0.4	0.3	0.2	1.1	13.9	4.5	2.8	11.8	28.5	4.3	0.9	5.6
Malacatoya	Las Banderas	894	1.0	0.7	0.4	0.3	2.3	9.5	5.2	4.4	11.7	28.6	5.1	1.5	5.7
Tamarindo	Tamarindo	198	0.3	0.2	0.2	0.1	0.4	2.8	1.2	0.7	4.1	19.6	3.7	0.6	2.9

1) Flow estimates by U.S. Corps of Engineers, 1940.

TABLE 5-56. NICARAGUA—FLOW CHARACTERISTICS OF SELECTED RIVERS

(Source: CEPAL, 1972)

River	Station	Basin area km²	Discharge, m³/s			% runoff between June to November	Period of record
			Mean	Max.	Min.		
Coco	Guanas	5,803	52.2	598.5	2.9	85	1958-66
Grande	Dario	771	3.2	311.5	0.0	98	1952-66
Tuma	El Dorado	579	12.7	386.5	0.0	80	1953-64
Tuma	Yacica	838	20.8	1,305.2	1.3	86	1952-65
Viejo	Santa Barbara	1,185	5.6	535.2	0.0	98	1953-63
Malacatoya	Las Banderas	894	5.7	783.0	0.0	95	1952-56
Tamarindo	Tamarindo	198	2.9	378.6	0.0	94	1954-64

TABLE 5-57. NICARAGUA–AVAILABLE WATER RESOURCES

(Source: CEPAL, 1972)

River basin	Basin area km2	Precipitation million m3		Surface water runoff m3/s			Ground water estimated dependable yield 2) m3/s
		Normal year	Dry year 1)	Normal year	Dry year 1)	95 % Flow	
Atlantic Slope							
Coco 3)	19,868	35,563	26,672	676	507	24	22
Wawa, Kukalaya, Prinzapolka	24,557	72,043	54,032	1,599	1,199	160	173
Grande de Matagalpa	19,668	36,975	27,731	762	571	23	31
Kurinwas, Escondido, Kukra, Punta Gorda	25,958	71,644	53,733	1,476	1,107	148	123
San Juan 3)	26,579	41,463	29,853	789	568	197	139
Total	116,630	257,688		5,302		552	488
Pacific Slope							
Choluteca 3)	538	592	444	5	3	1	–
Negro 3)	1,477	1,656	1,192	21	15	4	2
Estoro Real	4,100	6,152	4,306	78	55	1	16
Tamarindo, Brito and other coastal streams	7,255	11,972	8,140	114	77	6	21
Total	13,370	20,372		218		12	39
Total Nicaragua	130,000	278,060		5,519		564	527

1) Not totalled as droughts are usually limited to small areas.
2) Large-scale exploitation of ground water may result in decreased surface water flow.
3) International river basin; figures apply to Nicaragua only.

TABLE 5-58. NICARAGUA—WATER BALANCE ELEMENTS

(Source: CEPAL, 1972)

[Values in mm]

Element	Total Nation	Atlantic Slope	Pacific Slope
Precipitation	238	247	154
Total runoff	102	108	47
a) Direct runoff	85	90	40
b) Base flow	17	18	7
Ground water flow to sea	4	3	5
Total evapotranspiration	121	125	87
a) of ground water	11	12	04
b) of precipitation	110	113	83

TABLE 5-59. NICARAGUA—HYDROELECTRIC POWER PLANTS

(Source: CEPAL, 1972)

[Data as of 1970]

River	Plant	Capacity MW	Energy GWh	Head m	Water requirement, m³/s Total	Non-repetitive
EXISTING						
Tuma	Centroamerica	50	200	271	10.6	10.6
PROJECTED						
Grande de Matagalpa	Independencia	50	211	58	52.1	—
Grande de Matagalpa	Independencia de julio	50	220	45	69.8	—
Grande de Matagalpa	La Esperanza	100	520	49	151.8	151.8
Viejo	Santa Barbara	50	208	200	14.9	14.9

TABLE 5-60. PANAMA—MONTHLY DISCHARGE OF PRINCIPAL RIVERS
(Source: CEPAL, 1972)

River	Station	Basin area km²	Mean monthly discharge, m3/s												
			Jan.	Feb.	Mar.	Apr.	May	Jun.	Jul.	Aug.	Sep.	Oct.	Nov.	Dec.	Year
Changuinola	Bacon Bay	2,745	173.7	156.2	134.7	141.0	141.0	271.8	222.3	186.9	204.6	258.4	303.9	306.3	203.9
Cocle del Norte	El Torno	598	51.0	26.0	19.0	45.2	47.2	51.8	45.7	55.8	48.9	61.4	68.8	60.6	48.4
Indio	Limon	434	13.5	6.1	3.2	5.4	14.6	23.7	26.2	29.4	28.7	35.1	37.9	25.7	20.8
Gatun	Ciento	122	4.4	2.4	1.6	1.5	4.4	5.9	8.5	7.9	7.4	11.2	15.2	10.7	6.8
Chagres	Chico	414	26.5	16.0	12.7	14.6	26.1	28.7	31.8	33.7	32.5	37.6	51.0	51.6	30.2
Chiriqui Viejo	Paso Canoa	828	25.4	20.2	17.6	17.1	31.6	55.7	59.4	69.2	93.4	104.4	89.2	51.2	52.9
Chiriqui	David	1,392	72.6	48.0	35.3	38.4	67.4	171.5	135.9	144.3	191.6	283.0	193.4	108.5	123.7
Fonseca	San Lorenzo	667	33.2	21.7	13.6	11.2	25.4	74.4	63.6	70.9	104.3	150.3	104.5	56.2	60.8
Tabasara	Camaron	1,060	35.6	23.8	13.5	11.2	24.7	72.7	75.1	100.2	132.3	179.6	141.0	67.5	73.1
San Pablo	La Mesa	697	20.0	14.8	8.2	11.0	20.2	61.0	57.2	65.8	92.6	146.8	93.3	41.4	52.7
La Villa	Macaracas	513	10.0	5.5	3.5	3.3	7.2	15.2	15.4	16.2	22.9	42.8	36.6	21.0	16.6
Santa Maria	San Francisco	1,195	39.5	27.0	16.5	25.1	36.1	83.4	84.1	98.3	143.3	195.2	145.2	77.6	80.9
Gatun	San Juan	445	9.7	5.6	3.6	5.7	10.2	25.1	31.7	35.0	52.2	70.4	41.2	20.1	25.9
Grande	Grande	553	22.0	8.9	5.4	5.7	10.1	20.5	13.7	15.4	22.5	45.2	35.8	24.4	19.2
Calmito	Chorrera	313	4.4	2.6	1.4	1.1	1.9	3.8	5.5	7.5	9.6	15.2	17.2	11.3	6.8
Bayano	Canitas	3,941	90.8	35.3	21.5	72.6	188.4	174.6	251.9	256.3	178.7	313.7	341.2	281.0	182.4

TABLE 5-61. PANAMA—FLOW CHARACTERISTICS OF SELECTED RIVERS

(Source: CEPAL, 1972)

River	Station	Basin area km²	Discharge , m³/s			% Runoff July-Dec.	Period of record
			Mean	Max.	Min.		
Atlantic Slope							
Changuinola	Bacon Bay	2,745	203.9	3,766.2	65.0	59	1958-63
Cocle´del Norte	El Torno	598	48.4	1,019.9	5.0	59	1958-65
Chagres	Chico	414	30.2	—	4.7	66	1933-66
Ciri	Los Canones	186	10.3	—	0.4	76	1947-58
Trinidad	El Chorro	168	6.5	—	0.2	76	1947-66
Gatun	Ciento	122	6.8	—	0.1	81	1943-64
Boqueron	Peluca	91	8.0	—	0.5	68	1934-66
Pequeni	Candelaria	135	14.3	—	1.4	66	1934-66
Pacific Slope							
Bayano	Canitas	3,941	182.4	1,170.0	11.5	74	1958-62
Bayano	Maje	3,218	159.0	2,150.0	6.0	74	1958-66
Mamoni	Chepo	251	12.7	1,130.0	0.8	68	1957-66
Grande	Grande	553	19.1	1,900.9	1.8	69	1955-66
Santa Maria	San Francisco	1,195	80.9	2,373.0	5.5	76	1955-66
La Villa	Macaracas	513	16.6	730.6	1.5	78	1958-66
San Pablo	La Mesa	697	52.7	1,523.5	3.2	78	1956-63
Tabasara	Camaron	1,060	73.1	1,820.0	4.8	79	1956-65
Fonseca	San Lorenzo	667	60.8	4,086.1	2.6	75	1957-66
Chiriqui	David	1,392	123.7	2,330.0	9.0	71	1955-66

TABLE 5-62. PANAMA—AVAILABLE WATER RESOURCES
(Source: CEPAL, 1972)

River	Basin area km2	Precipitation million m3		Surface water runoff m3/s			Ground water estimated dependable yield 4) m3/s
		Normal year	Dry year 3)	Normal year	Dry year 3)	95 % Flow	
Atlantic Slope							
Sixaola, Home Creek, Changuinola 1)	3,655	9,776	8,212	217	182	43	3
Guarumo, Calovebora, Cricamola, Veraguas	7,547	30,037	25,532	667	567	100	11
Cocle, Miguel de la Borda, Indio, Chagres	7,650	23,026	19,572	475	403	148	—
Mandinga 2)	3,658	11,151	9,478	230	195	23	—
Total	22,520	73,990		1,588		314	14
Pacific Slope							
Chiriqui Viejo, Escarrea, Chico, Chiriqui	4,489	15,712	12,255	374	291	75	41
Fonseca, Tabasara, San Pablo, San Pedro 2)	11,704	34,292	26,063	659	496	66	15
La Villa, Parita, Sta. Maria, Grande 2)	9,667	19,237	14,620	396	301	39	19
Anton, Caimito 2)	2,224	4,159	3,327	86	69	4	6
Juan Diaz, Tocumen, Pacora	1,204	2,444	2,077	46	39	5	2
Bayano	4,632	9,959	8,465	216	161	22	—
Congo, Tucuti, Chucunaque, Tuira, Sambu	19,210	35,357	30,053	673	572	67	8
Total	53,130	121,160		2,450		278	91
Total Panama	75,650	195,150		4,038		592	105

1) International river basin; figures apply to Panama only.
2) Some smaller river basins included.
3) Not totalled as drought normally applies to small areas only.
4) Large-scale exploitation of ground water may diminish surface water flow.

TABLE 5-63. PANAMA—WATER BALANCE ELEMENTS

(Source: CEPAL, 1972)

[Values in mm]

Element	Total Nation	Atlantic Slope	Pacific Slope
Precipitation	258	329	232
Total runoff	152	195	136
Evapotranspiration	99	111	94

TABLE 5-64. PANAMA—HYDROELECTRIC POWER PLANTS

(Source: CEPAL, 1972)

[Data as of 1970]

River	Plant	Capacity MW	Energy GWh	Head m	Water use, m³/s Total	Water use, m³/s Non-repetitive
EXISTING						
Lake Gatun	Gatun [1]	22.5	80.0	24.6	46.0	—
Chagres	Madden	24.0	186.0	45.6	58.3	58.3
Macho Monte	Macho Monte	0.8	3.4	68.0	0.7	
Caldera	Caldera	5.3	23.2	90.0	3.6	
Cochea	Dolega I	0.8	3.5	41.6	1.2	8.0
Cochea	Dolega II	2.2	9.6	41.5	3.4	
David	Rovira	0.2	0.7	32.0	0.3	
Q. Grande	Boquete	0.3	1.1	33.5	0.5	
San Juan	La Yeguada	6.0	28.0	288.0	1.4	1.4
PROJECTED						
Changuinola [2]	Changuinola	200	1,770.0	230	110.0	110.0
Teribe [2]	Teribe	95	1,380.0	260	76.0	76.0
Hornito	Fortuna	170	1,350.0	800	24.0	—
Bayano	Bayano	22.5	556.0	45	176.0	176.0

[1] Power production dependent on Panama canal operation.
[2] International river basin; figures apply to Panama only.

TABLE 5-65. TRINIDAD & TOBAGO—DISCHARGE OF PRINCIPAL RIVERS

(Source: Trinidad Water Resources Survey, 1971)

River and station	Basin area Mi2	Mean monthly discharge, cfs												Instantaneous peak flow cfs	Date	Period of record
		Jan.	Feb.	Mar.	Apr.	May	Jun.	Jul.	Aug.	Sep.	Oct.	Nov.	Dec.			
Cunapo, Sangre Grande	31.7	85.5	39.3	8.69	3.03	13.0	19.4	20.5	270	249	124	167	214	2,250	Aug.9,'67	1967-70
North Oropuche, Toco Read	61.9	169	135	72.8	63.8	51.8	279	632	588	601	346	324	708	4,610	Nov.16,'69	1967-70
North Oropuche, W.A.S.A.	20.5	120	109	73.5	64.1	61.2	162	256	211	224	189	155	232	4,320	Nov.1,'66	1965-70
Navet, Cunapo S. Rd.	18.0	16.1	5.55	4.08	8.29	13.4	133	155	166	152	36.4	106	156	835	Aug.28,'70	1967-70
Ortoire, Rio Claro Guayaguayare Rd.	56.4	40.9	11.7	16.4	8.23	13.1	172	235	252	163	61.1	145	210	1,080	Nov.21,'69	1967-70
Pure River, Pascual Rd.	7.4	2.11	0.54	0.07	0.14	0.28	23.7	36.6	48.1	31.3	8.42	27.1	30.6	640	Nov.19,'67	1965-70
South Oropuche, Debe-Penal	55.5	12.7	1.42	1.30	0.99	1.62	102	294	399	159	57.7	229	240	1,730	Aug.13,'70	1967-70
Couva, Caroni Ltd.	24.3	7.44	1.19	0.52	0.29	0.14	34.8	229	127	120	27.3	31.9	60.7	7,950	Jul.24,'70	1967-70
Couva, Chickland	7.2	2.47	0.45	0.05	0.00	0.00	15.5	34.6	27.2	21.4	5.94	7.54	25.4	1,800	Aug.8,'67	1965-70
Caroni, Kelly Headworks	150	225	126	25.9	38.5	21.5	401	956	1,020	985	444	371	886	6,420	Aug.9,'67	1967-70
St. Joseph, E. Main Rd.	15.7	5.62	4.95	8.60	6.50	4.88	9.64	25.3	51.3	22.1	9.05	23.3	25.4	2,950	Aug.27,'70	1967-70
El Mamo, St. F9-4	4.9	3.16	2.83	1.58	1.75	2.50	12.2	30.9	29.2	28.4	9.67	7.23	35.8	1,730	Aug.7,'69	1967-70
Caura, Caura Rd.	10.0	8.12	6.90	6.62	4.55	3.67	6.95	13.9	24.6	17.3	13.2	14.3	16.4	1,462	Aug.27,'70	1968-70
Guanapo, E. Main Rd.	11.67	32.1	27.3	14.3	12.8	11.4	60.2	85.2	82.9	64.7	62.9	43.8	93.1	1,820	Oct.4,'70	1968-70

TABLE 5-66. TRINIDAD & TOBAGO—WITHDRAWAL OF GROUNDWATER BY MUNICIPAL WATER SUPPLY SYSTEMS ON TRINIDAD, 1970

(Source: Trinidad Water Resources Survey, 1971)

Aquifer system	Aquifer	Well field	Pumpage million gallons
Northern Gravels	El Socorro	El Socorro	2,167
	Valsayn	Valsayn	2,177
	Tacarigua	Tacarigua	1,029
North West Peninsula Gravels	Tucker Valley	Tucker Valley	531
	Diego Martin gravels	River Estate	334
		Four Roads	1,217
		Cocorite	580
Central Sands	Sum Sum Sands	Carlsen Field	592
		Freeport	711
		Freeport Todd's Road	224
		Total	9,562

TABLE 5-67. UNITED KINGDOM AMERICAN DEPENDENCIES—TORTOLA, VIRGIN ISLANDS— ESTIMATED WATER REQUIREMENT IN 1975

(Source: United Nations, 1964)

[Non-agricultural use only]

Consumer category	Number of persons	Per capita consumption		Daily water requirements	
		Imperial gallons	Liters	Imperial gallons	Cubic meters
Households	7,000	40	182	280,000	1,270
Industry (10% of households)				28,000	130
Tourism	3,500	80	364	280,000	1,270
Total				588,000	2,670
Public institutions (66% of sub-total)				388,000	1,760
Total				976,000	4,430

TABLE 5-68. UNITED KINGDOM AMERICAN DEPENDENCIES—ESTIMATED WATER REQUIREMENT FOR NEW PROVIDENCE, BAHAMA ISLANDS IN 1970

(Source: United Nations, 1964)

[Millions of units]

Category	Imperial gallons	Cubic meters
Households	730.0	3.32
Tourism	292.0	1.33
Industries, shipping, etc.	182.5	0.83
New connections to households	200.0	0.91
Total	1,404.5	6.39

TABLE 5-69. UNITED KINGDOM AMERICAN DEPENDENCIES—BERMUDA— AVAILABLE WATER RESOURCES

(Source: Stevens, Bermuda Public Works Dept., 1972)

Population (1969)	50,927
Total land area	20.6 mi^2 53.3 km^2
Mean annual precipitation	57.9 in. 1,470 mm
Evapotranspiration losses	42 in. 1,067 mm
Sources of water:	
Roof catchment	Typical catchment yields 20 gpd per capita, typical cistern capacity 10,000 gallons.
Artificial hillside catchments	Total area 50 acres; utilized by Government, U.S. Navy and Air-Force and hotels.
Ground water	Five fresh water lenses; largest is Pembroke-Devonshire lens tapped by infiltration galleries of Watlington Waterworks, Ltd. Supplied 74 million Imperial gallons in 1965 to 400 consumers in central Bermuda.
Ground water quality	Watlington Waterworks water contains 1,000-4,000 ppm TDS; corrosive; "Good" fresh water contains 80-250 ppm Cl and has hardness as CaCO$_3$ of 100-250 ppm.
Imported water...................................	Shipped in from Canada, U.S. and Britain by tankers during emergencies.
Desalination plants	Total distillation plant capacity 800,000 gpd (1972).

FIGURE 5-4. UNITED STATES — LARGE RIVERS
(Source: Iseri and Langbein, U.S. Geol. Survey, 1974)

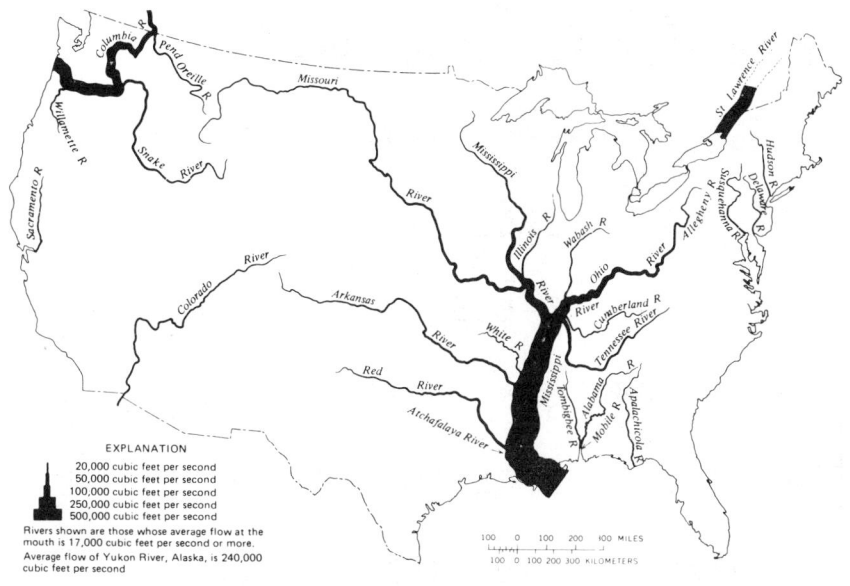

FIGURE 5-5. UNITED STATES — AVERAGE ANNUAL RUNOFF
(Source: Water Resources Council, 1968)

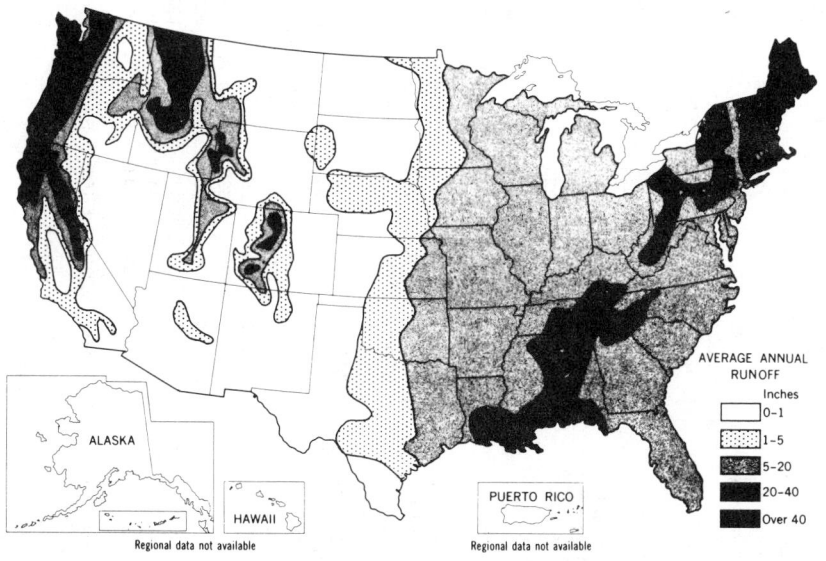

TABLE 5-70. UNITED STATES—MONTHLY DISCHARGE OF PRINCIPAL RIVERS

(Source: UNESCO, 1971)

River and station	Basin area km²	Mean monthly discharge, m³/s												Year	Period of record
		Jan.	Feb.	Mar.	Apr.	May	Jun.	Jul.	Aug.	Sep.	Oct.	Nov.	Dec.		
Penobscot, West Enfield (Me.)	17,000	214	186	290	809	685	345	221	181	176	224	328	282	327	1902-65
Kennebec, Bingham (Me.)	7,040	94.4	99.5	116	202	259	136	94.8	89.6	89.0	88.2	93.3	92.9	121	1907-10
Androscoggin, Auburn (Me.)	8,436	123	111	194	428	336	161	93.5	81.7	93.4	107	149	137	168	1929-65
Merrimack, Lowell (Mass.)	12,000	187	188	348	542	320	165	86.6	71.3	90.8	98.4	171	193	205	1923-65
Connecticut, Thompsonville (Conn.)	25,020	385	341	677	1,300	763	365	194	172	202	237	390	393	451 1)	1928-65
Delaware, Trenton (N.J.)	17,600	351	353	611	653	387	235	197	170	157	179	292	334	327	1912-65
Susquehanna, Harrisburg (Pa.)	62,400	1,070	1,100	2,210	2,090	1,290	703	422	332	306	457	696	877	963 1)	1890-1965
Potomac, Washington (D.C.)	29,940	369	461	642	552	379	221	132	143	110	150	161	245	296	1930-65
James, Richmond (Va.)	17,500	260	300	358	316	192	126	83.5	106	85.1	95.3	105	185	184	1934-65
Roanoke, Roanoke Rapids (N.C.)	21,800	304	329	362	315	233	182	175	183	158	155	153	221	230	1911-65
Cape Fear, Lillington (N.C.)	8,910	143	181	169	140	66.6	49.7	64.9	61.3	72.7	59.0	57.6	87.7	95.9	1923-65
Pee Dee, Pee Dee (S.C.)	22,900	316	431	471	396	241	179	190	183	210	182	171	224	265	1938-65
Santee, Pineville (S.C.)	38,100	58.6	123	161	139	47.8	24.8	22.2	18.8	63.6	72.9	31.6	43.0	67.0	1942-65
Savannah, Clyo (Ga.)	25,500	418	438	519	520	338	240	244	253	222	277	249	313	336	1929-33;37-65
Altamaha, Doctortown (Ga.)	35,200	477	636	795	782	388	239	231	214	177	177	162	295	385	1931-65
St. Johns, De Land (Fla.)	8,080	78.8	69.0	73.4	70.8	46.2	46.8	80.2	100	124	168	144	103	92.2	1933-65
Suwannee, Branford (Fla.)	20,000	169	203	287	316	207	145	148	180	185	175	141	142	191	1931-65
Apalachicola, Chattahoochee (Fla.)	44,300	752	876	1,150	1,040	652	465	477	429	341	361	359	550	620 1)	1928-65
Escambia, Century (Fla.)	9,886	223	246	339	337	166	108	120	123	95.6	74.6	90.5	160	173	1934-65
Alabama, Claiborne (Ala.)	57,000	1,280	1,550	1,830	1,770	899	531	519	484	369	367	433	834	902 1)	1930-65
Tombigbee, Leroy (Ala.)	49,500	1,190	1,580	1,750	1,530	714	318	330	208	175	162	347	661	742 1)	1928-60
Pascagoula, Merrill (Miss.)	17,000	381	492	561	515	279	140	170	124	100	78.5	133	269	269	1930-65
Pearl, Bogalusa (La.)	17,200	332	483	554	502	298	138	144	103	74.8	70.0	101	222	250	1938-65
Ohio, Louisville (Ky.)	236,100	4,880	5,470	7,210	6,020	3,730	2,280	1,610	1,190	743	738	1,500	2,780	3,120 1)	1928-65
Wabash, Mount Carmel (Ill.)	74,100	1,210	1,110	1,370	1,380	1,130	742	528	270	194	217	332	537	744 1)	1928-65
Cumberland, Smithland (Ky.)	46,395	1,210	1,580	1,740	1,390	718	446	348	272	217	188	383	809	779	1939-65
Tennessee, Paducah (Ky.)	104,000	2,890	3,590	3,160	2,100	1,520	1,170	1,140	1,080	1,050	987	1,370	2,130	1,820 1)	1939-65
Ohio, Metropolis (Ill.)	526,000	10,900	13,400	14,900	13,400	8,650	5,480	4,070	3,040	2,300	2,180	3,640	6,200	7,300 1)	1928-65
Fox, Wrightstown (Wis.)	15,900	109	112	132	199	171	143	95.4	76.2	73.1	86.4	99.8	105	117	1896-1965
Grand, Grand Rapids (Mich.)	12,700	90.7	111	213	185	128	88.9	54.4	41.0	46.0	54.0	64.8	70.9	95.2	1904-65
Maumee, Waterville (Ohio)	16,350	197	203	317	268	157	92.4	47.4	22.7	22.4	31.5	58.7	108	127	1899-1901; 1921-35;39-65
St. Lawrence, Ogdensburg (N.Y.)	764,600	6,230	6,150	6,430	6,970	7,230	7,340	7,290	7,100	6,860	6,630	6,510	6,450	6,760	1860-1965
Red of the North, Grand Forks (N.D.)	78,000	19.5	18.0	48.1	238	138	108	80.3	44.1	36.0	35.2	31.4	24.8	68.5	1882-1965
Mississippi, Clinton (Iowa)	222,000	677	730	1,330	2,420	2,280	1,990	1,560	1,030	1,040	1,100	1,030	708	1,330 1)	1874-1965
Mississippi, Alton (Ill.)	444,200	1,790	2,120	3,540	4,870	4,290	3,680	2,840	1,740	1,650	1,660	1,790	1,530	2,620 1)	1927-65
Missouri, Culbertson (Mont.)	237,130	187	171	212	259	205	184	236	363	389	377	249	197	252	1941-51;58-65

Station															
Yellowstone, Sidney (Mont.)	178,220	143	173	307	310	527	1,200	690	257	197	228	197	153	365[1]	1910-31;33-65
Missouri, Yankton (S.D.)	723,900	267	299	600	1,120	829	1,250	1,060	737	696	663	485	263	689[1]	1930-65
Platte, South Bend (Nebr.)	221,000	84.7	148	249	233	218	267	134	88.6	93.7	87.6	109	93.4	150	1953-65
Missouri, Nebraska City (Nebr.)	1,073,000	374	532	1,060	1,540	1,200	1,690	1,360	918	848	807	674	384	948[1]	1929-65
Kansas, Bonner Springs (Kans.)	155,100	67.0	106	161	221	261	426	332	169	180	136	99.1	67.5	185	1917-65
Missouri, Herman (Missouri)	1,368,000	1,080	1,420	2,430	3,360	3,020	4,160	3,390	1,930	1,820	1,600	1,440	1,050	2,220[1]	1897-1965
White, DeValls Bluff (Ark.)	60,686	850	943	1,140	1,270	1,230	829	505	339	291	305	402	542	720[1]	1928-45;50-65
Arkansas, Tulsa (Okla.)	193,250	73.6	90.9	118	261	375	368	277	151	164	185	116	82.6	188	1925-65
Canadian, Whitefield (Okla.)	123,220	61.5	102	147	255	430	291	164	69.0	88.6	117	87.0	76.9	157	1938-65
Arkansas, Little Rock (Ark.)	409,741	987	1,160	1,330	1,810	2,320	1,780	1,070	545	555	762	695	721	1,140[1]	1927-65
Mississippi, Vicksburg (Miss.)	2,964,300	16,200	21,100	24,800	27,900	23,600	18,500	14,600	9,340	7,320	7,090	8,220	11,100	15,800[1]	1928-65
Red, Alexandria (La.)	175,000	1,100	1,330	1,380	1,350	1,600	1,150	555	305	260	342	432	673	880[1]	1928-65
Ouachita, Monroe (La.)	39,622	627	879	1,010	1,039	940	524	272	103	90.1	109	174	316	507[1]	1932-65
Mississippi, Tarbert Landing	3,923,800	17,900	23,000	27,800	30,800	27,900	21,400	16,500	10,500	8,100	7,930	8,750	12,100	17,700[1]	1928-65
Sabine, Ruliff (Tex.)	24,160	399	416	406	366	381	252	120	82.9	57.5	55.5	116	221	239	1924-65
Neches, Evadale (Tex.)	20,590	273	297	308	308	325	188	73.4	37.3	28.2	36.6	81.1	164	176	1904-06;21-65
Trinity, Romayor (Tex.)	44,512	247	262	291	278	439	295	118	41.4	53.6	86.4	122	185	202	1924-65
Brazos, Richmond (Tex.)	114,000	200	234	231	257	459	303	145	68.9	99.2	153	140	188	207	1903-06;22-65
Colorado, Wharton (Tex.)	107,200	67.6	79.5	67.5	101	133	131	82.8	47.0	72.2	77.2	80.6	69.8	80.7	1919-25.38-65
Nueces, Mathis (Tex.)	43,150	8.6	12.0	11.4	15.0	43.5	38.9	29.5	6.9	42.7	38.2	10.8	4.2	21.9	1939-65
Pecos, Shumla (Tex.)	91,069	5.42	5.28	4.82	6.57	13.4	9.84	7.23	5.28	16.6	12.7	6.31	5.44	8.24	1954-65
Rio Grande, Laredo (Tex.)	352,178	73.7	72.2	61.9	69.7	121	158	136	143	269	188	188	82.5	123	1900-14;22-65
Green, Green River (Utah)	105,000	49.3	64.8	124	222	485	600	250	108	76.8	77.4	68.5	50.6	181	1895-99;1905-65
Colorado, Lees Ferry (Ariz.)	279,500	150	192	274	561	1,280	1,600	657	313	236	245	210	163	489[1]	1911-65
Colorado, Yuma (Ariz.)	629,100	222	262	293	367	665	1,040	586	306	230	214	200	213	383	1902-65
Sevier, Juab (Utah)	13,300	0.67	0.35	1.28	6.86	20.4	13.6	14.2	9.13	5.98	2.36	1.05	0.38	6.36	1911-65
Humboldt, Imlay (Nev.)	40,700	2.09	3.53	6.60	9.83	14.2	13.9	10.8	2.54	1.05	0.71	0.92	1.71	5.66	1935-41;45-65
San Joaquin, Vernalis (Calif.)	35,070	126	181	187	199	239	212	64.9	28.1	33.8	46.8	58.6	99.2	124	1922-65
Sacramento, Sacramento (Calif.)	60,940	940	1,180	1,020	960	850	501	299	289	323	320	419	681	647[1]	1948-65
Eel, Scotia (Calif.)	8,063	508	566	349	262	108	35.3	9.42	4.08	3.57	22.1	121	370	200	1910-65
Klamath, Klamath (Calif.)	31,300	806	1,050	749	793	648	381	162	94.2	93.6	148	348	629	486	1910-26;50-65
Chehalis, Porter (Wash.)	3,351	279	251	182	131	64.5	31.0	15.8	10.9	12.4	38.9	176	239	119	1952-65
Pend Oreille, Newport (Wash.)	62,700	377	382	439	723	1,540	2,060	1,210	505	339	340	402	391	726[1]	1903-41;52-65
Columbia, International Bdry.	155,000	1,080	1,010	1,200	1,930	5,040	8,140	5,920	3,130	1,940	1,600	1,430	1,190	2,810	1937-65
Snake, Clarkston (Wash.)	267,300	866	1,010	1,320	2,300	3,540	3,140	1,130	555	549	666	777	887	1,390[1]	1915-65
Columbia, Dalles (Ore.)	614,000	2,750	3,020	3,590	5,700	10,300	14,000	9,630	5,270	3,470	2,810	2,800	2,820	5,520[1]	1878-1965
Willamette, Salem (Ore.)	18,900	1,300	1,250	971	824	624	412	205	134	147	253	772	1,130	665[1]	1909-16;23-65
Cowlitz, Castle Rock (Wash.)	5,796	371	354	297	323	345	291	150	76.3	65.2	122	302	415	259	1927-65
Umpqua, Elkton (Ore.)	9,539	443	450	350	278	191	113	50.9	33.6	33.4	55.8	194	361	212	1905-65
Rogue, Agness (Ore.)	10,200	312	352	260	229	181	94.2	46.7	37.1	35.9	64.1	154	427	182	1960-65
Copper, Chitina (Alaska)	53,300	160	130	124	147	848	2,320	3,250	2,920	1,460	615	303	200	1,040[1]	1955-65
Kuskokwim, Crooked Creek (Alaska)	80,500	385	328	284	315	2,080	2,740	2,000	2,420	2,470	1,300	599	457	1,280	1951-65
Yukon, Eagle (Alaska)	294,000	475	433	394	408	5,420	6,160	5,130	3,980	2,980	1,930	980	598	2,220[1]	1911-14;50-65
Yukon, Rampart (Alaska)	516,400	674	591	502	517	5,420	11,500	7,950	6,480	5,090	2,780	1,220	799	3,630[1]	1955-65
Yukon, Kaltag (Alaska)	767,000	1,320	1,100	915	930	7,980	18,300	13,100	11,800	9,820	5,400	2,350	1,490	6,210[1]	1956-65

[1] Monthly and yearly averages rounded to three significant figures.

TABLE 5-71. UNITED STATES—MAXIMUM AND MINIMUM DAILY FLOW OF SELECTED RIVERS
(Source: UNESCO, 1971)

River and station	Daily discharge, m³/s			Period of record
	Maximum	Date	Minimum	
Penobscot, West Enfield, Me.	4,330	May 3, 1923	46	1902-67
Connecticut, Thompsonville, Conn.	7,990	Mar. 20, 1936	27	1928-67
Delaware, Trenton, N.J.	9,316	Aug. 20, 1955	5.4	1912-67
Susquehanna, Harrisburg, Pa.	21,000	Mar. 19, 1936	48	1890-1967
Potomac, Washington, D.C.	13,700	Mar. 3, 1936	4	1930-67
Savannah, Clyo, Ga.	3,620	Aug. 21, 22, 1940	55	1929-67
Altamaha, Doctortown, Ga.	5,040	Apr. 18, 1936	40	1931-67
Apalachicola, Chattahoochee, Fla.	8,300	Mar. 20, 1929	142	1928-67
Alabama, Claiborne, Ala.	7,561	Mar. 7, 1961	137	1930-67
Pascagoula, Merrill, Miss.	5,040	Feb. 27, 1961	20	1930-67
Pearl, Bogalusa, La.	2,498	Feb. 23, 1961	31	1938-67
Ohio, Metropolis, Ill.	50,404	Feb. 1, 1937	425	1928-67
Hudson, Green Island, N.Y.	5,130	Jul 31, 1948	28.6	1946-67
St. Lawrence, Ogdensburg, N.Y.	8,580	May 16, 1947	3,936	1918-67
Mississippi, Alton, Ill.	12,374	May 24, 1943	424	1927-67
Missouri, Yankton, S.D.	13,992	Apr. 13, 1952	76	1930-67
Missouri, Hermann, Mo.	19,142	Jun. 6, 7, 1903	119	1903-28-67
Arkansas, Little Rock, Ark.	15,178	May 27, 1943	24	1927-67
Red, Alexandria, La.	6,598	Apr. 17, 1945	25	1928-67
Rio Grande, Laredo, Tex.	20,300	Jun. 30, 1954	0	1904-14; 24-67
Colorado, Lees Ferry, Ariz.	6,230	Jun. 18, 1921	19.8	1921-67
Sacramento, Sacramento, Calif.	2,945	Nov. 21, 1950	158	1949-67
Klamath, Klamath, Calif.	15,773	Dec. 23, 1964	38	1910-26; 51-67
Snake, Clarkston, Wash.	10,449	May 29, 1948	300	1915-22; 29-67
Columbia, The Dalles, Oreg.	28,600	May 31, 1948	1,019	1933-67
Willamette, Salem, Oreg.	9,854	Jan. 8, 1923	70	1909-16; 23-67
Cowlitz, Castle Rock, Wash.	3,936	Jul. 23, 1933	30	1927-67
Copper, Chitina, Alaska	4,871	Jun. 10, 1958	—	1952-67
Yukon, Kaltag, Alaska	29,170	Jun. 22, 1964	—	1957-67

TABLE 5-72. UNITED STATES—LARGE RIVERS RANKED IN ORDER OF DISCHARGE AT MOUTH

(Source: U.S. Geol. Survey Circ. 686, 1974)

[Period of record 1931-60, 1941-70; order based on average discharge for 1941-70]

Rank	River	Drainage area (square miles)	Average discharge (1931-60) (cubic feet per second)	Average discharge (1941-70) (cubic feet per second)	Length (miles)	Most distant source	Mouth
1	Mississippi	1,247,266[1]	650,000[2]	640,000[2,3]	3,710[4]	Source of Red Rock River, Mont.	Gulf of Mexico.
2	Columbia	258,000	253,000	262,000	1,243	Columbia Lake, B.C.	Pacific Ocean.
3	Ohio	203,900	258,000	258,000	1,306	Source of Allegheny River, Potter Co., Pa.	Mississippi River.
4	St. Lawrence	302,000[5]	238,000[5]	243,000[5]	—		
5	Yukon	327,600	—	240,000[6]	1,770	Coast Mountains, B.C.	Bering Sea.
6	Atchafalaya[7]	95,105	161,000	183,000	135	Eastern edge of New Mexico.	Gulf of Mexico.
7	Mississippi above Missouri River	171,600	91,400	98,400	1,170	Lake Itasca, Minn.	Confluence with Missouri River.
8	Missouri	529,400	69,300	76,300	2,533	Source of Red Rock River, Mont.	Mississippi River.
9	Tennessee	40,910	64,000	(8)	900	Southwest Virginia, North Fork Holston River.	Ohio River.
10	Red	93,244	64,000	62,300	1,270	Eastern edge of New Mexico.	Atchafalaya River.
11	Mobile	43,800[9]	61,100	61,400	780	Northwest Georgia.	Mobile Bay.
12	Snake	109,000	49,500	50,000	1,038	Ocean Plateau, Teton Co., Wyo.	Columbia River.
13	Arkansas	160,600	41,900	45,100	1,450	Lake Co., Colo.	Mississippi River.
14	Susquehanna	27,570	38,200	37,190	444	Otsego Lake, Otsego Co., N.Y.	Chesapeake Bay.
15	Willamette	11,200	34,170	35,660	270	Tumblebug Creek, Douglas Co., Oreg.	Columbia River.
16	Alabama	22,600	32,000	32,400	735	Northwest Georgia.	Mobile River.
17	White	28,000	32,300	32,100	720	Madison Co., Ark.	Mississippi River.
18	Wabash	33,150	30,000	30,400	529	Darke Co., Ohio	Ohio River.
19	Pend Oreille	25,820	27,600	29,900	490	Near Butte, Mont.	Columbia River.
20	Tombigbee	20,100	27,400	27,300	525	Northeast Mississippi.	Mobile River.
21	Cumberland	18,080	26,900	(8)	720	Poor Fork, Letcher Co., Ky.	Ohio River.
22	Sacramento	27,100	—	—	377	Siskiyou Co., Calif.	Suisun Bay.
23	Apalachicola	19,600	24,200	24,700	524	Source of Chattahoochee River, Towns Co., Ga.	Gulf of Mexico.
24	Illinois	27,900	22,600	22,800	420	Source of Kankakee R., St. Joseph Co., Ind.	Mississippi River.
25	Colorado	242,900[10]	—	—	1,360[10]	Rocky Mountain National Park, Colo.	Gulf of California.
26	Hudson	13,370	21,300	19,500	306	Essex Co., N.Y.	Upper New York Bay.
27	Allegheny	11,700	19,800	19,290	325	Source of Allegheny River, Potter Co., Pa.	Ohio River.
28	Delaware	11,440[11]	19,200[12]	17,200	390[11]	Source of West Branch, Schoharie Co., N.Y.	Delaware Bay.

1 At Baptiste Collette Bayou, La.

2 About 25 percent of the flow of the Mississippi River system occurs in the Atchafalaya River.

3 Combined flow of Mississippi and Atchafalaya Rivers is 640,000 cubic feet per second. Flow of Mississippi River channel at mouth is 453,000 cubic feet per second.

4 Measured from the mouth of the Mississippi River and along its watercourse and that of the Missouri River to the source of Red Rock River in Montana. The length from mouth of Mississippi River to its source in Minnesota is 2,340 miles.

5 At international boundary, lat. 45°. Includes flow of St. Regis River.

6 Average is for 1957-70 period; station operated only since 1956.

7 Continuation of Red River.

8 Interbasin diversion beginning June 1966 between Lake Barkley on Cumberland River and Lake Kentucky on Tennessee River through Barkley-Kentucky Canal.

9 At Bankhead Tunnel.

10 At Arizona-Sonora boundary; natural flow not accurately known because of large depletions for irrigation.

11 At Liston Point on Delaware Bay.

12 Does not include flow of Chesapeake and Delaware canal.

FIGURE 5-6. UNITED STATES — HYDROLOGIC CYCLE AND WATER USE

(Source: Wolman, Publ. 1000-B, National Academy of Sciences — National Research Council, 1962)

*The same water may be reused at points spaced along a single stream.

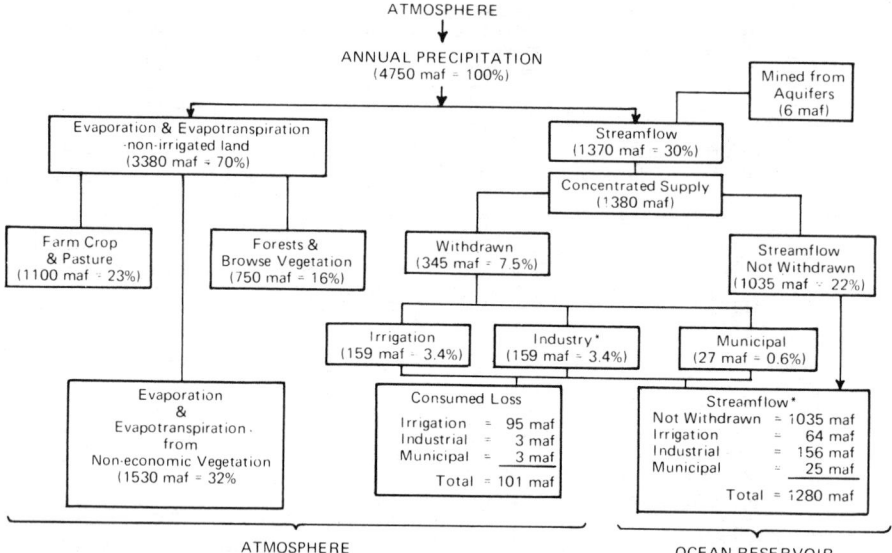

FIGURE 5-7. UNITED STATES — WATER RESOURCES REGIONS

(Source: Murray, 1972, U.S. Geol. Survey Circ. 676)

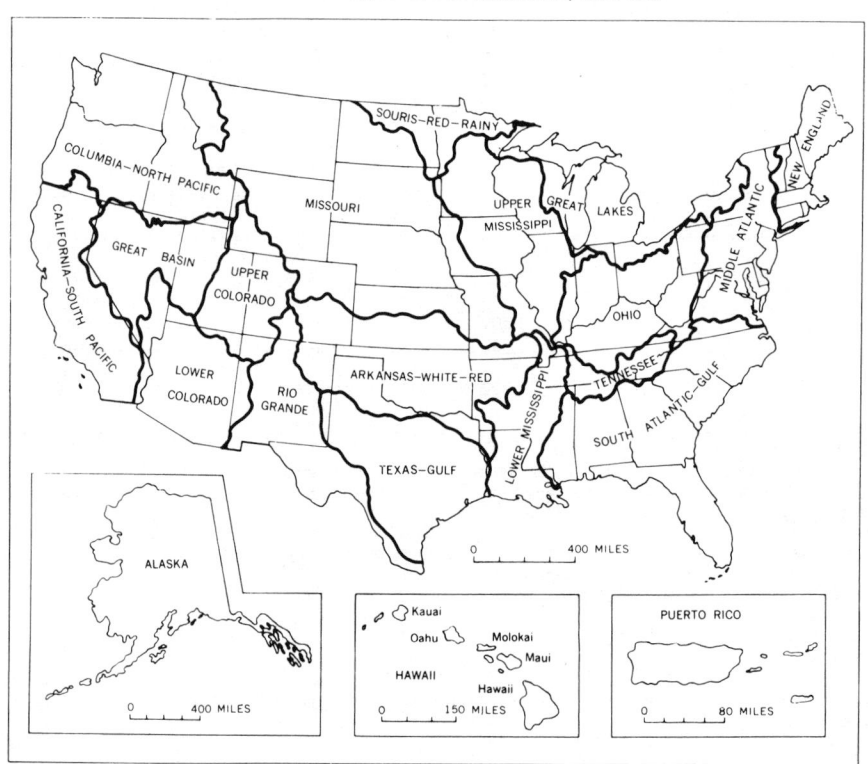

TABLE 5-73. UNITED STATES—AVAILABLE WATER RESOURCES, WITHDRAWAL AND CONSUMPTION

(Source: Murray, 1972, U.S. Geol. Survey Circ. 676)

[By Regions; see adjacent map]

Region	Area (1,000's sq mi)	Average runoff Inches per year	Average runoff Bgd	Estimated dependable supply, 1980 (bgd)	Withdrawals 1970[1] (bgd)	Fresh water consumed 1970 (bgd)	Annual flow in bdg. exceeded in 90 percent of years	Fresh surface water withdrawn, 1970 (bgd)
New England	59	24	67	22	9.7	0.41	49	4.1
Middle Atlantic	102	18	84	36	45	1.4	68	25
South Atlantic-Gulf	270	15	197	75	35	3.3	129	20
Great Lakes	126	12	75	69	39	1.2	54	38
Ohio	163	16	125	48	36	.92	75	34
Tennessee	41	21	41	14	7.9	.24	28	7.7
Upper Mississippi	190	7.2	65	31	16	.76	36	14
Lower Mississippi	96	17	79	25	13	3.6	38	8.5
Souris-Red-Rainy	59	2.2	6.2	3	.3	.07	2	.2
Missouri	515	2.2	54	30	24	12	29	18
Arkansas-White-Red	265	6.0	73	20	12	6.8	36	5.2
Texas-Gulf	175	3.9	32	17	21	6.2	11	7.4
Rio Grande	136	.8	5.0	3	6.3	3.3	2	3.8
Upper Colorado	110	2.5	13	13	8.1	4.1	8	8.0
Lower Colorado	137	.5	3.2	2	7.2	5.0	1	2.8
Great Basin	185	1.0	7.5	9	6.7	3.2	3	5.5
Columbia-North Pacific	271	16	210	70	30	11	148	26
California-South Pacific	120	9.0	62	28	48	22	30	21
United States (Conterminous)	3,020	8.3	1,200	515	365	87	747	249
Alaska	590				.2	.02		.2
Hawaii	6.4	44	13		2.7	.81		.8
Puerto Rico	3.4				3.0	.17		.4
Grand Total	3,620				371	88		250

1) Including some minor inter-regional diversions.

TABLE 5-74. UNITED STATES—TRENDS IN WATER USE, 1950-70

(Source: Murray and Reeves, 1972, U.S. Geol. Survey Circ. 676)

[in bgd]

	1950	1955	1960	1965	1970	Percent increase or decrease 1965-70
Total population (millions)	150.7	164	179.3	193.8	203.2	5
Withdrawals						
Public supplies	14	17	21	24	27	13
Rural domestic and livestock	3.6	3.6	3.6	4.0	4.5	13
Irrigation	110[1]	110	110	120	130	8
Thermoelectric power...........	40	72	100	130	170	33
Other self-supplied industrial use ...	37	39	38	46	47	2
Total withdrawals	200	240	270	310	370	19
Sources of Water						
Fresh ground water	34	47	50	60	68	13
Saline ground water	—	.65	.38	.47	1.0	113
Fresh surface water	160	180	190	210	250	19
Saline surface water	10	18	31	43	53	23
Reclaimed sewage	—	.2	.1	.7	.5	−29
Water consumed	—	—	61	77	87[2]	13
Water used for hydroelectric power ...	1,100	1,500	2,000	2,300	2,800	22

[1] Including an estimated 30 bgd in irrigation conveyance losses.
[2] Fresh water only.

TABLE 5-75. UNITED STATES—PROJECTED WATER USE, 1980-2020

(Source: National Water Commission, 1973)

[Billion gallons per day]

Type of Use	Projected Withdrawals			Projected Consumptive Use		
	1980	2000	2020	1980	2000	2020
Rural domestic	2.5	2.9	3.3	1.8	2.1	2.5
Municipal (public supplied)	33.6	50.7	74.3	10.6	16.5	24.6
Industrial (self-supplied)...........	75	127.4	210.8	6.1	10	15.6
Steam-electric power						
Fresh	134	259.2	410.6	1.7	4.6	8
Saline	59.3	211.2	503.5	.5	2	5.2
Agriculture						
Irrigation	135.9	149.8	161	81.6	90	96.9
Livestock.....................	2.4	3.4	4.7	2.2	3.1	4.2
U.S. Total......................	442.6	804.6	1,368.1	104.4	128.2	157.1

TABLE 5-76. UNITED STATES—WATER USED FOR PUBLIC SUPPLIES, 1970

(Source: Murray and Reeves, 1972, U.S. Geol. Survey Circ. 676)

[Partial figures may not add to totals because of independent rounding]

State	Population served (thousands)			Water withdrawn				Water delivered		Water consumed (mgd)
	Ground water (thousands)	Surface water (thousands)	All water (thousands)	Ground water (mgd)	Surface water (mgd)	All water (mgd)	Per capita (gpd)	Industrial and commercial uses (mgd)	Domestic use and losses (mgd)	
Alabama	765	1,420	2,190	100	360	470	213	260	210	36
Alaska	62	64	126	24	35	60	473	7.0	53	11
Arizona	989	509	1,500	190	120	310	208	47	260	160
Arkansas	605	637	1,240	71	95	170	133	63	100	36
California	8,000	10,700	18,700	1,600	1,800	3,400	181	620	2,800	1,400
Colorado	306	1,670	1,980	79	310	390	197	91	300	97
Connecticut	590	1,900	2,490	86	270	360	143	170	190	120
Delaware	217	193	410	30	46	76	185	32	44	20
Florida	4,820	592	5,410	760	120	880	163	170	720	230
Georgia	825	1,350	2,170	190	350	540	250	310	240	130
Hawaii	662	32	694	120	12	140	197	26	110	46
Idaho	407	63	470	96	15	110	237	5.8	110	29
Illinois	3,880	6,790	10,700	720	1,500	2,200	204	570	1,600	210
Indiana	1,500	1,990	3,480	210	280	490	141	140	350	120
Iowa	1,520	505	2,030	180	73	250	123	60	190	37
Kansas	817	823	1,640	130	120	250	155	61	190	74
Kentucky	304	1,860	2,160	24	160	180	83	78	100	26
Louisiana	1,290	1,370	2,670	140	240	380	144	58	330	200
Maine	153	599	752	20	89	110	146	38	72	22
Maryland	368	2,750	3,120	42	380	420	136	85	340	23
Massachusetts	1,460	3,940	5,400	170	590	750	140	320	440	38
Michigan	1,360	5,500	6,870	230	920	1,200	168	640	520	95
Minnesota	1,500	1,250	2,750	160	180	340	125	140	210	34
Mississippi	1,180	213	1,390	160	30	190	134	63	120	70
Missouri	942	3,020	3,970	92	420	510	128	260	250	95
Montana	152	354	507	26	85	110	219	32	79	55
Nebraska	908	211	1,120	150	34	190	168	46	140	38
Nevada	265	176	441	81	54	130	305	39	95	46
New Hampshire	261	283	544	32	38	70	128	22	48	3.6
New Jersey	3,030	3,390	6,420	340	560	900	139	110	780	180
New Mexico	645	67	712	130	16	150	204	41	100	66
New York	4,150	12,300	16,500	460	2,200	2,600	161	790	1,900	450
North Carolina	660	2,020	2,680	80	380	460	170	120	340	91
North Dakota	189	206	395	24	26	50	126	1.6	48	28
Ohio	2,480	5,980	8,460	320	1,000	1,300	157	340	990	170

TABLE 5-76. UNITED STATES—WATER USED FOR PUBLIC SUPPLIES, 1970 (continued)

State	Population served			Water withdrawn				Water delivered		Water consumed (mgd)
	Ground water (thousands)	Surface water (thousands)	All water (thousands)	Ground water (mgd)	Surface water (mgd)	All water (mgd)	Per capita (gpd)	Industrial and commercial uses (mgd)	Domestic use and losses [1] (mgd)	
Oklahoma	551	1,450	2,000	72	190	260	130	86	170	100
Oregon	355	842	1,200	67	160	230	188	90	140	48
Pennsylvania	1,350	8,300	9,650	250	1,500	1,800	181	880	870	180
Rhode Island	213	634	847	18	85	100	122	62	41	5.2
South Carolina	221	1,010	1,230	55	240	300	242	120	180	45
South Dakota	284	126	410	42	18	60	145	19	41	16
Tennessee	1,190	1,930	3,120	160	240	400	129	140	260	43
Texas	4,580	4,660	9,250	690	740	1,400	155	470	970	510
Utah	502	444	945	150	130	280	294	25	250	130
Vermont	100	189	289	14	29	42	146	13	30	1.7
Virginia	477	3,040	3,520	74	320	390	111	160	230	42
Washington	1,070	1,780	2,860	290	610	910	317	470	440	160
West Virginia	380	794	1,170	36	140	180	150	71	110	.2
Wisconsin	1,540	1,570	3,110	220	270	480	155	230	250	48
Wyoming	130	120	250	24	25	49	197	13	37	12
District of Columbia	0	757	757	0	160	160	211	50	110	16
Puerto Rico	396	1,940	2,330	34	170	200	86	77	130	43
United States [2]	60,600	104,000	165,000	9,400	18,000	27,000	166	8,800	19,000	5,900

1 Includes public use.
2 Including Puerto Rico.

TABLE 5-77. UNITED STATES—WATER FOR RURAL USE, 1970

(Source: Murray and Reeves, 1972, U.S. Geol. Survey Circ. 676)

[In million gallons per day; partial figures may not add to totals because of independent rounding]

State	Domestic use — Withdrawn — Ground water	Surface water	All water	Water consumed	Livestock use — Withdrawn — Ground water	Surface water	All water	Water consumed	Domestic and livestock uses — Withdrawn — Ground water	Surface water	All water	Water consumed
Alabama	63	0	63	63	13	15	27	27	76	15	90	90
Alaska	4.5	1.6	6.1	.2	0	.1	.1	.1	4.5	1.7	6.2	.4
Arizona	22	0	22	16	20	8.4	28	28	42	8.4	50	44
Arkansas	49	0	49	49	16	19	34	34	64	19	83	83
California	120	8.6	120	71	38	53	91	49	150	62	220	120
Colorado	9.3	1.4	11	2.1	20	15	35	31	29	16	45	33
Connecticut	38	.8	39	39	2.1	.4	2.5	2.5	40	1.2	42	42
Delaware	11	0	11	1.2	1.6	.1	1.7	1.4	13	.1	13	2.6
Florida	160	0	160	130	18	12	30	30	180	12	200	160
Georgia	72	1.3	73	8.8	31	.1	31	31	100	1.4	100	40
Hawaii	.1	.4	.5	.4	1.4	5.9	7.3	6.6	1.5	6.3	7.8	7.0
Idaho	22	2.5	24	6.1	10	12	22	19	32	14	46	25
Illinois	14	3.2	17	12	32	10	42	42	46	13	60	54
Indiana	76	11	87	61	29	17	46	46	100	28	130	110
Iowa	47	.1	47	19	110	25	130	130	160	25	180	150
Kansas	48	3.7	52	49	31	47	79	77	79	51	130	130
Kentucky	48	6.4	55	44	3.7	36	40	40	52	42	94	84
Louisiana	67	0	67	67	11	11	22	22	78	11	89	89
Maine	11	1.1	12	3.3	1.1	1.7	2.8	2.6	12	2.8	14	5.9
Maryland	46	0	46	46	10	.5	11	11	57	.5	57	57
Massachusetts	28	0	28	2.7	1.3	.8	2.1	2.3	30	.8	30	5.0
Michigan	160	0	160	26	24	6.8	31	28	180	6.8	190	54
Minnesota	110	0	110	110	59	9	68	68	170	9.0	170	170
Mississippi	25	0	25	22	16	24	39	39	40	24	64	61
Missouri	29	10	39	18	28	86	110	100	58	96	150	120
Montana	8.8	.7	9.5	9.5	17	16	33	33	26	17	43	43
Nebraska	22	0	22	22	80	20	100	95	100	20	120	120
Nevada	5.8	.2	6.0	2.3	1.6	2.8	4.4	2.4	7.4	3.0	10	4.7
New Hampshire	11	.2	11	1.1	.5	.8	1.3	1.3	12	1.0	13	2.4
New Jersey	80	0	80	40	1.5	.9	2.4	2.1	81	.9	82	42

TABLE 5-77. UNITED STATES—WATER FOR RURAL USE, 1970 (continued)

State	Domestic use				Livestock use				Domestic and livestock use			
	Withdrawn			Water consumed	Withdrawn			Water consumed	Withdrawn			Water consumed
	Ground water	Surface water	All water		Ground water	Surface water	All water		Ground water	Surface water	All water	
New Mexico	16	.6	17	8.1	12	32	44	43	29	33	61	51
New York	120	0	120	12	24	13	38	34	140	13	150	45
North Carolina	110	1.1	110	110	43	6.3	50	39	160	7.4	160	150
North Dakota	17	.1	17	17	9.6	6.2	16	16	26	6.3	32	32
Ohio	88	22	110	100	24	16	40	39	110	38	150	140
Oklahoma	24	4.1	28	25	6.7	46	52	52	31	50	80	78
Oregon	160	14	170	170	2.6	19	22	22	160	33	190	190
Pennsylvania	110	0	110	11	14	14	28	18	120	14	140	29
Rhode Island	4.6	0	4.6	.7	.1	.1	.2	.3	4.7	.1	4.8	1.0
South Carolina	46	0	46	46	4.0	4.9	8.9	9.0	50	4.9	55	55
South Dakota	15	1.0	16	11	81	27	109	93	96	28	120	100
Tennessee	39	0	39	9.9	5.4	29	34	34	45	29	73	44
Texas	95	0	95	95	96	52	150	150	190	52	240	240
Utah	23	.2	23	11	34	3.2	37	19	57	3.4	60	30
Vermont	10	.4	11	1.1	5.6	2.7	8.3	8.4	16	3.1	19	9.5
Virginia	73	1.6	74	45	12	17	29	23	84	19	100	68
Washington	43	12	55	19	4.2	2.1	6.3	4.4	48	14	62	24
West Virginia	17	.6	18	.2	.9	6.0	6.9	6.0	18	6.6	25	6.2
Wisconsin	74	0	74	7.3	56	15	71	71	130	15	140	78
Wyoming	5.7	.7	6.4	5.7	3.8	15	19	18	9.5	16	25	24
District of Columbia	0	0	0	0	0	0	0	0	0	0	0	0
Puerto Rico	.4	3.2	3.6	3.0	1.3	7.5	8.8	8.0	1.7	11	12	11
United States[1]	2,500	110	2,600	1,700	1,100	790	1,900	1,700	3,600	910	4,500	3,400

1 Including Puerto Rico.

TABLE 5-78. UNITED STATES—WATER USED FOR IRRIGATION, 1970

(Source: Murray and Reeves, 1972, U.S. Geol. Survey Circ. 676)

[Partial figures may not add to totals because of independent roundings]

State	Acres irrigated (1,000 acres)	Total water withdrawn (1,000 acre-feet per year)				Water consumed[1] (1,000 ac-ft/yr)	Convey-ance loss (1,000 ac-ft/yr)	Total water withdrawn (million gallons per day)				Water consumed[1] (mgd)	Conveyance loss (mgd)
		Ground water	Surface water	Re-claimed sewage	All water			Ground water	Surface water	Re-claimed sewage	All water		
Alabama	27	6.0	14	0	20	20	0	5.4	12	0	18	18	0
Alaska	2.2	.5	.5	0	1.0	.9	0	.4	.4	0	.9	.8	0
Arizona	1,200	4,300	2,700	0	7,000	5,000	270	3,800	2,400	0	6,300	4,500	240
Arkansas	1,100	1,200	260	0	1,400	1,000	100	1,100	230	0	1,300	890	90
California	8,700	18,000	19,000	140	37,000	23,000	5,500	16,000	17,000	120	33,000	20,000	4,900
Colorado	4,600	2,100	12,000	90	14,000	7,400	1,600	1,900	11,000	80	13,000	6,600	1,400
Connecticut	14	.7	6.2	0	6.9	6.9	0	.5	5.4	0	5.9	5.9	0
Delaware	17	2.6	.7	0	3.3	3.3	0	2.2	.5	0	2.7	2.7	0
Florida	1,700	1,400	1,100	0	2,500	1,500	160	1,300	970	0	2,200	1,300	150
Georgia	150	7.6	45	0	53	53	0	6.6	40	0	47	47	0
Hawaii	160	610	760	64	1,400	840	250	550	680	57	1,300	750	220
Idaho	3,700	2,300	15,000	2.8	17,000	5,200	4,800	2,100	13,000	2.5	15,000	4,700	4,300
Illinois	36	17	7.2	0	24	24	0	15	6.0	0	21	21	0
Indiana	34	20	8.9	0	29	29	0	18	7.4	0	25	25	0
Iowa	54	26	3.8	0	30	30	0	23	3.1	0	26	26	0
Kansas	1,800	3,100	230	0	3,300	2,600	48	2,800	200	0	3,000	2,300	43
Kentucky	25	.4	7.8	0	8.2	8.2	0	.4	6.7	0	7.1	6.8	0
Louisiana	670	870	880	0	1,700	1,300	480	770	780	0	1,600	1,100	430
Maine	22	.2	9.9	0	10	10	0	.2	8.7	0	8.9	8.8	0
Maryland	16	2.5	4.9	.2	7.6	7.5	0	2.1	4.3	.2	6.6	6.6	0
Massachusetts	34	21	44	0	65	49	0	18	40	0	58	43	0
Michigan	100	25	41	0	66	66	0	22	36	0	58	58	0
Minnesota	50	14	9.5	0	23	23	0	12	8.0	0	20	20	0
Mississippi	200	240	180	0	420	210	42	220	160	0	370	190	37
Missouri	180	79	8.2	0	87	62	6.7	70	6.9	0	77	55	5.5
Montana	2,200	71	8,500	.1	8,600	6,000	2,500	63	7,600	.1	7,600	5,400	2,200
Nebraska	4,100	3,000	2,300	0	5,300	3,900	1,400	2,700	2,100	0	4,700	3,500	1,300
Nevada	830	430	2,900	4.8	3,400	1,600	1,800	380	2,600	4.2	3,000	1,400	1,600
New Hampshire	3.3	0	3.3	0	3.3	2.5	0	0	2.8	0	2.8	2.0	0
New Jersey	170	63	22	0	85	78	0	56	20	0	76	70	0

TABLE 5-78. UNITED STATES—WATER USED FOR IRRIGATION, 1970 (continued)

State	Acres irrigated (1,000 acres)	Total water withdrawn (1,000 acre-feet per year)				Water consumed[1] (1,000 ac-ft/yr)	Conveyance loss (1,000 ac-ft/yr)	Total water withdrawn (million gallons per day)				Water consumed[1] (mgd)	Conveyance loss (mgd)
		Ground water	Surface water	Reclaimed sewage	All water			Ground water	Surface water	Reclaimed sewage	All water		
New Mexico	1,100	1,500	1,700	25	3,200	1,500	170	1,300	1,500	22	2,800	1,300	160
New York	75	16	15	0	31	31	0	14	13	0	27	27	0
North Carolina	470	56	36	0	92	92	0	50	32	0	82	82	0
North Dakota	74	30	190	0	210	150	66	26	170	0	190	130	59
Ohio	32	11	25	0	35	32	0	9.0	22	0	31	28	0
Oklahoma	620	810	110	0	920	640	22	720	99	0	820	570	20
Oregon	1,900	710	4,700	3.6	5,400	2,600	1,700	630	4,200	3.1	4,800	2,300	1,500
Pennsylvania	35	.9	11	0	12	12	0	.8	9.4	0	10	10	0
Rhode Island	3.8	.5	4.7	0	5.2	3.9	0	.4	4.1	0	4.5	3.4	0
South Carolina	42	10	22	0	32	32	0	8.9	20	0	29	29	0
South Dakota	150	35	230	0	260	147	88	31	200	0	230	130	79
Tennessee	9.3	1.8	3.6	0	5.4	4.6	0	1.3	2.9	0	4.2	3.8	0
Texas	8,300	8,800	2,800	17	12,000	9,100	540	7,800	2,500	15	10,000	8,100	480
Utah	1,300	470	3,500	58	4,100	2,200	710	420	3,200	52	3,600	2,000	630
Vermont	.3	0	.1	0	.1	.1	0	0	.1	0	.1	.1	0
Virginia	80	6.1	34	0	40	38	0	5.2	30	0	35	34	0
Washington	1,400	390	5,900	0	6,300	2,500	1,200	350	5,300	0	5,600	2,200	1,000
West Virginia	2.6	0	1.6	0	1.6	1.6	0	0	1.3	0	1.3	1.3	0
Wisconsin	100	37	22	0	60	45	1.2	33	20	0	52	40	.7
Wyoming	1,700	140	5,900	8.7	6,000	2,600	1,700	130	5,200	7.7	5,400	2,300	1,500
District of Columbia	0	0	0	0	0	0	0	0	0	0	0	0	0
Puerto Rico	91	75	82	0	160	110	47	67	73	0	140	98	42
United States[2]	50,000	51,000	91,000	410	140,000	82,000	25,000	45,000	81,000	370	130,000	73,000	22,000

[1] Excluding conveyance losses by evapotranspiration.
[2] Including Puerto Rico.

TABLE 5-79. UNITED STATES—SELF-SUPPLIED INDUSTRIAL WATER USE, 1970

(Source: Murray and Reeves, 1972, U.S. Geol. Survey Circ. 676)

[In million gallons per day; partial figures may not add to totals because of independent rounding]

State	Thermoelectric power (electric utility) use						Other uses									All industrial uses		
	Fresh ground water	Water withdrawn — Surface water Fresh	Surface water Saline	Total fresh water[1]	Water consumed Fresh	Water consumed Saline	Water withdrawn — Ground water Fresh	Ground water Saline	Surface water Fresh	Surface water Saline	Reclaimed sewage	All water Fresh	All water Saline	Water consumed Fresh	Water consumed Saline	Water withdrawn Fresh[1]	Saline	Fresh water consumed
Alabama	2.6	4,800	200	4,800	25	.4	93	5.0	910	96	0	1,000	100	59	6.1	5,800	300	84
Alaska	1.4	68	1.0	70	0	0	8.0	0	100	0	0	110	0	4.2	0	180	1.0	4.2
Arizona	40	2.0	0	42	32	0	150	0	24	0	0	180	0	86	0	220	0	120
Arkansas	4.0	950	0	950	3.0	0	330	0	200	0	0	520	0	210	0	1,500	0	220
California	300	1,200	8,200	1,500	24	66	410	180	28	440	3.5	440	620	170	31	2,000	8,800	190
Colorado	30	83	0	110	8.0	0	55	7.0	130	9.0	0	180	16	45	3.3	300	16	53
Connecticut	0	1,100	1,800	1,100	2.0	0	20	1.0	55	130	0	75	140	6.0	0	1,100	2,000	8.0
Delaware	.7	1.4	720	2.1	0	0	22	0	64	300	0	87	300	1.4	0	89	1,000	1.4
Florida	13	1,700	9,300	1,700	20	86	710	87	190	46	0	900	130	140	1.3	2,600	9,500	160
Georgia	6.7	3,900	140	3,900	39	1.1	330	0	300	0	0	630	0	160	0	4,500	140	200
Hawaii	82	46	860	130	0	0	160	13	100	3.2	9.4	270	16	4.6	0	400	870	4.6
Idaho	7.0	0	0	0	0	0	340	0	110	0	0	450	0	16	0	450	0	16
Illinois	7.0	11,000	0	11,000	5.0	0	250	40	1,700	0	0	1,900	40	76	0	13,000	40	81
Indiana	1.0	4,800	0	4,800	5.0	0	140	4.4	3,100	0	0	3,200	4.4	130	0	8,000	4.4	130
Iowa	0	1,400	0	1,400	20	0	150	0	130	0	0	280	0	5.3	0	1,700	0	25
Kansas	38	220	0	260	32	0	120	0	37	0	0	160	0	57	0	420	0	89
Kentucky	1.9	3,800	0	3,800	21	0	87	15	280	0	0	370	15	40	0	4,200	15	61
Louisiana	36	2,700	140	2,800	170	25	460	41	2,900	740	0	3,400	780	600	78	6,100	920	770
Maine	1.0	20	180	21	0	0	3.0	0	400	24	0	400	24	24	1.4	420	200	24
Maryland	0	510	2,700	510	0	0	43	0	450	820	130	620	820	52	0	1,100	3,500	52
Massachusetts	0	580	2,100	580	1.0	1.0	140	0	390	160	0	530	160	53	16	1,100	2,300	54
Michigan	0	9,800	0	9,800	0	0	70	400	1,600	0	0	1,700	400	120	120	11,000	400	120
Minnesota	280	1,400	0	1,700	.2	0	160	0	1,100	0	0	1,200	0	85	0	2,900	0	85
Mississippi	35	410	490	450	35	0	310	0	190	49	0	500	49	53	0	950	540	88
Missouri	6.0	2,400	0	2,500	13	0	200	.2	110	0	130	310	.2	29	0	2,800	.2	42
Montana		60	0	60	0	0	34	0	120	0	0	150	0	22	0	210	0	22
Nebraska	270	620	0	890	8.4	0	100	0	.7	0	0	100	0	3.7	0	990	0	12
Nevada	7.1	49	0	57	7.5	0	39	6.2	31	0	0	70	6.2	41	.4	130	6.2	48
New Hampshire	0	250	160	250	0	0	12	0	180	0	0	190	0	9.6	0	440	160	9.6
New Jersey	8.0	1,000	3,200	1,000	26	0	550	0	450	0	0	1,000	0	70	0	2,000	3,200	96

TABLE 5-79. UNITED STATES—SELF-SUPPLIED INDUSTRIAL WATER USE, 1970 (continued)

State	Thermoelectric power (electric utility) use						Other uses									All industrial uses		
	Fresh ground water	Surface water Fresh	Surface water Saline	Total fresh water[1]	Water consumed Fresh	Water consumed Saline	Ground water Fresh	Ground water Saline	Surface water Fresh	Surface water Saline	Reclaimed sewage	All water Fresh	All water Saline	Water consumed Fresh	Water consumed Saline	Water withdrawn Fresh[1]	Water withdrawn Saline	Fresh water consumed
New Mexico .	8.6	20	0	29	27	0	72	0	14	0	0	86	0	51	9.5	110	0	78
New York...	130	4,700	8,500	4,800	9.6	17	140	1.7	1,300	64	2.3	1,400	66	120	0	6,200	8,600	130
North Carolina	1.0	4,300	170	4,300	30	0	140	0	480	0	0	620	0	120	2.8	5,000	170	150
North Dakota	1.0	350	.9	350	1.2	.8	4.7	9.6	5.2	0	0	9.9	9.6	1.9	0	360	10	3.1
Ohio	49	13,000	0	13,000	14	0	390	0	3,300	0	0	3,700	0	110	0	17,000	0	120
Oklahoma ...	3.8	180	0	180	28	0	32	37	95	13	0	130	50	48	50	310	50	76
Oregon	0	22	0	22	.1	0	110	0	590[2]	0	0	700[2]	0	26	0	720[2]	0	26
Pennsylvania	0	12,000	0	12,000	8.9	0	400	0	5,000	50	0	5,400	50	220	0	18,000	50	230
Rhode Island	0	0	310	0	0	0	15	.4	23	50	0	38	.4	3.8	0	38	310	3.8
South Carolina	.7	2,600	68	2,600	14	0	53	0	300	32	0	350	32	33	0	2,900	100	47
South Dakota	.7	2.7	0	3.4	.6	0	14	170	2.6	0	0	17	170	1.9	17	20	170	2.5
Tennessee ...	0	4,900	0	4,900	62	0	88	0	960	0	6.0	1,000	0	47	0	5,900	0	110
Texas	60	5,800	3,800	5,900	120	24	480	0	1,200	3,600	0	1,700	3,600	680	840	7,500	7,400	800
Utah	0	87	0	87	3.8	0	56	3.5	140	5.5	0	190	9.0	45	3.0	280	9.0	48
Vermont ...	1.0	4.0	0	5.0	0	0	12	0	34	0	0	46	0	2.3	0	51	0	2.3
Virginia ...	0	3,100	770	3,100	.8	0	120	0	920	80	0	1,000	80	7.4	0	4,200	850	8.2
Washington ..	0	4.3	0	4.3	0	0	150	0	410	37	0	560	37	100	4.1	560	37	100
West Virginia	0	5,000	0	5,000	1.1	0	25	0	630	0	0	660	0	57	0	5,600	0	58
Wisconsin. ...	0	5,300	0	5,300	0	0	110	0	220	0	0	330	0	10	0	5,600	0	10
Wyoming ...	1.0	200	0	200	5.3	0	67	21	20	0	0	87	21	6.0	.3	290	21	11
District of Columbia	0	1,100	0	1,100	0	0	.8	0	.6	0	0	1.4	0	.3	0	1,100	0	.3
Puerto Rico..	.4	0	2,100	.4	.4	0	40	1.5	180	360	0	220	360	18	0	220	2,400	19
United States[3]	1,400	120,000	46,000	120,000	820	220	8,000	1,000	31,000	7,100	150	39,000	8,100	4,100	1,200	160,000	54,000	4,900

1 Included 670 mgd from public supplies.
2 Excludes 600 mgd used for log ponds.
3 Including Puerto Rico.

TABLE 5-80. UNITED STATES—WATER USED FOR THERMOELECTRIC POWER GENERATION, 1970

(Source: Murray and Reeves, 1972, U.S. Geol. Survey Circ. 676)

[In million gallons per day; partial figures may not add to totals because of independent rounding]

State	Condenser cooling						Other uses						Water consumed	
	Self-supplied			Public supplies	Self-supplied and public supplies		Fresh ground water	Self-supplied		Public supplies	Self-supplied and public supplies		Fresh	Saline
	Fresh ground water	Surface water						Surface water						
		Fresh	Saline					Fresh	Saline					
Alabama	0	4,600	190	0	4,800		2.6	250	4.0	0	260		25	0.4
Alaska	.9	2.5	1.0	0	4.4		.5	66	0	0	66		0	0
Arizona	40	2.0	0	0	42		0	0	0	0	0		32	0
Arkansas	3.0	950	0	0	950		1.0	1.0	0	0	2.0		3.0	0
California	300	1,200	8,200	130	9,900		0	1.0	0	10	11		24	66
Colorado	30	83	0	0	110		0	.1	0	.1	.2		8.0	0
Connecticut	0	1,000	1,800	0	2,900		0	5.0	0	3.0	8.0		2.0	0
Delaware	.6	1.4	720	2.0	730		.1	0	0	.3	.4		0	0
Florida	12	1,700	9,300	2.6	11,000		1.0	0	0	2.5	3.5		20	86
Georgia	4.4	3,900	140	0	4,000		2.3	28	1.1	9.5	40		39	1.1
Hawaii	82	46	860	0	980		0	0	0	0	0		0	0
Idaho	0	0	0	0	0		0	0	0	0	0		0	0
Illinois	0	11,000	0	1.0	11,000		7.0	320	0	3.0	320		5.0	0
Indiana	1.0	4,700	0	26	4,700		0	110	0	1.0	110		5.0	0
Iowa	0	1,300	0	7.7	1,400		0	42	0	.2	42		20	0
Kansas	38	220	0	0	260		0	0	0	0	0		32	0
Kentucky	1.0	3,600	0	0	3,600		.9	230	0	.7	240		21	0
Louisiana	34	2,600	140	0	2,700		1.8	140	7.2	0	140		170	25
Maine	0	20	180	0	200		1.0	0	0	1.0	2.0		0	0
Maryland	0	500	2,700	0	3,200		0	12	3.0	1.0	16		0	0
Massachusetts	0	550	2,100	0	2,700		0	29	0	2.5	32		1.0	0
Michigan	0	9,700	0	0	9,700		0	49	0	0	49		0	0
Minnesota	280	1,400	0	0	1,700		0	.2	0	0	.2		.2	0
Mississippi	34	400	490	0	920		1.0	12	0	0	13		35	0
Missouri	4.0	2,400	0	2.0	2,500		2.0	5.0	0	3.0	10		13	0
Montana	0	60	0	0	60		0	0	0	0	0		0	0
Nebraska	270	620	0	84	970		0	0	0	0	0		8.4	0
Nevada	7.1	49	0	0	56		0	.1	0	0	.1		7.5	0
New Hampshire	0	250	160	0	410		0	0	0	0	0		0	0
New Jersey	0	1,000	3,200	0	4,200		8.0	0	0	18	26		26	0

TABLE 5-80. UNITED STATES—WATER USED FOR THERMOELECTRIC POWER-GENERATION, 1970 (continued)

State	Condenser cooling					Other uses					Water consumed	
	Self-supplied			Public supplies	Self-supplied and public supplies	Fresh ground water	Self-supplied		Public supplies	Self-supplied and public supplies	Fresh	Saline
	Fresh ground water	Surface water					Surface water					
		Fresh	Saline				Fresh	Saline				
New Mexico	8.6	19	0	.1	28	0	.9	0	0	.9	27	0
New York	0	4,400	8,500	22	13,000	130	260	0	3.9	390	9.6	17
North Carolina	0	4,300	170	0	4,500	1.0	41	0	0	42	30	0
North Dakota	1.0	350	.9	0	350	0	0	0	0	0	1.2	.8
Ohio	46	13,000	0	300	13,000	3.0	590	0	16	610	14	0
Oklahoma	2.8	180	0	7.8	190	1.0	.1	0	.5	1.6	28	0
Oregon	0	22	0	0	22	0	0	0	0	0	.1	0
Pennsylvania	0	12,000	0	1.9	12,000	0	270	0	7.0	280	8.9	0
Rhode Island	0	0	310	0	310	0	0	0	0	0	0	0
South Carolina	0	2,500	68	0	2,600	.7	25	0	0	26	14	0
South Dakota	.7	2.7	0	.2	3.6	0	0	0	0	0	.6	0
Tennessee	0	4,500	0	3.7	4,500	0	340	0	0	340	62	0
Texas	58	5,800	3,800	.5	9,600	1.9	2.1	.3	.1	4.4	120	24
Utah	0	87	0	0	87	0	.2	0	.4	.6	3.8	0
Vermont	0	0	0	0	0	1.0	4.0	0	0	5.0	0	0
Virginia	0	3,100	770	0	3,900	0	0	0	0	0	.8	0
Washington	0	4.0	0	0	4.0	0	.3	0	0	.3	0	0
West Virginia	0	4,800	0	0	4,800	0	120	0	0	120	1.1	0
Wisconsin	0	5,300	0	0	5,300	0	0	0	0	0	0	0
Wyoming	.4	200	0	0	200	.6	1.0	0	0	1.6	5.3	0
District of Columbia	0	1,000	0	0	1,000	0	60	0	0	60	0	0
Puerto Rico	.2	0	2,100	.5	2,100	.2	0	0	1.3	1.5	.4	0
United States[1]	1,300	120,000	46,000	590	170,000	170	3,000	16	85	3,300	820	220

1 Including Puerto Rico.

TABLE 5-81. UNITED STATES—SUMMARY OF WATER WITHDRAWN, EXCEPT FOR HYDROELECTRIC POWER, 1970

(Source: Murray and Reeves, 1972, U.S. Geol. Survey Circ. 676)

[In million gallons per day; partial figures may not add to totals because of independent rounding]

State	Population (thousands)	Per capita use (gpd)	Ground water Fresh	Ground water Saline	Ground water Fresh and saline	Surface water Fresh	Surface water Saline	Surface water Fresh and saline	Reclaimed sewage	All sources Fresh	All sources Saline	All sources Fresh and saline	Conveyance losses	Fresh water consumed[1]
Alabama	3,444	1,900	280	5.0	280	6,100	290	6,400	0	6,400	300	6,700	0	230
Alaska	302	830	39	0	39	210	1.0	210	0	250	1.0	250	0	16
Arizona	1,772	3,900	4,200	0	4,200	2,600	0	2,600	0	6,800	0	6,800	240	4,800
Arkansas	1,923	1,600	1,500	0	1,500	1,500	0	1,500	0	3,000	0	3,000	90	1,200
California	19,953	2,400	18,000	180	18,000	20,000	8,700	29,000	130	39,000	8,800	48,000	4,900	22,000
Colorado	2,207	6,000	2,100	7.0	2,100	11,000	9.0	11,000	80	13,000	16	13,000	1,400	6,800
Connecticut	3,032	1,200	150	1.0	150	1,400	2,000	3,300	0	1,500	2,000	3,500	0	180
Delaware	548	2,200	68	0	68	110	1,000	1,100	0	180	1,000	1,200	150	26
Florida	6,789	2,300	2,900	87	3,000	3,000	9,400	12,000	0	5,900	9,500	15,000	150	1,900
Georgia	4,590	1,200	630	0	630	4,500	140	4,700	0	5,200	140	5,300	0	420
Hawaii	770	3,500	910	13	920	850	860	1,700	66	1,800	870	2,700	220	810
Idaho	713	22,000	2,500	0	2,500	13,000	0	13,000	2.5	16,000	0	16,000	4,300	4,700
Illinois	11,114	1,400	1,000	40	1,100	15,000	0	15,000	0	16,000	40	16,000	0	360
Indiana	5,194	1,700	470	4.4	470	8,200	0	8,200	0	8,600	4.4	8,600	0	390
Iowa	2,825	750	500	0	500	1,600	0	1,600	0	2,100	0	2,100	0	240
Kansas	2,249	1,700	3,100	0	3,100	640	0	640	0	3,800	0	3,800	43	2,600
Kentucky	3,219	1,400	170	15	180	4,300	0	4,300	0	4,500	15	4,500	0	180
Louisiana	3,643	2,500	1,500	41	1,500	6,700	880	7,500	0	8,100	920	9,100	430	2,200
Maine	994	760	36	0	36	520	200	720	0	560	200	760	0	61
Maryland	3,922	1,300	140	0	140	1,300	3,500	4,900	130	1,600	3,500	5,200	0	140
Massachusetts	5,689	740	350	0	350	1,600	2,300	3,900	0	2,000	2,300	4,200	0	140
Michigan	8,875	1,500	500	400	900	12,000	0	12,000	0	13,000	400	13,000	0	330
Minnesota	3,805	900	780	0	780	2,700	0	2,700	0	3,400	0	3,400	0	310
Mississippi	2,217	950	760	0	760	810	540	1,300	0	1,600	540	2,100	37	410
Missouri	4,677	750	430	.2	430	3,100	.9	3,100	0	3,500	1.1	3,500	5.5	310
Montana	694	12,000	150	0	150	7,900	0	7,900	0	8,000	0	8,000	2,200	5,500
Nebraska	1,484	4,100	3,300	0	3,300	2,700	0	2,700	0	6,000	0	6,000	1,300	3,700
Nevada	489	6,700	520	6.2	530	2,700	0	2,700	4.2	3,300	6.2	3,300	1,600	1,500
New Hampshire	738	940	55	0	55	470	160	630	0	530	160	700	0	18
New Jersey	7,168	870	1,000	0	1,000	2,000	3,200	5,200	0	3,100	3,200	6,300	0	380

TABLE 5-81. UNITED STATES—SUMMARY OF WATER WITHDRAWN, EXCEPT FOR HYDROELECTRIC POWER, 1970 (continued)

State	Population (thousands)	Per capita use (gpd)	Ground water			Surface water			Reclaimed sewage	All sources			Conveyance losses	Fresh water consumed[1]
			Fresh	Saline	Fresh and saline	Fresh	Saline	Fresh and saline		Fresh	Saline	Fresh and saline		
New Mexico	1,016	3,100	1,600	0	1,600	1,600	0	1,600	22	3,200	0	3,200	160	1,500
New York	18,191	970	890	1.7	890	8,200	8,600	17,000	2.3	9,100	8,600	18,000	0	660
North Carolina	5,082	1,200	430	0	430	5,300	170	5,400	0	5,700	170	5,900	0	480
North Dakota	618	1,000	82	9.6	92	550	.9	550	0	630	10	650	59	200
Ohio	10,652	1,700	880	0	880	17,000	0	17,000	0	18,000	0	18,000	0	460
Oklahoma	2,559	590	860	37	890	610	13	620	0	1,500	50	1,500	20	830
Oregon	2,091	2,800	970	0	970	5,000	0	5,000	3.1	5,900	0	5,900	1,500	2,600
Pennsylvania	11,794	1,700	770	0	770	19,000	50	19,000	0	20,000	50	20,000	0	450
Rhode Island	950	490	38	.4	39	110	310	430	0	150	310	460	0	13
South Carolina	2,591	1,300	170	0	170	3,100	100	3,200	0	3,300	100	3,400	0	180
South Dakota	666	910	180	170	350	250	0	250	0	440	170	600	79	250
Tennessee	3,924	1,600	290	0	290	6,100	0	6,100	0	6,400	0	6,400	0	200
Texas	11,197	2,400	9,200	0	9,200	10,000	7,400	18,000	21	19,000	7,400	27,000	480	9,600
Utah	1,059	4,000	680	3.5	680	3,500	5.5	3,500	52	4,200	9.0	4,200	630	2,200
Vermont	445	250	42	0	42	69	0	69	0	110	0	110	0	13
Virginia	4,648	1,200	280	0	280	4,400	850	5,300	0	4,700	850	5,500	0	150
Washington	3,409	2,100	840	0	840	6,300	37	6,300	0	7,100	37	7,200	1,000	2,500
West Virginia	1,744	3,300	80	0	80	5,700	0	5,700	0	5,800	0	5,800	0	66
Wisconsin	4,418	1,400	490	0	490	5,800	0	5,800	0	6,300	0	6,300	.7	180
Wyoming	332	17,000	230	21	250	5,500	0	5,500	7.7	5,700	21	5,800	1,500	2,400
District of Columbia	757	1,700	.8	0	.8	1,300	0	1,300	0	1,300	0	1,300	0	16
Puerto Rico	2,712	1,100	140	1.5	140	430	2,400	2,900	0	580	2,400	3,000	42	170
United States[2]	205,897	1,800	68,000	1,000	69,000	250,000	53,000	300,000	520	320,000	54,000	370,000	22,000	87,000

1 Excluding irrigation conveyance losses by evapotranspiration.
2 Including Puerto Rico.

TABLE 5-82. UNITED STATES—WATER USED FOR HYDROELECTRIC POWER, 1970

(Source: Murray and Reeves, 1972, U.S. Geol. Survey Circ. 676)

State	Mgd	1,000 acre-feet per year	State	Mgd	1,000 acre-feet per year	State	Mgd	1,000 acre-feet per year
Alabama	130,000	150,000	Maine	80,000	90,000	Oregon	350,000	390,000
Alaska	780	870	Maryland	23,000	25,000	Pennsylvania	30,000	34,000
Arizona	20,000	23,000	Massachusetts	17,000	19,000	Rhode Island	58	65
Arkansas	25,000	28,000	Michigan	59,000	66,000	South Carolina	41,000	46,000
California	84,000	94,000	Minnesota	18,000	20,000	South Dakota	59,000	66,000
Colorado	4,000	4,500	Mississippi	0	0	Tennessee	130,000	150,000
Connecticut	6,600	7,400	Missouri	9,000	10,000	Texas	9,300	10,000
Delaware	0	0	Montana	72,000	80,000	Utah	3,000	3,300
Florida	11,000	13,000	Nebraska	29,000	32,000	Vermont	15,000	16,000
Georgia	45,000	50,000	Nevada	4,200	4,800	Virginia	16,000	18,000
Hawaii	330	370	New Hampshire	22,000	24,000	Washington	720,000	810,000
Idaho	84,000	94,000	New Jersey	120	130	West Virginia	25,000	28,000
Illinois	10,000	12,000	New Mexico	430	480	Wisconsin	66,000	74,000
Indiana	18,000	21,000	New York	270,000	300,000	Wyoming	6,500	7,200
Iowa	35,000	39,000	North Carolina	88,000	98,000	District of Columbia	6.0	6.7
Kansas	520	590	North Dakota	14,000	15,000	Puerto Rico	610	680
Kentucky	79,000	88,000	Ohio	370	410			
Louisiana	0	0	Oklahoma	20,000	22,000	United States[1]	2,800,000	3,100,000

1 Including Puerto Rico.

TABLE 5-83. UNITED STATES—NATURAL FRESH-WATER LAKES OF 100 SQUARE MILES OR MORE

(Source: Bue, 1963, U.S. Geol. Survey Circ. 476)

[The Great Lakes are excluded]

Name	Location	Area sq mi	Name	Location	Area sq mi
Lake of the Woods	Minnesota and Ontario	1,485	Mille Lacs	Minnesota	207
Iliamna	Alaska	1,000	Flathead	Montana	197
Okeechobee	Florida	700	Tahoe	California and Nevada	193
Champlain	New York, Vermont, and Quebec	490	Leech	Minnesota	176
St. Clair	Michigan and Ontario	460	Pend Oreille	Idaho	148
Becharof	Alaska	458	Upper Klamath	Oregon	142
Upper and Lower Red	Minnesota	451	Utah	Utah	140
			Yellowstone	Wyoming	137
Rainy	Minnesota and Ontario	345	Tustumena	Alaska	117
			Moosehead	Maine	117
			Clark	Alaska	110
Teshekpuk	Alaska	315	Bear	Idaho and Utah	110
Naknek	Alaska	242	Winnibigoshish	Minnesota	109
Winnebago	Wisconsin	215	Dall	Alaska	100

TABLE 5-84. UNITED STATES—NATURAL FRESH-WATER LAKES OF 250 FEET DEEP OR MORE

(Source: Bue, 1963, U.S. Geol. Survey Circ. 476)

[The Great Lakes are excluded]

Name	Location	Depth feet	Name	Location	Depth feet
Crater	Oregon	1,932	Cooper	Alaska	>400
Tahoe	California and Nevada	1,645	Champlain	New York, Vermont, and Quebec	400
Chelan	Washington	1,605	Kasnyku	Alaska	393
Pend Oreille	Idaho	1,200	Chakachamna	Alaska	380
Nuyakuk	Alaska	930	Ozette	Washington	331
Deer	Alaska	877	Aleknagik	Alaska	330
Chauekuktuli	Alaska	700	Sebago	Maine	316
Crescent	Washington	624	Swan	Alaska	>314
Seneca	New York	618	Baranoff	Alaska	303
Clark	Alaska	606	Payette	Idaho	>300
Beverley	Alaska	500	Quinault	Washington	About 300
Nerka	Alaska	475	Crescent	Alaska	291
Tokatz	Alaska	474	Wallowa	Oregon	283
Long	Alaska	470	Chilkoot	Alaska	282
Lower Sweetheart	Alaska	459	Odell	Oregon	279
Cayuga	New York	435	Silver	Alaska	278
Crater	Alaska	414	Grant	Alaska	>250

FIGURE 5-8. UNITED STATES – PROFILE OF THE GREAT LAKES –
ST. LAWRENCE RIVER DRAINAGE SYSTEM

(Source: Internat. Great Lakes Levels Board, 1973)

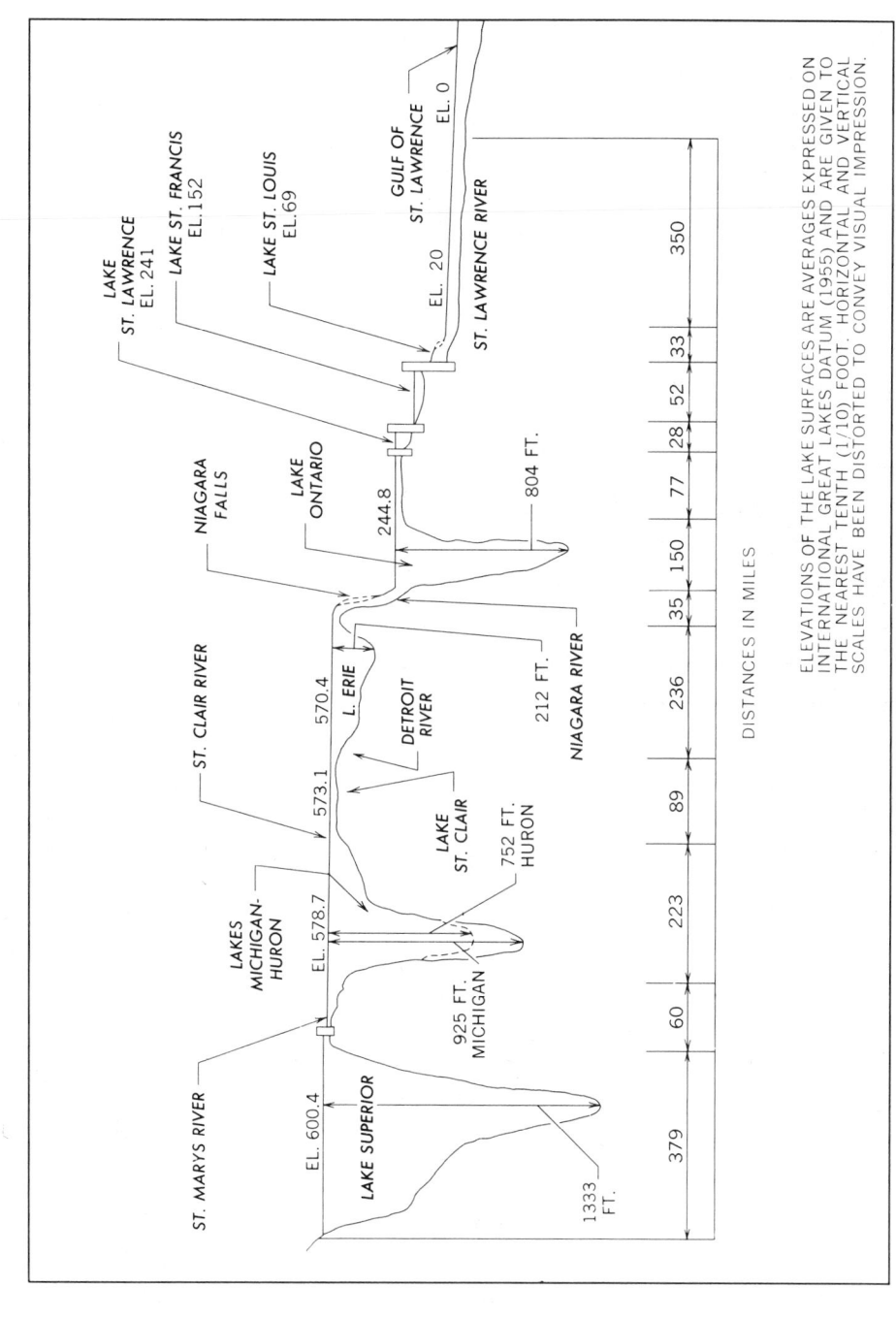

DISTANCES IN MILES

ELEVATIONS OF THE LAKE SURFACES ARE AVERAGES EXPRESSED ON
INTERNATIONAL GREAT LAKES DATUM (1955) AND ARE GIVEN TO
THE NEAREST TENTH (1/10) FOOT. HORIZONTAL AND VERTICAL
SCALES HAVE BEEN DISTORTED TO CONVEY VISUAL IMPRESSION.

TABLE 5-85. UNITED STATES—PHYSICAL AND HYDROLOGIC DATA OF THE GREAT LAKES

(Source: Internat. Great Lakes Levels Board, 1973)

[Historical Record: 1860-1972; all elevations in feet IGLD (1955)]

	Lake Superior	Lake Michigan	Lake Huron	Lake St. Clair a)	Lake Erie	Lake Ontario
Elevation of Low Water Datum	600.0	576.8	576.8	571.7	568.6	242.8
Monthly Elevations						
Average	600.39	578.70	578.70	573.09	570.41	244.77
Maximum	602.06	581.94	581.94	575.70b)	572.76b)	248.06
Minimum	598.23	575.35	575.35	569.86	567.49	241.45
Range of Stage	3.8	6.6	6.6	5.8	5.3	6.6
Range, Winter Low to Summer High(Monthly)						
Average	1.1	1.1	1.1	1.8	1.5	1.9
Maximum	1.9	2.2	2.2	3.3	2.7	3.5
Minimum	0.4	0.1	0.1	0.9	0.5	0.7
Recorded Monthly Outflows (cfs) Outlet c)	St. Marys	Str. of Mackinac	St. Clair	Detroit	Niagara	St. Lawrence
Average	75,400	52,000d)	187,900	188,900	202,300	239,700
Maximum	127,000	—	242,000b)	— e)	256,000b)	314,000b)
Minimum	40,900	—	99,000	100,000e)	116,000	154,000
Average Outflow in Inches on Total Drainage Basin f)	12.4	11.1	11.1	11.1	10.2	10.5
Drainage Areas (sq mi)						
Land Area g)	49,300	45,600	51,800	6,100	23,600	27,200i)
Water Surface Area h)	31,700	22,300	23,000	400	9,900	7,600i)
Storage Capacity per ft. Depth (CFS-months)	337,000	481,000		5,000	105,000	80,000

a) Lake St. Clair elevations are available only for the period 1898 to date.
b) New maximums set in 1973: Lake St. Clair 576.23 (June), Lake Erie 573.51 (June).
   Outflows: St. Clair River 245,000; Niagara River 265,000; St. Lawrence River 350,000.
c) Outflows include the effects of diversions.
d) Approximate.
e) Insufficient records available.
f) Drainage basin includes land and water surface areas.
g) Land areas include the total drainage area to the outlet of the upstream lake.
h) Water areas do not include areas of connecting channels.
i) Includes area down to the St. Lawrence Power Project at Cornwall.

TABLE 5-86. UNITED STATES—PRINCIPAL SALINE LAKES

(Source: Bue, 1963, U.S. Geol. Survey Circ. 476)

Lake	Present area square miles	Remarks
Salton Sea (California)	350	About 650 sq mi at highest stage in 1905-07.
Goose (in California and Oregon ..... About	100	Maximum, 186 sq mi, 125 in California and 61 in Oregon; overflowed into Pitt River in 1869 and 1881; dry in 1930; 150 sq mi in 1958.
Pontchartrain, Louisiana	625	These lakes are connected with the Gulf of Mexico, and are subject to tidal fluctuation.
Pyramid, Nevada	180	Maximum size, 220 sq mi. Low until 1860; reached extreme high level in 1862 and 1868 or 1869; nearly as high in 1890; began to drop in 1917 (Hardman and Venstrom, 1941).
Walker, Nevada	107	Maximum size, 125 sq mi.
Great Salt, Utah .... About	1,000	Maximum size since 1851, 2,400 sq mi in 1870's; minimum, 950 sq mi in October 1961; seasonal high in 1962 was 1,050 sq mi in June.

TABLE 5-87. UNITED STATES—MAJOR RESERVOIRS

(Source: International Commission on Large Dams, 1967)

Name of Dam	Name of Reservoir	Reservoir capacity acre-feet	Name of Dam	Name of Reservoir	Reservoir capacity acre-feet
Hoover	Lake Mead	29,827,000	Buford	Lake Lanier	2,554,000
Glen Canyon	Glen Canyon	28,040,000	Texarkana	Lake Texarkana	2,509,000
Oahe	Oahe	23,600,000	Trinity	Trinity	2,500,000
Garrison	Garrison	23,000,000	Tuttle Creek	Tuttle Creek	2,280,000
Fort Peck	Fort Peck	19,400,000	Marshall Ford	Lake Travis	2,200,200
Grand Coulee	Franklin D. Roosevelt Lake	9,402,000	Elephant Butte	Elephant Butte	2,185,400
Rockland	Rockland	6,809,700	Center Hill	Center Hill	2,092,000
Fort Randall	Fort Randall	6,300,000	Canyon Ferry	Canyon Ferry	2,051,000
Wolf Creek	Lake Cumberland	6,089,000	Whitney	Whitney	2,017,500
Kentucky	Kentucky	6,002,600	Kingsley	McConaughy Lake	2,000,000
Denison	Lake Texoma	5,530,000	Pensacola	Lake O'the Cherokees	2,000,000
Bull Shoals	Bull Shoals	5,408,000	Norfolk	Norfolk	1,983,000
Shasta	Shasta Lake	4,500,000	Bagnell	Lake of the Ozarks	1,973,000
McGee Bend	McGee Bend	4,478,800	Keystone	Keystone	1,879,000
Falcon	Falcon	4,085,000	Davis	Lake Mohave	1,818,330
Eufaula	Eufaula	3,848,000	Dale Hollow	Dale Hollow	1,706,000
Oroville	N.A.	3,500,000	American Falls	American Falls	1,700,000
Hungry Horse	Hungry Horse	3,468,000	Martin	Martin	1,630,000
Table Rock	Table Rock	3,462,000	Monticello	Lake Berryessa	1,600,000
Clark Hill	Clark Hill	2,900,000	Sardis	Sardis	1,569,900
Hartwell	Hartwell	2,858,100	Cherokee	Cherokee	1,565,400
Greers Ferry	N.A.	2,844,000	Oologah	Oologah	1,519,000
John H. Kerr	John H. Kerr	2,808,400	Dougals	Douglas	1,514,100
Blakeley Mountain	Lake Ouachita	2,768,500	Brownlee	Brownlee	1,470,000
Norris	Norris	2,567,000	Fontana	Fontana	1,444,300

TABLE 5-87. UNITED STATES—MAJOR RESERVOIRS (continued)

Name of Dam	Name of Reservoir	Reservoir capacity acre-feet	Name of Dam	Name of Reservoir	Reservoir capacity acre-feet
Ross	Ross Lake	1,405,000	Pathfinder	Pathfinder	1,010,900
Palisades	Palisades	1,402,000	Folsom	Folsom	1,010,000
Roosevelt	Roosevelt	1,381,600	Buchanan	Lake Buchanan	1,000,000
Santee	Lake Marion	1,350,000	Pine Flat	Pine Flat	1,000,000
McNary	McNary	1,345,000	Possum Kindgom	Possum Kingdom	990,000
Grenada	Grenada	1,337,400	Boysen	Boysen	970,000
Tiber	Tiber	1,337,000	Winnigoshish	Winnigoshish	967,940
Lake Almanor	Lake Almanor	1,308,000	Chamita	Chamita	945,000
Fort Gibson	Fort Gibson	1,287,000	Iron Bridge	Iron Bridge	930,000
Dry Falls	Equalizing	1,275,000	Fort Gaines	N.A.	925,000
Winsor	Quabbin	1,265,000	Link River	Upper Klamath	873,000
Tenkiller Ferry	Tenkiller Ferry	1,230,000	Harlan County	Harlan	850,000
Kerr	Flathead Lake	1,220,000	Jackson Lake	Jackson Lake	847,000
Coolidge	San Carlos	1,208,800	Watauga	Watauga	778,800
Swift	N.A.	1,190,000	South Holston	South Holston	744,000
Albeni Falls	Pend Oreille Lake	1,153,000	Federal	Leech Lake	743,330
Wheeler	Wheeler	1,150,400	Canyon	Canyon	740,900
Watts Bar	Watts Bar	1,132,000	Glendo	Glendo	735,000
Owyhee	Owyhee	1,120,000	Lake Tahoe	Lake Tahoe	732,000
Pinopolis	Lake Moultrie	1,110,000	Parker	Havasu Lake	716,600
Belton	Belton	1,097,600	Chickamauga	Chickamauga	705,300
Pickwick Landing	Pickwick Landing	1,091,400	Cascade	Cascade	704,100
Guntersville	Guntersville	1,018,700	Conklingville	Sacandaga	690,000
Lewisville	Garza-Little Elm	1,016,200	Ripogenus	Chesuncook Lake	690,000
Seminoe	Seminoe	1,012,000	Lake Chelan	Lake Chelan	676,100

N.A. — Not available.

TABLE 5-88. UNITED STATES—MUNICIPAL WATER SUPPLY SYSTEMS

(Source: American Water Works Association, 1970)

[Operating data of selected water utilities as of 1970]

City or utility	Total population served	Quantity available, 10⁶ gallons/year				Average production mgd	Delivered to distribution system 10⁶ gallons/year	Consumption gallons per capita per day
		Production		Purchased	Total			
		Surface water	Ground water					
Alabama								
Birmingham	525,000	20,368	522	8,258	29,148	79.86	27,983	152
Huntsville	160,000	2,383	4,413	719	7,515	20.59	7,515	129
Mobile	200,000	—	—	—	7,289	19.97	7,289	100
Montgomery	160,000	3,550	3,803	—	7,353	20.15	7,253	126
Alaska								
Anchorage	60,000	3,120	1,323	—	4,443	12.17	—	203
Arizona								
Phoenix	633,000	35,153	17,575	—	52,728	144.46	52,728	228
Arkansas								
Little Rock	159,728	12,065	—	—	12,065	33.05	11,792	207
California								
Bakersfield 1)	130,600	—	17,004	—	17,004	46.59	17,004	357
Fresno	180,000	—	21,147	—	21,147	57.94	21,147	322
Long Beach	361,000	—	8,854	12,407	21,261	58.25	21,254	161
Los Angeles	2,862,000	147,988	24,062	17,005	189,055	517.96	190,185	181
Los Angeles 2)	663,700	2,130	22,500	18,450	43,080	118.03	39,650	178
Oakland 3)	1,080,000	77,053	128	—	77,181	211.45	77,181	196
Pasadena	137,187	990	4,157	6,270	11,417	31.28	11,417	228
Riverside	122,000	—	12,141	7	12,148	33.28	12,148	273
Sacramento	270,000	21,000	3,000	—	24,000	65.75	24,000	244
San Diego	723,000	2,788	—	37,519	40,307	110.43	40,307	153
San Francisco	2,000,000	274,754	—	—	274,754	752.75	88,927	376
San Jose Waterworks	525,000	5,052	21,377	8,525	34,954	95.76	34,954	182
Colorado								
Colorado Springs	150,000	13,900	—	1,600	15,500	42.47	13,930	283
Denver	772,790	70,976	—	—	70,976	194.45	59,571	252
Connecticut								
Bridgeport 4)	340,000	24,160	891	—	25,051	68.63	25,051	202
Hartford 5)	393,000	21,304	—	—	21,304	58.37	20,740	149
New Haven	400,000	19,996	744	—	20,740	56.82	20,740	142
District of Columbia		73,008	—	—	73,008	200.02	73,007	185
Florida								
Fort Lauderdale	139,000	—	13,728	—	13,728	37.61	13,728	271
Miami	500,000	—	58,199	—	58,199	159.45	51,176	319
Orlando	177,000	—	11,820	—	11,820	32.38	11,820	183
St. Petersburg	235,000	—	10,067	—	10,067	27.58	9,901	117

TABLE 5-88. UNITED STATES—MUNICIPAL WATER SUPPLY SYSTEMS (continued)

City or utility	Total population served	Quantity available, 10⁶ gallons/year				Average production mgd	Delivered to distribution system 10⁶ gallons/year	Consumption gallons per capita per day
		Production		Purchased	Total			
		Surface water	Ground water					
Georgia								
Atlanta	700,000	38,000	—	—	38,000	104.11	32,000	149
Columbus	145,000	9,020	—	—	9,020	24.71	8,805	170
Hawaii								
Honolulu	535,000	—	40,197	54	40,251	110.28	40,251	206
Illinois								
Chicago	4,506,000	377,636	—	—	377,636	1,034.62	377,636	230
Peoria	163,040	456	7,896	—	8,352	22.88	8,352	140
Indiana								
Fort Wayne	194,670	10,741	—	—	10,741	29.43	10,741	151
Gary 6)	250,000	11,260	—	—	11,260	30.85	10,968	123
Indianapolis	680,000	30,996	2,250	—	33,246	91.08	33,805	134
Iowa								
Cedar Rapids	110,640	—	6,372	—	6,372	17.46	6,297	158
Des Moines	243,770	12,534	—	—	12,534	34.34	12,124	141
Kansas								
Kansas City	180,000	11,561	—	—	11,561	31.67	—	176
Topeka	133,000	6,010	—	—	6,010	16.47	6,010	124
Wichita	284,720	4,160	9,732	—	13,892	38.06	13,760	134
Kentucky								
Louisville	673,000	42,377	—	—	42,377	116.10	40,214	173
Louisiana								
Shreveport	200,000	10,022	—	—	10,022	27.46	10,021	137
Maine								
Augusta	22,000	1,311	125	—	1,436	3.93	1,436	179
Portland	146,480	7,984	200	—	8,184	22.42	8,184	153
Maryland								
Baltimore	1,500,000	93,803	—	—	93,803	256.99	89,556	171
Hyattsville	1,200,000	44,678	524	—	45,202	123.84	44,414	103
Massachusetts								
Boston	608,000	—	—	51,714	51,714	141.68	51,714	233
Springfield	200,000	14,612	—	—	14,612	40.03	14,612	200
Michigan								
Detroit	3,790,500	244,562	—	—	244,562	670.03	244,562	177
Lansing	146,360	—	8,002	—	8,002	21.92	7,946	150
Saginaw	92,570	10,226	—	—	10,226	28.02	10,015	303
Minnesota								
Duluth	100,000	6,565	—	—	6,565	17.99	6,504	180
St. Paul	404,380	20,122	—	—	20,122	55.13	19,737	136

Mississippi								
Jackson	170,000	7,068	—	—	7,068	19.36	5,931	114
Missouri								
Kansas City	632,724	35,739	—	—	35,739	97.92	34,469	155
St. Louis	622,500	68,660	—	—	68,660	188.11	68,078	302
St. Louis County	764,512	38,566	—	—	38,566	105.66	37,960	138
Montana								
Billings	73,800	5,365	—	—	5,365	14.70	5,365	199
Nebraska								
Lincoln	149,520	—	11,604	—	11,604	31.79	11,604	213
Omaha	425,000	19,576	10,867	—	30,443	83.41	25,484	196
Nevada								
Las Vegas 7)	200,000	—	15,638	4,387	20,025	54.86	20,025	274
Reno 8)	109,000	10,522	1,839	—	12,361	33.87	12,361	311
New Jersey								
Elizabeth 9)	1,251,670	30,000	8,000	—	38,000	104.11	34,000	83
Short Hills 10)	232,000	—	—	—	10,110	27.70	10,110	119
Weehawken 11)	1,000,000	32,370	5,287	2,099	39,756	108.92	39,756	109
New Mexico								
Albuquerque	250,000	—	18,000	—	18,000	49.32	18,000	197
New York								
Lynbrook 12)	262,400	—	10,637	—	10,637	29.14	10,511	111
New Rochelle	160,900	1,084	—	7,106	8,190	22.44	8,190	139
New York City	7,898,000	795,708	—	—	795,708	2,180.02	486,371	276
Oakdale 13)	650,000	—	24,277	—	24,277	66.51	24,277	102
Rochester	291,270	18,599	—	—	18,599	50.96	18,599	175
Syracuse	197,210	18,826	—	53	18,879	51.72	18,005	262
North Carolina								
Asheville	125,000	7,360	—	—	7,360	20.16	7,306	161
Greensboro	150,000	7,100	—	—	7,100	19.45	7,100	130
North Dakota								
Fargo	53,460	2,663	—	—	2,663	7.30	2,482	136
Ohio								
Akron	380,000	18,053	—	—	18,053	49.46	17,684	130
Dayton	350,000	—	25,698	—	25,698	70.41	25,372	201
Oklahoma								
Tulsa	395,000	22,626	—	—	22,626	61.99	22,290	157
Oregon								
Eugene	73,350	8,191	—	—	8,191	22.44	8,191	306
Portland	650,000	70,805	—	—	70,805	193.99	33,444	298
Pennsylvania								
Bryn Mawr 14)	808,000	19,649	7,157	27,701	54,507	149.33	27,292	185
Chester	115,000	10,509	—	—	10,509	28.79	10,357	250
Erie	200,000	—	—	—	16,764	45.93	16,764	230
Pittsburgh 15)	513,240	23,902	—	—	23,902	65.48	23,902	128
South Carolina								
Charleston	175,900	11,040	—	—	11,040	30.25	10,700	172
Greenville	200,000	12,500	—	—	12,500	34.25	12,500	171

TABLE 5-88. UNITED STATES—MUNICIPAL WATER SUPPLY SYSTEMS (continued)

City or utility	Total population served	Quantity available, 10^6 gallons/year				Average production mgd	Delivered to distribution system 10^6 gallons/year	Consumption gallons per capita per day
		Production		Purchased	Total			
		Surface water	Ground water					
South Dakota								
Sioux Falls	72,440	56	4,050	—	4,106	11.25	4,106	155
Tennessee								
Knoxville	185,000	10,728	—	—	10,728	29.39	10,210	159
Memphis	623,530	—	32,959	—	32,959	90.30	32,959	145
Nashville	360,000	21,011	—	—	21,011	57.56	20,607	160
Texas								
Amarillo	127,010	6,513	2,429	—	8,942	24.50	8,942	193
Austin	250,000	18,397	—	—	18,397	50.40	17,614	202
Corpus Christi	274,720	26,959	—	—	26,959	73.86	26,959	269
Dallas	1,130,000	63,575	—	2,739	66,314	181.68	60,744	161
El Paso	330,000	3,109	19,811	—	22,920	62.79	22,920	190
Houston	1,076,000	46,436	51,708	—	98,144	268.89	98,114	250
Lubbock	170,000	8,519	1,660	—	10,179	27.89	9,936	164
San Antonio	578,860	—	34,633	—	34,633	94.88	34,633	164
Utah								
Salt Lake City	400,000	20,105	—	—	20,105	55.08	4,949	138
Virginia								
Annandale	477,000	16,470	406	1,174	18,050	49.45	18,050	104
Norfolk	500,000	21,885	—	—	21,885	59.96	21,379	120
Richmond	250,000	14,961	—	506	15,467	42.38	14,967	170
Washington								
Seattle	573,150	60,247	—	—	60,247	165.06	49,706	288
Spokane	171,800	—	20,716	—	20,716	56.76	20,716	330
Tacoma	175,000	23,020	4,840	—	27,860	76.33	27,860	436
West Virginia								
Morgantown	43,500	2,300	—	—	2,300	6.30	2,218	145
Wisconsin								
Madison	184,000	—	10,615	—	10,615	29.08	10,615	158
Milwaukee	933,000	59,243	—	—	59,243	162.31	59,243	174
Wyoming								
Cheyenne	50,000	3,036	1,011	—	4,047	11.09	4,047	222

1) California Water Service.
2) Southern California Water Co.
3) East Bay Municipal Utilities.
4) Bridgeport Hydraulic Co.
5) Hartford Metropolitan District.
6) Gary Hobart Water Co.
7) Las Vegas Valley Water District.
8) Sierra Pacific Power Co.
9) Elizabethtown Water Co.
10) Commonwealth Water Co.
11) Hackensack Water Co.
12) Long Island Water Co.
13) Suffolk Water Authority.
14) Philadelphia Suburban Water Co.
15) South Pittsburgh Water Co.

NORTH AMERICA

REFERENCES CITED

References cited in more than one country:

United Nations, 1964, Water Desalination in Developing Countries, ST/ECA/82.

UNESCO, Discharge of Selected Rivers of the World; Vol. I. General and Regime Characteristics of Stations Selected (1969); Vol. II. Monthly and Annual Discharges Recorded at Various Selected Stations (1971); Vol. III. Mean Monthly and Extreme Discharges 1965-1969 (1971).Copyright UNESCO/IASH/WMO. Reprinted by permission.

NORTH AMERICA GENERAL

Comision Economica para America Latina (CEPAL), 1972, Aprovechamiento de los Recursos Hidraulicos en Centroamerica 1970 a 1980, Mexico, D.F.

U.S. Federal Power Commission, 1968, World Power Data, Publ. P. 40. Washington, D.C.

BARBADOS

Barbados Water Works Department, 1972, (Private Communication).

BRITISH HONDURAS

Smith, S.F., 1972, Ministry of Power and Communications, Belmopan (Private Communication).

CANADA

Canadian National Committee International-Hydrological Decade, 1972, Discharge of Selected Rivers of Canada, Ottawa.

Dominion Bureau of Statistics, Canada Year Book, 1970-71, Ottawa.

Brown, I.C., (Ed.), 1966, Groundwater in Canada, Geological Survey of Canada, Economic Geology Report, No. 24.

Conner, R.A., 1967, Fluoridation of Water Supplies in Canada and Related Problems, Proc. of the International Conference on Water for Peace, Washington, D.C.

COSTA RICA

CEPAL, 1971, Istmo Centroamericano. Programa de Evaluacion de Recursos Hidraulicos I. Costa Rica, Mexico, D.F.

CUBA

Comite Nacional Cubano para el Decenio Hidrologico Internacional, 1969, Memoria 1965-68, Datos Basicos de la Red de Estaciones de Observacion, Vol. II.

Keshishev, V.N., 1967, The Runoff in Eastern Cuba, Soviet Geography: Review and Translation, Vol. VIII.

DOMINICAN REPUBLIC

Organization of the American States, 1969, Survey of the Natural Resources of the Dominican Republic, Washington, D.C.

EL SALVADOR

CEPAL, 1971, Istmo Centroamericana. Programa de Evaluacion de Recursos Hidraulicos-II. El Salvador, Mexico, D.F.

Lemus Serrano, F., 1972, Planificacion de los Recursos Naturales para el Aprovechamiento Racional y Multiple de los Recursos Hidraulicos en El Salvador-Servicio Hidrologico, San Salvador.

FRENCH AMERICAN DEPENDENCIES

GUADELOUPE and MARTINIQUE—Office de la Recherche Scientifique et Technique Outre-Mer (ORSTOM), Annuaire Hydrologique de la France d'Outre-Mer, Annees 1963-1964-1965, Paris, 1968.

GUATEMALA

CEPAL, 1971, Istmo Centroamericano. Programa de Evaluacion de Recursos Hidraulicos-III. Guatemala, Mexico, D.F.

HAITI

Organisation des Etats Americains et Conseil National de Developpement et de Planification de la Republique d'Haiti, 1972, Haiti Mission d'Assistance Technique Integree, Washington, D.C.

HONDURAS

CEPAL, 1973, Istmo Centroamericano. Programa de Evaluacion de Recursos Hidraulicos-IV. Honduras, Mexico, D.F.

JAMAICA

Williams, J.B., 1969, The Water Resources of Jamaica - A Review, Bull. Scientific Research Council Jamaica, Vol. 9, Nos. 1-4, March.

MEXICO

del Arenal C., Rodolfo, 1969, Water Resources of Mexico, Water Resources Bull., Vol. 5, March.

Oriver Alba, Adolfo, 1969, Water Resources Administration in Mexico, U.N. Interregional Seminar on Current Issues of Water Resources Administration, New Delhi, 1973.

Lamadrid, Arturo and Camhaji Samra, Salomon, 1973, La Demanda de Agua en la Industria de Transformacion, International Symposium on Water Resources Planning, Mexico, D.F.

NETHERLANDS AMERICAN DEPENDENCIES

ARUBA, BONAIRE, CURACAO — Pijpers, F.J., 1972, Ministry of Economic Development, Government of Netherlands Antilles, Willemstad, Curacao (Private Communication).

NICARAGUA

CEPAL, 1972, Istmo Centroamericano. Programa de Evaluacion de Recursos Hidraulicos-V. Nicaragua, Mexico, D.F.

PANAMA

CEPAL, 1972, Istmo Centroamericano. Programa de Evaluacion de Recursos Hidraulicos-VI. Panama, Mexico, D.F.

TRINIDAD & TOBAGO

Trinidad Water Resources Survey, 1971, Annual Data Report - 1970.

UNITED KINGDOM AMERICAN DEPENDENCIES

BERMUDA

Stevens, M.R., 1972, Water Statistics, Public Works Department, Hamilton (Private Communication).

Bermuda Public Works Department, 1965, Report on the Sources of Water Supply Available in Bermuda Together with Related Problems.

UNITED STATES

National Water Commission, 1973, Water Policies for the Future, U.S. Government Printing Office, Washington, D.C.

Iseri, K.T. and Langbein, W.B., 1974, Large Rivers of the United States, U.S. Geological Survey Circular 686.

Murray, C.R., and Reeves, E.B., 1972, Estimated Use of Water in the United States in 1970, U.S. Geological Survey Circular 676.

International Great Lakes Levels Board, 1973, Regulation of Great Lakes Water Levels - Report to the International Joint Commission, Washington, D.C.

American Water Works Association, 1973, Operating Data for Water Utilities 1970 and 1965, AWWA No. 20112, Denver, Colo.

Water Resources Council, 1968, The Nation's Water Resources, U.S. Government Printing Office, Washington, D.C.

Wolman, Abel, 1962, Water Resources; A Report to the Committee on Natural Resources of the National Academy of Sciences — National Research Council, Washington, D.C. (NAS-NRC Publ. 1000-B).

MEXICO

CUBA

JAMAICA

PUERTO RICO

HAITI

DOMINICAN
REPUBLIC

Caribbean Sea

BELIZE

HONDURAS

GUATEMALA

EL SALVADOR

NICARAGUA

COSTA
RICA

PANAMA

BARBADOS

TRINIDAD AND TOBAGO

VENEZUELA

Orinoco

GUYANA

SURINAM

FRENCH GUIANA

COLOMBIA

Magdalena

ECUADOR

Negro

Galapagos Is.

Amazon

P E R U

Marañon

Madeira

Tapajos

B R A Z I L

Xingu

Tocantins

São Francisco

Lake
Titicaca

BOLIVIA

Pacific Ocean

PARAGUAY

Paraná

Paraguay

Paraná

C H I L E

A R G E N T I N A

URUGUAY

Atlantic Ocean

Salado

Falkland Is.

MILES
0   200   400   600

0   600
KILOMETERS

mp

TABLE 6-1. SOUTH AMERICA—STATUS OF WATER SUPPLY AND SEWERAGE SYSTEM SERVICES IN LATIN AMERICA

(Source: Pan American Health Organization, 1973)

[Population in thousands] a

| | | Water Supply | | | | | | | | | | | | | | | Sewage disposal | | | | |
| | | Total | | | | | Urban | | | | | Rural | | | | | Urban | | Rural | Total | |
Country or other political unit	Date of data	Population	House connections	Easy access	Total	%	Population	House connections	Easy access	Total	%	Population	House connections	Easy access	Total	%	No.	%	No.	No.	%
Argentina	Dec. 72	24,210	14,100	1,200	15,300	63	18,400	13,200	1,000	14,200	77	5,810	900	200	1,100	19	6,560	36	—	6,560	27
Barbados	Oct. 71	241	135	106	241	100	110	105	5	110	100	131	30	101	131	100	—	—	—	—	—
Bolivia	Nov. 72	5,190	894	515	1,409	27	1,650	790	485	1,275	77	3,540	104	30	134	4	390	24	68	458	9
Brazil	Dec. 70	94,317	38,500	14,800	53,300	57	53,789	28,700	12,600	41,300	77	40,528	9,800	2,200	12,000	30	15,600	29	1,384	16,984	18
British Honduras	Dec. 72	127	36	21	57	45	69	33	19	52	75	58	3	2	5	9	3	4	—	3	2
Chile	Dec. 72	10,120	4,890	1,930	6,820	67	6,950	4,750	1,800	6,550	94	3,170	140	130	270	9	2,630	38	185	2,815	28
Colombia	May 72	22,800	11,293	2,680	13,973	61	13,300	9,293	2,000	11,293	85	9,500	2,000	680	2,680	28	7,817	59	3,060	10,877	48
Costa Rica	Dec. 72	1,867	1,293	143	1,436	77	635	603	32	635	100	1,232	690	111	801	65	255	40	0	255	14
Cuba	Jun. 66	7,950	5,610	650	6,260	79	5,020	3,840	650	4,490	89	2,930	1,770	—	1,770	60	1,700	34	—	1,700	21
Dominican Republic	Dec. 72	4,406	1,226	655	1,881	43	1,936	1,064	450	1,514	78	2,470	162	205	367	15	307	16	—	307	7
Ecuador	Dec. 72	6,600	1,750	300	2,050	31	2,630	1,550	150	1,700	65	3,970	200	150	350	9	1,560	59	40	1,600	24
El Salvador	Dec. 72	3,684	715	1,126	1,841	50	1,452	581	479	1,060	73	2,232	134	647	781	35	450	31	11	461	13
Guatemala	Dec. 71	5,309	795	1,290	2,085	39	1,836	739	897	1,636	89	3,473	56	393	449	13	769	42	—	769	14
Guyana	Dec. 71	735	374	40	414	56	225	206	15	221	98	510	168	25	193	38	67	30	—	67	9
Haiti	Dec. 72	5,200	160	264	424	8	971	160	264	424	44	4,229	—	—	—	—	—	—	—	—	—
Honduras	Dec. 72	2,682	648	354	1,002	37	805	522	268	790	98	1,877	126	86	212	11	367	46	1	368	14
Jamaica	Mar. 72	1,925	780	869	1,649	86	520	500	9	509	98	1,405	280	860	1,140	81	139	27	29	168	9
Mexico	Dec. 72	53,320	28,360	1,700	30,060	56	32,300	21,810	1,700	23,510	73	21,020	6,550	—	6,550	31	15,600	48	32	15,632	29
Nicaragua	Oct. 71	1,951	781	252	1,033	53	942	663	192	855	91	1,009	118	60	178	18	398	42	—	398	20
Panama	Dec. 72	1,499	722	385	1,107	74	733	662	70	732	99	766	60	315	375	49	497	68	5	502	33
Paraguay	Dec. 72	2,329	180	226	406	17	877	180	135	315	36	1,452	—	91	91	6	129	15	0	129	6
Peru	Dec. 72	14,020	3,630	2,035	5,665	40	6,410	3,470	1,200	4,670	73	7,610	160	835	995	13	4,170	65	12	4,182	30
Surinam	Dec. 72	393	163	104	267	68	201	147	54	201	100	192	16	50	66	34	85	42	—	85	22
Trinidad and Tobago	Dec. 70	1,060	562	460	1,022	96	358	297	59	356	99	702	265	401	666	95	181	51	2	183	17
Uruguay	Dec. 72	2,956	2,155	292	2,447	83	2,389	2,065	222	2,287	96	567	90	70	160	28	960	40	262	1,222	41
Venezuela	Dec. 71	10,700	6,893	1,963	8,856	83	7,300	5,570	1,730	7,300	100	3,400	1,323	233	1,556	46	3,400	47	121	3,521	33
Eastern Caribbean countries and territories	Dec. 70	504	131	232	363	72	168	74	55	129	77	336	57	177	234	70	14	8	—	14	3
TOTAL		286,095	126,776	34,592	161,368	56	161,976	101,574	26,540	128,114	79	124,119	25,202	8,052	33,254	27	64,048	40	5,212	69,260	24

a Current estimates of population served as received from countries by the Department of Engineering and Environmental Sciences, PASB.

TABLE 6-2. SOUTH AMERICA—FUNDS ALLOCATED FOR CONSTRUCTION OF WATER SUPPLY AND SEWERAGE SYSTEMS IN LATIN AMERICA

(Source: Pan American Health Organization, 1973)

[January 1961-December 1972; amounts in thousands of U.S. dollars]

Country	International loans and grants									Estimated national matching funds
	IDB [1]		IBRD [2]		AID [3]		EXIMBANK [4]		CIDA [5]	
	Water	Sewerage	Water	Sewerage	Water	Sewerage	Water	Sewerage	Water	
Argentina	45,730	2,270	—	—	1,400	—	—	—	—	56,030
Barbados	—	0,150	—	—	—	—	—	—	2,600	0,070
Bolivia	10,600	4,800	—	—	1,145	—	—	—	—	9,397
Brazil	151,060	16,650	22,000	15,000	30,695	33,900	—	—	—	423,164
Chile	27,945	1,700	—	—	2,000	0,840	0,188	—	—	23,654
Colombia	36,751	10,733	127,700	3,900	3,800	9,600	1,261	—	—	124,670
Costa Rica	3,900	3,940	—	—	5,900	0,500	4,000	—	—	8,024
Dominican Republic	9,060	1,090	—	—	3,000	—	—	—	—	5,925
Ecuador	38,900	19,168	—	—	—	—	—	—	—	22,023
El Salvador	9,180	1,520	—	—	0,075	—	—	—	—	5,340
Guatemala	24,318	2,000	—	—	1,369	—	—	—	—	14,625
Guyana	—	—	—	—	2,650	—	—	—	—	1,200
Haiti	7,510	0,088	—	—	—	—	—	—	—	1,600
Honduras	3,300	0,400	—	—	1,050	—	—	—	—	1,470
Jamaica	—	—	5,000	—	3,700	—	—	—	—	5,900
Mexico	27,974	2,550	—	—	—	—	0,036	—	—	20,296
Nicaragua	9,000	10,385	9,900	—	0,143	—	—	—	—	14,628
Panama	12,042	1,670	—	—	26,140	10,851	—	—	—	19,597
Paraguay	3,895	4,670	—	—	—	—	—	—	—	3,550
Peru	25,024	10,836	—	—	5,700	2,900	5,123	1,500	—	43,079
Trinidad and Tobago	7,900	—	—	—	—	—	—	9,000	—	15,013
Uruguay	12,943	3,300	—	—	—	—	1,900	—	—	23,768
Venezuela	46,000	7,200	21,300	—	—	—	7,500	—	—	121,131
Eastern Carribbean countries and territories	—	—	—	—	—	—	—	—	6,132	—
Total	513,032	105,120	185,900	18,900	88,767	58,591	20,008	10,500	8,732	964,154

Summary (in U.S. dollars)

International loans . . . . $ 1,009,550,000

Water.   816,439,000

Sewerage   193,111,000

National matching funds   964,154,000

Other national funds 1,097,770,000

Total funds . . . . $ 3,071,474,000

[1] Inter-American Development Bank
[2] International Bank for Reconstruction and Developemnt (World Bank)
[3] Agency for International Development
[4] Export-Import Bank (USA)
[5] Canadian International Development Agency

TABLE 6-3. SOUTH AMERICA—HYDROELECTRIC AND THERMAL POWER GENERATING
CAPACITY AND PRODUCTION, 1968

(Source: U.S. Federal Power Commission, 1968)

Country	Installed capacity (MW.) [1]			Energy production (GWh.) [2]			Population (1,000's)	Kwh. per capita
	Hydro	Thermal	Total	Hydro	Thermal	Total		
Argentina	544	5,293	5,837	1,481	15,705	17,186	23,617	728
Bolivia	144	78	222	576	87	663	4,680	142
Brazil	6,183	2,372	8,555	30,550	7,631	38,181	88,209	433
Chile	858	862	1,720	3,529	3,278	6,807	9,351	728
Columbia	1,336	830	2,166	4,723	2,120	6,843	19,825	345
Ecuador	122	168	290	372	452	824	5,695	145
French Guiana	0	5	5	0	20	20	39	513
Guyana	0	81	81	0	257	257	719	357
Paraguay	45	63	108	6	158	164	2,213	74
Peru	836	722	1,558	2,970	1,805	4,775	12,772	374
Surinam	180	52	232	762	294	1,056	375	2,816
Uruguay	236	245	481	1,138	780	1,918	2,818	681
Venezuela	912	1,852	2,764	1,807	8,639	10,446	9,686	1,078
Subtotal	11,396	12,623	24,019	47,914	41,226	89,140	180,017	495

[1] MW.—Megawatts=thousand kilowatts.
[2] GWh.—Gigawatt-hours=million kilowatt-hours.

TABLE 6-4. ARGENTINA—MONTHLY DISCHARGE OF PRINCIPAL RIVERS

(Source: UNESCO, 1971)

River and station	Basin area km²	Mean monthly discharge, m³/s												Year	Period of record
		Jan.	Feb.	Mar.	Apr.	May	Jun.	Jul.	Aug.	Sep.	Oct.	Nov.	Dec.		
Parana, Posadas	975,375	4,300	16,400	16,100	13,800	11,600	11,300	9,440	8,140	8,400	10,100	10,300	11,200	11,800	1901-63
Pilcomayo, F.N. Pilcomayo	130,000	363	583	412	230	105	52.0	34.0	20.0	14.0	21.0	83.0	157	167	1950-62
Pescado, Col. Colpana	5,150	220	277	240	113	53.0	31.0	25.0	20.0	17.0	24.0	49.0	113	98.0	1945-62
Bermejo, Zan. del Tigre	25,000	669	947	732	360	160	95.0	70.0	52.0	44.0	63.0	146	339	307	1940-62
San Francisco, Urundel	25,800	198	390	308	112	48.0	34.0	28.0	24.0	17.0	16.0	29.0	74.0	105	1946-62
Juramento, Miraflores	34,500	73.0	119	74.0	33.0	18.0	15.0	13.0	12.0	11.0	11.0	13.0	25.0	34.0	1929-62
Pasaje, El Tunal	38,000	80.0	132	91.0	42.0	23.0	17.0	15.0	13.0	10.0	9.0	13.0	24.0	38.0	1943-62
Salado, El Arenal	40,000	43.1	85.9	61.8	30.2	11.3	5.4	3.4	1.5	0.4	0.6	1.1	6.8	20.2	1929-62
Yabebiry, Col. Martires	650	3.3	5.0	3.8	11.5	11.6	18.2	14.3	8.5	20.9	25.3	14.8	7.8	12.1	1951-65
Dulce, El Sauce	20,200	152	183	203	124	75.0	42.0	30.0	22.0	14.0	19.0	43.0	80.0	82.0	1927-62
Soto, P. de la Corriente	449	9.9	7.4	7.4	2.4	1.3	0.6	0.4	0.4	0.6	2.7	5.3	8.0	3.9	1953-60
Tercero, Embalse	3,300	55.0	46.0	47.0	28.0	18.0	10.0	8.0	6.0	7.0	21.0	34.0	45.0	27.0	1913-62
Manso, Los Moscos	580	34.0	23.0	20.0	18.0	37.0	42.0	45.0	39.0	33.0	39.0	51.0	43.0	35.0	1946-62
San Juan, La Puntilla	26,000	110	80.0	54.0	41.0	40.0	40.0	38.0	36.0	38.0	49.0	82.0	119	60.0	1909-62
Mendoza, Guido	8,180	70.0	63.0	43.0	26.0	21.0	18.0	17.0	17.0	18.0	26.0	41.0	64.0	35.0	1956-62
Colorado, Buta Ranquil	15,300	218	142	99.0	80.0	77.0	80.0	78.0	78.0	95.0	187	366	350	154	1942-62
Colorado, Pichi Mahuida	22,300	211	132	89.0	70.0	74.0	79.0	75.0	76.0	85.0	147	268	292	132	1918-62
Neuquen, Paso Indios	30,200	248	135	101	105	239	344	354	322	342	500	640	484	319	1902-62
Limay, Paso Limay	26,400	590	373	299	317	651	947	1,070	999	896	1,030	1,060	865	759	1903-62
Negro, Ia. Angostura	95,000	730	425	290	281	546	1,120	1,240	1,250	1,130	1,340	1,530	1,240	934	1927-62
Futaleufu, Balsa Garzon	4,650	299	217	193	178	311	374	343	297	254	289	379	355	291	1948-62
Chubut, Los Altares	16,400	23	13	12	12	31	64	72	72	72	82	79	44	48	1943-62
Senguerr, Nacimiento	1,300	34	22	16	19	25	34	34	29	28	38	59	48	31	1949-62
La Leona, Paso La Leona	7,450	366	533	611	485	338	254	186	141	94	84	132	214	286	1956-62
Santa Cruz, Charles Fuhr	15,550	827	1,120	1,270	1,130	847	636	433	315	222	316	425	591	675	1956-62

TABLE 6-5. ARGENTINA—FLOW CHARACTERISTICS OF PRINCIPAL RIVERS

(Source: CEPAL, 1972)

[Rivers with mean flow of less than 10m3/s omitted]

River	Gaging station	Basin area 1) km2	Flow, m3/s						Runoff l/s/km2
			Mean	Max.	Min.	5%	50%	95%	
I. Plata Basin									
Parana	Rosario	2,302,000	14,900.0	22,380	6,500				6.5
Parana	Parana	2,039,000	13,920.0	25,130	6,300				6.8
Parana	Corrientes	1,925,000	15,420.0	29,300	5,100				8.0
Parana	R. Apipe	886,000	11,800.0	29,400	3,100				13.3
Iguazu	Iguazu	70,000	1,470.0	8,300	230				
Pilcomayo	Fortin Nuevo Pilcomayo	110,000	166.0	1,150 2)	1.0				1.5
Bermejo	Zanja del Tigre y Manuel Elordi	25,000	305.0	13,190	22	1,070.0	114.0	34.0	12.2
Bermejo	Junta de San Antonio	15,300	195.0	8,650	10	670.0	70.0	10.0	12.7
Bermejo	Aguas Blancas	4,450	84.0	5,360	6	280.0	25.0	15.0	18.9
a) Pescado	Colonia Colpana	5,150	91.0	5,790	9.5	285.0	35.0	15.0	17.7
San Francisco	Urundel	25,800	106.0	4,700	4.4				4.1
a) Grande	Puente Perez	7,650	24.7	358 2)	3.0	95.0	11.0	3.0	3.3
b) Lavayen	Bajada de Pinto	4,100	13.2	625	1.0	49.0	5.0	3.0	3.2
c) Mojotoro	El Angosto	850	16.1	860	0.7	58.0	5.0	1.0	18.9
Juramento-Pasaje-Salado	Suncho Corral	44,000	15.0	400	0.0	77.0	3.0	0.0	0.3
Salado	El Arenal	40,000	19.6	400	0.0	97.0	3.0	0.0	0.5
Pasaje	El Tunal	38,000	39.2	1,092 2)	0.5	140.0	17.0	3.0	1.0
Pasaje	Miraflores	34,500	34.0	1,200	5.6	128.0	13.0	7.0	1.0
a) Arias	San Gabriel	7,100	24.7	417	5.0	79.0	11.0	7.0	3.5
b) Tercero	Bell Ville	8,500	16.7	250 2)	0.4				2.0
c) Tercero	Embalse	3,300	26.4	2,000	1.4				8.0
Uruguay	Salto Grande	239,000	4,660.0	37,000	92.0				19.7
Uruguay	Paso Hervidero	231,000	3,840.0	18,200	380.0				16.6
Uruguay	Santo Tome	127,500	2,260.0	17,800	112.0				17.7
II. Atlantic Ocean									
Jachal	Pachimoco	25,500	11.5	173 2)	2.5	25.0	9.0	5.0	0.5
San Juan	Gobernador I. de la Roza	26,000	61.0	1,100 2)	15.0	125.0	39.0	27.0	2.3
San Juan	Km. 47	25,667	69.0	462 2)	24.0	178.0	33.0	12.0	2.7
a) Patos	La Plateada	8,500	55.0		3.0	46.0	11.6	5.5	6.5
b) Patos	Alvarez Condarco		18.3						
Mendoza	T. Usina Cacheuta	9,050	52.0	2,800	9.0	118.0	32.0	17.0	5.7
Mendoza	Punta de Vacas		31.3						
a) Tupungato	Punta de Vacas	1,800	20.0	95 2)	4.0				11.1
Tunuyan	Valle de Uco	2,380	26.1	112 2)	4.0	65.0	18.0	9.0	10.9
Diamante	Los Reyunos	4,150	36.5	210 2)	8.0	100.0	27.0	13.0	8.8

Atuel	Rincon del Atuel y Angostura	3,800	32.0	152 2)	6.0	71.0	25.0	15.0	8.4
a) Salado del Atuel	Canada Ancha	812	11.5	58 2)	2.0	25.5	8.5	4.5	14.2
Colorado	Pichi Mahuida	22,300	134.0	830	32.0	365.0	95.0	55.0	6.0
Colorado	Buta Ranquil	15,300	148.8						9.7
Negro	Primera Angostura	95,000	930.0	3,420	75.0	2,080.0	830.0	180.0	9.3
Negro	Paso Roca	89,000	1,020.0	6,500	87.0	2,220.0	880.0	220.0	11.5
Neuquen	Paso de los Indios	30,200	303.0	5,340	47.0	820.0	216.0	64.0	10.1
Limay	Paso Limay	26,400	725.0	5,120	69.0	1,630.0	630.0	130.0	27.5
Limay	Paso Flores	9,800	282.0	1,120	23.0	535.0	255.0	75.0	28.8
Limay	Nahuel Huapi	3,900	211.0	658 2)	29.0	355.0	250.0	85.0	54.1
Chubut	Los Altares	16,400	48.6	540	4.0	138.0	33.0	8.0	3.0
Alto Chubut	El Maiten	1,200	18.1	340	1.0	49.0	15.0	3.0	15.1
Gualjaina	Gualjaina	2,800	11.0	170	0.9				3.9
Senguerr	Vuelta del Senguerr y Dique Toma	23,500	49.0	290	4.7	133.0	43.0	8.0	2.1
Senguerr	En Nacimiento	1,300	31.9	187 2)	6.0				24.6
Mayo	Paso rio Mayo	5,450	10.0	110	0.6				1.8
Santa Cruz	Charles Fuhr	15,550	748.0	2,090	194.0				48.1
La Leona	La Leona	7,450	300.0	910	65.0				40.2
III. Pacific Ocean									
Manso	Lago Steffen	1,260	67.0	400	10.2				53.2
Manso	Lago Los Alerces	750	43.9	240	5.0				58.7
Manso	Los Moscos	580	35.2	240	5.0	67.5	32.5	12.5	60.7
Epuyen	Angostura	500	16.0	140	2.5	33.0	15.0	3.0	32.0
Futaleufu	Balsa Garzon	4,650	296.0	1,900	6.6	560.0	260.0	112.5	63.7
Carrenleufu	La Elena	1,500	31.0	260	11.2				20.7
Carrenleufu	Lago Vintter	790	23.9	120	7.6				30.3
IV. Interior Basins									
Sali-Dulce	El Sauce	20,200	51.0	3,200	0.0	288.0	34.0	2.0	4.0
Sali-Dulce	La Escuela	19,700	97.0	1,950	1.0	355.0	55.0	5.0	4.9
Sali	El Cadillal	4,700	15.0	800	0.8	55.0	5.0	1.0	3.2

1) Non-Argentine portion included in international river basins.
2) Mean daily maximum flow.

TABLE 6-6. ARGENTINA–DISTRIBUTION OF SURFACE WATER RESOURCES

(Source: CEPAL, 1972)

[By drainage basin]

Drainage basin	Area 1,000 km2	% of Country	Population (1,000's)	Density of population per km2	River discharge m3/s	% of total discharge	Runoff l/s/km2	Runoff l/s/person
Plata River.	918.9	33.1	16,000	17.4	18,360	84.7	19.8	1.15
Atlantic Ocean	1,051.3	37.8	2,700	2.6	2,349	10.8	2.2	0.87
Pacific Ocean.	37.5	1.3	15	0.4	795	3.7	21.2	53.00
Interior	771.8	27.8	2,600	3.4	182	0.8	0.2	0.07
Total Argentina	2,779.5	100.0	21,300	7.7	21,686	100.0	7.8	1.02

TABLE 6-7. ARGENTINA—MAJOR RESERVOIRS
(Source: CEPAL, 1972)

[Listed according to year of completion]

Year	Province	River	Name of reservoir	Type of dam	Usable storage Hm3	Dam volume m3	Purpose
1890	San Luis	Cuchi-Corral	Chorrillos	—	0.8	—	—
1890	Cordoba	Primero	San Roque	G	—	790,000	I-FC-W
1923	Jujuy	Perico	La Cienaga	E	26.0	2,140	—
1927	San Luis	Los Molles	P. de Funes	A	8.5	272,000	—
1931	La Rioja	La Rioja	Los Sauces	RF	19.0	300,000	I-E
1936	Cordoba	Tercero	Rio III No. 1	RF	560.0	91,000	—
1938	La Rioja	Anzulon	Anzulon	RF	36.0	11,500	FC
1939	Mendoza	Arroyo Frias	Frias	RF	—	34,000	—
1941	San Luis	Conlara	San Felipa	B	81.0	24,000	W-E
1941	San Luis	Cuchi-Corral	Cruz de Piedra	B	12.0	10,000	W
1942	Cordoba	San Jeronimo	San Jeronimo	A	0.2	75,000	FC
1943	Mendoza	A° Papagallos	Papagallos	—	0.8	40,000	FC
1944	Mendoza	A° Maure	Maure	RF	0.6	230,000	E
1944	Cordoba	Rio Tercero	Rio III No. 2	RF	10.5	89,000	I-E-W
1944	Cordoba	Rio Primero	San Roque	G	201.0	184,000	E
1944	Cordoba	Los Sauces	La Vina	A	230.0	230,000	I-E-W
1944	Cordoba	Cruz del Eje	Cruz del Eje	LG	129.0	3,500	W
1944	Cordoba	Los Alazanes	Los Alazanes	A	0.2	68,000	I-E
1947	Mendoza	Atuel	Nihuil I	G	259.0	64,300	I-W
1949	Catamarca	Tala	Juncal	RF	1.4	216,000	I-W
1951	Tucuman	Marapa	Escaba	B	126.0	96,000	I-E
1953	Cordoba	Segundo	Los Molinos I	A	307.0	68,000	I-E
1954	Cordoba	Segundo	Los Molinos II	LG	4.0	20,000	I-E
1954	San Luis	Quinto	La Florida	LG	105.0	17,000	I-E
1955	Tucuman	Marapa	Batiruana	LG	0.5	—	I-E
1956	Catamarca	—	Ipizca	A	10.0	30,000	I-E
1957	Cordoba	Los Sauces	La Vina II	G	2.0	—	I-E
1958	Catamarca	—	La Canada	A	10.0	—	I-W
1958	San Luis	Lujan	Lujan	LG	6.0	9,700	—
1960	La Rioja	Olta	Olta	A	8.0	600,000	I-E-W
1963	Catamarca	Del Valle	Las Pirquitas	E	65.0	500,000	I-FC-E
1963	Chubut	Chubut	F. Ameghino	LG	1,900.0	—	I-FC-E
1966	Tucuman	Sali	Cadillal	E	390.0	—	I-FC
1966	Mendoza	Atuel	Valle Grande	B	290.0	—	I-FC-E
1967	Santiago del Estero	Rio Dulce	Rio Hondo	E-G	1,050.0	8,000,000	I-FC-E
		Total (1890-1967)	35 dams		5,829.5		

Abbreviations used:

Dam types:
G: Gravity
B: Buttressed
A: Arch
LG: Gravity, light
RF: Rock-fill
E: Earth

Purpose:
I: Irrigation
W: Drinking water
FC: Flood control
E: Energy

TABLE 6-8. ARGENTINA—ESTIMATED WATER DEMAND, 1961-80
(Source: CEPAL, 1972)

Category	Water demand, million m^3		
	1961	1970	1980
Public water supply	—	1,750 (1967)	—
Industry	702	1,144	1,931
Thermoelectric power generation	1,691	3,517	4,266
Irrigation	15,000 (1960)	16,100 (1973)	19,700 (1985)
Irrigated area (1,000 ha)	1,150	1,341	1,643

TABLE 6-9. ARGENTINA—INDUSTRIAL WATER DEMAND, 1961-80
(Source: CEPAL, 1972)

Industry	Estimated water demand, millions of m^3		
	1961	1970	1980
Slaughterhouses and refrigerated storage	38	36	36
Sugar	2	3	4
Fruit and vegetable canning	2	3	4
Beer and non-alcoholic beverages			
Wine and cider	12	15	19
Textiles (wool, cotton and synthetics)	73	123	155
Pulp and paper	59	91	130
Leather	5	9	10
Sulfuric and acetic acid, synthetic rubber, carbon black, calcium carbide, caustic soda, solvay soda, soap and detergents	9	32	53
Steel	148	290	780
Portland cement	10	15	26
Aluminum	—	57	74
Petroleum production and refining	319	437	593
Mining (lead, iron, sulfur and coal)	3	8	16
Total	702	1,144	1,931

TABLE 6-10. BOLIVIA–DISCHARGE OF SELECTED RIVERS

(Source: CEPAL, 1964)

Basin and river	Period of record	Jan.	Feb.	Mar.	Apr.	May	Jun.	Jul.	Aug.	Sep.	Oct.	Nov.	Dec.	Max.	Mean	Min.	Coefficient of irregularity	Location
Bermejo River basin:																		
Pajonal	48-49	2.80	4.80	3.08	1.32	0.92	0.53	0.48	0.37	0.32	0.55	0.58	0.69	19.22	1.37	0.12	0.40	Pte. Carretero
Santa Ana	48-49	3.44	6.89	1.24	2.10	1.24	0.78	0.63	0.49	0.47	0.67	0.79	0.98	36.92	1.64	0.23	0.36	Sifon entre Rios
Guapore River basin:																		
Parapeti	43-49	59.60	79.50	70.60	37.10	16.20	9.80	7.20	5.10	6.30	10.50	17.40	30.50	882.60	29.15	1.80	0.38	Choreti-Camiri
Mamore River basin:																		
Corani	53-60	11.30	12.60	8.30	2.90	1.20	0.60	0.80	0.60	1.40	2.20	2.70	4.80	—	4.12	—	0.42	Sitio Presa
Grande	45-59	—	—	—	—	—	—	—	—	—	—	—	—	—	158.10	—	0.40	Pte. Abapo
Mairana	46-49	12.19	2.65	1.67	1.34	0.79	0.59	0.34	0.20	0.25	0.40	0.28	0.77	77.00	1.79	0.00	0.53	Sitio Presa
Piray	46-49	10.64	17.02	13.25	8.16	3.32	3.21	2.37	1.79	3.67	4.39	3.16	5.75	352.00	6.39	0.70	0.31	Pte. Carretero
Rocha	41-47	3.64	5.15	2.23	0.18	0.00	0.00	0.00	0.00	0.49	0.13	0.24	2.01	328.00	1.17	0.00	0.59	Cochabamba
Sulti	40-45	10.20	12.63	4.66	1.55	0.00	0.00	0.00	0.00	0.00	0.03	0.86	5.19	90.00	2.93	0.00	0.60	La Angostura
Pilcomayo River basin:																		
Pilcomayo	42-49;52-56	428.50	695.00	451.90	206.90	144.30	56.40	43.40	33.70	34.10	31.20	78.30	155.00	2,420.50	196.56	6.30	0.42	Pte. Ustariz
Lake Titicaca basin:																		
Contador	45-55	3.03	3.85	2.80	1.38	0.74	0.51	0.42	0.38	0.42	0.62	1.14	1.98	11.64	1.44	0.18	0.34	Aguas Abajo
Desaguadero	41-50	83.20	145.80	63.00	20.30	6.90	6.10	5.50	4.80	4.30	4.50	3.00	30.00	753.20	31.45	0.80	0.52	Chuquina
Hichuceta	45-55	1.89	2.40	1.88	1.15	0.72	0.50	0.38	0.33	0.38	0.54	0.91	1.48	6.65	1.05	0.22	0.28	Boquilla Presa
Pallina	47-51	3.37	14.01	2.64	1.10	0.14	0.08	0.08	0.06	0.05	0.07	0.10	1.29	212.83	1.92	0.01	0.62	Pte. Carretero
Tacagua	42-49	7.82	12.40	3.60	2.80	0.60	0.57	0.51	0.26	0.22	0.28	0.29	2.07	247.50	3.62	0.02	0.51	Sitio Presa
Vizcachini	42-59	2.51	4.55	1.92	1.03	0.61	0.53	0.53	0.43	0.43	0.49	0.43	1.19	107.67	1.22	0.02	0.36	Pte. F.C. Sitio Presa
Oruro System *																		
Azeruni	31-59	3.38	3.06	2.13	1.20	0.45	0.42	0.40	0.47	0.78	1.34	2.00	2.80	—	1.54	—	0.31	Sta. 10-C
Carabuco	51-58	4.40	4.50	3.10	1.40	0.80	0.50	0.40	0.40	0.90	1.20	1.90	2.80	—	1.90	—	0.32	Sta. 4
Carabuco en Arriba	51-60	3.02	3.56	2.00	1.01	0.40	0.26	0.30	0.24	0.57	0.75	1.22	1.80	—	1.26	—	0.35	Sta. 6
Choquetanga	52-57	2.89	3.15	1.74	0.62	0.48	0.39	0.42	0.31	0.62	0.67	1.14	1.70	—	1.18	—	0.34	Sta. 1
Choquetanga	30-60	5.10	3.50	3.90	2.10	1.30	0.90	0.80	0.80	1.30	1.50	2.10	3.30	—	2.20	—	0.26	Sta. 16-A
La Paz System *																		
Cuticucho	30-31;40-48	5.15	4.30	3.56	2.43	0.90	0.38	0.36	0.32	0.50	1.03	1.44	3.85	—	2.02	—	0.38	Sta. 9-D
Chununi	30-32	1.31	1.09	1.09	1.04	1.64	1.37	1.69	0.94	0.85	1.15	1.12	1.24	—	1.21	—	0.07	Sta. 7-G
Hankchuma	30-31;38-39	1.52	1.46	1.48	1.42	1.20	0.93	0.94	1.00	1.14	1.13	1.15	1.23	—	1.22	—	0.07	Sta. 7-G
Milluni (lake)	24-59	4.50	5.35	2.70	1.19	0.64	0.43	0.43	0.33	0.35	0.62	1.20	2.09	—	1.65	—	0.41	—
Zongo en Botijlaca	30-32;38-42	3.64	3.10	3.12	2.07	1.54	1.10	0.95	1.03	1.03	1.64	2.33	2.43	—	2.00	—	0.20	Sta. 7-D
Zongo en Canaviri	30-32;39-48	15.90	12.50	13.60	3.50	2.50	2.30	2.00	1.80	1.90	2.70	3.40	7.20	—	5.80	—	0.38	Sta. 9-C

* Rivers regulated by Bolivian Power Company.

TABLE 6-11. BOLIVIA—MAJOR HYDROELECTRIC POWER PLANTS

(Source: CEPAL, 1964)

[Data as of 1960; capacity of 1,000 kW or more]

Plant	Year of installation	River	Type	Installed capacity kW	Energy produced in 1959 million kWh	Head m	Mean flow m³/s
EXISTING							
Achachicala	1909-53	Milluni	R	4,600	12.94	449	0.50
Zongo	1929-48	Zongo	R	4,800	6.79	384	0.13
Botijlaca	1938-41	Zongo	RR	3,600	17.17	382	0.37
Cuticucho	1942-55	Zongo	RR	8,700	36.35	662	0.32
Santa Rosa I	1952	Zongo	RR	2,800 }	46.81	183	0.83
Santa Rosa II	1955	Coscapa	RR	6,800 }		835	0.32
Sainani	1956	Zongo	RR	9,900	26.06	274	1.00
Miguilla	1931	Miguilla	R	2,600	6.96	489	0.28
Angostura	1936	Miguilla	R	3,900	17.45	533	0.68
Choquetanga	1939-44	Choquetanga	RR	6,700	37.25	488	1.55
Carabuco I	1958	Carabuco	RR	6,400	39.07	352	1.79
Pongo	–	Caracoles	RR }	1,000	5.90	–	–
Calatranca	–	Caracoles	RR }				
Rea-Rea	–	Colquiri	RR	2,300	19.50 1)	460	–
Lupi-Lupi	–	El Tranque	R	3,400	8.00 1)	128	–
Incachaca	1914-49	Malaga	RR	2,160	8.70	188	0.80
Angostura	1954	Sulti	R	2,120	0.44	58	1.20
Cayara	–	–	R	1,600	2.99	110	0.29
Quilpani	–	Yura	R	6,000	2)	260	–
Punutuma	–	Yura	RR	2,500	2)	102	–
Landara	–	Yura	RR	2,000	2)	104	–
Yocalla	–	Pilcoyo	–	1,100	2.9 1)	48	–
UNDER CONSTRUCTION							
Chururaqui		Zongo	R	20,000			
Carabuco II		Miguilla	R	6,200			
Corani		Corani	R	32,000		620	2.40
Chapisirca		Titiri	R	18,000		1,340	0.45

1) 1958
2) Installation under repair in 1959
R— Reservoir
RR—Run-of-the-river

TABLE 6-12. BOLIVIA—ESTIMATED WATER DEMAND, 1959-71

(Source: CEPAL, 1964)

Category	Estimated water demand, million m^3	
	1959	1971
Public water supply	35 (1964)	130
Irrigation (220,000 ha)........................	2,625	— 1)
Industry (total)	37	79
Manufacturing	4	7
Production and refining of petroleum	20	40
Thermoelectric power generation	2	2
Tin mining	11	30

1) Projected demand for 1990 is 12,335 million m^3.

TABLE 6-13. BOLIVIA—WATER BALANCE OF LAKE TITICACA

(Source: CEPAL, 1964)

[Million m^3; by various hydrologists]

	Bucher	Rudolph	Forti	Monheim
Inflow:				
Precipitation on lake	4,528	5,337	5,000	5,062
Tributaries	11,997	6,274	15,230	7,718
Total	16,507	11,611	20,230	12,780
Outflow:				
Rio Desaguadero	1,416	1,724	4,403	630
Evaporation	15,091	9,350	15,827	12,150
Ground water recharge		537		
Total	16,507	11,611	20,230	12,780

TABLE 6-14. BOLIVIA—MUNICIPAL WATER SUPPLY SYSTEMS

(Source: CEPAL, 1964)

[Existing (1961) and projected (1971)]

City	Estimated population in 1961 (1,000's)	Percentage of population served		Per Capita Use l/d	
		1961	1971	1961	1971
La Paz	450	55	100	151	280
Cochabamba	114	60	100	101	330
Oruro	88	49	100	40	240
Potosi	64	60	100	50	280
Santa Cruz	60	10	100	—	180
Sucre	57.5	52	100	42	170
Tarija	23.5	54	100	200	200
Trinidad	14.7	—	100	70	—
Cobija......................	2.3	—	100	—	100
Other cities	260	—	100	—	100
Total	1,134	—	100	—	225

TABLE 6-15. BRAZIL—DISCHARGE OF SELECTED RIVERS
(Source: UNESCO, 1971)

River and station	Basin area km2	Mean monthly discharge, m3/s												Year	Max. flow m3/s	Date	Min. flow m3/s	Period of record
		Jan.	Feb.	Mar.	Apr.	May	Jun.	Jul.	Aug.	Sep.	Oct.	Nov.	Dec.					
Amazon, Obidos	4,688,000	114,300	142,900	169,000	192,200	207,400	205,800	193,700	168,000	128,100	95,000	89,300	95,000	150,000	224,000	May 23,'35	72,100	1928-47
Sao Francisco, Juazeiro	470,770	4,830	5,310	5,010	4,200	2,770	1,870	1,560	1,340	1,180	1,230	1,920	3,490	2,890	13,630	Mar. 13,'49	577	1929-64
Jequitinhonha, Jacinto	63,900	856	606	559	397	233	185	163	140	131	252	521	1,100	429	10,120	Dec. 24,'43	40.3	1938-65
Paraiba, Campos	55,770	1,570	1,690	1,520	1,050	721	590	496	424	400	466	675	1,210	900	4,837	Jan. 28,'61	102	1929-64
Parana, Guaira	806,000	11,500	13,400	13,500	10,700	7,910	6,950	5,650	4,850	4,770	5,380	6,510	8,120	8,260	29,624	Mar. 3,'29	2,291	1921-65
Iguacu, Salto Osorio	46,400	669	922	910	701	758	882	911	887	1,060	1,260	1,080	763	900	11,880	Aug. 20,'57	138	1941-65

TABLE 6-16. BRAZIL—LENGTH OF NAVIGABLE RIVERS
(Source: Brazil Yearbook, 1968)

River	Basin	Distance navigable km	River	Basin	Distance navigable km
Amazon	Amazon	3,100	Parnaiba	Northeast	668
Purus	Amazon	2,853	Das Velhas	Sao Francisco	467
Sao Francisco	Sao Francisco	2,712	Jequitinhonha	East	614
Tocantins	Amazon	1,372	Uruguay	Uruguay	530
Araguaia	Amazon	1,300	Parana	Parana	550
Rondonia	Amazon	1,329	Ribeira do Iguape	Southeast	300
Madeira	Amazon	1,090	Doce	East	220
Itapicuru	Northeast	826	Jacui	Southeast	220
Paraguay	Paraguay	722	Itajai-Acu	Southeast	180

TABLE 6-17. BRAZIL—HYDROELECTRIC POTENTIAL OF RIVER BASINS

(Source: Brazil Ministry of Mines and Energy, 1973)

[As of June 30, 1970]

State or region	River basin	Total	in oper- ation	under con- struction	under study	under future study
			Hydroelectric Potential (Megawatts)			
			Stage of Development			
NORTH						
Amazonas	Amazon	55.1	—	—	—	55.1
Para	Amazon, Tocantins-Araguaia and Northeast	5,325.0	—	20.0	70.0	5,235.0
Amapa	Amazon	100.0	—	40.0	20.0	40.0
NORTHEAST						
Maranhao	Tocantins-Araguaia and Northeast	57.0	—	—	37.0	20.0
Piaui	Northeast	216.0	108.0	108.0	—	—
Ceara	Northeast	80.0	5.0	15.0	60.0	—
Paraiba	Northeast	11.8	4.4	—	7.4	—
Pernambuco	Sao Francisco	661.3	—	—	661.3	—
Sergipe	Sao Francisco	4,255.0	—	—	4,255.0	—
Bahia	Sao Francisco and East	6,744.3	787.0	1,428.0	4,129.3	400.0
CENTRAL-SOUTH						
Minas Gerais	Sao Francisco, East and Parana	15,855.7	2,113.6	759.9	5,410.8	7,571.4
Espirito Santo	East	736.8	46.8	118.0	409.0	163.0
Rio de Janeiro	East	2,777.5	799.9	210.0	732.6	1,035.0
Sao Paulo	East, Parana and Southeast	11,608.9	1,716.5	6,706.0	2,786.4	400.0
SOUTH						
Parana	Parana and Southeast	23,440.3	144.9	250.0	73.1	22,972.3
Santa Catarina	Southeast and Uruguay	1,229.8	76.4	22.1	21.3	1,110.0
Rio Grande do Sul	Southeast and Uruguay	2,530.0	208.5	366.5	1,155.0	800.0
CENTRAL-WEST						
Mato Grosso	Parana and Paraguay	120.5	—	16.3	72.2	32.0
Goias	Tocantins-Araguaia and Parana	3,151.8	139.7	305.0	547.6	2,159.5
Rondonia	Amazon	377.0	—	—	—	377.0
Distrito Federal	Parana	25.5	25.5	—	—	—
SUMMARY						
North		5,480.1	—	60.0	90.0	5,330.1
Northeast		12,025.4	904.4	1,551.0	9,150.0	420.0
Central-South		30,978.9	4,676.8	7,793.9	9,338.8	9,169.4
South		27,200.1	429.8	638.6	1,249.4	24,882.3
Central-West		3,674.8	165.2	321.3	619.8	· 2,568.5
Brazil		79,359.3	6,176.2	10,364.8	20,448.0	42,370.3

TABLE 6-18. BRAZIL—WATER QUALITY OF AMAZON RIVER AND ITS TRIBUTARIES

(Source: Oltman, 1964, U.S. Geol. Survey Circ. 486)

River	Date 1963	Location		Temperature °F	Specific conductance (micromhos at 25°C)	pH	Bicarbonate ppm	Dissolved oxygen ppm
		Lat. S.	Long. W.					
Baia de Guajara	July 10	1°24'	48°30'	84.0	33	—	—	—
Rio Para	July 10	1°50'	50°09'	85.0	35.0	6.7	—	—
Estraito de Boiucu	July 11	1°47'	50°28'	83.0	37.0	6.65	—	—
Furo do Tajapuru	July 11	1°21'	50°49'	83.0	38.0	6.70	—	—
Furo do Ituquara	July 11	1°02'	51°03'	83.5	37.8	6.55	—	—
Rio Tapajos	July 13	2°25'	54°43'	86.0	15.0	7.15	—	7.7
Rio Negro	July 20	3°09'	60°01'	83.0	9.5	4.9	—	5.5
Parana do Careiro	July 22	3°13'	59°50'	83.0	50.2	6.65	—	—
Rio Solimoes	July 23	3°11'	59°54'	83.0	52.8	6.70	26.5	6.0
Rio Madeira	July 26	3°24'	58°46'	84.0	52.0	6.9	21.0	7.5
Amazon River (south channel)	July 11	1°19'	51°27'	84.0	37.2	6.60	—	—
Amazon River	July 12	1°26'	52°06'	82.5	39.0	6.40	—	—
	July 17	2°35'	57°17'	82.0	42.5	6.4	15.6	6.0
	July 20	3°08'	59°54'	83.5	10.0	5.10	—	5.2
	July 27	1°58'	55°23'	83.0	37.5	6.35	12.3	5.7

TABLE 6-19. BRAZIL—WATER WELL DRILLING 1960-61

(Source: Brazil Yearbook, 1968)

[By State]

State	Year	No. of wells	Number of meters drilled		Number of successful wells		Hourly yield (liters)	
			Total	Average per well	Total	%	Total	Average per well
Piaui	1960	110	7,432	68	103	94	534,112	5,186
	1961	126	11,725	93	109	87	842,327	7,728
Ceara	1960	68	3,749	55	55	81	185,920	3,380
	1961	56	3,714	66	47	84	152,500	3,245
Rio Grande do Norte	1960	37	3,262	88	33	89	115,000	3,485
	1961	51	3,080	60	46	90	151,000	3,283
Paraiba	1960	30	719	24	28	93	84,150	3,005
	1961	12	642	54	10	83	48,800	4,880
Pernambuco	1960	18	1,040	58	16	89	54,870	3,429
	1961	16	859	54	13	91	59,240	4,557
Alagoas	1960	2	139	70	2	100	5,460	2,730
	1961	3	120	40	3	100	4,400	1,467
Sergipe	1960	13	547	42	7	54	24,400	3,486
	1961	16	696	44	10	63	54,050	5,405
Bahia	1960	7	331	47	7	100	21,480	3,069
	1961	14	741	53	13	93	42,500	3,269
Minas Gerais	1960	58	5,736	99	52	90	359,600	6,915
	1961	99	4,845	49	95	96	313,000	3,295
Total	1960	343	22,955	67	303	88	1,384,992	4,571
	1961	393	26,422	67	346	88	1,667,817	4,820

TABLE 6-20. BRAZIL—INDUSTRIAL WATER USE IN GREATER SAO PAULO REGION, 1967

(Source: Departamento de Aguas e Energia Eletrica, 1972)

Type of industry	Water use, m^3/d			
	River water	Municipal water	Ground water	Total
Textile	22,888	10,197	29,137	62,222
Pharmaceutical and chemical	156,694	5,265	14,527	176,486
Petroleum derivatives and gas	6,002	1,074	5,523	12,599
Automobile	5,958	4,545	10,840	21,343
Meat processing and canning	5,538	—	7,482	13,020
Electric instruments, radio and television, illumination	—	4,507	4,917	9,424
Iron and steel	53,350	1,761	9,926	65,037
Metallurgical	3,844	3,754	4,180	11,778
Rubber (tires)	3,360	910	4,573	8,843
Ceramic and cement	1,772	48	3,593	5,413
Plastic utensils	450	1,036	1,650	3,136
Glass	600	358	1,510	2,468
Oil, fat and detergents	10,380	848	4,506	15,734
Paper	126,742	2,766	1,187	130,695
Food (milk)	—	450	4,208	4,658
Food (wheat, sugar, coffee, etc.)	1,918	348	810	3,076
Food (beer, whisky, beverages)	—	9,150	4,450	13,600
Total	399,496	47,017	113,019	559,532

TABLE 6-21. CHILE—DISCHARGE OF PRINCIPAL RIVERS

(Source: Donoso, Water for Peace, 1967)

[Arranged by region from north to south]

Region and river	Gaging station	Length of river km	Elevation above sea level m	Basin area km2	Mean annual flow m3/s	Mean runoff l/s/km2
Norte Grande						
Lauca	Est. Lago	11	4,450	492	0.98	2
Norte Chico						
Copiapo	La Puerta	169	857	5,068	2.19	0.4
Huasco	Carmen en Junta	150	825	3,010	4.66	1.5
Limari	Grande en Palena	106	342	6,254	8.0	1.3
Choapa	Puente Negro	150	226	3,648	10.8	3
Chile Central						
Aconcagua	Chacabuquito	70	1,030	2,084	32.9	16
Maipo	Cabimbao	233	150	14,820	103	7
Rapel	Corneche	214	16	13,186	160	12
Maule	Armerillo	88	450	5,454	225	41
Bio-Bio	Rucalhue	221	310	7,044	485	69
Tolten	Villarrica	118	230	3,050	334	109
Sur Chico						
Valdivia	Trafun	154	150	4,416	450	102
Bueno	Bueno	122	100	3,710	365	99
Petrohue	Desague	77	300	2,210	275	125
Puelo	Basilio	149	8	8,620	670	78
Cisnes	Pto. Cisnes	141	7	5,160	250 E	49
Sur Grande						
Aisen	Coyhaique	110	20	3,110	45	14
Baker	Colonia	312	105	23,460	1,000	43
San Juan	Fte. Bulnes	72	8	864	10 E	12

E-Estimated.

TABLE 6-22. CHILE—SEASONAL FLOW OF SELECTED RIVERS

(Source: CEPAL, 1960)

Province	River	Basin area km2	Mean flow (million m3)			Period of record
			October to April	January to April	Year	
Coquimbo	Elqui	9,570	290.4	151.5	447.4	1928-56
Aconcagua.............	Aconcagua	2,640	1,099.2	469.8	1,368.9	1940-56
Santiago	Maipo-Mapocho	15,400	2,745.0	1,461.4	3,576.4	1914-52
O'Higgins-Colchagua	Rapel	13,520	2,679.3	1,332.3	4,050.4	1946-56
Talca-Maule-Linares	Maule E	21,700	6,350.8	1,847.4	12,760.5	1947-56
Nuble	Itata E	11,500	2,607.6	895.3	5,921.1	1947-56
Malleco-Bio-Bio-Concepcion .	Bio-Bio	23,900	9,880.6	3,671.2	25,225.0	1949-56

E— Flow measurements based in part on estimates.

TABLE 6-23. CHILE—HYDROELECTRIC PLANTS

(Source: CEPAL, 1960)

[Plants of less than 10,000 KW not listed individually]

Region and plant	River	Type 1)	Capacity 1,000 KW	Power generated in 1957 million KWH	Elevation m	Meanflow m3/s	Meanflow million m3/yr
Existing in 1957							
Norte Chico							
Molles	Molles	RR	16.0	39.3	1,154	0.7	22
Chile Central							
Florida	Maipo	RR	13.5	77.0	96	17.5	550
Maitenes	Colorado	RR	26.0	118.0	176	10.7	340
Volcan	Volcan	RR	13.0	87.3	172	7.8	250
Queltehues	Maipo	RR	36.4	285.8	202	14.4	450
Sauzal	Cachapoal	RR	76.8	262.2	120	45.0	1,420
Coya	Cachapoal	RR	33.0	231.6	138	2.8	90
Pangal	Pangal	RR	21.6	149.4	474	6.4	200
Cipreses	Maule	R	101.4	297.3	350	18.5	580
Abanico	Laja	RR	86.0	430.3	146	56	1,760
Various			63.4	397.4			
Sur Chico							
Pilmaiquen	Pilmaiquen	RR	24.2	107.7	32	87.0	2,740
Small capacity plants			9.2	22.6			
Total			520.7	2,508.0			9,500
Under construction and projected to 1973							
Norte Chico							
Cuncumen	Choapa	RR	30.0	—	600 E	4.0	130
Chile Central							
Isla	Maule	RR	68.0	—	95	68.0	2,100
Garzas	Maule	RR	200.0	—	200 E	73.8	2,300
Rapel	Rapel	R	260.0	—	100 E	111.0	3,500
Abanico (Expansion)	Laja	RR	50.0	—	146	8.5	270
Lago Laja	Laja	R	240.0	—	340	35.0	1,100
Antuco	Laja	RR, R	200.0	—	180	97.1	3,100
Sur Chico							
Pullinque	Calle Calle	RR	49.0		47	81.5	2,600
Pilmaiquen (Expansion)	Pilmaiquen	RR	10.6		32	26.1	800
Small capacity plants			18.0				
Total			1,124.6				18,330

1) R—Reservoir
RR—Run-of-the-river
E—Estimated

TABLE 6-24. CHILE—WATER DEMAND FOR COPPER AND NITRATE PRODUCTION

(Source: Wollman, The Water Resources of Chile, Resources for the Future, Inc., The Johns Hopkins Press, Baltimore,Md.)

COPPER

Province	Production (1,000 tons)			Water requirements (1,000 m^3/yr)		
	1963	1985		1963	1985	
		Low	High		Low	High
Tarapaca	Less than .5	Less than .5	4	Less than .5	Less than .5	160
Antofagasta	304.3	592	775	27,029	51,652	67,550
Atacama	107.7	220	437	38,391	72,035	166,326
Coquimbo	10.9	33	74	2,761	7,564	20,898
Aconcagua	17.6	54	188	807	2,477	14,168
Valparaiso	3.3	10	22	1,108	3,402	72,992
Santiago	15.5	48	76	3,925	12,050	19,294
O'Higgins	144.4	264	429	12,006	21,927	35,607
Total	603.7	1,221	2,005	86,027	171,107	396,995

NITRATE

Province	Production (1,000 tons)		Water use (1,000 m^3/year)	
	1963	1985	1963	1985
Tarapaca	118.3	164	890	1,230
Antofagasta	1,026.3	1,436	7,290	10,765
Total	1,144.6	1,600	8,180	11,995

TABLE 6-25. CHILE—WATER USE, 1970

(Source: Court, University of Chile, 1971)

Category	Water use in 1970 million m^3	Equivalent stream flow m^3/s	%
Public supply	495	15.7	1.2
Industrial	450	14.3	1.1
Mining	100	3.2	0.2
Irrigation	15,900	505.0	38.5
Hydroelectric power generation	24,400	775.0	59.0
Total	41,345	1,313.2	100.0

TABLE 6-26. CHILE—IRRIGATION WITH WATER FROM RESERVOIRS

(Source: CEPAL, 1960)

	Total irrigated area	Area irrigated with water from reservoirs	Percentage
	Hectares		
A. Area			
Projects in operation	1,360,000	139,000	10.2
Projects under construction or study			
New	607,000	364,000	59.9
Improvements	(627,000)	614,000	
Total	1,967,000	1,117,000	56.7

	Area irrigated with water from reservoirs (hectares)	Reservoir capacity	
		million m^3	m^3/ha
B. Reservoir capacity			
Projects in operation	139,000	580	4,170
Projects under construction or study	1,117,000	8,030	7,180

TABLE 6-27. CHILE—MUNICIPAL WATER DEMAND, 1957-73

(Source: CEPAL, 1960)

[By region[1]]

Region and city	Urban Population (1,000's) 1957	Urban Population (1,000's) 1973	Water Available l/d capita 1957	Water Available l/d capita 1973	Total Consumption 1957 (Million m³)	Estimated Consumption 1973 (Million m³)	Waterworks under Construction 1958 (Million m³)	Surplus (+) or Deficit (−) 1973 (Million m³)
Norte Grande								
Arica	230		205	250	1.6	3.2	—	− 1.6
Antofagasta			90	300	2.7	11.6	9.7	+ 0.8
Other towns of over 5,000 pop.			150	160	6.3	9.8	—	− 3.5
Population at nitrate plants			160	200	2.2	3.6	2.3	+ 0.9
Others			60	100	0.15	0.5	—	− 0.35
Total		380	190	200	13.0	28.7	12.0	− 5.4
Norte Chile								
Copiapo	175		170	250	1.35	3.3	—	− 1.95
La Serena			200	250	3.05	6.4	—	− 3.35
Coquimpo			145	220	1.45	3.8	—	− 2.35
Other towns of over 5,000 pop.			130	220	3.1	8.9	3.7	− 2.1
Others			80	130	0.55	1.4	0.6	− 0.25
Total		300	145	220	9.5	23.8	4.3	− 10.0
Chile Central								
Valparaiso y Vina del Mar	3,390		250	380	31.0	81.5	39.9	− 10.6
San Antonio			185	250	2.2	5.0	5.9	+ 3.1
Santiago			300	440	165.0	420.0	125.0	− 130.0
Talca			270	300	6.0	11.3	—	− 5.3
Chillan			210	250	4.5	8.8	—	− 4.3
Concepcion			210	370	10.0	30.0	—	− 20.0
Other towns of over 5,000 pop.			125	300	40.8	165.0	26.9	− 97.3
Others			40	250	5.5	46.4	3.4	− 37.5
Total		5,700	210	370	264.0	768.0	201.0	− 305.0
Sur Chico								
Valdivia	230		180	260	3.25	8.1	—	− 4.85
Osorno			160	300	2.6	8.4	8.4	+ 2.6
Puerto Montt			160	200	1.85	4.0	—	− 2.15
Other towns of over 5,000 pop.			75	200	1.8	8.8	0.7	− 6.3
Others			35	180	0.5	4.5	0.5	− 3.5
Total		400	115	240	10.0	33.8	9.7	− 16.6
Sur Grande								
Punta Arenas	65		220	270	3.0	6.4	—	− 3.4
Other towns of over 5,000 pop.			45	220	0.3	2.5	—	− 2.2
Others			60	200	0.2	1.1	0.2	− 0.7
Total		120	155	250	3.5	10.0	0.2	− 6.3
Grand Total Chile	4,090	6,900	200	350	300.0	864.3	227.3	− 343.3

1 Description of regions:

	Area km2	South Latitudes
Norte Grande	178,000	18°–26°
Norte Chico	120,000	26°–33°
Chile Central	147,000	33°–39°
Sur Chico	72,000	39°–45°
Sur Grande	225,000	45°–56°

TABLE 6-28. COLOMBIA—DISCHARGE OF PRINCIPAL RIVERS
(Source: CEPAL, 1964)

River	Gaging station	Period of record years	Mean monthly discharge million m^3	Mean annual discharge million m^3
I. Magdalena River basin:				
Chicamocha	Termopaipa	3	13.6	163.6
Tuta	Tuta	4	6.2	73.9
Fonce	San Gil	3	159.0	1,907.7
Lebrija	Rocas	2	52.1	625.0
Negro	Colorados	4	293.5	3,522.1
Sogamoso	El Tablazo	1	874.1	10,489.0
Nevado	Puente Guican	3	11.6	139.7
Sumapaz	San Bartolo	2	90.6	1,086.9
Coello	Chicoral	5	113.4	1,362.1
Lagunilla	El Bosque	3	6.9	83.2
Prado	Puente Casabianca	4	105.3	1,263.8
II. Cauca River basin:				
Sonson	El Bosque	2	15.6	186.9
Amaime	Ingenio Manuelita	8	24.0	287.8
Bugalagrande	Bugalagrande	10	46.8	561.9
Cali	Acueducto	13	10.7	128.3
Cauca	Juanchito	25	734.3	8,811.1
Cauca	La Balsa	13	522.1	6,265.2
Cauca	La Virginia	12	1,478.4	17,741.1
Cauca	Suarez	12	370.9	4,451.3
Fraile	El Penon	7	14.4	172.5
Guabas	Puentepiedra	11	12.2	146.2
Jamundi	Planta Electrica	12	13.6	163.6
La Paila	Hacienda El Medio	11	12.8	154.0
La Vieja	Caicedonia	12	166.8	2,001.6
La Vieja	Cartago	13	263.3	3,159.4
Lili	Lili	5	2.1	25.0
Mondomo	Puente Carretera	5	17.8	213.4
Palo	Arriba	13	48.3	579.6
Palo	Puerto Tejada	3	65.1	781.7
Pescador	Bolivar	12	3.5	41.6
Quilichao	Santander de Quilichao	10	1.6	19.2
Rioclaro	Rioclaro	13	18.7	223.9
Riofrio	Riofrio	13	42.9	514.8
Risaralda	La Suiza	3	72.3	868.2
Timba	Timba	11	66.7	800.5
Tulua	Mateguadua	13	38.8	465.3
Cauca	Cocunuco	3	25.6	307.1
Consota	Pereira	8	8.3	99.7
III. Pacific Ocean				
Calima	Madronal	10	29.7	356.5
Bravo	La Esperanza	3	10.3	124.0
Anchicaya	Central Hidroelectrica	12	233.4	2,801.1
Sajandi	Planta Hidroelectrica	5	29.0	347.6
Blanco	Carlosana	3	13.7	163.9
Chiguacos	San Pedro	3	4.6	55.8
Guabo	Ricaurte	3	72.0	864.1
Mayo	San Pablo	3	21.9	262.7
Patia	El Bordo	4	214.8	2,577.4
Pasto	Pandiaco	3	8.7	104.9
Sapuyes	Puente Carretera	3	9.0	108.3
IV. Orinoco River basin:				
Chitaga	Chorro Colorado	4	105.1	1,261.3
Bata	Santa Maria	4	178.7	2,144.4
Garagoa	Las Juntas	4	87.0	1,044.4
Somondoco	Las Juntas	3	51.0	611.5
Macheta	Tibirita	2	31.7	380.5
Teatinos	Planta Municipal	3	4.3	51.1
V. Amazon River basin:				
Balsayaco	San Andres	3	73.6	883.1
VI. Sierra Nevada de Santa Marta basin:				
Tucorinca	El Trebol	1	33.8	405.1
V. Lake Maracaibo basin:				
Zulia	Puente Ospina Perez	2	145.8	1,750.2

TABLE 6-29. COLOMBIA—HYDROELECTRIC POWER PLANTS
(Source: CEPAL, 1964)

Plant	River	Capacity (MW)	
		Present	Future
Tucurinca	Tucurinca	—	38.0
Guatapuri	Guatapuri	—	2.4
Tuamica	—	—	10.0
Anori	Anori	3.2	—
Guadalupe	Guadalupe	50.0	240.0
Troneros	Guadalupe	—	36.0
Mocorongo	Grande	75.0	—
Piedras Blancas	Force	11.5	—
Sonson	Sonson	—	7.0
S. Juan	S. Juan	3.2	—
La Esmeralda	S. Eugenio	—	26.6
La Insula	S. Eugenio	17.0	—
S. Francisco	S. Eugenio	—	80.0
Belmonte	Otun	3.8	—
Dos Quebradas	Dos Quebradas	4.0	—
El Bosque	Quindio	3.7	—
Quindio	Quindio	2.5	—
Guadalajara	Guadalajara	—	3.0
Nima 1 and 2	Nima	9.7	—
Cali 1 and 2	Cali	2.0	—
Rio Timba	Timba	—	60.0
Calima	Calima	—	120.0
Calima	Calima	—	40.0
Anchicaya	Anchicaya	64.0	—
Sajandi	Patia	—	2.4
Rio Mayo	Mayo	—	20.0
Rio Pasto	Pasto	2.9	—
Lebrija	Lebrija	9.0	18.0
Caracoli	—	1.6	—
La Cascada	Fonce	2.1	—
Guatape	Nare	—	300.0
Rio Negro	Negro	—	10.0
Rio Neusa	Neusa	—	25.0
Rio Recio	Recio	—	10.0
R. Bogota 5 and 6	Bogota	—	229.5
Canoas	Bogota	—	24.0
El Charquito	Bogota	5.5	16.0
El Salto	Bogota	54.0	66.0
Laguneta	Bogota	54.0	18.0
Salsipuedes	Bogota	—	6.0
Mirolindo	Conveima	3.9	—
Rio Prado	Prado	—	45.0
Iquira	Iquira	2.9	6.1
Diamante	—	4.4	—
Sueva	—	10.2	—
Lo Vuelta	—	4.0	—

TABLE 6-30. COLOMBIA—WATER DEMAND FOR IRRIGATION, 1960-90
(Source: CEPAL, 1964)

Department	Irrigated area 1,000 ha 1960	Water demand million m³ 1960	1990
Atlantico	1.0	15	315
Bolivar	—	—	600
Boyaca	3.2	35	540
Cauca	1.2	15	550
Caldas	—	—	130
Cundinamarca	15.0	150	600
Huila	10.0	100	390
Magdalena	15.0	225	3,600
Norte de Santander	7.0	105	400
Santander	12.0	120	350
Tolima	—	1,100	1,900
Valle	50.0	600	2,800
Various	15.6	160	160
Total Colombia	220.0	2,625	12,335

TABLE 6-31. COLOMBIA—PUBLIC WATER DEMAND, 1960-75
(Source: CEPAL, 1964)

[By Department]

Department	Water Consumption, millions of m³ 1960*	1970*	1975
Antoiquia			171
Atlantico			180
Bolivar			34
Boyaca			17
Caldas			99
Cauca			17
Cordoba			16
Cundinamarca			254
Choco			4
Huila			16
Magdalena			30
Meta			5
Narino			17
Norte de Santander			26
Santander			61
Tolima			39
Valle			227
Intendencies and commissaries			7
Total Colombia	410	850	1,220
Per Capita Consumption l/day	150	195	220

* No breakdown by Department available.

TABLE 6-32. COLOMBIA—MUNICIPAL WATER AND SEWER CONNECTIONS

(Source: CEPAL, 1964)

Department	Urban Population (1,000's)	Number of houses (1,000's)	Water service		Sewerage service	
			Number of connections (1,000's)	Percentage houses connected	Number of sewer connections (1,000's)	Percentage houses connected
Antioquia	1,095	159.7	129.3	61.0	100.1	62.7
Atlantico	595	61.4	40.9	66.6	15.9	38.8
Bolivar	410	63.2	18.7	29.5	10.6	16.7
Boyaca	180	29.1	19.0	65.5	14.7	50.6
Caldas	664	98.9	79.6	81.0	67.5	68.2
Cauca	160	24.2	11.2	46.2	8.5	35.3
Cordoba	135	20.8	7.1	34.2	2.2	10.8
Cundinamarca	1,482	160.3	139.5	87.0	126.8	79.1
Choco	38	7.4	3.3	46.1	2.7	36.7
Huila	149	26.1	17.2	66.0	6.9	26.3
Magdalena	293	49.2	19.3	39.2	2.9	5.8
Meta	61	9.4	4.7	50.0	4.1	43.0
Narino	189	31.8	17.2	54.0	10.3	32.4
Norte de Santander	209	32.9	26.2	79.8	15.9	48.5
Santander	391	56.3	43.8	77.7	33.2	59.0
Tolima	302	45.9	30.7	66.0	19.8	43.1
Valle	1,062	134.3	97.4	72.6	84.1	62.6
Intendencies and commissaries	101	16.2	7.2	44.2	1.0	6.1
Total	7,518	1,027.1	712.4	69.4	527.2	51.3
In addition:						
Bogota	1,166	115.6	104.9	90.7	101.0	87.4
Medellin	562	75.0	60.1	80.0	50.0	66.7
Cali	501	51.0	40.8	80.0	40.8	80.0
Barranquilla	431	408.0	31.4	79.3	15.0	36.7
Subtotal	2,660	282.4	237.2	82.5	206.8	72.0
Nation	4,858	744.7	475.2	64.2	320.4	43.3

TABLE 6-33. ECUADOR–DISCHARGE OF SELECTED RIVERS
(Source: UNESCO, 1971)

River and station	Basin area km2	Mean monthly discharge, m3/s												Year	Max. flow m3/s	Date	Min. flow m3/s	Year or period of record
		Jan.	Feb.	Mar.	Apr.	May	Jun.	Jul.	Aug.	Sep.	Oct.	Nov.	Dec.					
Mira, D.J. Lita	–	105	181	194	218	123	92	118	70	93	168	146	75	132	1,170	Feb. 20	42	1968
Quevedo, Quevedo	–	70	424	735	677	286	242	124	62	46	36	49	33	229	1,928	Feb. 11'66	11	1965-67
Vinces, Vinces	5,380	51	304	501	441	241	178	107	60	41	32	39	31	167	868	Feb. 2'66	20	1965-68
Pastaza, Banos	–	119	79	115	99	70	119	214	149	96	107	60	52	107	918	Apr. 1'66	23	1965-68
Zapotal, Lechugal	2,300	175	382	410	388	258	97	52	26	18	18	14	18	154	–	–	–	1965-66
Chimbo, Bucay	2,250	31	43	56	83	65	34	22	15	12	14	19	12	34	–	–	–	1965-66

TABLE 6-34. ECUADOR–AVAILABLE SURFACE-WATER RESOURCES
(Source: Mancheno, Servicio National de Meteorologia e Hidrologia, 1973)

River	Basin area km2	Annual flow million m3
Pacific Slope		
Mira	7,200[1]	9.052
Santiago-Cayapas	6,190	8.695
Esmeraldas	21,186	25.911
Guayas	35,245	47.078
Jama	1,607	1.290
Chone	2,597	1.595
Portoviejo	2,230	1.292
Naranjal	3,324	3.688
Jubones	5,350	4.028
Puyango-Tumbes	3,705[1]	4.410
Catamayo-Chira	7,010[1]	2.934
Amazon River Basin		
Napo-Aquarico	–	49.245
Pastaza	–	29.000
Santiago	–	20.000
Estimated total annual runoff Ecuador		314.000

1) International river basin; Ecuador portion only.

TABLE 6-35. ECUADOR—WATER WELLS AND GROUND WATER WITHDRAWAL

(Source: Mancheno, Servicio Nacional de Meteorologia e Hidrologia, 1973)

Number of dug wells	2,915
Average depth, m	28
Average yield, l/s	0.22
Number of drilled wells	154
Average depth, m	96
Average yield, l/s	4
Groundwater withdrawal in Ecuador, l/s	1,257
from dug wells	(641)
from deep wells	(616)

TABLE 6-36. FRENCH AMERICAN DEPENDENCIES—GUIANA—DISCHARGE OF PRINCIPAL RIVERS

(Source: ORSTOM, 1968)

River and station	Basin area km²	Mean monthly discharge, m³/s												Year	Max. flow m³/s	Date	Period of record
		Jan.	Feb.	Mar.	Apr.	May	Jun.	Jul.	Aug.	Sep.	Oct.	Nov.	Dec.				
Lawa-Maroni, Maripasoula	28,285	477	758	965	972	1,321	1,334	923	604	335	189	153	245	689	2,906	1955	1951-65
Tampoc, Degrad Roche	7,655	98	185	255	253	350	305	214	126	62	30	22	38	161	788	1963	1950-65
Oyapock, Camopi	17,120	336	582	774	832	1,026	887	651	444	241	138	108	150	513	2,490	1953	1952-65
Camopi, Camopi	5,920	126	202	267	293	359	306	219	153	86	52	46	54	180	1,087	1953	1952-65

TABLE 6-37. GUYANA—MONTHLY DISCHARGE OF ESSEQUIBO RIVER

(Source: UNESCO, 1971)

River and station	Basin area km²	Jan.	Feb.	Mar.	Apr.	May	Jun.	Jul.	Aug.	Sep.	Oct.	Nov.	Dec.	Year	Period of record
Essequibo, Plantain Island	68,600	1,040	1,120	998	1,180	2,790	4,770	5,320	4,450	2,270	979	698	889	2,190	1950-66

TABLE 6-38. GUYANA—FLOW CHARACTERISTICS OF SELECTED RIVERS

(Source: Potter, Min. of Works, Hydraulics and Supply, 1970)

River	Station	Drainage area mi²	Max. daily flow			Min. daily flow			Mean flow		Period of record
			cfs	cfs/mi²	Date	cfs	cfs/mi²	Date	cfs	cfs/mi²	
Essequibo	Plantain Is.	25,700	283,000	11	Jul. 1'51	5,130	.2	May 5-8'61	78,570	3.1	1950-69
Cuyuni	Kamaria Falls	20,600	190,500	9.3	Sep. 2'54	350	.02	Apr. 18-19 '47	37,560	1.8	1946-68
Mazaruni	Apaikwa	5,420	92,150	17	Jun. 22'54	1,500	.28	Apr. 25'61	25,990	4.8	1950-68
Mazaruni	Hillfoot	8,000	146,350	18	May 20'63	2,000	.25	May 1'51	40,460	5.1	1961-68
Potaro	Kaieteur Fall	1,020	39,600	39	Jun. 18'68	400	.39	Apr. 16-18 '61	7,224	7.1	1950-68
Potaro	Tumatumari	2,395	78,550	33	Jul. 1'47	1,550	.65	Apr. 6,14 '47	18,427	7.7	1946-54
Demerara	Great Falls	950	18,100	19	Jul. 27'55	150	.16	May 11-16 '64	2,585	2.7	1949-67
Demerara	Saka	1,560	15,790	11	Jul. 2'51	410	.27	May 13'64	3,938	2.5	1950-67
Berbice	Itabru Falls	1,970	14,740	7.5	Jun. 16'63	60	.03	Apr. 23-27 '62	1,412	.7	1960-68
Canje	Reynold's Bridge	107	304	2.3	Jul. 16'69	51	.48	Apr. 7-9'67	94	.9	1969

TABLE 6-39. GUYANA—PRINCIPAL AQUIFERS IN COASTAL ARTESIAN BASIN
(Source: Gibson, 1971)

Aquifer	Depth below land surface ft	Thickness ft	Transmissivity gpd/ft^2	Remarks
Upper Sand	100-200	50-400	—	Little used due to high Fe content (more than 5 mg/l) and salinity (to 1,200 mg/l).
"A" Sand	300-1,000	50-200	200,000	Yields 90% of country's drinking water; treated for Fe.
"B" Sand	1,200-2,600	50-200	—	Excellent quality but high temperature (105° F);trace H_2S.

TABLE 6-40. GUYANA—WATER USE, 1971
(Source: Gibson, 1971)

Category	Quantity, mgd[1]	Source of water
Agriculture (mainly for flood irrigation of sugar and rice)	3,240	Surface water; shallow reservoirs (conservancies) and streams.
Domestic	50	90% obtained from wells; 10% from surface sources.
Industrial (mainly bauxite industry)	1	

1) Imperial gallons.

TABLE 6-41. NETHERLANDS AMERICAN DEPENDENCIES—DISCHARGE OF SELECTED RIVERS IN SURINAM
(Source: ICID, 1969)

River	Basin area km^2	Mean annual discharge m^3/s
Marowijne	66,000	2,000
Suriname	15,900	440
Saramacca	10,200	240
Coppename	17,900	470
Nickerie	9,750	200
Commewijne	6,700	120
Corantijn	66,700	2,000

TABLE 6-42. PARAGUAY–MONTHLY DISCHARGE OF PARAGUAY RIVER AT ASUNCION

(Source: Administracion Nacional de Navegacion y Puertos, 1972)

Year	Mean monthly discharge, m3/s 1)												
	Jan.	Feb.	Mar.	Apr.	May	Jun.	Jul.	Aug.	Sep.	Oct.	Nov.	Dec.	Year
1965	2,396	2,677	3,426	2,713	4,824	6,503	3,983	2,444	1,907	1,898	2,166	1,980	3,076
1966	3,691	3,235	3,621	3,815	3,206	2,565	2,193	1,706	1,372	1,301	1,595	1,403	2,475
1967	1,712	1,887	2,261	2,291	1,865	1,561	1,644	1,372	1,030	872	947	1,063	1,542
1968	1,341	2,244	1,816	1,572	1,770	1,717	1,538	1,406	1,181	1,238	1,075	1,144	1,504
1969	2,307	1,606	1,390	1,963	2,616	3,468	2,085	1,229	909	1,743	2,534	1,144	2,033
1970	1,494	1,337	1,472	1,721	1,522	1,613	1,723	1,428	1,189	1,258	992	1,036	1,402

1) Mean daily maximum flow (period 1965-70):   7,501 m3/s on May 31, 1965.
   Mean daily minimum flow (period 1965-70):   818 m3/s on Sep. 24-26, 1969.

TABLE 6-43. PERU–DISCHARGE OF SELECTED RIVERS
(Source: CEPAL, 1968)

River and station	Basin area km2	Mean monthly discharge, m3/s												Year	Number of years of record
		Jan.	Feb.	Mar.	Apr.	May	Jun.	Jul.	Aug.	Sep.	Oct.	Nov.	Dec.		
Pacific Slope															
Tumbes, Pte. Tumbes	1,940	140.5	217.1	314.3	318.4	171.1	86.9	45.7	29.6	42.1	41.1	44.2	43.0	124.6	13
Chira, Pte. Sullana	10,100	79.2	205.1	288.4	298.2	137.1	80.6	55.9	39.6	32.6	31.9	27.9	35.3	109.4	21
Piura, Pte. Piura	12,640	3.6	59.9	108.2	89.7	30.0	12.6	6.1	3.4	1.8	1.1	0.7	0.5	26.4	47
La Leche, Puchaca	1,740	5.4	10.0	17.4	14.5	7.6	4.9	3.1	2.3	3.6	4.0	2.9	3.0	6.6	44
Chancay (Lamb.), Carhuaquero	5,180	24.7	44.0	66.4	69.5	38.2	20.0	10.8	7.3	8.4	15.3	16.3	18.0	28.2	50
Zana, El Batan	2,000	5.5	10.2	15.3	17.2	11.2	0.8	4.1	2.8	2.7	3.8	3.7	3.7	8.3	41
Jequetepeque, Ventanillas	4,200	24.8	51.5	93.4	81.0	32.0	13.2	7.1	4.8	3.9	7.9	10.1	13.0	28.6	38
Chicama, Salinar	4,770	33.1	66.9	102.5	78.2	29.6	12.4	7.6	5.5	5.9	6.4	5.7	8.7	30.2	49
Moche, Quirihuae	2,130	9.9	16.8	34.2	29.6	10.1	2.5	1.2	0.7	0.8	1.6	2.3	3.9	9.6	47
Viru, Huacapongo	1,960	4.3	10.4	14.8	10.0	4.0	0.7	0.3	0.1	0.1	0.6	1.0	1.3	3.9	23
Santa, Pte. Carretera	11,700	203.5	270.1	350.7	274.1	146.1	91.5	60.3	53.9	53.8	67.3	85.3	125.0	149.1	23
Nepena, San Jacinto	1,810	2.1	6.1	9.0	5.1	1.9	1.0	0.8	0.6	0.4	0.4	0.4	0.7	2.4	23
Casma, Pte. Carretera	2,900	5.6	16.4	23.2	14.6	3.7	1.0	0.5	0.3	0.2	0.3	0.8	1.5	5.7	27
Huarmey, Pte. Carretera	2,100	2.7	9.7	16.5	12.2	2.2	0.2	0.1	0.1	0.0	0.0	0.7	0.6	3.7	19
Pativilca, Alpas	4,700	63.5	102.3	123.4	77.6	36.4	20.4	15.5	14.1	15.2	22.3	28.0	35.0	46.0	24
Huaura, Casa Blanca	5,470	40.6	57.8	65.1	43.1	22.8	15.6	12.7	10.9	10.5	14.0	17.9	24.8	27.9	31
Chancay (Huar.), Santo Domingo	3,030	23.3	36.1	40.1	24.6	10.4	6.3	4.6	4.0	4.3	5.1	6.4	9.5	14.6	35
Chillon, Pte. Magdalena	2,020	13.3	24.3	31.1	16.8	6.7	3.3	2.5	2.0	1.9	2.0	2.4	5.2	9.3	37
Rimac, Chosica	3,540	39.5	64.6	78.1	41.4	21.8	14.3	12.0	11.8	12.8	13.4	15.8	22.5	29.0	50
Lurin, Manchay	2,500	6.3	16.0	9.3	8.2	2.6	0.8	0.2	0.0	0.0	0.0	0.5	0.6	3.7	20
Mala, La Capilla	2,130	37.1	50.8	62.1	20.6	7.2	2.8	2.1	1.8	1.6	2.8	6.9	13.8	18.0	23
Canete, Toma Imperial	6,750	94.8	144.0	160.9	87.1	36.2	20.8	14.7	12.0	11.2	13.1	24.4	42.2	55.1	46
San Juan, Conta	3,910	26.7	52.5	67.8	21.5	4.3	1.3	0.7	0.3	1.4	2.1	2.1	6.6	15.7	33
Pisco, Letrayoc	4,440	52.2	86.7	93.3	36.6	10.6	4.3	2.4	2.1	1.9	3.5	7.8	16.2	26.5	46

Station															
Ica, Huemani	7,390	17.4	39.7	45.6	14.2	2.2	0.7	0.3	0.2	0.3	0.8	0.8	5.2	10.6	47
Grande, Pte. Carretera	12,694	30.6	75.8	68.1	27.3	5.2	2.9	0.4	0.3	0.3	0.0	0.0	0.7	17.6	27
Acari, T. Bella Union	4,131	20.8	50.6	48.6	16.7	6.7	3.1	2.0	0.9	0.7	0.7	1.3	3.2	12.9	12
Yauca, Pte. Yaqui	4,541	13.9	35.5	33.4	10.7	3.5	1.5	0.8	0.9	0.8	0.8	1.0	2.3	8.7	14
Camana o Majes, Huatiapa	17,300	170.6	260.2	267.5	109.0	64.5	41.1	35.9	30.4	27.7	27.1	28.4	61.1	93.6	15
Chili, Charcani	12,643	24.6	38.9	34.3	15.4	6.6	5.0	5.0	4.6	4.4	4.4	4.3	7.2	12.9	30
Tambo, Chucarapi	12,910	44.5	100.4	86.1	36.0	24.3	20.7	18.7	16.5	12.5	12.5	9.9	16.7	33.2	19
Moquegua, Tumilaca	3,441	2.5	4.7	5.0	1.7	1.4	1.1	1.1	1.0	1.1	1.1	1.2	1.2	1.9	11
Caplina, Calientes	2,246	1.6	2.9	2.1	1.0	0.7	0.8	0.8	0.8	0.8	0.8	0.8	0.8	1.2	23
Uchusuma, Piedra Blanca	–	0.4	1.0	0.5	0.3	0.4	0.4	0.4	0.4	0.3	0.4	0.2	0.3	0.4	20
Titicaca Basin															
Ramis, Pte. Carretera	14,640	150.8	217.0	206.1	112.6	55.8	32.3	21.8	16.2	14.0	18.0	21.3	68.1	77.8	7
Amazon Basin															
Huancabamba	–	17.2	23.2	24.0	24.1	20.4	18.2	14.8	15.3	12.5	14.8	16.3	14.4	17.9	–
Mantaro, La Mejorada	17,500	216.7	348.0	389.3	239.9	159.4	124.3	117.1	105.7	105.3	107.9	122.9	173.8	184.0	16
Chamaya, La Savila	2,000	62.0	73.4	105.4	105.7	81.4	59.7	49.4	41.2	43.4	55.9	49.2	37.0	63.7	11

TABLE 6-44. PERU—RUNOFF, WATER CONSUMPTION AND DISCHARGE TO SEA OF RIVERS ON PACIFIC SLOPE

(Source: CEPAL, 1968)

River basin		Basin area km^2	Mean annual runoff million m^3	Mean annual water consumption million m^3	Mean annual discharge to sea million m^3
Tumbes	M	1,940	3,920	165	3,755
Chira	M	10,140	3,420	206	3,214
Piura	M	12,640	835	444	391
La Leche	M	1,740	210	210	0
Chancay (Lam.)	M	5,180	889	841	48
Zana	M	2,000	260	228	32
Jequetepeque	M	4,200	902	219	683
Chicama	M	4,770	950	555	395
Moche	M	2,130	302	209	93
Viru	M	1,960	125	78	47
Santa	M	11,700	4,700	120	4,580
Nepana	M	1,810	75	51	24
Casma	M	2,900	179	39	140
Culebras	E	680	50	20	30
Huarmey	M	2,100	118	25	93
Fortaleza	E	2,220	110	25	85
Pativilca	M	4,700	1,450	366	1,084
Supe	E	1,000	50	50	0
Huaura	M	5,470	883	414	469
Chancay (Huar.)	M	3,030	460	204	256
Chillon	M	2,020	292	107	185
Rimac	M	3,540	915	366	549
Lurin	M	2,500	117	29	88
Mala	M	2,130	568	60	508
Omas	E	1,440	80	15	65
Canete	M	6,750	1,735	233	1,502
Topara	E	650	100	15	85
San Juan	M	3,910	495	150	345
Pisco	M	4,440	835	259	576
Ica	M	7,390	334	158	176
Grande	M	12,690	555	129	426
Acari	M	4,130	406	57	349
Yauca	M	4,540	275	35	240
Chala	E	1,240	60	10	50
Chaparra	E	1,310	60	10	50
Atico	E	1,370	60	10	50
Caraveli	E	1,940	90	15	75
Ocana	E	15,590	1,240	200	1,040
Majes	M	17,300	2,950	159	2,791
Chili	M	12,640	406	42	364
Yura	E	4,400	268	20	248
Tambo	M	12,910	1,050	177	873
Moquegua	M	3,440	60	41	19
Locumba	E	6,240	69	50	19
Sama	E	4,620	54	25	29
Caplina	M	2,250	38	24	14
Uchusuma	M		13	9	4
Total		223,690	32,741	6,874	26,139

M = Measured
E = Estimated

TABLE 6-45. PERU—DATA ON RIVER BASINS

(Source: CEPAL, 1968)

Basin	Area km2	% Total area Peru	% Total population Peru	Mean annual precipitation		
				mm	109 m3	% Total Peru
Pacific Ocean	283,600	22	55	198	56	2.6
Amazon River	952,800	74	39	2,185	2,080	95.8
Lake Titicaca	48,800	4	6	722	35	1.6
Total	1,285,200	100	100	1,691	2,171	100

TABLE 6-46. PERU—HYDROELECTRIC POWER AND THERMOELECTRIC POWER CAPACITY, 1965

(Source: CEPAL, 1968)

[In MW; by Departamento]

Departamento	Public Service		Other		Total capacity
	Hydropower plants	Thermopower plants	Hydropower plants	Thermopower plants	
Amazonas	0.28	0.97	—	1.31	2.56
Ancash	51.50	6.26	2.66	3.76	64.18
Apurimac	1.42	0.21	1.33	—	2.96
Arequipa	10.89	14.50	2.92	4.37	32.68
Ayacucho	0.87	0.84	2.05	1.04	4.80
Cajamarca	1.09	1.51	1.83	0.90	5.33
Callao	Included under Lima				
Cuzco	20.80	4.04	2.42	0.58	27.84
Huancavelica	0.66	0.14	4.85	—	5.65
Huanuco	0.71	1.20	1.40	—	3.31
Ica	Very small	11.75	—	30.81	42.56
Junin	4.53	1.48	83.79	8.53	98.33
La Libertad	0.46	4.09	12.67	31.32	48.54
Lambayeque	—	10.86	1.31	18.78	30.95
Lima	400.39	78.29	8.05	45.66	532.39
Loreto	—	3.84	—	8.84	12.68
Madre de Dios	—	0.30	—	1.31	1.61
Moquehua	0.15	1.07	—	179.69	180.91
Pasco	0.61	0.10	167.05	3.62	171.38
Piura	0.20	10.99	—	19.06	30.25
Puno	0.21	6.12	1.47	6.20	14.00
San Martin	—	0.91	—	1.31	2.22
Tacna	—	2.11	—	7.78	9.89
Tumbes	—	0.12	—	1.51	1.63
Total Peru	494.77	161.70	293.80	376.38	1,326.65

TABLE 6-47. PERU—GROUNDWATER WITHDRAWAL IN COASTAL ZONE
(Source: CEPAL, 1968)

Coastal Zone	Length of coastal zone km	Number of wells	Average yield per well l/s	Groundwater withdrawn per km of coastline m3/s 1)	m3/year
Costa Norte	1,000	1,411	25	0.036	245,000
Costa Central	350	831	23	0.055	415,000
Costa Sur	950	1,685	40	0.071	495,000
Total Coastal Zone	2,300	3,947	31	0.054	376,000

1) Mean pumping rate.

TABLE 6-48. PERU—POTABLE WATER DEMAND, 1965-80
(Source: CEPAL, 1968)

Geographic Zone	1965 Population	Water demand l/cap/d	km3/yr	1980 Population	Estimated water demand l/cap/d	km3/yr
Total Country 1)	11,600,000	160		18,000,000	240	
Urban population	5,800,000	275	0.58	10,000,000	300	1.10
Rural population	5,800,000	70	0.15	8,000,000	160	0.47
Livestock	41,000,000	17	0.25	53,000,000	17	0.33
Total water demand			0.98			1.90
Coastal Zone	4,700,000	250		7,300,000	320	
Urban population	3,600,000	300	0.40	5,800,000	350	0.74
Rural population	1,100,000	75	0.03	1,500,000	200	0.11
Livestock	10,000,000	10	0.04	15,000,000	10	0.06
Total water demand			0.47			0.91
Mountain Region (excl. Titicaca)	5,400,000	110		8,400,000	190	
Urban population	1,800,000	200	0.13	3,400,000	250	0.31
Rural population	3,600,000	65	0.09	5,000,000	150	0.27
Livestock	25,000,000	20	0.18	30,000,000	20	0.22
Total water demand			0.40			0.80
Lake Titicaca basin	700,000	115		1,100,000	175	
Urban population	125,000	200	0.01	350,000	250	0.03
Rural population	575,000	65	0.02	750,000	150	0.04
Livestock	5,000,000	20	0.04	6,000,000	20	0.04
Total water demand			0.07			0.11

1) The difference between country total and total of three listed zones represents the Amazon basin.

TABLE 6-49. URUGUAY–DISCHARGE OF SELECTED RIVERS
(Source: UNESCO, 1971)

River and station	Basin area km2	Mean monthly discharge, m3/s												Year	Period of record
		Jan.	Feb.	Mar.	Apr.	May	Jun.	Jul.	Aug.	Sep.	Oct.	Nov.	Dec.		
Uruguay, Salto	238,900	2,160	2,050	2,370	4,980	5,390	6,390	6,110	5,400	6,350	7,440	4,600	3,310	4,660	1898-1960
Negro, Palmar	63,000	170	196	323	584	836	870	1,000	1,070	1,030	785	329	358	637	1909-44
Cebollati, Averias	18,010	39	29	31	73	102	168	173	164	215	126	56	55	102	1934-58

TABLE 6-50. URUGUAY–FLOW CHARACTERISTICS OF SELECTED RIVERS
(Source: CEPAL, 1972)

River	Station	Basin area km2	Flow, m3/s		
			Mean	Max	Min
Negro	Paso Mazangano	6,670	71.0	2,100[1]	0.3[1]
Negro	Paso Aguiar	8,225	124.0	5,670[1]	
Negro	Paso de los Toros	40,300	536.0	17,300	1.0
Negro	Rincon de Baygorria	43,600	572.0	11,500	
Negro	Paso Palmar	62,560	725.0		
Tacuarembo	Paso del Borracho	6,590	130.0	4,840[1]	0.3[1]
Tacuarembo	Paso de la Laguna	14,050		4,080	
Arroyo Corrales	Paso de la Compania	1,030	23.5	882	0.0
Arroyo Yaguari	Paso de Coelho	2,460	41.0	1,760[1]	0.1
Arroyo Cunapiru	Cunapiru	1,900	33.7	359	
San Jose	Puente Carretero	3,000	35.0	4,000	0.1
Santa Lucia	Picada de Almeyda	2,600	30.0	1,200	0.2
Santa Lucia	San Ramon	3,180	39.0	1,300	0.2
Uruguay	Salto Grande	239,000	4,660.0	37,000	92.0

1) Mean daily flow.

TABLE 6-51. URUGUAY—EXISTING AND PROJECTED HYDROPOWER PLANTS, 1969
(Source: CEPAL, 1972)

River basin and plant	River	Type	Capacity MW	Mean annual production Gwh	Height m	Mean flow m3/s
Rio Negro Basin		**EXISTING**				
Rincon del Bonete	Negro	Reservoir	128.0	600	22.0	526
Rincon de Baygorria	Negro	Reservoir	108.0	450	15.0	580
Total Uruguay			236.0	1,050		
Rio Negro Basin		**PROJECTED**				
Los Cuervos	Cunapiru	Reservoir	12.0	40	25.0	34
Palmar	Negro	Reservoir	270.0	1,090	27.8	725
Tacuarembo-Rivera[1]	Various		18.0	54	—	—
Rio Uruguay Basin						
Salto Grande[2]	Uruguay	Run-of-the-river	720.0	2,860	24.0	2,330[2]
Barra Viraro	Queguay	Reservoir	15.0	64	19.3	81
San Salvador[1]	San Salvador		0.9	—	—	—
Rio de la Plata Basin						
Piedra Alta[1]	Santa Lucia Chico		1.4	—	—	—
Rosario[1]	Rosario		0.5	—	—	—
Laguna Merin Basin						
Sierra del Tigre	Cebollati	Reservoir	11.2	34	16.0	40
La Cachocira[1]	Taouari		2.5	—	—	—
Paso del Centurion[1]	Yaguaron		80.0	—	—	—
Total Projected			1,231.6			

1) Not under study; potential site only.
2) Information pertains to Uruguyan plant only.

TABLE 6-52. URUGUAY—REGULATING CAPACITY OF SELECTED RESERVOIRS
(Source: CEPAL, 1972)

Basin and river	Dam	Status[1]	Storage Capacity Gross Hm3	Net Hm3
Uruguay				
Uruguay	Salto Grande	UC	—	—
Negro	Rincon del Bonete	E	13,000	10,400
	Rincon de Baygorria	E	570	143
	Palmar	UC	2,854	1,054
Rio Negro-Cunapiru	Cerro Los Cuervos	UC	1,190	970
	Subtotal		17,614	12,567
De la Plata				
Canelon	Canelon Grande	E	29	—
Santa Lucia Chico	Santa Lucia	UC	—	—
Santa Lucia Grande	Picada de Almeida	UC	280	140
	Subtotal		309	140
Laguna Merin				
Cebollati	Sierra del Tigre	UC	1,420	1,300
Yi	Paso del Bote	UC	—	—
Uaguaron	Paso del Centurion	UC	—	—
Olimar	Cerro La Bolsa	UC	900	500
	Subtotal		2,320	1,800
	Total		20,243	—

1) E — Existing
   UC — Under construction
   — Not available

TABLE 6-53. URUGUAY—WATER DEMAND, 1965-90

(Source: CEPAL, 1972)

Category	1965 Hm3	1965 %	1974 Hm3	1974 %	1990 Hm3	1990 %
Drinking water	103[1]	15.7	160	16.3	215	10.7
Industrial water	49[1]	7.7	67	6.9	111	5.5
Irrigation	277	42.0	502	50.1	1,391	69.2
Stock watering	225	34.6	250	25.7	300	14.6
Total	654	100.0	979	100.0	2,017	100.0

[1] National public water supply systems provide 110 million Hm3 of which 103 million are for human consumption and 7 million are supplied to industries. The remainder of industrial water demand (42 million) is self-supplied or furnished by others.

TABLE 6-54. URUGUAY—INDUSTRIAL WATER DEMAND, 1968-85

(Source: CEPAL, 1972)

[Hydroelectric power generation excluded]

Industry	Consumption 1968 Hm3	Rate of annual increase 1974	Rate of annual increase 1985	Estimated consumption 1974 Hm3	Estimated consumption 1985 Hm3
Slaughterhouses and refrigerated storage	3,920	3	4.8	4,800	7,950
Dairy products	1,000	3	4.8	1,194	1,980
Canned goods	870	3	4.5	1,045	1,700
Sugar plantations and refineries	7,400	—	—	9,000	9,800
Beverage	880	3	4.5	1,042	1,700
Alcoholic beverages	900	3	4.8	1,076	1,790
Beer	1,190	3	4.5	1,420	2,300
Non-alcoholic beverages	410	3	4.5	490	795
Textile	4,190	3	5	5,000	8,550
Paper	8,200	3	5	9,800	16,700
Leather	960	5	8	1,005	2,390
Rubber	960	3	5	1,145	1,960
Chemical industries	1,740	4	7	2,200	4,650
Petroleum and coal	14,800	3.6	5	18,450	31,500
Non-metallic minerals	950	3	7	1,125	2,350
Metal products	510	3	5	560	955
Gas	140	3	5	167	286
Others	2,000	3	3	2,388	3,320
Total	51,020			61,957	100,676

TABLE 6-55. URUGUAY—MUNICIPAL WATER CONNECTIONS

(Source: CEPAL, 1972)
[Cities of over 50,000 pop.]

City	Population (1963)	Number of connections	Population served at home	Number of public fountains
Montevideo	1,154,465	162,424	1,011,600	175
Paysandu	51,645	9,324	—	50
Salto	50,714	8,815	—	66
Total	1,256,824	180,563	—	291

TABLE 6-56. VENEZUELA—DISCHARGE OF SELECTED RIVERS

(Source: UNESCO, 1971)

River and station	Basin area km2	Mean monthly discharge, m3/s												Year	Period of record
		Jan.	Feb.	Mar.	Apr.	May	Jun.	Jul.	Aug.	Sep.	Oct.	Nov.	Dec.		
Apure, San Fernando	115,000	567	338	248	298	775	2,340	3,410	3,930	3,960	3,320	2,140	1,260	1,890	1962-65
Aragua, La Chorrera	1,881	0.7	0.8	0.2	0.1	0.4	1.0	2.8	6.9	5.5	3.9	3.4	1.5	2.3	1945-66
Bocono, Pena Larga	1,578	27.5	21.5	22.1	51.3	97.9	136	137	108	94.3	77.2	63.7	34.4	72.5	1952-65
Caroni, Carhuachi	95,000	2,480	1,980	1,420	1,710	4,260	8,180	9,140	8,890	6,760	4,620	4,040	3,480	4,750	1949-65
Guarico, La Puerta	538	1.3	0.9	0.6	0.7	1.6	4.3	6.7	7.1	7.8	7.7	5.0	2.7	3.9	1940-66
Motatan, Agua Viva	4,454	23.7	18.1	16.1	23.1	39.7	37.8	32.4	30.5	33.0	46.7	57.5	40.8	33.8	1941-66
Orinoco, Cdad. Bolivar	850,000	12,600	8,920	7,520	9,270	18,400	30,800	40,500	46,200	44,400	35,400	27,100	19,700	25,200	1923-64
Paguey, El Paso	810	12.2	11.0	10.1	28.1	59.6	85.9	87.9	85.6	83.6	64.0	46.2	22.5	49.9	1950-65
Pao, La Balsa	2,730	5.8	3.4	2.4	4.1	13.7	29.5	46.7	62.5	49.8	33.3	25.1	12.5	24.1	1950-66
Portuguesa, Pte. Portuguesa	762	6.5	3.1	2.3	3.9	21.3	57.4	72.9	58.2	43.2	37.6	20.7	11.1	28.4	1950-66
Tirgua, Paso Viboral	1,563	8.8	7.1	5.5	6.4	12.1	16.6	22.9	25.0	23.5	22.1	18.7	12.5	15.2	1947-66
Tocuyo, Pte. Torres	3,590	5.4	4.8	3.0	11.5	21.2	17.9	16.8	10.9	10.2	16.0	15.7	7.2	11.7	1940-66
Tuy, El Vigia	3,620	12.3	9.8	7.4	5.7	10.9	28.2	36.4	38.7	31.8	27.3	20.3	18.5	20.7	1947-66
Yaracuy, Pte. Penon	1,206	5.9	5.2	4.0	4.7	9.4	12.4	16.7	14.9	10.8	10.5	10.5	8.7	9.4	1942-64

FIGURE 6-1. VENEZUELA. WATER RESOURCES REGIONS AND URBAN WATER DEMAND
(Source: COPLANARH, 1972)

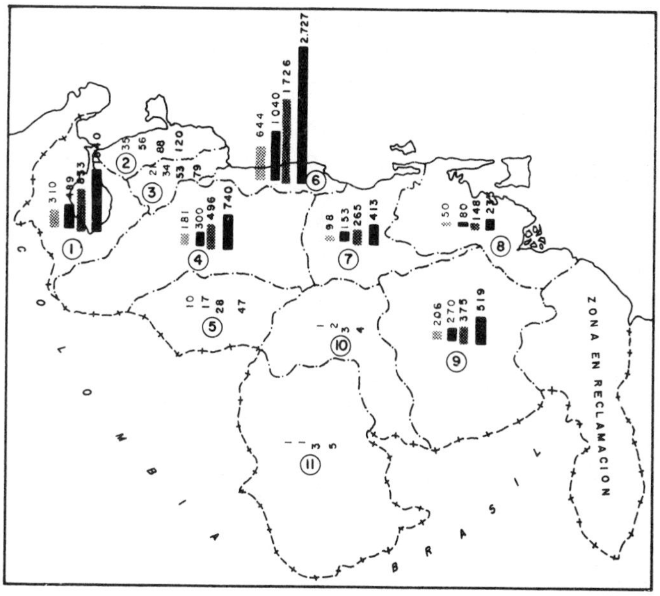

EXPLANATION

URBAN WATER DEMAND
(million m³/yr.)

From left to right

Column	Year
1	1970
2	1980
3	1990
4	2000

TABLE 6-57. VENEZUELA—PRECIPITATION AND RUNOFF BY REGION
(Source: COPLANARH, 1972)
[For location of regions see adjacent map]

Region	Area km²	Mean annual precipitation mm	Mean annual runoff, million m³	
			Groundwater	Surface water
1. Lago de Maracaibo	61,900	1,373	5,074 [1]	17,894
2. Costa Noroccidental	24,970	738	373 [1]	2,458
3. Centro Occidental	20,660	845	412 [1]	1,831
4. Llanos Centrales and Llanos Occidentales	140,360	1,566	11,282 [1]	48,810
5. Llanos Meridionales	68,650	1,770	Inundated [1]	22,778
6. Central	18,540	1,341	871	5,168
7. Centro Oriental	71,020	1,210	1,256 [1]	9,117
8. Oriental	78,990	1,247	1,052 [1]	10,257
9. Guayana Oriental	150,270	2,357	1,206	164,511
10. Guayana Occidental	88,020	2,722	615	121,956
11. Amazonica	176,770	3,250	171	295,359
Guayana Esequiba (French Guiana-claimed)	159,500	2,592	1,600 E	132,302
Subtotal left bank Orinoco River	468,250	1,373	20,113	111,577
Subtotal right bank Orinoco River	431,900	2,771	2,199	588,562
Total Venezuela	1,059,650	2,126	23,912	832,441
(less Guayana Esequiba)	900,150	2,044	22,312	700,139

[1] Inundated zones excluded.
E— Estimated.

TABLE 6-58. VENEZUELA—AVAILABLE WATER RESOURCES BY REGION
(Source: COPLANARH, 1972)

| Region | Surface water runoff | | Volume of water available per year | | | | | |
| | | | Actual 1970[2] | | Practical | | Potential | |
	% of total	million m³/yr	% of total runoff	million m³	% of total runoff	million m³	% of total runoff	million m³
1. Lago de Maracaibo	16	17,894	6	1,140	19	3,470	44	7,800
2. Costa Noroccidental	2	2,458	2	40	17	410	45	1,130
3. Centro Occidental	2	1,831	9	155	46	990	65	1,230
4. Llanos Centrales and Llanos Occidentales	43	48,810	4	1,980	31	15,370	49	23,910
5. Llanos Meridionales	21	22,778	0	0	–	–	10	2,280
6. Central	5	5,168	9	460	36	1,875	47	2,420
7. Centro Oriental	8	9,117	5	492	21	1,942	35	3,172
8. Oriental (left bank)	3	3,521	*	3	10	400	39	1,390
Oriental (right bank)	1	6,736	–	–	–	–	30	2,020
9. Guayana Oriental 1)	28	164,511	43	70,000	70	115,000	83	137,300
10. Guayana Occidental	21	121,956	0	0	–	–	70	85,564
11. Amazonica	50	295,359	0	0	–	–	70	206,748
Subtotal left bank Orinoco River	16	111,577	4	4,270	22	24,457	39	43,332
Subtotal right bank Orinoco River	84	588,562	12	70,000	20	115,000	73	431,632
TOTAL	100	700,139	11	74,270	20	139,457	68	474,964
French Guiana (claimed)	19	132,302	–	–	–	–	68	90,000
Total Venezuela	119	832,441	9	74,270	17	139,457	68	564,964

* No information.
1) Region of French Guiana excluded.
2) Includes only water available from storage reservoirs.

TABLE 6-59. VENEZUELA—WATER RESOURCES AND WATER DEMAND, 1970-2000

(Source: COPLANARH, 1972)

[By region]

Water resources region	1970					2000				
	Available water resources 1) million m3	Urban	Agriculture	Industry 2)	Surplus (+) or Deficit (−) million m3	Available water resources 1) million m3	Urban	Agriculture	Industry 2)	Surplus (+) or Deficit (−) million m3
1. Lago de Maracaibo	1,160	260	499	40	+ 361	4,532	1,193	3,202	150	− 13
2. Costa Noroccidental	40	35	60	*	− 55	762	129	618	*	+ 15
3. Centro Occidental	0	15	228	*	− 243	477	54	476	*	− 53
4. Llanos Centrales y Occidentales	655	161	460	*	+ 34	5,321	643	4,678	*	0
5. Llanos Meridionales	0	10	40	*	− 50	266	44	222	*	0
6. Central	1,064	613	965	38	− 552	4,735	2,695	3,078	88	− 1,126
7. Centro Oriental	492	98	160	*	+ 234	1,568	413	1,191	*	− 36
8. Oriental	3	56	125	*	− 178	2,474	235	2,929	*	− 690
9. Guayana Oriental	70,000	48	*	158	+ 70,000	70,000	330	326	189	+ 70,000
10. Guayana Occidental	0	1	*	*	0	77	4	73	*	0
11. Amazonica	0	1	*	*	0	49	5	44	*	0

* Insignificant.
1) Water from storage reservoirs.
2) Not served by urban water systems.

TABLE 6-60. VENEZUELA–IRRIGATED AREA AND WATER DEMAND FOR IRRIGATION, 1970-2000
(Source: COPLANARH, 1972)

Planning region	1970		1980		2000	
	Irrigated area 1,000 ha	Water demand million m3/yr	Irrigated area 1,000 ha	Water demand million m3/yr	Irrigated area 1,000 ha	Water demand million m3/yr
1. Lago Maracaibo	70	470	133	930	563	3,200
2. Costa Noroccidental	4	60	34	520	43	640
3. Centro Occidental	26	340	55	490	86	810
4. Llanos Centrales and Llanos Occidentales	89	990	276	2,500	841	6,900
5. Llanos Meridionales	3	40	3	40	15	230
6. Central	54	310	58	330	66	380
7. Centro Oriental	11	150	76	860	99	1,150
8. Oriental	9	80	16	120	229	2,930
9. Guayana Oriental	–	–	15	100	45	330
10. Guayana Occidental	–	–	–	–	10	70
11. Amazonica	–	–	–	–	–	40
Total	266	2,440	666	5,950	1,997	16,680

TABLE 6-61. VENEZUELA–MAJOR STORAGE RESERVOIRS
(Source: CEPAL, 1962)

Reservoir	Capacity million m3	Surface area of reservoir ha	Mean depth m	Remarks
Suata	44	860	5.1	Existing
Taiguaiguay	90	1,800	5.0	Existing
Guarico	1,840	23,150	7.9	Existing
Pao	287	4,275	6.7	Projected
Tinaco	981	13,000	7.5	Projected
Cojedes	810	2,325	34.8	Projected
Majaguas	320	4,050	7.9	Under construction
Bocono-Tucupido	4,500	17,500	25.7	–

TABLE 6-62. VENEZUELA—PRESENT AND PROJECTED ELECTRIC POWER GENERATING CAPACITY

(Source: COPLANARH, 1972)

[In 1,000 Kw]

Year	Thermal electric plants	Hydroelectric plants	Total generating capacity
1970	1,690	1,340	3,030
1980	2,800	3,255	5,635
1990	5,500	5,500	11,000
2000	10,350	10,350	20,700

TABLE 6-63. VENEZUELA—NUMBER AND DEPTH OF WATER WELLS

(Source: Direccion de Geologia, Div. de Hidrogeologia, 1969)

[By state]

State	Number of active water wells [1]	Mean depth m
Distrito Federal	121	54
Anzoategui	311	61
Apure	87	51
Aragua	1,031	65
Barinas	396	29
Bolivar	225	29
Carabobo	839	64
Cojodes	235	39
Falcon	237	63
Guarico	650	45
Lara	598	61
Merida	29	49
Miranda	862	59
Monagas	198	55
Nueva Esparta	42	23
Portuguesa	526	38
Sucre	83	48
Tachira	81	40
Trujillo	104	64
Yaracuy	401	64
Zulia	549	100
Territorio Amazonas	—	—
Territorio D. Amacuro	27	64
Dependencias Federales	—	—
Total	7,632	57

[1] Depth of wells known; wells constructed by oil companies omitted.

TABLE 6-64. VENEZUELA—MUNICIPAL WATER DEMAND, 1970-2000
(Source: COPLANARH, 1972)

City	Population to be served year 2000 1,000	Water consumption in 1969 million m^3	% Population served in 1969	Water demand, million m^3			
				1970	1980	1990	2000
Caracas	5,547	204	75	395	592	904	1,331
Maracaibo [1]	2,209	33	54	127	203	333	530
Valencia [1]	1,277	30	79	63	63	187	306
C. Guayana	916	16	60	23	62	128	220
Valles del Tuy	865	—	—	13	31	95	206
Maracay	862	23	78	36	73	127	207
Barquisimeto	664	19	80	55	80	117	159
San Cristobal	432	10	73	26	42	70	104
Pto. La Cruz [1] Barcelona	419	12	74	23	41	65	98
Cabimas	364	3	37	24	39	63	87
C. Bolivar	349	7	60	18	31	55	84
Maiquetia	325	9	47	20	33	50	78
Pto. Cabello	313	7	77	11	24	42	75
El Tablazo	295	—	—	4	13	34	71
C. Ojeda	292	3	24	13	27	44	70
Total 15,129		576	67	851	1,398	2,314	3,626

[1] Includes population within metropolitan area.

SOUTH AMERICA

REFERENCES CITED

References cited in more than one country:

UNESCO, Discharge of Selected Rivers of the World; Vol. I. General and Regime Characteristics of Stations Selected, (1969); Vol. II. Monthly and Annual Discharges Recorded at Various Selected Stations (1971); Vol. III. Mean Monthly and Extreme Discharges 1965-1969, (1971). Copyright UNESCO/IASH/WMO. Reprinted by permission.

SOUTH AMERICA GENERAL

Pan American Health Organization, 1973, Annual Report of the Director of the Pan American Sanitary Bureau, Doc. No. 124, Washington, D.C.

U.S. Federal Power Commission, 1968, World Power Data, FPC P - 40, Washington, D.C.

ARGENTINA

CEPAL, 1972, Los Recursos Hidraulicos de America Latina, Argentina, U.N. S. 72. II. 6. 2.

BOLIVIA

CEPAL, 1964, Los Recursos Hidraulicos de America Latina, III Bolivia y Colombia, United Nations, New York.

BRAZIL

Oltman, R.E., and others, 1964, Amazon River Investigations Reconnaissance Measurements of July 1963, U.S. Geological Survey Circular 486.

Departamento de Aguas e Energia Eletrica, 1972, Estudio de Aguas Subterraneas, Geopesquisadora Brasileira/TAHAL.

Ministerio das Minas e Energia, 1973, Relatorio das atividades de 1972, Brasilia.

Conselho Nacional de Estatistica, 1969, Brazil Yearbook — 1968, Rio de Janeiro.

CHILE

Donoso, Jaime and Matte, Augusto, 1967, Country Report: Chile, International Conference Water for Peace, Vol. I. Washington, D.C.

CEPAL, 1960, Los Recursos Hidraulicos de America Latina, I Chile, Santiago.

Court, Luis, 1971, El Problema de Aguas en Chile, Universidad de Chile, Santiago, Publ. 71/11/C.

Wollman, Nathaniel, 1968, The Water Resources of Chile, Resources for the Future, Inc., The Johns Hopkins Press, Baltimore, Md.

COLOMBIA

CEPAL, 1964, Los Recursos Hidraulicos de America Latina, III Bolivia y Colombia, United Nations, New York.

ECUADOR

Mancheno, E., 1973, Estado de la Utilizacion de los Recursos Hidricos y su Desarrollo en el Ecuador, First World Congress on Water Resources, Internat. Water Resources Association, Chicago, Ill.

FRENCH AMERICAN DEPENDENCIES

GUIANA

Office de la Recherche Scientifique et Technique Outre-Mer (ORSTOM), 1968, Annuaire Hydrologique de la France d'Outre-Mer, Annees 1963-1964-1965, Paris.

GUYANA

Potter, K.E.D., 1970, An Appraisal of the Hydrology and Climate of Guyana, Hydrometeorological Service, Min. of Works, Hydraulics and Supply.

Gibson, U.P., 1971, Water Resources Management in Guyana, United Nations Interregional Seminar on Current Issues of Water Resources Administration, New Delhi, 1973.

NETHERLANDS AMERICAN DEPENDENCIES

SURINAM

Framji, K.K., and Mahajan, I.K., 1969, Irrigation and Drainage in the World, Internat. Commission on Irrigation and Drainage (ICID), New Dehli.

PARAGUAY

Administracion Nacional de Navegacion y Puertos, 1972, Annuario Hidrografico Ano 1965-1970, Asuncion.

PERU

ECLA, 1968, Los Recursos Hidraulicos de America Latina, IV Peru, Santiago, Chile.

URUGUAY

ECLA, 1972, Los Recursos Hidraulicos de America Latina, Uruguay, Santiago, Chile, Vols. I and II.

VENEZUELA

Comision del Plan Nacional de Aprovechamiento de los Recursos Hidraulicos (COPLANARH), 1972, Tomo I - El Plan, Tomo II - Documentacion Basica, Caracas.

Division de Hidrogeologia, Direccion de Geologia, 1969, Explicacion del Mapa de Posibilidades de Abastecimiento de Aguas Subterraneas de Venezuela, Caracas.

TABLE 7-1. DATA ON LAND, PEOPLE AND ECONOMY, BY COUNTRY

(Source: U.S. Department of State, 1972)

Country	LAND Total sq mi (000)	Agriculture [1] Percent of total	Agriculture [1] Acres per capita	PEOPLE Population 1970 (mill)	Population Growth rate (%)[1]	Population Density per sq mi 1969	Urban (%)	Infant deaths per 1,000 live births [1]	Life expectancy (yrs)	Average daily caloric intake (1969-70)	ECONOMY Gross National Product 1970 ($ bill)	Gross National Product 1970 per capita ($)	Exports 1967-69 primary (%)
WESTERN EUROPE													
EEC													
Belgium	12	53	0.4	9.7	0.3	822	69	22	74	3,150	25	2,595	Various
France	211	56	1.6	50.8	0.9	239	67	16	75	3,270	146	2,874	"
Germany, West (FRG)	96	56	0.5	61.7	1.2	636	81	23	74	2,940	186	3,007	"
Italy	116	68	0.9	54.5	0.6	460	53	30	73	2,950	92	1,684	"
Luxembourg	1	52	1.0	0.3	0.6	341	67	17	73	3,150	1	2,847	"
Netherlands	16	56	0.4	13.0	1.2	819	71	13	76	3,030	31	2,402	"
EFTA													
Austria	32	47	1.3	7.4	0.2	229	54	25	73	2,950	14	1,861	"
Denmark	17	72	1.5	4.9	0.6	296	80	15	75	3,140	16	3,197	"
Finland [2]	130	8	1.6	4.7	0.4	36	61	14	73	2,960	10	2,166	"
Iceland	40	22	27.5	0.2	1.0	5	72	12	76	2,900	0.5	2,319	"
Norway	125	5	0.6	3.9	0.7	31	53	14	77	2,900	11	2,879	"
Portugal	36	54	1.3	9.6	1.2	270	37	57	69	2,730	6	659	"
Sweden	174	9	1.1	8.1	1.1	47	80	13	76	2,750	31	3,896	"
Switzerland	16	52	0.9	6.3	0.7	393	58	15	75	2,990	21	3,266	"
United Kingdom	94	80	0.9	55.7	0.3	593	81	19	74	3,180	117	2,092	"
OTHER													
Cyprus	4	57	2.0	0.6	1.6	177	44	26	66	2,460	0.5	844	"
Greece	51	67	2.4	8.9	0.7	174	49	32	74	2,900	9	1,035	"
Ireland	27	68	4.0	2.9	0.8	109	47	21	72	3,450	4	1,314	"
Malta	0.12	50	0.1	0.3	0.9	2,657	68	24	70	2,680	0.2	694	"
Spain	195	68	2.7	33.3	1.1	169	61	30	72	2,750	32	974	"
Turkey	301	68	3.7	35.2	2.5	114	35	153 [3]	55	2,760	9	247	"
EASTERN EUROPE													
COMECON													
Bulgaria	43	52		8.5	0.8	198	51	31	73	3,070	12	1,375	"
Czechoslovakia	49	56		14.5	0.4	294	52	22	74	3,030	33	2,240	"
East Germany (DRG)	42	58		17.1	0.0	411	81	19	74	3,040	40	2,320	"
Hungary	36	74		10.3	0.3	289	43	36	72	3,050	16	1,530	"
Poland	121	63		32.5	0.9	270	53	33	74	3,140	46	1,420	"
Romania	92	63		20.3	1.1	218	42	50	70	3,010	24	1,200	"
USSR	8,649	26	314	242.8	0.9	29	57	25	70	3,180	486	2,095	"
OTHER													
Albania	11	43		2.1	2.7	187	38	87	66	2,370	1	570	"
Yugoslavia	99	57	1.8	20.5	0.9	208	37	55	69	3,130	19	900	"

1) 1969 and 1970.
2) Associate member of EFTA.
3) Represents 1967.

EEC – European Economic Community.
EFTA – European Free Trade Association.
COMECON – Council for Mutual Economic Assistance.

TABLE 7-1. DATA ON LAND, PEOPLE AND ECONOMY, BY COUNTRY (continued)

Country	Total sq mi (000)	Agriculture [1] Percent of total	Agriculture [1] Acres per capita	Population 1970 (mill)	Population Growth rate (%)	Population Density per sq mi	Urban (%)	Infant deaths per 1,000 live births	Life expectancy (yrs)	Average daily caloric intake	GNP [2] 1970 ($ bill)	GNP [2] 1970 per capita ($)	Exports 1967-69 primary (%)
MIDDLE EAST AND NORTH AFRICA													
I. ISLAMIC—ARAB													
Algeria	920	19	8	13.8	3.2	15	43	103	51	1,950	3.7	275	65 crude petrol / 12 wine
Bahrain	.23	N.A.	N.A.	215	3.1	927	75	146	N.A.	—	N.A.	420	petrol
Egypt	387	3	0.2	34.2	2.5	88	43	120	53	2,960	6.1	188	45 cotton / 14 rice
Iraq	168	27	3.0	9.8	3.4	58	47	104	N.A.	2,050	2.7	288	93 petrol
Jordan	38	14	1.4	2.4	3.3	64	44	115	52	2,400	.6 (1967)	286	41 fruits/veg / 33 phosph
Kuwait	6	8	0.4	.77	3.3	125	80	41	N.A.	N.A.	2.4 (1969)	3,360	97 petrol
Lebanon	4	31	0.3	2.9	2.4	736	41	N.A.	N.A.	2,360	1.4 (1969)	511	21 fruits/veg
Libya	679	2	4.6	1.9	3.7	3	27	N.A.	37	2,340	3.0 (1969)	1,601	99 petrol
Morocco	172	35	2.5	15.8	3.2	92	35	145	55	2,180	3.1 (1969)	203	17 foods / 24 phosph
Oman	82	N.A.	N.A.	.66	3.1	8	5	N.A.	N.A.	N.A.	N.A.	210 E (1969)	petrol
Qatar	8	N.A.	N.A.	.79	3.1	9	68	N.A.	N.A.	N.A.	0.05 (1966)	1,550	petrol
Saudi Arabia	830	40	38	5.5	2.8	7	25	157	30-40	2,080	2.8 (1969)	592	96 petrol
Sudan	967	12	4.9	15.8	2.9	16	10	121	50	1,940	1.7 (1969)	113	58 cotton
Syria	71	61	4.4	6.3	3.4	88	44	98	30-40	2,450	1.6 (1969)	273	40 cotton
Tunisia	63	50	3.9	5.2	2.8	78	44	120	52	2,190	1.2 (1969)	242	26 phosph / 18 petrol
United Arab Emirates	32	N.A.	N.A.	193	3.1	6	55	N.A.	N.A.	N.A.	N.A.	1,590 E (1969)	petrol / pearls
Yemen (San'a)	75	N.A.	0.2	5.9	2.9	.78	6	160	30-40	1,910	0.4 (1969)	80	90 cotton/coffee
Yemen (Aden)	111	32	17.5	1.3	2.9	12	33	160	N.A.	2,080	0.13 (1969)	110	70 petrol prod
II. ISLAMIC—NON-ARAB													
Iran	636	11	1.5	29.5	3.0	46	41	160	N.A.	2,030	9.9	345	90 petrol
Turkey	301	69	3.7	36.1	2.5	120	35	119	54	2,760	8.2	232	24 cotton / 19 tobacco

													Principal Products
III. NON-ISLAMIC													
Cyprus	4	57	2.0	.65	1.5	182	44	25	66	2,460	.53	828	28 copper pyrites / 20 citrus
Greece	51	67	2.4	9.0	0.8	176	49	32	69	2,900	9.0	1,011	22 tobacco / 15 fruits & nuts
Israel	8	59	1.0	3.0	2.4	370	80	19	71	2,930	5.1	1,751	35 diamonds / 18 fruits
SUB-SAHARAN AFRICA													
Angola	481	24	13.4	5.7	1.3	11	14	230	35	1,870	1.1 (1969)	210	43 coffee / 18 diamonds
Botswana	232	69	152	.62	3.0	3	22	175	N.A.	N.A.	.054 (1969)	94	96 animals & prod
Burundi	11	64	1.1	3.6	2.0	330	3	161	39	2,020	.207 E (1969)	60	80 coffee
Cameroon	183	35	5.3	5.8	2.1	32	20	110	49	2,130	.85 (1969)	150	24 cocoa / 24 coffee / 8 aluminum
Central African Republic	241	10	10	1.5	2.2	6	26	160–170	35 (1960)	2,120	N.A.	130	49 diamonds / 24 cotton / 16 coffee
Chad	496	40	35.9	3.7	1.6	7	8	179	35	2,180	.24 (1968)	70 (1968)	77 cotton
Comoro Islands	.84	49	N.A.	.27	3.9	N.A.	N.A.	N.A.	N.A.	N.A.	N.A.	N.A.	vanilla / copra
Congo (Brazzaville)	132	44	39	.94	1.9	7	39	148	37	2,120	.2	220	52 wood / 27 diamonds
Dahomey	43	18	2	2.5	2.1	58	17	149	38	2,170	182 (1968)	70 (1968)	42 palm prod
Equatorial Guinea	11	12	2.8	.28	1.7	27	15	N.A.	N.A.	N.A.	N.A.	290	60 cocoa
Ethiopia	472	67	8	25.3	2.3	54	8	162	35	2,050	1.69	67	57 coffee / 10 hides
Fr. Terr. of Afars & Issas	9	11	N.A.	.13	N.A.	N.A.	N.A.	N.A.	N.A.	N.A.	N.A.	N.A.	N.A.
Gabon	103	20	27	.49	1.2	5	21	184	39	2,200	.16	325	33 oil / 28 wood / 32 manganese
Gambia	4	53	4.1	.36	2.0	83	10	125	43	2,320	.032 (1969)	100	75 peanuts
Ghana	92	12	0.8	9.0	2.7	96	31	122	45	2,130	2.32	262	55 cocoa
Guinea	95	N.A.	N.A.	3.9	3.0	43	11	N.A.	43	2,060	.397 (1969)	104	63 aluminum
Ivory Coast	125	52	9.9	4.2	2.8	34	21	154	39	2,440	1.26 (1969)	308	30 coffee / 27 wood / 20 cocoa
Kenya	225	10	1.2	11.2	3.4	50	10	126	47	2,240	1.54	137	24 coffee / 15 tea / 9 oil
Lesotho	12	94	7.3	1.0	2.8	82	2	137	48	N.A.	.09 (1969)	100	44 wool / 25 livestock
Liberia	43	37	7.0	1.5	3.0	28	26	143	41	2,260	.289 (1969)	196	68 iron ore / 16 rubber

See footnotes at end of table.

TABLE 7-1. DATA ON LAND, PEOPLE AND ECONOMY, BY COUNTRY (continued)

Country	LAND Total sq mi (000)	Agriculture[1] Percent of total	Agriculture[1] Acres per capita	Population 1970 (mill)	Population Growth rate (%)	Population Density per sq mi	Urban (%)	Infant deaths per 1,000 live births	Life expectancy (yrs)	Average daily caloric intake	ECONOMY GNP[2,3] 1970 ($ bill)	ECONOMY GNP[2,3] 1970 per capita ($)	Exports 1967-69 primary (%)
SUB-SAHARAN AFRICA													
Malagasy Republic	227	63	12.5	7.3	2.2	32	13	161	40	2,360	.755 (1969)	106	31 coffee / 9 vanilla
Malawi	45	30	2.0	4.4	2.5	98	5	119	47	2,110	.285	64	25 tobacco / 23 tea
Mali	479	34	20	5.1	2.4	11	12	190	50	2,130	.447 (1969)	90	26 livestock / 26 cotton
Mauritania	398	38	84	1.2	2.2	3	2	137	40	1,980	.170 (1969)	140	86 iron
Mauritius	0.72	70	.3	.84	1.8	1,166	48	70	64	2,300	.158 (1969)	230	91 sugar
Mozambique	302	60	15.4	7.7	1.6	25	5	92	45	2,050	1.2	158	17 cotton / 12 cashew / 11 sugar
Namibia (South-West Africa)	318	65	212	0.75	1.7	2	34	111	N.A.	N.A.	.4 (GDP)	600 (GDP)	4)
Niger	489	11	9.7	3.8	2.9	8	3	148-159	37	2,170	.321 (1969)	90	66 nuts / 9 livestock
Nigeria	357	52	2.1	55.1	2.8	158	23	157	39	2,170	4.83 (1969)	90	32 oil / 20 cocoa / 17 nuts
Portuguese Guinea	14	N.A.	N.A.	.55	0.2	39	N.A.	N.A.	N.A.	N.A.	.10 (1969)	200	66 nuts
Reunion	1	33	0.5	.46	2.3	464	40	59	57	N.A.	N.A.	610	84 sugar
Rhodesia, Southern	150	17	3.2	5.3	3.3	35	22	65	56	2,510	1.30 (1969)	255	33 tobacco
Rwanda	10	71	1.3	3.6	3.0	354	0	120-130	46	1,900	.15 (1969)	40	51 coffee / 28 tin
Senegal	76	58	7.1	3.9	2.4	52	26	156	45	2,300	.71 (1969)	186	68 nuts
Seychelles	0.16	54	N.A.	.05	2.0	N.A.	N.A.	49	N.A.	N.A.	N.A.	N.A.	50 cinnamon/vanilla / 40 copra
Sierra Leone	28	82	5.5	2.6	2.1	95	14	136	N.A.	2,120	.43 (1969)	164	63 diamonds / 9 iron
Somali Republic	246	34	19	2.8	2.2	11	26	N.A.	N.A.	1,770	.175 (1969)	63	54 livestock / 29 bananas
South Africa[5]	471	84	12.6	21.5	2.3	43	48	138	51	2,870	16.83	740	35 gold
Spanish Sahara	103	N.A.	N.A.	0.06	N.A.	N.A.	N.A.	N.A.	N.A.	N.A.	N.A.	N.A.	fish / skins
Swaziland	7	88	8.9	.42	2.9	63	4	168	48	N.A.	N.A.	180	24 sugar / 23 iron

Tanzania	363	60	10.5	13.3	2.7	37	7	165	45	2,140	1.27 (1969)	98	16 cotton / 15 coffee
Togo	22	42	3.1	1.9	2.6	86	16	163	40	2,220	.225 (1969)	124	33 phosphates / 30 cocoa
Uganda	91	42	2.5	9.8	2.5	107	6	124	46	2,160	1.13	116	54 coffee
Upper Volta	106	85	11.3	5. 1	1.8	48	5	181	36	2,040	.25 (1969)	50	45 livestock / 22 cotton
Zaire (Congo-Kinshasa)	906	31	10	17.8	2.4	20	16	115	45	1,920	1.80	101	68 copper
Zambia	291	51	22.1	4.2	3.1	15	22	159	44	2,290	1.67 (1969)	398	94 copper
EAST ASIA													
China, People's Rep. of	3,700	11	0.34	800	2.2	215	15	N.A.	50	2,000	120	145	agr prod/text/light mfg (% N.A.)
China, Rep. of	14	25	0.2	14.6	2.4	1,045	64	28	68	2,650	6.2	410	19 textiles / 11 metals
Hong Kong	.4	13	0.01	3.9	2.4	9,850	92	21	70	2,370	3.4	820	32 clothing / 14 textiles
Japan	143	18	.2	103.5	1.2	724	72	14	72	2,300	196.0	1,898	44 mach / 13 iron/steel / 12 textiles
Korea, North	47	17	N.A.	14.5	2.8	286	34	N.A.	N.A.	N.A.	2.9 (1967)	230 (1967)	min/chem/metals (% N.A.)
Korea, South	38	24	0.2	31.8	2.2	837	38	53	58	2,430	7.8	245	13 woods / 6 fish
Mongolian Rep.	604	0.5	0.9	1.3	3.4	2	N.A.	N.A.	64	2,540	.75	577	wool, hides (% N.A.)
SOUTH ASIA				1970 (000)									
Afghanistan	250	22	2.0	17.4	2.7	70	8	190	N.A.	2,060	1.4 (1969)	85	37 fruits/nuts / 17 karakal
Bangladesh	55.1	64	0.3	74	2.7	1,340	5	135 E	47	2,350 E	7.6	105	jute
Bhutan	18	15	N.A.	0.87	2.5	48	N.A.	N.A.	N.A.	N.A.	N.A.	100	rice/dolomite
Ceylon (Sri Lanka)	25.3	37	0.5	12.8	2.4	505	20	50	62	2,210	2.0	165	61 tea / 19 rubber / 13 coconut
India	1,262	55	.8	547,000	2.6	448	21	118	50	1,940	53.0	96	23 jute / 12 tea / 7 iron
Maldives	.12	N.A.	N.A.	0.11	1.9	N.A.	N.A.	N.A.	N.A.	N.A.	.01	80 E	90 fish
Nepal	54	30	0.9	11.3	2.2	208	5	162	25-40	2,540	0.9	80	70-80 rice/foods
Pakistan	310.4	24	0.8	60	2.7	190	22	135 E	51	2,350 E	9.6	150	cotton

See footnotes at end of table.

TABLE 7-1. DATA ON LAND, PEOPLE AND ECONOMY, BY COUNTRY (continued)

Country	LAND	Agriculture[1]		Population			PEOPLE Urban (%)	Infant deaths per 1,000 live births	Life expectancy (yrs)	Average daily caloric intake	ECONOMY Gross National Product[2]		Exports
	Total sq mi (000)	Percent of total	Acres per capita	1970 (mill)	Growth rate (%)	Density per sq mi					1970 ($ bill)	1970 per capita ($)	1967-69 primary (%)
SOUTHEAST ASIA													
Brunei	2.2	8	1.4	0.1	3.7	54	44	42	—	—	.11 (1969)	950	96 oil
Burma	262	29	1.5	28	2.2	108	19	75	47	1,940 (1966)	2.0	74	50 rice
Indonesia	736	12	.3	120 E	2.6 E	163	17 E	150	40	1,870	12.8 E	108 E	40 oil / 23 rubber
Khmer Republic (Cambodia)	70	16	1.3	7	2.2	98	12	127	44	2,170	.74 (1969)	111	34 rice / 24 rubber
Laos	91	8	1.6	3 E	2.4	32	13	N.A.	35	2,000	.2 (1969)	73	57 tin
Malaysia	128	18	0.8	11	2.9	85	40	60	63	2,190	3.9	360	36 rubber / 20 tin
Philippines	116	41	0.7	38	3.4	332	34	83	55	2,010	8.5	222	24 wood / 17 sugar
Singapore	0.23	21	.02	2	2.0	9,165	100	21	62	2,440	1.7 (1969)	843	22 rubber / 19 oil
Thailand	198	24	0.8	36	3.3	178	15	67	56	2,100 (1965)	6.8	181	27 rice / 14 rubber
Viet-Nam, North	61	14	.3	20	2.0 E	328	15	N.A.	N.A.	1,700	1.6 E (1969)	90	coal
Viet-Nam, South	67	33	0.8	18	2.6	277	24	37	35	2,140	3.1 (1969)	175	80 rubber
OCEANIA	sq mi			(000)									
American Samoa	76	30	N.A.	30	3.2	N.A.	N.A.	28.6	N.A.	N.A.	N.A.	N.A.	95 fish
Australia	3 mill	64	N.A.	12,550	2.1	4.1	84	18.3	68	3,300	31,600	2,542	24 wool / 17 machinery
Brit. Solomon Islands	11,500	6	N.A.	161	2.0	N.A.	N.A.	N.A.	N.A.	N.A.	N.A.	N.A.	68 copra / 27 timber
Cook Islands	93	N.A.	N.A.	20	1.3	N.A.	N.A.	N.A.	N.A.	N.A.	N.A.	N.A.	N.A.
Fiji	7,000	30	N.A.	520	2.7	67	23	24.5	N.A.	N.A.	200	370	50 sugar
French Polynesia	1,543	21	N.A.	98	3.1	N.A.	N.A.	N.A.	N.A.	N.A.	N.A.	N.A.	14 copra / 7 vanilla
Gilbert and Ellice Islands	342	46	N.A.	55	3.6	N.A.	N.A.	22.7	N.A.	N.A.	N.A.	N.A.	N.A.
Guam	209	36	N.A.	85	3.2	N.A.	N.A.	20.3	N.A.	N.A.	220 (1969)	N.A.	N.A.

	(000)		(mill)										
Micronesia * (Pacific Isl.)	712	40	N.A.	102	2.2	N.A.	22	32.6	N.A.	N.A.	.38 (1969)	430	63 copra / 10 fish
Nauru	8	neg	N.A.	7	3.8	722	0	55	N.A.	N.A.	.25	4,500	phosphates
New Caledonia	8,548	28	N.A.	107	2.6	5	55	32.8	N.A.	N.A.	19 (1969)	1,900	98 nickel
New Hebrides	5,700	N.A.	N.A.	84	5.0	N.A.	N.A.	N.A.	N.A.	N.A.	N.A.	N.A.	N.A.
New Zealand	103,736	51	N.A.	2,820	1.4	27	64	16.9	71	3,500	6	2,100	32 meat / 20 wool
Niue Island	100	N.A.	N.A.	6	N.A.	N.A.	N.A.	N.A.	N.A.	N.A.	N.A.	N.A.	N.A.
Papua New Guinea	183,540	18	N.A.	2,420	2.5	5	N.A.	25.8	N.A.	N.A.	55 (1969)	225	26 copra / 25 cocoa
Pitcairn Group	18	N.A.	N.A.	**	3.4	N.A.	N.A.	N.A.	N.A.	N.A.	N.A.	N.A.	N.A.
Tokelau Group	4	N.A.	N.A.	2	N.A.	N.A.	N.A.	N.A.	N.A.	N.A.	N.A.	N.A.	N.A.
Tonga Islands	270	80	N.A.	87	3.0	116	16	9	N.A.	N.A.	N.A.	310	46 bananas / 38 copra
Wallis Futuna	93	N.A.	N.A.	9	2.0	N.A.	N.A.	N.A.	N.A.	N.A.	N.A.	N.A.	N.A.
Western Samoa	1,133	N.A.	N.A.	146	2.4	139	23	25.5	N.A.	N.A.	17 (1967)	250	49 copra / 32 cocoa
WESTERN HEMISPHERE													
North America													
United States	3,600	46	5.0	205	1.1	57	74	21	70	3,200	977	4,754	varied
Canada	3,800	6	5.0	22	1.8	6	74	19.3	70	3,200	81	3,770	varied
Latin America													
Argentina	1,072	64	18.0	24.1	1.4	23	79	58	67	2,920	21.0	871	26 meat / 22 grain
Barbados	0.2	70	0.3	.26	1.3	1,290	45		65	N.A.	0.13 (1969)	523 (1969)	41 sugar / 12 petrol
Bolivia	424	13	8.0	4.7	2.4	11	36	108	50	1,980	0.9	201	52 tin / 13 petrol
Brazil	3,286	16	3.7	91.9	2.7	28	54	92	57	2,690	33.7	366	39 coffee / 7 cotton
British Honduras	9	3	1.3	.12	2.5	14	51	54	60	N.A.	.05 E (1969)	410 E (1969)	40 sugar / 19 citrus
Chile	292	19	3.9	9.3	1.9	33	74	100	61	2,900	6.3	682	76 copper
Colombia	440	17	2.3	21.1	3.2	48	55	80	60	2,200	6.6	313	61 coffee / 9 petrol
Costa Rica	20	30	2.2	1.7	3.2	89	34	87	65	2,610	0.9	509	33 coffee / 26 bananas
Cuba	44	55	N.A.	8.4	2.0	195	60	N.A.	N.A.	N.A.	3.5	409	85 sugar
Dominican Republic	19	40	1.2	4.1	3.0	230	37	80	58	1,900	1.3	332	54 sugar / 11 coffee

Oceania — * Includes whole western Pacific Island area: Mariana District, except Guam, Marshall District, Palau District, Ponape District, Truk District and Yap District.
** Pitcairn population is only 124.
See footnotes at end of table.

TABLE 7-1. DATA ON LAND, PEOPLE AND ECONOMY, BY COUNTRY (continued)

Country	LAND Total sq mi (000)	Agriculture 1 Percent of total	Agriculture 1 Acres per capita	Population 1970 (mill)	Population Growth rate (%)	Population Density per sq mi	Population Urban (%)	Infant deaths per 1,000 live births	Life expectancy (yrs)	Average daily caloric intake	GNP 2 1970 ($ bill)	GNP 2 1970 per capita ($)	Exports 1967-69 primary (%)
Latin America													
Ecuador	109	17	1.9	6.1	34	56	46	87	52	2,020	1.8	294	52 bananas 17 coffee
El Salvador	83	59	0.9	3.4	3.4	410	38	59	58	1,840	1.0	288	45 coffee
Guatemala	42	23	1.2	5.3	2.9	130	34	94	49	2,220	1.7	326	33 coffee 16 cotton
Guyana	83	15	10	.76	3.0	9	30	43	61	2,180	.23 (1969)	317	44 bauxite 32 sugar
Haiti	11	31	0.4	4.9	2.0	490	17	N.A.	48	1,780 E	.42 (1968)	91 (1968)	39 coffee 13 bauxite
Honduras	43	38	3.9	2.7	3.4	63	28	48	49	2,010	0.7	249	46 bananas 11 coffee
Jamaica	4	42	0.6	2.0	1.9	470	36	33	65	2,280	1.1	556	51 bauxite 18 sugar
Mexico	762	52	5.0	50.1	3.3	66	60	66	61	2,570	31.6	630	13 cotton
Nicaragua	50	14	2.3	1.9	3.2	38	45	55	50	2,350	0.7	390	35 cotton 14 coffee
Panama	29	18	2.4	1.4	3.0	50	50	56	66	2,500	1.0	693	56 bananas
Paraguay	157	27	11.0	2.4	3.1	15	36	84	58	2,520 E	0.6	245	29 meat 19 lumber
Peru	496	23	5.0	13.6	3.1	27	49	62	54	2,340	5.4	396	28 copper 26 fish
Surinam	63	1	0.3	0.40	3.5	6	38	48	65	2,350	.24 (1969)	617	88 bauxite & aluminum
Trinidad & Tobago	2	28	0.4	1.0	1.2	530	53	36	66	2,450 E	.77 (1969)	751 (1969)	77 petrol & prou
Uruguay	72	84	13.0	2.9	1.3	40	80	50	69	3,170	2.0	705	42 wool 30 meat
Venezuela	352	24	5.0	10.4	3.5	30	72	46	66	2,490	10.1	974	92 petrol

1) Consists of (a) arable land under permanent crops and (b) permanent meadows and pastures.
2) GNP data in constant 1969 prices; unadjusted for inequalities in purchasing power between countries.
3) Gross Domestic Product (GDP) figures used whenever GNP figures unavailable.
4) Included in data for South Africa.
5) Includes South-West Africa.

TABLE 7-2. TEMPERATURE AND PRECIPITATION DATA FOR REPRESENTATIVE WORLD-WIDE STATIONS

(Source: U.S. Department of Commerce, 1969)

COUNTRY AND STATION	LATITUDE ° '	LONGITUDE ° '	ELEVATION FEET	Length of record YEAR	JAN. Max °F	JAN. Min °F	APR. Max °F	APR. Min °F	JULY Max °F	JULY Min °F	OCT. Max °F	OCT. Min °F	Extreme Max °F	Extreme Min °F	Length of record YEAR	Jan IN.	Feb IN.	Mar IN.	Apr IN.	May IN.	June IN.	July IN.	Aug IN.	Sep IN.	Oct IN.	Nov IN.	Dec IN.	YEAR IN.
United States:																												
Albuquerque, N. Mex.	35 03N	106 37W	5,311	30	46	24	69	42	91	66	71	45	104	-16	30	0.4	0.4	0.5	0.5	0.8	0.6	1.2	1.3	1.0	0.8	0.4	0.5	8.4
Asheville, N. C.	35 26N	82 32W	2,140	30	48	28	67	42	84	61	68	45	99	-7	30	4.2	4.0	4.8	4.0	3.7	3.5	5.9	4.9	3.6	3.1	2.8	3.6	48.1
Atlanta, Ga.	33 39N	84 26W	1,010	30	52	37	70	50	87	71	72	52	103	-9	30	4.4	4.5	5.4	4.5	3.2	3.8	4.7	3.6	3.6	3.3	3.0	4.4	47.2
Austin, Tex.	30 18N	97 42W	597	30	60	41	78	57	95	74	82	60	109	-2	30	2.4	2.6	2.1	3.6	3.7	3.2	2.2	1.9	3.4	2.8	2.1	2.5	32.5
Birmingham, Ala.	33 34N	86 45W	620	30	57	36	76	50	93	71	79	52	107	-10	30	5.0	5.3	6.0	4.5	3.4	4.0	5.2	4.9	3.3	3.0	3.5	5.0	53.1

NORTH AMERICA

COUNTRY AND STATION	LATITUDE ° '	LONGITUDE ° '	ELEVATION FEET	Length of record YEAR	JAN. Max °F	JAN. Min °F	APR. Max °F	APR. Min °F	JULY Max °F	JULY Min °F	OCT. Max °F	OCT. Min °F	Extreme Max °F	Extreme Min °F	Length of record YEAR	Jan IN.	Feb IN.	Mar IN.	Apr IN.	May IN.	June IN.	July IN.	Aug IN.	Sep IN.	Oct IN.	Nov IN.	Dec IN.	YEAR IN.
Bismarck, N. Dak.	46 46N	100 45W	1,647	30	20	0	55	32	86	58	59	34	114	-45	30	0.4	0.4	0.8	1.2	2.0	3.4	2.2	1.7	1.2	0.9	0.6	0.4	15.2
Boise, Idaho	43 34N	116 13W	2,838	30	36	22	63	37	91	59	65	38	112	-28	30	1.3	1.3	1.3	1.2	1.3	0.9	0.2	0.2	0.4	0.8	1.2	1.3	11.4
Brownsville, Tex.	25 54N	97 26W	16	30	71	52	82	66	93	76	85	67	104	12	30	1.4	1.0	1.0	1.6	2.4	2.8	1.7	2.8	5.0	3.5	1.3	1.7	26.9
Buffalo, N.Y.	42 56N	78 44W	705	30	31	18	53	34	80	59	60	41	99	-21	30	2.8	2.7	3.2	3.0	3.0	2.5	2.6	3.1	3.1	3.0	3.6	3.0	35.6
Cheyenne, Wyo.	41 09N	104 49W	6,126	30	37	14	56	30	85	55	63	32	100	-38	30	0.5	0.6	1.2	1.9	2.5	2.1	1.8	1.4	1.1	0.8	0.6	0.5	15.0
Chicago, Ill.	41 47N	87 45W	607	30	33	19	57	41	84	67	63	47	105	-23	30	1.9	1.6	2.7	3.0	3.7	4.1	3.4	3.2	2.7	2.8	2.2	1.9	33.2
Des Moines, Iowa	41 32N	93 39W	938	30	29	11	59	38	87	65	66	43	110	-30	30	1.3	1.1	2.1	2.5	4.1	4.7	3.1	3.7	2.9	2.1	1.8	1.1	30.5
Dodge City, Kans.	37 46N	99 58W	2,582	30	42	20	66	41	93	68	71	46	109	-26	30	0.6	0.7	1.2	1.8	3.2	3.0	2.3	2.4	1.5	1.4	0.6	0.5	19.2
El Paso, Tex.	31 48N	106 24W	3,918	30	56	30	78	49	95	69	79	50	109	-8	30	0.5	0.4	0.4	0.3	0.4	1.3	1.3	1.2	1.1	0.9	0.3	0.5	8.0
Indianapolis, Ind.	39 44N	86 17W	792	30	37	21	61	40	86	64	67	44	107	-25	30	3.1	2.3	3.4	3.7	4.0	4.6	3.5	3.0	3.2	2.6	3.1	2.7	39.2
Jacksonville, Fla	30 25N	81 39W	20	30	67	45	80	58	92	73	80	62	105	10	30	2.5	2.9	3.5	3.6	3.5	6.3	7.7	6.9	7.6	5.2	1.7	2.2	53.6
Kansas City, Mo.	39 07N	94 36W	742	30	40	23	66	46	92	71	72	49	113	-22	30	1.4	1.2	2.5	3.6	4.4	4.6	3.2	3.8	3.3	2.9	1.8	1.5	34.2
Las Vegas, Nev.	36 05N	115 10W	2,162	30	54	32	78	51	104	76	80	53	117	8	30	0.5	0.4	0.4	0.2	0.1	0.1	0.5	0.5	0.3	0.2	0.3	0.4	3.8
Los Angeles, Calif.	33 56N	118 23W	97	30	64	45	67	52	76	62	73	57	110	23	30	2.7	2.9	1.8	1.1	0.1	0.1	*	*	0.3	0.4	1.1	2.4	12.8
Louisville, Ky.	38 11N	85 44W	477	30	44	27	66	43	89	67	70	46	107	-20	30	4.1	3.3	4.6	3.8	3.9	4.0	3.4	3.0	2.6	2.3	3.2	3.2	41.4
Miami, Fla	25 48N	80 16W	7	30	76	58	83	66	89	75	85	71	100	28	30	2.0	1.9	2.3	3.9	6.4	7.4	6.8	7.0	9.5	8.2	2.8	1.7	59.9
Minneapolis, Minn.	44 53N	93 13W	834	30	22	2	56	33	84	61	61	37	108	-34	30	0.7	0.8	1.5	1.9	3.2	4.0	3.3	3.2	2.4	1.6	1.4	0.9	24.9
Missoula, Mont.	46 55N	114 05W	3,190	30	28	10	57	31	85	49	58	30	105	-33	30	0.9	0.9	0.7	1.0	1.9	1.9	0.9	0.7	1.0	1.0	0.9	1.1	12.9
Nashville, Tenn.	36 07N	86 41W	590	30	49	31	71	48	91	70	74	49	107	-15	30	5.5	4.5	5.2	3.7	3.7	3.3	3.7	2.9	1.0	2.3	3.3	4.2	45.2
New Orleans, La.	29 59N	90 15W	3	30	64	45	78	58	91	73	80	61	102	7	30	3.8	4.0	5.3	4.6	4.4	4.4	6.7	5.3	5.0	2.8	3.3	4.1	53.7
New York, N.Y.	40 47N	73 58W	132	30	40	27	60	43	85	68	66	50	106	-15	30	3.3	2.8	4.0	3.4	3.7	3.3	3.7	4.4	3.9	3.1	3.4	3.3	42.3
Oklahoma City, Okla.	35 24N	97 36W	1,285	30	46	28	71	49	93	72	74	52	113	-17	30	1.3	1.4	2.0	3.1	5.2	4.5	2.4	2.5	3.0	2.5	1.6	1.4	30.9
Phoenix, Ariz.	33 26N	112 01W	1,117	30	64	35	84	50	105	75	87	55	118	16	30	0.7	0.9	0.7	0.3	0.1	0.1	0.8	1.1	0.7	0.5	0.5	0.9	7.3
Pittsburgh, Pa.	40 00N	80 00W	747	30	40	25	63	42	86	65	60	45	103	-20	30	2.8	2.3	3.5	3.4	3.8	4.0	3.6	3.5	2.7	2.5	2.3	2.5	36.9
Portland, Maine	43 39N	70 19W	47	30	32	12	53	32	80	57	60	37	103	-39	30	4.4	3.8	4.3	3.7	3.4	3.2	2.9	2.4	3.5	3.2	4.2	3.9	42.9

See footnotes at end of table.

TABLE 7-2. TEMPERATURE AND PRECIPITATION DATA FOR REPRESENTATIVE WORLD - WIDE STATIONS (continued)

COUNTRY AND STATION	LATI-TUDE	LONGI-TUDE	ELE-VATION FEET	TEMP Length of record YEAR	JAN Max °F	JAN Min °F	APR Max °F	APR Min °F	JULY Max °F	JULY Min °F	OCT Max °F	OCT Min °F	Extreme Max °F	Extreme Min °F	PRECIP Length of record YEAR	January IN.	February IN.	March IN.	April IN.	May IN.	June IN.	July IN.	August IN.	September IN.	October IN.	November IN.	December IN.	YEAR IN.
Portland, Oreg.	45 36N	122 36W	21	30	44	33	62	42	79	56	63	45	107	-3	30	5.4	4.2	3.8	2.1	2.0	1.7	0.4	0.7	1.6	3.6	5.3	6.4	37.2
Reno, Nev.	39 30N	119 47W	4,404	30	45	16	65	31	89	46	69	29	106	-19	30	1.2	1.0	0.7	0.5	0.5	0.4	0.3	0.2	0.2	0.5	0.6	1.1	7.2
Salt Lake City, Utah	40 46N	111 58W	4,220	30	37	18	63	36	94	60	65	38	107	-30	30	1.4	1.2	1.6	1.8	1.4	1.0	0.6	0.9	0.5	1.2	1.3	1.2	14.1
San Francisco, Calif.	37 37N	122 23W	8	30	55	42	64	47	72	54	71	51	106	20	30	4.0	3.5	2.7	1.3	0.5	0.1	*	*	0.2	0.7	1.6	4.1	18.7
Sault Ste. Marie, Mich.	46 28N	84 22W	721	30	23	8	46	30	76	54	55	38	98	-37	30	2.1	1.5	1.8	2.2	2.8	3.3	2.5	2.9	3.8	2.8	3.3	2.3	31.3
Seattle, Wash.	47 27N	122 18W	400	30	44	33	58	40	76	54	60	44	100	0	30	5.7	4.2	3.8	2.4	1.7	1.6	0.8	1.0	2.1	4.0	5.4	6.3	39.0
Sheridan, Wyo.	44 46N	106 58W	3,964	30	34	9	56	31	87	56	62	33	106	-41	30	0.6	0.7	1.4	2.2	2.6	2.6	1.2	0.9	1.2	1.1	0.8	0.6	15.9
Spokane, Wash.	47 38N	117 32W	2,356	30	31	19	59	36	86	55	60	38	108	-30	30	2.4	1.9	1.5	0.9	1.5	1.5	0.6	0.4	1.2	1.6	2.4	2.4	17.2
Washington, D.C.	38 51N	77 03W	14	30	44	30	66	46	87	69	68	50	106	-15	30	3.0	2.5	3.2	3.2	4.1	3.2	4.2	4.9	3.8	3.1	2.8	2.8	40.8
Wilmington, N.C.	34 16N	77 55W	28	30	58	37	74	51	89	71	76	55	104	5	30	2.9	3.4	4.0	2.9	3.5	4.3	7.7	6.9	6.3	3.0	3.1	3.4	51.4
United States, Alaska:																												
Anchorage	61 13N	149 52W	85	30	21	4	44	28	65	50	42	28	86	-38	30	0.8	0.7	0.5	0.4	0.5	1.0	1.9	2.6	2.5	1.9	1.0	0.9	14.7
Annette	55 02N	131 34W	110	30	38	30	50	37	63	51	51	42	90	-4	30	11.4	8.5	9.6	9.1	7.1	5.7	6.0	7.5	9.9	16.9	14.7	12.1	118.5
Barrow	71 18N	156 47W	31	30	-9	-23	7	-7	45	33	21	12	78	-56	30	0.2	0.2	0.1	0.1	0.1	0.4	0.8	0.9	0.6	0.5	0.2	0.1	4.3
Bethel	60 47N	161 48W	125	30	11	-4	34	18	62	48	38	25	90	-52	30	1.1	1.1	1.0	0.6	1.0	1.2	2.0	4.2	2.6	1.5	1.1	1.0	18.4
Cold Bay	55 12N	162 43W	96	30	33	23	38	28	54	45	45	36	78	-9	30	2.3	3.2	1.8	1.5	2.3	2.0	1.8	4.3	4.3	4.6	3.8	2.6	34.5
Fairbanks	64 49N	147 52W	436	30	-1	-21	42	17	72	48	35	17	99	-66	30	0.9	0.5	0.4	0.3	0.7	1.4	1.8	2.2	1.1	0.9	0.6	0.5	11.3
Juneau	58 22N	134 35W	12	30	30	20	45	31	63	48	47	37	89	-21	30	4.0	3.1	3.3	2.9	3.2	3.4	4.5	5.0	6.7	8.3	6.1	4.2	54.7
King Salmon	58 41N	156 39W	49	30	21	6	41	25	63	47	43	29	88	-40	30	1.1	1.0	0.9	0.6	0.7	1.4	2.1	3.4	3.1	2.2	1.5	1.0	19.4
Nome	64 30N	165 26W	13	30	12	-3	28	14	55	44	35	24	84	-47	30	1.0	0.9	0.9	0.8	0.7	0.9	2.3	3.8	2.7	1.7	1.2	1.0	17.9
St. Paul Island	57 09N	170 13W	22	30	30	21	33	24	49	42	41	33	64	-26	30	1.8	1.2	1.1	1.0	1.3	1.2	2.3	3.3	3.1	3.2	2.5	1.8	23.8
Shemya	52 43N	174 06E	122	30	34	29	38	33	49	44	42	38	63	16	30	2.5	2.3	2.6	2.1	2.4	1.3	2.2	2.1	2.3	2.8	2.7	2.1	27.4
Yakutat	59 31N	139 40W	28	30	34	20	45	29	61	48	49	35	86	-24	30	10.9	8.2	8.7	7.2	8.0	5.1	8.4	10.9	16.6	19.6	16.1	12.3	132.0
Canada:																												
Aklavik, N.W.T.	68 14N	135 00W	30	22	-10	-26	19	-2	66	47	25	15	93	-62	22	0.5	0.3	0.4	0.5	0.5	0.8	1.4	1.4	0.9	0.9	0.8	0.4	9.0
Alert, N.W.T.	82 31N	62 20W	95	10	-19	-29	-8	-18	44	36	2	-7	67	-53	10	0.2	0.3	0.3	0.3	0.5	0.6	0.5	1.1	1.0	0.9	0.2	0.4	6.3
Calgary, Alta.	51 06N	114 01W	3,540	55	24	2	53	27	76	47	54	29	97	-49	55	0.5	0.5	1.0	1.0	2.3	3.1	2.5	2.3	1.5	0.7	1.4	0.6	16.7
Charlottetown, P.E.I.	46 17N	63 08W	181	65	26	10	43	30	73	58	54	41	97	-27	65	3.8	3.0	3.2	2.8	2.7	2.6	3.4	3.4	3.4	4.1	3.8	4.0	39.8
Chatham, N.B.	47 00N	65 27W	109	50	23	2	47	28	77	56	55	37	102	-43	50	3.4	2.7	3.3	3.0	3.2	3.6	3.9	4.0	3.1	4.0	3.4	3.2	40.8
Churchill, Man.	58 45N	94 04W	94	30	-11	-27	24	4	64	43	34	20	96	-57	30	0.5	0.6	0.9	0.9	0.9	1.9	2.2	2.7	2.3	1.4	1.0	0.7	16.0
Edmonton, Alta.	53 34N	113 31W	2,219	71	16	-3	52	28	74	50	51	30	99	-57	71	0.9	0.7	0.7	1.0	1.9	3.2	3.3	2.4	1.3	0.8	0.9	0.9	18.0
Fort Nelson, B.C.	58 50N	122 35W	1,253	12	1	-15	47	25	74	51	43	25	98	-61	13	0.9	1.2	0.7	0.8	1.4	2.5	2.4	1.5	1.3	1.0	1.4	1.2	16.3
Fort Simpson, N.W.T.	61 45N	121 14W	554	42	-10	-27	38	14	74	50	36	21	97	-70	42	0.7	0.7	0.5	0.7	1.4	1.5	2.0	1.5	1.3	1.1	0.9	0.8	13.1
Frobisher Bay, N.W.T.	63 45N	68 33W	110	18	-9	-23	16	-1	53	39	29	18	76	-49	10	0.7	0.9	0.8	0.8	0.7	0.9	1.5	2.0	1.8	1.1	1.1	1.0	13.3

Station	Lat	Long	Elev																								Ann
Gander, Nfld.	48 57N	54 34W	496	14	27	13	40	27	71	52	51	37	96	-17	14	2.6	3.3	2.8	2.6	2.6	2.8	3.6	3.6	3.7	4.2	3.7	39.6
Halifax, N.S.	44 39N	63 34W	83	75	32	15	47	31	74	55	57	41	99	-21	71	5.4	4.4	4.9	4.1	4.5	4.0	3.8	4.4	4.1	5.3	5.4	55.7
Kapuskasing, Ont.	49 25N	82 28W	743	19	10	-14	43	19	75	50	47	31	101	-53	19	2.0	1.1	1.6	2.1	1.8	2.3	3.4	2.9	3.5	2.4	1.9	27.5
Knob Lake, Que.	54 48N	66 49W	1,712	30	-3	-21	30	12	64	46	37	25	88	-59	10	1.9	1.9	1.4	1.7	1.6	3.3	3.3	4.4	3.4	2.4	1.5	29.7
Montreal, Que.	45 30N	73 34W	187	67	21	6	30	33	78	61	54	40	97	-35	77	3.8	3.0	3.5	3.1	2.6	3.4	3.7	3.5	3.7	3.5	3.6	40.8
North Bay, Ont.	46 21N	79 25W	1,216	17	22	2	48	28	78	56	49	36	99	-46	23	2.0	1.5	1.8	2.5	2.5	3.2	2.7	2.7	3.7	3.2	2.1	30.8
Ottawa, Ont.	45 19N	75 40W	374	65	21	3	51	31	81	58	54	37	102	-38	65	2.9	2.2	2.8	2.5	2.6	3.5	3.4	3.2	3.2	3.0	2.6	34.3
Penticton, B.C.	49 28N	119 36W	1,129	32	32	21	61	35	84	53	59	38	105	-16	32	1.0	0.7	0.7	1.1	1.1	1.2	0.8	1.0	1.0	0.9	1.1	10.8
Port Arthur, Ont.	48 22N	89 19W	644	62	17	-4	44	26	74	52	50	34	104	-42	59	0.9	0.8	1.0	2.1	1.4	2.8	3.6	3.4	3.4	1.5	0.9	23.8
Prince George, B.C.	53 53N	122 41W	2,218	27	23	3	54	27	75	44	52	30	102	-58	27	1.8	1.2	1.4	1.3	1.3	2.1	2.8	2.8	3.4	2.0	1.9	19.9
Prince Rupert, B.C.	54 17N	130 23W	170	26	39	30	50	37	62	49	53	42	90	-3	26	9.8	7.6	8.4	5.3	4.1	4.8	5.1	7.7	12.2	12.3	11.3	95.3
Quebec, Que.	46 48N	71 23W	239	72	18	2	44	29	76	57	51	37	97	-34	72	3.5	2.7	3.0	3.1	3.7	4.0	4.0	3.6	3.6	3.2	3.2	39.8
Regina, Sask.	50 26N	104 40W	1,884	55	10	-11	50	26	79	51	52	27	110	-56	49	0.5	0.3	0.7	1.8	3.3	2.4	1.8	1.3	0.9	0.6	0.4	14.7
Resolute, N.W.T.	74 43N	94 59W	220	13	-20	-33	-1	-16	45	35	11	0	61	-61	7	0.1	0.1	0.2	0.5	0.8	0.9	1.1	0.8	0.5	0.2	0.1	5.5
St. John, N.B.	45 17N	66 04W	119	61	28	11	43	32	69	54	54	41	93	-24	61	4.1	3.1	3.7	3.1	3.2	3.1	3.6	3.7	4.1	3.9	3.8	42.6
St. Johns, Nfld.	47 32N	52 44W	211	68	30	18	41	29	69	51	53	40	93	-21	58	5.3	5.1	4.6	3.8	3.9	3.1	4.0	3.7	4.8	5.7	6.0	53.1
Saskatoon, Sask.	52 08N	106 38W	1,690	38	9	-11	49	26	77	52	51	27	104	-55	38	0.9	0.5	0.7	1.4	1.4	2.6	1.9	1.5	0.9	0.5	0.6	14.6
The Pas, Man.	53 49N	101 15W	890	27	1	-18	45	21	75	54	45	26	100	-54	27	0.6	0.5	0.7	0.8	1.4	2.2	2.2	2.1	2.0	1.2	0.8	15.5
Toronto, Ont.	43 40N	79 24W	379	105	30	16	50	34	79	59	56	40	105	-26	105	2.7	2.4	2.6	2.5	2.9	3.0	2.9	2.7	2.8	2.4	2.6	32.2
Vancouver, B.C.	49 17N	123 05W	127	43	41	32	58	40	74	54	57	44	92	2	41	8.6	5.8	5.0	3.3	2.8	2.5	1.7	1.7	3.6	5.8	8.3	57.4
Whitehorse, Y.T.	60 43N	135 04W	2,303	10	13	-3	41	22	67	45	41	28	91	-62	10	0.6	0.5	0.6	0.4	0.6	1.0	1.0	1.6	1.3	0.7	1.0	10.6
Winnipeg, Man.	49 54N	97 14W	783	66	7	-13	48	27	79	55	51	31	108	-54	66	0.9	0.9	1.2	1.4	2.3	3.1	2.5	2.3	1.5	1.1	0.9	21.2
Yellow Knife, N.W.T.	62 28N	114 27W	674	13	8	-23	29	9	69	52	36	26	90	-60	13	0.8	0.6	0.7	0.4	0.7	1.2	1.4	1.0	1.3	1.0	0.8	10.8
Greenland:																										
Angmagssalik	65 36N	37 33W	95	30	23	10	35	16	54	37	35	25	77	-26	38	2.9	2.4	2.6	2.1	1.8	1.5	2.1	3.3	4.7	3.0	2.7	31.1
Denmarkshaven	76 46N	19 00W	7	2	1	-15	6	-13	47	34	13	2	63	-42	22	1.2	0.7	0.7	0.2	0.5	0.6	0.4	0.5	0.6	1.0	0.7	6.0
Eismitte	70 53N	40 42W	9,843	1	-33	-53	-14	-37	19	1	-23	-42	27	-85	1	0.6	0.2	0.3	0.1	0.1	0.1	0.4	0.3	0.5	1.0	1.0	4.3
Godthaab	64 10N	51 43W	66	40	19	10	31	20	52	38	35	26	76	-20	45	1.4	1.7	1.6	1.2	1.4	2.2	3.1	3.3	2.5	1.9	1.5	23.5
Ivigtut	61 12N	48 10W	98	48	24	12	38	24	57	42	40	29	86	-20	50	3.3	2.6	3.4	3.2	3.1	3.5	3.7	5.7	5.9	4.6	3.1	44.6
Jacobshavn	69 13N	51 10W	104	32	24	-7	24	6	51	40	31	20	71	-46	52	0.4	0.4	0.5	0.6	0.8	1.2	1.3	0.9	1.3	0.7	0.5	9.2
Nord	81 36N	16 40W	118	8	-15	-28	-5	-18	44	35	3	-6	61	-60	8	0.8	0.8	0.5	0.3	0.1	0.3	1.0	1.4	1.2	0.6	0.5	8.9
Scoresbysund	70 29N	21 58W	56	12	12	-3	22	6	49	36	25	15	63	-42	12	0.8	1.4	0.9	0.1	0.4	0.4	1.7	1.4	0.7	1.4	1.9	15.0
Thule	76 31N	68 44W	251	12	-4	-17	10	-7	46	38	19	8	63	-44	12	0.4	0.3	0.2	0.2	0.4	0.7	0.6	0.7	1.1	0.5	0.2	4.9
Upernivik	72 47N	56 07W	59	40	1	-13	15	-1	48	35	29	21	69	-44	50	0.4	0.5	0.6	0.5	0.6	0.9	1.1	1.1	0.6	0.7	0.6	9.2
Mexico:																										
Acapulco	16 50N	99 56W	10	8	85	70	87	71	89	75	88	74	97	60	40	0.3	*	0.0	0.0	*	1.4	12.8	9.1	9.3	13.9	6.7	55.1
Chihuahua	28 42N	105 57W	4,429	9	65	36	81	51	89	66	79	51	102	12	22	0.2	0.2	0.2	0.1	0.2	1.7	3.6	3.7	3.3	1.7	0.9	15.4
Guadalajara	20 41N	103 20W	5,194	26	73	45	85	53	79	60	78	56	101	26	33	0.4	0.4	0.2	0.2	1.1	8.8	9.4	8.5	7.2	2.2	0.8	39.7
Guaymas	27 57N	110 55W	58	9	74	57	84	65	96	82	96	73	117	41	41	0.2	0.1	*	0.1	0.1	1.7	2.2	2.1	1.9	0.7	0.8	9.4
La Paz	24 07N	110 17W	85	9	74	54	86	58	96	73	90	68	108	31	12	0.2	0.1	0.0	0.0	0.0	0.4	1.2	1.4	0.6	0.5	1.1	5.7
Lerdo	25 30N	103 32W	3,740	10	72	45	86	57	90	68	82	58	105	23	14	0.4	0.1	0.1	0.2	0.3	0.8	1.5	2.0	0.8	0.8	0.5	10.2
Manzanillo	19 04N	104 20W	26	17	86	68	87	67	93	76	91	76	103	54	17	0.1	0.2	*	0.0	0.1	4.7	5.7	6.4	14.5	5.1	0.9	39.5
Mazatlan	23 11N	106 25W	256	10	81	61	76	65	86	77	85	71	93	52	46	0.8	0.5	0.2	0.1	0.1	1.5	5.9	8.3	8.0	2.6	0.9	30.2
Merida	20 58N	89 38W	72	22	83	62	92	73	92	73	85	71	106	51	40	1.2	0.8	0.7	0.8	3.2	5.6	5.2	5.6	6.8	3.8	1.3	36.5
Mexico City	19 26N	99 04W	7,340	42	66	42	78	52	74	54	70	50	92	24	48	0.2	0.3	0.5	0.7	1.9	4.1	4.5	4.3	4.1	1.6	0.5	23.0

TABLE 7-2. TEMPERATURE AND PRECIPITATION DATA FOR REPRESENTATIVE WORLD - WIDE STATIONS (continued)

COUNTRY AND STATION	LATITUDE (° ')	LONGITUDE (° ')	ELEVATION (FEET)	TEMP Length of record (YEAR)	JAN Max °F	JAN Min °F	APR Max °F	APR Min °F	JULY Max °F	JULY Min °F	OCT Max °F	OCT Min °F	Extreme Maximum °F	Extreme Minimum °F	PRECIP Length of record (YEAR)	Jan IN.	Feb IN.	Mar IN.	Apr IN.	May IN.	Jun IN.	Jul IN.	Aug IN.	Sep IN.	Oct IN.	Nov IN.	Dec IN.	YEAR IN.
Mexico cont'd:																												
Monterrey	25 40N	100 18W	1,732	11	68	48	84	62	90	71	80	64	107	25	33	0.6	0.7	0.8	1.3	1.3	3.0	2.3	2.4	5.2	3.0	1.5	0.8	22.9
Salina Cruz	16 12N	95 12W	184	10	85	72	88	76	89	76	87	75	98	62	22	*	0.4	0.6	0.5	3.3	11.6	4.5	5.5	7.1	4.0	0.9	0.1	38.5
Tampico	22 16N	97 51W	78	12	75	59	83	69	85	71	85	73	104	34	12	1.5	1.2	1.0	1.5	1.9	8.7	4.1	4.8	10.8	5.0	3.0	1.6	44.9
Vera Cruz	19 12N	96 08W	52	10	77	66	83	72	87	74	85	73	98	53	40	0.9	0.6	0.6	0.8	2.6	10.4	4.1	11.1	13.9	6.9	3.0	1.0	65.7
CENTRAL AMERICA																												
British Honduras:																												
Belize	17 31N	88 11W	17	27	81	67	86	74	87	75	86	72	97	49	33	5.4	2.4	1.5	2.2	4.3	7.7	6.4	6.7	9.6	12.0	8.9	7.3	74.4
Canal Zone:																												
Balboa Heights	08 57N	79 33W	118	34	88	71	90	74	87	74	85	74	97	63	46	1.0	0.4	0.7	2.9	8.0	8.4	7.1	7.9	8.2	10.1	10.2	4.8	69.7
Cristobal	09 21N	79 54W	35	36	84	76	86	77	85	76	86	75	97	66	73	3.4	1.5	1.5	4.1	12.5	13.9	15.6	15.3	12.7	15.8	22.3	11.7	130.3
Costa Rica:																												
San Jose	09 56N	84 08W	3,760	8	75	58	79	62	77	62	77	60	92	49	34	0.6	0.2	0.8	1.8	9.0	9.5	8.3	9.5	12.0	11.8	5.7	1.6	70.8
El Salvador:																												
San Salvador	13 42 N	89 13W	2,238	39	90	60	93	65	89	65	87	65	105	45	39	0.3	0.2	0.4	1.7	7.7	12.9	11.5	11.7	12.1	9.5	1.6	0.4	70.0
Guatemala:																												
Guatemala City	14 37N	90 31W	4,855	6	73	53	82	58	78	60	76	60	90	41	29	0.3	0.1	0.5	1.2	6.0	10.8	8.0	7.8	9.1	6.8	0.9	0.3	51.8
Honduras:																												
Tela	15 46N	87 27W	41	4	82	67	87	72	88	73	86	71	96	58	20	8.9	5.1	2.6	3.3	4.3	5.0	6.4	9.4	7.7	13.5	15.9	14.0	96.1
WEST INDIES																												
Bridgetown, Barbados	13 08N	59 36W	181	35	83	70	86	72	86	74	86	73	95	61	22	2.6	1.1	1.3	1.4	2.3	4.4	5.8	5.8	6.7	7.0	8.1	3.8	50.3
Camp Jacob, Guadaloupe	16 01N	61 42W	1,750	19	77	64	79	65	81	68	81	68	92	54	21	9.2	6.1	8.1	7.3	11.5	14.1	17.6	15.3	16.4	12.4	12.3	10.1	140.4
Ciudad Trujillo, Dom. Rep.	18 29N	69 54W	57	26	84	66	85	69	88	72	87	72	98	59	25	2.4	1.4	1.9	3.9	6.8	6.2	6.4	6.3	7.3	6.0	4.8	2.4	55.8
Fort-de-France, Martinique	14 37N	61 05W	13	22	83	69	86	71	86	74	87	73	96	56	31	4.7	4.3	2.9	3.9	4.7	7.4	9.4	10.3	9.3	9.7	7.9	5.9	80.4
Hamilton, Bermuda	32 17N	64 46W	151	59	68	58	71	59	85	73	79	69	99	40	62	4.4	4.7	4.8	4.1	4.6	4.4	4.5	5.4	5.2	5.8	5.0	4.7	57.6
Havana, Cuba	23 08N	82 21W	80	25	79	65	84	69	89	75	85	73	104	43	72	2.8	1.8	1.8	2.3	4.7	6.5	4.9	5.3	5.9	6.8	3.1	2.3	48.2
Kingston, Jamaica	17 58N	76 48W	110	33	86	67	87	70	90	73	88	73	97	56	59	0.9	0.6	0.9	1.2	4.0	3.5	1.5	3.6	3.9	7.1	2.9	1.4	31.5
La Guerite, St. Christopher (St. Kitts)	17 20N	62 45W	157	19	80	71	83	73	86	76	85	75	91	61	21	4.1	2.0	2.3	2.3	3.8	3.6	4.4	5.2	6.0	5.4	7.3	4.5	50.9
Nassau, Bahamas	25 05N	77 21W	12	35	77	65	81	69	88	75	85	73	94	54	57	1.4	1.5	1.4	2.5	4.6	6.4	5.8	5.3	6.9	6.5	2.8	1.3	46.4
Port-au-Prince, Haiti	18 33N	72 20W	121	72	87	68	90	69	94	74	89	71	101	58	97	1.3	2.3	3.4	6.3	9.1	7.6	2.9	5.7	6.9	6.7	3.4	1.3	53.3
Saint Clair, Trinidad	10 40N	61 31W	67	49	87	69	90	69	88	71	89	71	101	52	52	2.7	1.6	1.8	2.1	3.7	7.6	8.6	9.7	7.6	6.7	7.2	4.9	64.2
Saint Thomas, Virgin Is.	18 20N	64 58W	11	9	82	71	85	74	88	77	87	76	92	63	9	2.5	1.9	1.7	2.2	4.6	3.2	3.2	4.1	6.9	5.6	3.9	3.9	43.7
San Juan, Puerto Rico	18 26N	66 00W	13	30	81	67	84	69	87	74	87	73	94	60	30	4.7	2.9	2.2	3.7	7.1	5.7	6.3	7.1	6.8	5.8	6.5	5.4	64.2

SOUTH AMERICA

Station	Lat	Long	Elev (ft)													Jan	Feb	Mar	Apr	May	Jun	Jul	Aug	Sep	Oct	Nov	Dec	Ann
Argentina:																												
Bahia Blanca	38 43S	62 16W	95	33	88	62	71	51	57	39	71	48	109	18	46	1.7	2.2	2.5	2.3	1.2	0.9	1.0	1.0	1.6	2.2	2.1	1.9	20.6
Buenos Aires	34 35S	58 29W	89	23	85	63	72	53	57	42	69	50	104	22	70	3.1	2.8	4.3	3.5	3.0	2.4	2.2	2.4	3.1	3.4	3.3	3.9	37.4
Cipolletti	38 57S	67 59W	889	9	89	56	72	40	55	29	72	43	107	9	24	0.4	0.4	0.4	0.7	0.6	0.6	0.5	0.5	0.6	0.9	0.5	0.5	6.4
Corrientes	27 28S	58 50W	177	39	93	71	81	63	71	53	82	60	112	30	40	4.7	4.5	5.3	5.6	3.3	1.9	1.7	2.8	4.7	5.2	5.2	5.2	46.4
La Quiaca	22 06S	65 36W	11,345	23	70	41	69	32	60	16	71	32	95	0	25	3.5	2.6	1.8	0.3	*	0.0	*	*	0.1	0.3	1.0	2.7	12.3
Mendoza	32 53S	68 49W	2,625	23	90	60	73	47	59	35	76	50	109	15	46	0.9	1.2	1.1	0.5	0.4	0.3	0.2	0.3	0.5	0.7	0.7	0.7	7.5
Parana	31 44S	60 31W	210	12	91	67	77	58	62	45	75	54	113	21	23	3.1	3.1	3.9	2.6	2.0	1.2	1.2	1.6	2.4	2.8	3.7	4.5	35.0
Puerto Madryn	42 47S	65 01W	26	50	81	57	70	46	55	28	68	45	104	10	50	0.4	0.4	0.7	0.5	0.6	0.9	0.6	0.6	0.4	0.6	0.4	0.6	7.0
Santa Cruz	50 01S	68 32W	39	12	70	48	57	39	41	28	58	39	94	1	20	0.4	0.3	0.3	0.6	0.4	0.5	0.4	0.5	0.3	0.3	0.4	0.7	5.3
Santiago del Estero	27 46S	64 18W	653	28	97	69	82	59	70	44	87	59	116	19	20	3.4	3.0	3.0	1.3	0.6	0.3	0.2	0.5	1.4	2.5	2.5	4.1	20.4
Ushuaia	54 50S	68 20W	26	16	57	41	48	33	39	25	52	35	85	-6	21	2.0	2.6	1.9	2.1	1.5	1.2	1.1	1.3	1.3	1.6	1.5	1.9	19.9
Bolivia:																												
Concepcion	16 15S	62 03W	1,607	5	85	66	86	62	81	54	88	62	101	32	16	7.2	4.7	4.4	2.0	1.5	1.1	0.9	1.2	2.9	5.0	5.0	5.9	38.6
La Paz	16 30S	68 08W	12,001	31	63	43	65	40	62	33	66	40	80	26	50	4.5	4.2	2.6	1.3	0.5	0.3	0.4	0.5	1.1	1.6	1.9	3.7	22.6
Sucre	19 03S	65 17W	9,344	5	63	48	63	45	61	37	65	46	88	25	52	7.3	4.9	3.7	0.2	0.1	0.2	0.3	1.0	1.6	2.6	4.3	7.8	27.8
Brazil:																												
Barra do Corda	05 35S	45 28W	266	9	89	71	89	71	92	64	94	72	103	45	9	6.7	8.7	8.0	6.1	2.3	1.0	0.7	1.0	1.0	2.5	3.9	5.7	47.2
Bela Vista	22 06S	56 22W	525	13	91	67	85	61	77	49	87	61	108	20	20	6.6	4.9	5.0	4.3	5.0	2.8	1.3	1.8	2.9	5.4	5.8	7.0	52.2
Belem	01 27S	48 29W	42	16	87	72	87	73	88	51	89	71	98	61	20	12.5	14.1	14.1	12.6	10.2	6.7	5.9	4.4	3.5	2.9	2.3	6.7	96.0
Brasilia	15 51S	47 56W	3,481	3	80	65	82	62	78	51	82	64	93	46	3	9.0	7.8	7.8	4.8	1.4	0.4	*	0.4	1.9	3.4	*	5.9	54.0
Conceicao do Araguaia	08 15S	49 12W	53	5	88	70	91	68	95	63	93	68	102	55	5	14.9	12.1	10.8	4.1	1.9	0.4	*	0.5	1.5	4.9	6.6	8.6	66.2
Corumba	19 00S	57 39W	381	8	94	73	92	73	84	64	93	70	106	33	11	7.3	5.9	5.1	2.9	1.9	1.2	0.3	1.2	2.6	4.0	5.6	7.1	48.5
Florianopolis	27 35S	48 33W	96	17	83	72	74	64	68	57	73	63	102	32	25	7.6	5.6	6.3	4.1	3.6	3.5	2.2	3.7	4.3	5.1	3.5	4.3	53.1
Goias	15 58S	50 04W	1,706	10	86	63	91	63	89	56	94	63	104	41	11	12.5	9.9	9.9	4.6	0.4	0.3	0.0	0.3	2.3	5.3	9.4	9.5	64.8
Guarapuava	25 16S	51 30W	3,592	10	79	61	73	55	66	47	74	53	94	23	5	8.7	5.8	5.4	4.5	4.6	6.5	3.6	4.6	6.9	6.6	6.6	6.1	65.8
Manaus	03 08S	60 01W	144	11	88	75	87	75	89	75	92	76	101	63	25	9.8	9.1	10.3	8.7	6.7	3.3	2.3	1.8	4.2	4.6	6.1	6.9	71.3
Natal	05 46S	35 12W	52	18	87	76	86	73	82	69	85	75	100	61	18	1.9	4.8	7.0	7.1	9.2	7.0	7.1	3.8	2.6	1.4	0.8	0.7	54.2
Parana	12 26S	48 06W	853	19	90	58	74	64	68	57	75	58	105	37	19	11.3	9.3	11.2	4.0	0.5	0.1	0.2	1.1	4.5	5.0	9.1	12.2	62.3
Porto Alegre	30 02S	51 13W	33	22	87	67	78	60	66	49	74	57	105	25	13	3.5	3.2	3.9	4.1	4.5	5.0	5.2	5.0	5.0	3.4	3.1	3.5	49.1
Quixeramobim	05 12S	39 18W	653	9	88	74	88	74	88	74	93	69	94	63	13	0.7	5.0	7.0	7.0	5.1	1.7	0.7	0.4	0.6	0.4	0.6	0.7	29.6
Recife	08 04S	34 53W	97	27	86	77	85	75	80	71	84	75	94	50	56	2.1	3.3	6.3	8.7	10.5	10.9	10.0	6.0	2.5	1.0	1.0	1.1	63.4
Rio de Janeiro	22 55S	43 12W	201	38	84	73	80	69	75	63	77	66	102	46	84	4.9	4.8	5.1	4.2	3.1	2.1	1.6	1.7	2.6	3.1	4.0	4.1	42.6
Salvador (Bahia)	13 00S	38 30W	154	25	86	74	84	74	79	69	83	71	100	20	20	2.6	5.3	6.1	10.8	9.4	7.2	4.8	3.3	4.0	4.5	5.6	5.6	74.8
Santarem	02 30S	54 42W	66	22	86	73	85	73	87	71	91	73	99	65	22	6.8	10.9	13.2	11.3	6.9	4.1	1.7	1.5	1.9	2.3	4.1	4.1	77.9
Sao Paulo	23 37S	46 39W	2,628	44	77	63	73	59	66	53	68	57	100	32	17	8.8	7.8	6.0	4.1	2.4	2.2	1.5	2.1	3.5	4.6	6.0	9.4	57.3
Sena Madureira	09 04S	68 39W	443	12	92	69	91	68	91	63	93	69	100	41	10	11.2	11.3	10.2	9.4	4.1	2.2	1.5	2.1	4.0	7.0	7.5	11.7	81.2
Uaupes	00 08S	67 05W	272	15	88	72	88	72	85	70	89	71	100	52	10	10.3	7.7	10.0	12.0	10.6	9.2	8.8	7.2	5.1	6.9	7.2	10.4	105.4
Uruguaiana	29 46S	57 07W	246	15	91	69	78	59	66	48	77	55	108	27	12	3.6	3.6	5.1	5.6	3.7	3.2	2.8	3.6	4.1	2.9	4.1	4.1	46.6
Chile:																												
Ancud	41 47S	73 52W	184	30	62	51	57	47	50	42	55	45	82	46	30	3.1	3.7	5.3	7.4	9.9	11.0	10.3	9.4	6.5	4.2	4.7	4.6	80.1
Antofagasta	23 42S	70 24W	308	22	76	63	70	58	63	51	66	55	86	32	37	0.0	0.0	0.0	0.0	0.0	0.1	0.2	0.1	*	0.0	0.0	*	0.5

TABLE 7-2. TEMPERATURE AND PRECIPITATION DATA FOR REPRESENTATIVE WORLD - WIDE STATIONS (continued)

COUNTRY AND STATION	LATI-TUDE ° '	LONGI-TUDE ° '	ELE-VATION FEET	Length of record (YEAR)	JAN. Max °F	JAN. Min °F	APR. Max °F	APR. Min °F	JULY Max °F	JULY Min °F	OCT. Max °F	OCT. Min °F	Extreme Maximum °F	Extreme Minimum °F	Length of record (YEAR)	January IN.	February IN.	March IN.	April IN.	May IN.	June IN.	July IN.	August IN.	September IN.	October IN.	November IN.	December IN.	YEAR IN.
Chile cont'd:																												
Arica	18 28S	70 20W	95	15	78	64	74	60	66	54	69	58	93	39	25	*	0.0	0.0	0.0	0.0	0.0	0.0	*	0.0	0.0	0.0	*	*
Cabo Raper	46 50S	75 38W	131	8	58	46	54	44	47	38	51	40	72	28	10	7.8	5.8	7.1	7.7	7.5	7.9	9.5	7.5	5.6	7.0	6.7	7.0	87.1
Los Evangelistas	52 23S	75 07W	190	16	50	44	48	41	43	36	45	39	66	19	27	11.7	10.0	11.3	11.4	9.6	9.4	9.4	8.6	9.2	8.8	9.9	10.1	119.4
Potrerillos	26 30S	69 27W	9,350	7	65	49	63	47	57	40	61	44	75	20	7	*	0.1	0.3	*	0.7	*	0.5	0.3	0.2	0.2	0.0	*	2.2
Puerto Aisen	42 24S	72 42W	33	8	63	50	55	43	45	37	55	42	93	18	11	7.8	7.8	8.3	7.5	14.7	10.4	11.1	11.1	6.5	7.8	7.0	7.9	107.9
Punta Arenas	53 10S	70 54W	26	15	58	45	50	39	40	31	51	38	86	11	15	1.5	0.9	1.3	1.4	1.3	1.6	1.1	1.2	0.9	1.1	0.7	1.4	14.4
Santiago	33 27S	70 42W	1,706	14	85	53	74	45	59	37	72	45	99	24	58	0.1	0.1	0.2	0.5	2.5	3.3	3.0	2.2	1.2	0.6	0.3	0.2	14.2
Valdivia	39 48S	73 14W	16	29	73	52	62	46	52	41	63	44	97	19	60	2.6	2.9	5.2	9.2	14.2	17.7	15.5	12.9	8.2	5.0	4.9	4.1	102.4
Valparaiso	33 01S	71 38W	135	30	72	56	67	52	60	47	65	50	94	32	41	0.1	*	0.3	0.6	4.1	5.9	3.9	2.9	1.3	0.4	0.2	0.2	19.9
Colombia:																												
Andagoya	05 06N	76 40W	197	8	90	75	90	75	89	74	90	74	97	62	15	25.0	21.4	19.5	26.1	25.5	25.8	23.3	25.3	24.6	22.7	22.4	19.5	281.1
Bogota	04 42N	74 08W	8,355	10	67	48	67	51	64	50	66	50	75	30	49	2.3	2.6	4.0	5.8	4.5	2.4	2.0	2.0	2.4	4.7	2.6	2.6	41.8
Cartagena	10 28N	75 30W	39	6	84	73	87	76	88	78	87	77	98	61	10	0.3	0.0	0.4	0.9	3.4	3.4	3.0	0.6	0.5	10.8	8.9	4.5	36.8
Ipiales	00 50N	77 42W	9,680	9	61	50	63	49	57	42	62	49	77	32	13	3.1	2.3	3.5	3.5	2.8	1.9	1.3	1.1	1.4	3.1	3.3	2.6	29.9
Tumaco	01 49N	78 45W	7	10	82	75	84	76	82	75	82	75	90	64	10	16.9	11.7	9.6	14.6	17.4	12.0	7.7	7.3	7.3	5.9	4.9	7.0	122.3
Ecuador:																												
Cuenca	02 53S	78 39W	8,301	7	69	50	69	50	65	47	70	49	81	29	10	2.0	1.8	3.2	4.3	4.3	1.7	0.9	1.1	1.6	3.1	1.8	2.5	28.3
Guayaquil	02 10S	79 53W	20	5	87	72	88	72	84	67	86	68	98	52	10	8.3	11.4	11.5	8.1	2.1	0.4	0.2	*	*	*	0.1	1.1	43.2
Quito	00 08S	78 29W	9,222	54	67	46	69	47	71	44	71	46	86	25	33	3.9	4.4	5.6	6.9	5.4	1.7	0.8	1.2	2.7	4.4	3.8	3.1	43.9
French Guiana:																												
Cayenne	04 56N	52 27W	20	38	84	74	86	75	88	73	91	74	97	65	51	14.4	12.3	15.8	18.9	21.7	15.5	6.9	2.8	1.2	1.3	4.6	10.7	126.1
Guyana:																												
Georgetown	06 50N	58 12W	6	54	84	71	85	76	85	75	87	76	93	68	35	8.0	4.5	6.9	5.5	11.4	11.9	10.0	6.9	3.2	3.0	6.1	11.3	88.7
Lethem	03 24N	59 38W	270	3	91	73	91	74	87	73	92	76	97	63	9	1.2	1.4	1.3	5.7	11.5	11.9	14.8	9.4	3.4	2.3	4.3	1.3	68.5
Paraguay:																												
Asuncion	25 17S	57 30W	456	15	95	71	84	65	74	53	86	65	110	29	30	5.5	5.1	4.3	5.2	4.6	2.7	2.2	1.5	3.1	5.5	5.9	6.2	51.8
Bahia Negra	20 14S	58 10W	318	20	92	74	87	68	79	61	90	69	106	35	20	5.4	5.3	4.9	2.9	2.3	1.6	1.5	0.6	2.3	4.2	5.3	4.3	40.6
Peru																												
Arequipa	16 21S	71 34W	8,460	13	67	49	67	48	67	47	68	47	82	25	37	1.3	1.8	0.7	0.2	*	*	*	*	0.0	*	*	0.4	4.4
Cajamarca	07 09S	78 30W	8,662	9	71	48	70	47	70	41	71	47	79	25	9	3.6	4.2	4.6	3.4	1.7	0.6	0.2	0.4	2.3	2.3	1.9	3.2	28.2
Cusco	13 33S	71 59W	10,866	13	68	45	71	40	70	31	72	43	86	16	12	6.4	5.9	4.3	2.0	0.6	0.2	0.2	0.4	1.0	2.6	3.0	5.4	32.0
Iquitos	03 45S	73 13W	384	5	90	71	89	71	88	68	90	70	100	54	16	9.1	10.4	9.4	13.6	10.7	5.7	6.4	5.2	10.5	7.3	9.1	10.3	107.7
Lima	12 05S	77 03W	394	15	82	66	80	63	67	57	71	58	93	49	15	0.1	*	*	*	0.2	0.2	0.3	0.3	0.3	0.1	0.1	*	1.6
Mollendo	17 00S	72 07W	80	10	79	66	76	63	67	57	70	59	90	50	10	*	0.1	*	*	0.1	0.1	*	0.2	0.2	0.1	0.1	*	0.9

Climatic data table. Temperatures in °F (mean daily maximum and minimum for January, April, July, October; extreme maximum and minimum), elevation in feet, precipitation in inches. "Yrs" = length of record. (*) denotes a trace or less than the smallest unit shown. Column headings are inferred; the printed headings appear at the beginning of the table.

Temperature (°F) and position

Station	Lat.	Long.	Elev.	Yrs	Jan max	Jan min	Apr max	Apr min	Jul max	Jul min	Oct max	Oct min	Ext max	Ext min
Surinam: Paramaribo	05 49N	55 09W	12	35	85	72	86	73	87	73	91	73	99	62
Uruguay: Artigas	30 24S	56 23W	384	13	91	65	77	55	65	45	75	54	107	24
Montevideo	34 52S	56 12W	72	56	83	62	71	53	58	43	68	49	109	25
Venezuela: Caracas	10 30N	66 56W	3,418	30	75	56	81	60	78	61	79	61	91	45
Ciudad Bolivar	08 07N	63 32W	197	10	90	72	93	75	90	75	93	75	100	64
Maracaibo	10 39N	71 36W	20	12	90	73	92	76	94	76	92	76	102	66
Merida	08 36N	71 10W	5,293	14	73	56	75	60	76	59	75	60	90	48
Santa Elena	04 36N	61 07W	2,976	10	82	61	82	63	81	61	84	61	95	48
PACIFIC ISLANDS														
Easter Is. (Isla de Pascua)	27 10S	109 26W	98	4	77	64	78	63	70	58	73	58	88	46
Mas a Tierra (Juan Fernandez)	33 37S	78 52W	20	25	72	60	68	57	60	50	61	51	86	39
Seymour Is. (Galapagos Is.)	00 28S	90 18W	36	3	86	72	87	75	81	69	81	67	93	58
ATLANTIC ISLANDS														
Fernando de Noronha	03 50S	32 25W	148	32	84	75	82	75	81	73	82	75	93	63
Cumberland Bay, South Georgia	54 16S	36 30W	8	23	48	35	42	29	34	23	41	28	84	-3
Laurie Is., South Orkneys	60 44S	44 44W	13	48	35	29	31	21	20	4	30	19	54	-40
Stanley, Falkland Isles	51 42S	57 51W	6	25	56	42	49	37	40	31	48	35	76	12
EUROPE														
Albania: Durres	41 19N	19 28E	23	10	51	42	63	55	83	74	68	58	95	21
Andorra: Les Escaldes	42 30N	01 31E	3,543	5	43	29	59	39	78	55	61	42	91	0
Austria: Innsbruck	47 16N	11 24E	1,909	34	34	20	60	39	78	55	58	40	97	-16
Vienna	48 15N	16 22E	664	50	34	26	57	41	75	59	55	44	98	-14
Bulgaria: Sofia	42 42N	23 20E	1,805	30	34	22	62	41	82	57	63	42	99	-17
Varna	43 12N	27 55E	115	30	40	30	59	43	84	63	67	50	98	-12
Cyprus: Nicosia	35 09N	33 17E	716	40	58	42	74	50	97	69	81	58	116	23
Czechoslovakia: Prague	50 05N	14 25E	662	40	34	25	55	40	74	58	54	44	98	-16
Prerov	49 27N	17 27E	702	20	34	25	57	38	77	55	56	40	100	-23
Denmark: Copenhagen	55 41N	12 33E	43	30	36	29	50	37	72	55	53	42	91	-3
Aarhus	56 08N	10 12E	161	21	35	27	51	37	70	54	53	42	87	-12

Precipitation (inches)

Station	Yrs	Jan	Feb	Mar	Apr	May	Jun	Jul	Aug	Sep	Oct	Nov	Dec	Year
Paramaribo	75	8.4	6.5	7.9	9.0	12.2	11.9	9.1	6.2	3.1	3.0	4.9	8.8	91.0
Artigas	50	4.3	3.9	4.7	5.1	4.1	4.1	2.8	4.0	3.0	4.7	3.8	4.1	48.6
Montevideo	56	2.9	2.6	3.9	3.9	3.3	3.2	2.9	3.0	3.1	2.6	2.9	3.1	37.4
Caracas	46	0.9	0.4	0.6	1.3	3.1	4.0	4.3	4.2	4.3	4.3	3.7	1.8	32.9
Ciudad Bolivar	10	1.4	0.8	0.7	1.0	3.8	5.5	7.1	6.3	3.6	4.0	2.8	1.3	38.3
Maracaibo	36	0.1	*	0.3	0.8	2.7	2.2	2.2	1.8	2.8	5.9	3.3	0.6	22.7
Merida	14	2.5	1.5	3.6	6.7	9.8	7.3	4.7	5.7	6.7	9.5	8.2	3.4	69.7
Santa Elena	10	3.2	3.2	3.2	5.7	9.6	9.5	9.1	7.6	5.3	4.9	4.9	4.5	70.7
Easter Is.	10	4.8	3.7	4.6	4.2	4.6	4.3	3.5	3.0	2.7	3.7	4.6	4.9	48.6
Mas a Tierra	29	0.8	1.2	1.6	3.4	5.9	6.4	5.8	4.4	2.9	1.9	1.6	1.0	36.9
Seymour Is.	3	0.8	1.4	1.1	0.7	*	*	*	*	*	*	*	*	4.0
Fernando de Noronha	32	1.7	4.7	7.4	10.5	10.5	7.3	5.4	1.9	0.7	0.3	0.4	0.5	51.3
Cumberland Bay	24	3.3	4.3	5.3	5.4	5.2	4.9	5.5	5.3	3.5	2.6	3.4	3.0	51.7
Laurie Is.	46	1.4	1.5	1.9	1.6	1.2	1.0	1.3	1.3	1.1	1.3	1.3	1.0	15.7
Stanley	41	2.8	2.3	2.5	2.6	2.6	2.1	2.0	2.0	1.5	2.0	2.1	2.8	26.8
Durres	10	3.0	3.3	3.9	2.2	4.7	3.1	2.2	1.7	3.1	2.4	2.8	2.9	42.9
Les Escaldes	9	1.5	1.7	2.9	2.4	2.9	2.2	2.6	3.1	2.0	2.6	3.5	2.2	34.3
Innsbruck	35	2.1	1.8	1.5	2.2	2.9	4.1	5.1	4.5	3.1	2.2	2.0	1.8	33.8
Vienna	100	1.5	1.4	1.8	2.0	2.8	2.7	3.0	2.7	2.0	1.9	1.8	1.8	25.6
Sofia	27	1.3	1.1	1.7	2.3	3.3	3.2	2.4	2.0	2.1	1.9	1.2	1.4	25.0
Varna	20	1.5	0.9	1.2	1.2	1.8	2.6	1.9	1.5	1.9	1.9	1.5	2.0	19.6
Nicosia	64	2.9	2.0	1.3	0.8	1.1	0.4	*	0.2	0.9	1.7	1.2	3.0	14.6
Prague	70	0.9	0.8	1.1	1.5	2.4	2.8	2.6	1.7	1.2	1.2	1.2	1.4	19.3
Prerov	21	1.3	1.1	1.1	2.0	2.4	2.9	3.5	2.0	2.4	1.5	1.5	2.0	24.8
Copenhagen	30	1.6	1.3	1.2	1.7	1.7	2.1	2.2	1.9	2.1	2.2	2.2	2.1	23.3
Aarhus	21	2.3	1.5	1.4	1.8	1.2	2.2	2.5	2.6	2.6	2.5	2.5	2.1	26.6

See footnotes at end of table.

TABLE 7-2. TEMPERATURE AND PRECIPITATION DATA FOR REPRESENTATIVE WORLD - WIDE STATIONS (continued)

COUNTRY AND STATION	LATITUDE ° '	LONGITUDE ° '	ELEVATION FEET	TEMPERATURE — Length of record YEAR	JAN. Max °F	JAN. Min °F	APR. Max °F	APR. Min °F	JULY Max °F	JULY Min °F	OCT. Max °F	OCT. Min °F	Extreme Max °F	Extreme Min °F	PRECIP. Length of record YEAR	Jan IN.	Feb IN.	Mar IN.	Apr IN.	May IN.	Jun IN.	Jul IN.	Aug IN.	Sep IN.	Oct IN.	Nov IN.	Dec IN.	Year IN.
Finland:																												
Helsinki	60 10N	24 57E	30	20	27	17	43	31	71	57	45	37	89	-23	20	2.2	1.7	1.7	1.7	1.9	2.0	2.3	3.3	2.8	2.9	2.7	2.4	27.6
Kuusamo	65 57N	29 12E	843	20	17	2	35	18	68	50	36	27	90	-40	20	1.1	1.1	0.8	1.1	1.4	2.3	2.8	3.0	2.1	2.1	1.6	1.1	20.8
Vaasa	63 05N	21 36E	13	18	26	16	41	28	69	55	44	36	89	-29	19	1.1	0.8	0.8	1.0	1.4	1.8	2.4	2.5	2.7	2.3	1.7	1.1	19.6
France:																												
Ajaccio (Corsica)	41 52N	08 35E	243	46	56	40	66	48	85	64	72	55	103	23	86	3.0	2.3	2.6	2.2	1.6	0.9	2.8	0.7	1.7	3.8	4.4	3.1	29.1
Bordeaux	44 50N	00 43W	157	51	48	35	63	44	80	58	66	47	102	9	47	2.7	2.8	2.9	2.6	2.5	2.3	2.0	1.9	2.2	3.0	3.9	3.9	32.7
Brest	48 19N	04 47W	56	56	49	40	57	44	70	56	61	49	95	7	56	3.5	3.0	2.5	2.5	1.9	2.0	2.0	2.2	2.3	3.6	4.2	4.4	34.1
Cherbourg	49 39N	01 38W	30	47	47	40	54	43	67	57	59	50	91	14	47	3.3	2.9	2.7	2.0	1.9	1.8	1.9	3.0	2.9	4.6	5.1	5.2	37.3
Lille	50 35N	03 05W	141	40	42	33	58	40	75	55	59	45	96	0	40	2.5	1.9	2.5	2.0	2.4	2.2	2.8	2.3	2.6	3.0	3.0	3.2	30.3
Lyon	45 42N	04 47E	938	70	41	30	61	42	80	58	61	45	105	-13	70	1.4	1.4	1.8	2.1	2.8	2.9	3.1	2.9	3.1	3.1	2.6	1.9	28.8
Marseille	43 18N	05 23E	246	72	53	38	59	41	78	58	76	57	101	9	72	1.9	1.5	1.5	2.0	2.6	1.0	0.6	0.9	2.0	3.7	3.1	2.2	23.2
Paris	48 49N	02 29E	164	66	42	32	60	41	76	55	59	44	105	1	66	1.5	1.3	1.5	1.7	2.0	2.1	2.1	2.0	2.0	2.2	2.0	1.9	22.3
Strasbourg	48 35N	07 46E	465	20	40	31	59	41	78	57	58	43	101	-8	118	1.6	1.4	1.7	2.6	2.6	3.1	3.4	3.4	3.1	2.7	2.0	1.9	29.5
Toulouse	43 33N	01 23E	538	47	47	35	62	43	82	59	66	48	111	1	20	1.9	1.7	2.3	2.7	2.9	2.4	1.5	2.1	2.3	2.2	2.4	2.3	26.7
Germany:																												
Berlin	52 27N	13 18E	187	50	35	26	55	38	74	55	55	41	96	-15	40	1.9	1.3	1.5	1.7	1.9	2.3	3.1	2.2	1.9	1.7	1.7	1.9	23.1
Bremen	53 05N	08 47E	52	50	37	30	53	38	71	55	54	43	94	-7	80	1.9	1.6	1.8	1.5	2.1	2.6	3.2	2.8	2.1	2.2	2.0	2.2	26.0
Frankfurt A/M	50 07N	08 40E	338	50	37	29	58	41	75	56	56	43	100	-7	80	1.7	1.3	1.6	1.5	2.0	2.5	2.8	2.6	1.9	2.2	2.0	2.0	24.1
Hamburg	53 33N	09 58E	66	50	35	28	51	39	69	56	53	44	92	-4	80	2.1	1.9	2.0	1.8	2.0	2.7	3.4	3.2	3.2	2.6	2.0	2.5	28.9
Munich	48 09N	11 34E	1,739	50	33	23	54	37	72	54	53	40	92	-14	80	1.7	1.4	1.9	2.7	3.7	4.6	4.7	4.2	3.2	2.2	1.9	1.9	34.1
Munster	51 58N	07 38E	207	50	39	29	56	38	73	54	56	42	96	-17	80	2.6	1.9	2.2	2.0	2.2	2.7	3.3	3.1	2.5	2.7	2.4	2.9	30.5
Nurnberg	49 27N	11 03E	1,050	50	35	26	56	38	74	55	55	41	99	-18	40	1.5	1.2	1.3	1.7	2.2	2.5	3.1	3.1	2.1	2.1	1.9	1.7	24.4
Gibraltar:																												
Windmill Hill	36 06N	05 21W	400	12	58	50	64	55	77	66	70	61	97	35	12	4.6	3.4	3.7	2.5	1.4	0.2	*	0.1	0.8	3.5	4.1	5.4	29.7
Greece:																												
Athens	37 58N	23 43E	351	72	54	42	67	52	90	72	74	60	109	20	80	2.2	1.6	1.4	0.8	0.8	0.6	0.2	0.4	0.6	1.7	2.8	2.8	15.8
Iraklion (Crete)	35 20N	25 08E	98	21	60	48	70	54	85	72	77	62	114	32	22	3.7	3.0	1.6	0.9	0.7	0.1	*	0.1	0.7	1.7	2.7	4.0	19.2
Rhodes	36 26N	28 15E	289	10	59	51	61	59	83	74	76	68	104	30	6	5.7	3.9	2.6	1.7	0.5	0.3	0.0	*	0.4	1.7	5.2	6.7	28.5
Thessaloniki	40 37N	22 57E	78	9	49	37	66	49	90	70	73	56	107	15	26	1.5	1.5	1.6	1.9	2.0	1.2	1.0	0.7	1.2	2.4	2.1	1.9	19.0
Hungary:																												
Budapest	47 31N	19 02E	394	50	35	26	62	44	82	61	61	45	103	-10	50	1.5	1.5	1.7	2.0	2.7	2.6	2.0	1.9	1.8	2.1	2.4	2.0	24.2
Debrecen	47 36N	21 39E	430	50	33	21	61	39	81	57	60	41	102	-22	80	1.2	1.1	1.4	1.8	2.4	2.8	2.5	2.3	1.8	2.2	2.0	1.6	23.1
Iceland:																												
Akureyri	65 41N	18 05W	16	23	34	26	40	30	57	47	43	34	83	-8	26	1.7	1.5	1.7	1.3	0.6	0.9	1.3	1.6	1.9	2.3	1.9	1.9	18.6
Reykjavik	64 09N	21 56W	92	25	36	28	43	33	58	48	44	36	74	4	30	4.0	3.1	3.0	2.1	1.6	1.7	2.0	2.6	3.1	3.4	3.6	3.7	33.9

Place	Lat	Long	Elev (ft)	Yrs(T)	Jan max	Jan min	Apr max	Apr min	Jul max	Jul min	Oct max	Oct min	Rec hi	Rec lo	Yrs(P)	Jan	Feb	Mar	Apr	May	Jun	Jul	Aug	Sep	Oct	Nov	Dec	Ann
Ireland:																												
Cork	51 54N	08 29W	56	27	48	38	55	41	68	53	58	44	85	15	35	4.9	3.6	3.3	2.6	2.9	2.0	2.9	3.1	2.9	3.9	4.5	4.7	41.3
Dublin	53 22N	06 21W	155	30	47	35	54	38	67	51	57	43	86	8	35	2.7	2.2	2.0	1.9	2.3	2.0	2.8	3.0	2.8	2.7	2.7	2.6	29.7
Shannon Airport	52 41N	08 55W	8	9	46	36	55	41	66	53	58	45	87	12	12	3.8	3.0	2.0	2.2	2.4	2.1	3.1	3.0	3.1	3.4	4.2	4.3	36.5
Italy:																												
Ancona	43 37N	13 32E	52	30	46	36	62	50	83	68	67	55	102	18	30	2.6	1.7	1.6	2.3	2.1	1.9	0.7	1.5	2.0	3.7	2.5	3.0	28.0
Cagliari (Sardinia)	39 15N	09 03E	3	30	56	43	66	50	86	67	72	62	102	25	25	2.2	1.5	1.5	1.2	1.2	0.5	0.4	0.4	1.0	3.0	1.8	2.3	17.0
Genoa	44 24N	08 55E	318	10	50	41	65	53	82	70	73	58	100	18	10	3.9	4.0	3.3	3.4	4.6	1.4	2.3	2.3	4.7	6.1	7.2	4.1	46.6
Naples	40 51N	14 15E	82	30	54	40	65	52	84	70	71	60	101	24	30	3.7	3.2	3.0	2.6	1.8	0.6	0.7	0.7	2.8	5.1	4.5	5.4	35.2
Palermo (Sicily)	38 07N	13 19E	354	10	58	47	67	53	86	71	75	62	113	31	30	3.8	3.4	2.4	1.9	1.1	0.6	0.6	0.6	1.9	3.7	4.1	4.5	28.3
Rome	41 48N	12 36E	377	10	54	39	68	46	88	64	73	53	104	20	30	3.3	2.0	2.0	2.0	1.9	0.7	0.4	0.7	2.8	4.3	4.4	4.1	29.5
Taranto	40 28N	17 17E	56	10	55	43	59	50	89	70	73	58	108	26	10	1.6	0.9	1.3	0.8	1.0	0.6	0.4	0.7	1.7	2.2	1.8	1.9	14.2
Venice	45 26N	12 23E	82	10	43	33	63	49	82	67	65	52	97	14	30	2.0	2.1	2.4	2.8	3.2	3.3	2.6	2.6	2.8	3.7	3.5	2.6	33.4
Luxembourg:																												
Luxembourg	49 37N	06 03E	1,096	7	36	29	58	40	74	55	56	43	99	-10	100	2.3	2.0	1.9	2.1	2.4	2.5	2.8	2.6	2.4	2.7	2.7	2.8	29.2
Malta:																												
Valletta	35 54N	14 31E	233	90	59	51	66	56	84	72	76	66	105	34	90	3.3	2.3	1.5	0.8	0.4	0.1	*	0.2	1.3	2.7	3.6	3.9	20.3
Monaco:																												
Monaco	43 44N	07 25E	180	60	54	46	61	53	77	70	67	60	93	27	60	2.4	2.3	3.1	2.2	2.1	1.4	0.7	1.1	2.3	4.7	4.3	3.5	30.1
Netherlands:																												
Amsterdam	52 23N	04 55E	5	29	40	34	52	43	69	59	56	48	95	3	29	2.0	1.4	1.3	1.6	1.8	1.8	2.6	2.7	2.8	2.8	2.6	2.2	25.6
Norway:																												
Bergen	60 24N	05 19E	141	49	43	27	55	34	72	51	57	38	89	3	75	7.9	6.0	5.4	4.4	3.9	4.2	5.2	7.3	9.2	9.2	8.0	8.1	78.8
Kristiansand	58 10N	07 59E	175	11	32	25	50	35	71	53	53	39	90	-14	56	5.0	3.6	3.6	2.7	2.5	2.8	3.5	5.3	4.7	6.2	5.7	6.4	52.0
Oslo	59 56N	10 44E	308	44	30	20	50	34	73	56	49	37	93	-21	56	1.7	1.3	1.4	1.6	1.8	2.4	2.9	3.8	2.5	2.9	2.3	2.3	26.9
Tromso	69 39N	18 57E	335	47	27	22	37	27	59	48	40	33	83	-1	75	4.1	3.8	3.3	2.4	2.1	2.1	2.3	2.9	4.7	4.5	4.0	3.9	40.1
Trondheim	63 25N	10 27E	417	44	30	22	45	32	66	51	46	36	95	-22	65	3.1	2.7	2.6	2.0	1.7	1.9	3.0	3.0	3.7	3.7	2.8	2.8	32.1
Vardo	70 22N	31 06E	43	40	27	19	34	26	53	44	38	32	80	-11	56	2.5	2.5	2.3	1.5	1.5	1.3	1.7	1.7	1.9	2.5	2.1	2.4	23.5
Poland:																												
Danzig	54 24N	18 40E	36	36	33	25	49	37	70	56	53	42	94	-16	35	1.2	1.0	1.3	1.8	1.8	2.3	2.8	2.6	2.7	1.8	1.8	1.5	21.7
Krakow	50 04N	19 57E	723	33	32	22	55	38	76	57	54	41	97	-28	35	1.1	1.1	1.4	2.8	4.0	4.0	4.5	3.8	2.8	2.2	2.2	1.3	28.6
Warsaw	52 13N	21 02E	294	25	30	21	55	38	75	56	54	41	98	-22	113	1.2	1.1	1.3	1.5	2.6	2.6	3.0	3.0	1.9	1.7	1.4	1.4	22.0
Wroclaw (Breslau)	51 07N	17 05E	482	50	35	25	55	39	74	57	55	42	98	-26	40	1.5	1.1	1.5	1.7	2.4	2.4	2.7	2.7	1.5	1.7	1.5	1.5	23.2
Portugal:																												
Braganca	41 49N	06 47W	2,395	11	46	31	59	39	80	54	62	42	103	10	11	11.9	6.9	7.7	3.7	3.0	1.6	0.6	0.6	1.5	3.0	6.3	7.1	53.8
Lagos	37 06N	08 38W	46	21	61	47	67	52	83	64	73	58	107	28	17	3.2	2.6	2.8	1.4	0.8	0.2	*	*	0.4	1.5	2.6	2.8	18.3
Lisbon	38 43N	09 08W	313	75	56	46	64	52	79	63	69	57	103	29	75	3.3	3.2	3.1	2.4	1.7	0.7	0.2	0.2	1.4	3.1	4.2	3.6	27.0
Romania:																												
Bucharest	44 25N	26 06E	269	41	33	20	63	41	86	61	65	44	105	-18	41	1.5	1.1	1.7	1.6	2.5	2.3	1.8	2.3	1.5	1.6	1.9	1.5	22.8
Cluj	46 47N	23 40E	1,286	15	31	18	58	38	79	56	60	41	100	-26	16	1.3	1.2	1.0	2.1	3.3	2.6	3.3	3.3	2.0	1.7	1.0	1.2	24.0
Constanta	44 11N	28 39E	13	20	37	25	55	42	79	63	62	49	101	-13	39	1.2	1.2	1.1	1.1	1.3	1.3	1.1	1.3	1.1	1.4	1.2	1.4	15.1
Spain:																												
Almeria	36 51N	02 28W	213	20	61	47	69	54	85	69	76	62	108	34	20	0.9	1.0	0.7	0.9	0.7	0.2	0.1	0.1	0.6	0.9	1.1	1.1	8.6
Barcelona	41 24N	02 09E	312	20	56	42	64	51	81	69	71	58	98	24	30	1.2	2.1	1.9	1.8	1.8	1.3	1.7	2.3	2.6	3.4	2.7	1.8	23.5
Burgos	42 20N	03 42W	2,825	29	42	30	57	38	77	53	61	43	99	0	29	1.5	1.5	2.1	1.9	2.4	1.7	0.7	0.3	1.4	2.0	2.2	2.0	20.2
Madrid	40 25N	03 41W	2,188	30	47	33	64	44	87	62	66	48	102	14	30	1.1	1.7	1.7	1.7	1.5	1.2	0.3	0.1	1.1	1.9	1.9	1.6	16.5
Sevilla	37 29N	05 59W	98	26	59	42	73	51	96	67	78	57	117	27	29	2.2	2.9	3.3	2.3	1.7	0.4	0.1	0.1	2.2	2.6	3.7	2.8	23.3
Valencia	39 28N	00 23W	79	26	58	41	67	51	83	68	73	57	107	20	29	0.9	1.5	0.9	1.2	1.1	1.3	0.5	0.4	2.2	1.6	2.5	1.3	15.4

TABLE 7-2. TEMPERATURE AND PRECIPITATION DATA FOR REPRESENTATIVE WORLD - WIDE STATIONS (continued)

COUNTRY AND STATION	LATITUDE ° '	LONGITUDE ° '	ELEVATION FEET	Length of record YEAR	JAN. Max	JAN. Min	APR. Max	APR. Min	JULY Max	JULY Min	OCT. Max	OCT. Min	Extreme Max °F	Extreme Min °F	Length of record YEAR	Jan IN.	Feb IN.	Mar IN.	Apr IN.	May IN.	Jun IN.	Jul IN.	Aug IN.	Sep IN.	Oct IN.	Nov IN.	Dec IN.	YEAR IN.
Sweden:																												
Abisko	68 21N	18 49E	1,273	11	20	6	33	19	61	45	35	24	82	-30	11	0.7	0.6	0.5	0.5	0.7	1.8	1.8	1.8	1.2	1.0	0.6	0.6	11.7
Goteberg	57 42N	11 58E	55	39	35	27	48	36	69	56	51	42	88	-13	61	2.5	2.0	1.9	1.7	1.9	2.2	2.8	3.7	3.1	3.1	2.7	2.8	30.5
Haparanda	65 50N	24 09E	30	20	22	10	38	23	71	53	39	30	89	-34	20	2.2	1.6	1.2	1.5	1.4	1.7	2.1	2.8	2.6	2.8	2.5	2.0	24.4
Karlstad	59 23N	13 30E	164	30	30	20	49	32	73	56	49	38	93	-21	30	1.9	1.2	1.2	1.4	1.9	1.9	2.6	3.1	2.9	2.4	2.4	1.9	24.8
Sarna	61 41N	13 07E	1,504	20	19	4	42	23	69	46	42	28	91	-51	20	1.6	0.8	0.9	1.2	1.6	2.8	3.6	3.3	2.6	2.3	1.8	1.8	24.3
Stockholm	59 21N	18 04E	146	30	31	23	45	32	70	55	48	39	97	-26	30	1.5	1.1	1.1	1.5	1.6	1.9	2.8	3.1	2.1	2.1	1.9	1.9	22.4
Visby (Gotland)	57 39N	18 18E	36	30	35	28	44	33	67	55	50	41	88	1	30	1.7	1.1	1.2	1.4	1.1	1.4	2.0	2.7	1.7	1.9	2.1	2.0	20.3
Switzerland:																												
Berne	46 57N	07 26E	1,877	30	35	26	56	39	74	56	55	42	96	-9	77	1.9	2.0	2.6	3.0	3.7	4.4	4.4	4.3	3.5	3.5	2.7	2.5	38.5
Geneva	46 12N	06 09E	1,329	30	39	29	58	41	77	58	58	44	101	-1	125	1.9	1.8	2.2	2.5	3.0	3.1	2.9	3.6	3.6	3.8	3.1	2.4	33.9
Zurich	47 23N	08 33E	1,617	23	38	28	57	39	76	55	57	42	98	-12	23	2.3	1.9	2.9	3.4	4.0	4.9	5.0	4.6	3.3	3.2	2.5	2.9	40.9
Turkey:																												
Edirne	41 39N	26 34E	154	18	41	28	66	44	88	63	70	49	107	-8	18	2.2	1.9	1.7	1.9	1.7	2.1	1.5	1.1	1.1	2.1	2.9	3.0	23.2
Istanbul	40 58N	28 50E	59	18	45	36	61	45	81	65	67	54	100	17	18	3.7	2.3	2.6	1.9	1.4	1.3	1.7	1.5	2.3	3.8	4.1	4.9	31.5
United Kingdom:																												
Belfast	54 35N	05 56W	57	7	42	34	53	38	65	52	55	44	82	14	30	4.2	2.8	2.3	2.4	2.3	2.5	3.5	3.5	3.4	3.8	3.6	3.9	38.2
Birmingham	52 29N	01 56W	535	30	42	35	53	40	69	54	55	45	92	11	30	2.9	2.1	1.7	2.2	2.5	1.8	2.8	2.7	2.3	2.9	3.2	2.6	29.7
Cardiff	51 28N	03 10W	203	30	45	36	55	41	69	54	57	45	91	2	30	4.6	3.0	2.3	2.5	3.0	2.2	3.4	3.9	3.6	4.5	4.6	4.3	41.9
Dublin	53 22N	06 21W	155	35	47	35	54	38	67	51	57	43	86	8	35	2.7	2.2	2.0	1.9	2.3	2.0	2.8	3.0	2.8	2.7	2.7	2.6	29.7
Edinburgh	55 55N	03 11W	441	30	43	35	50	39	65	52	53	44	83	15	30	2.5	1.6	1.6	1.6	2.2	1.9	3.1	3.1	2.6	2.9	2.4	2.1	27.6
London	51 29N	00 00	149	30	44	35	56	40	73	55	58	44	99	9	30	2.0	1.5	1.4	1.8	1.8	1.6	2.0	2.2	1.8	2.3	2.5	2.0	22.9
Liverpool	53 24N	03 04W	198	30	44	36	52	41	66	55	55	46	87	15	30	2.7	1.9	1.5	1.6	2.2	2.0	2.6	3.1	2.6	3.0	2.5	2.0	28.9
Perth	56 24N	03 27W	77	30	43	32	53	38	68	51	55	41	89	0	30	3.1	2.2	1.9	1.7	2.3	2.0	3.1	2.9	2.8	3.3	2.7	2.7	30.7
Plymouth	50 21N	04 07W	87	30	47	40	54	43	66	55	58	49	88	16	30	4.3	3.0	2.6	2.3	2.5	2.0	2.6	2.9	2.8	3.8	4.6	4.4	37.8
Wick	58 26N	03 05W	119	30	42	35	48	38	59	50	52	43	80	8	30	2.9	2.1	1.8	2.1	1.8	2.0	2.6	2.6	2.9	3.2	3.1	2.9	30.0
U.S.S.R.:																												
Arkhangelsk	64 33N	40 32E	22	23	9	2	36	23	64	51	36	30	91	-49	25	1.2	1.1	1.1	0.7	1.3	1.9	2.6	2.7	2.2	1.9	1.6	1.3	19.8
Astrakhan	46 21N	48 02E	45	10	23	14	57	40	85	69	56	40	99	-22	25	0.5	0.5	0.4	0.6	0.6	0.7	0.5	0.4	0.6	0.4	0.6	0.6	6.4
Dnepropetrovsk	48 27N	35 04E	259	18	25	16	53	39	80	62	56	40	101	-25	25	1.4	1.3	1.2	1.4	1.8	3.0	1.9	1.6	1.0	1.8	1.6	1.6	19.4
Kaunas	54 54N	23 53E	118	19	26	18	49	34	72	53	50	38	96	-23	19	1.6	1.3	1.3	1.8	2.0	3.3	3.5	3.5	1.9	1.9	1.6	1.6	25.0
Kirov	58 36N	49 41E	594	20	6	-2	41	27	72	55	37	29	92	-43	29	1.2	1.0	0.9	0.9	1.9	2.5	2.1	2.9	2.3	2.0	1.6	1.3	20.6
Kursk	51 45N	36 12E	773	15	19	11	47	35	74	58	48	36	91	-23	20	1.5	1.3	1.2	1.5	2.2	2.5	3.2	2.3	1.6	1.8	1.5	1.7	22.3
Leningrad	59 56N	30 16E	16	23	23	12	45	31	71	57	45	37	91	-36	95	1.0	0.9	0.9	1.0	1.6	2.0	2.5	2.8	2.1	1.8	1.4	1.2	19.2
Lvov	49 50N	24 01E	978	9	31	22	53	38	77	59	55	43	97	-29	35	1.3	1.5	1.8	2.0	2.8	3.7	4.1	3.1	2.4	2.1	0.8	1.6	28.2
Minsk	53 54N	27 33E	738	12	22	13	47	33	70	54	47	36	92	-27	11	1.4	1.5	1.3	1.5	2.0	2.8	3.0	3.1	1.9	1.5	1.5	1.7	22.9
Moscow	55 46N	37 40E	505	15	21	9	47	31	76	55	46	34	96	-27	11	1.5	1.4	1.1	1.9	2.2	2.9	3.0	2.9	1.9	2.7	1.7	1.6	24.8

Station	Lat.	Long.	Elev.	Yrs									Hi	Lo	Yrs	Jan	Feb	Mar	Apr	May	Jun	Jul	Aug	Sep	Oct	Nov	Dec	Ann.
Odessa	46 29N	30 44E	214	20	28	22	52	41	57	47	79	65	99	-13	15	1.0	0.7	0.7	1.1	1.1	1.6	1.9	1.1	1.4	1.4	1.1	1.1	14.3
Riga	56 57N	24 06E	67	30	29	20	48	35	49	39	72	56	93	-20	57	1.3	1.0	1.1	1.2	1.7	3.0	2.4	3.0	3.0	2.0	1.9	1.5	22.2
Saratov	51 32N	46 03E	197	14	15	7	50	35	48	36	82	64	102	-27	15	1.0	0.8	1.1	1.0	1.8	1.8	1.8	1.2	1.3	1.4	1.4	1.2	14.5
Sevastopol	44 37N	33 31E	75	20	39	30	55	42	63	50	79	65	97	-4	30	1.1	1.1	1.1	0.9	0.6	1.1	1.1	0.8	1.3	1.5	1.2	1.1	12.2
Stalingrad	48 42N	44 31E	136	8	15	4	52	36	53	37	84	65	106	-30	12	0.9	1.0	0.6	0.6	1.0	0.9	1.9	0.7	1.0	1.0	1.5	1.3	12.2
Stavropol	45 02N	41 58E	1,886	18	26	17	50	37	55	42	76	60	95	-22	41	1.4	1.1	1.5	2.4	3.0	3.0	4.1	2.5	2.3	2.0	1.8	1.8	26.9
Tallin	59 26N	24 48E	146	15	27	18	42	31	47	38	70	55	89	-19	63	1.1	1.0	0.9	1.1	1.7	2.1	1.9	2.3	2.1	1.9	1.9	1.5	20.2
Tbilisi	41 43N	44 48E	1,325	10	39	26	61	44	64	48	83	65	95	6	10	0.7	0.8	1.3	1.6	3.6	3.1	1.7	1.9	2.0	1.3	1.4	1.2	21.4
Ust'Shchugor	64 16N	57 34E	279	15	4	-14	35	17	33	23	65	49	90	-67	15	1.1	0.8	0.8	1.4	2.2	3.0	3.2	2.4	2.2	1.9	1.3	1.3	20.6
Ufy	54 43N	55 56E	571	20	6	-3	44	30	41	31	75	58	99	-42	23	1.6	1.3	1.2	1.6	2.4	2.6	2.2	1.8	2.3	2.2	1.5	2.3	22.5
Yugoslavia:																												
Belgrade	44 48N	20 28E	453	16	37	27	64	45	65	47	84	61	107	-14	16	1.6	1.3	1.6	2.2	2.6	2.8	1.9	2.5	1.7	2.7	1.8	1.9	24.6
Skopje	41 59N	21 28E	787	10	40	26	67	42	65	43	88	60	105	-11	10	1.5	1.2	1.3	1.5	1.9	1.9	1.3	1.1	1.1	2.6	2.3	1.8	19.5
Split	43 31N	16 26E	420	14	51	29	65	50	69	55	87	68	100	17	51	3.1	2.5	3.2	3.0	2.1	1.2	1.6	2.9	2.9	4.4	4.2	4.4	35.1
OCEAN ISLANDS																												
Bjornoya, Bear Island	74 31N	19 01E	49	10	26	17	27	16	36	29	44	36	71	-25	25	1.6	1.3	1.3	0.9	0.8	0.7	0.8	1.8	1.2	1.7	1.4	1.6	15.1
Gronfjorden, Spitzbergen	78 02N	14 15E	23	19	10	-4	15	-3	25	17	46	38	60	-57	15	1.4	1.1	1.1	0.5	0.5	0.4	0.6	1.0	0.9	1.2	0.9	1.5	11.7
Horta, Azores	38 32N	28 38W	200	30	62	54	76	65	77	62	82	70	88	38	51	4.5	4.2	4.1	3.0	2.9	2.0	1.9	2.5	3.2	4.4	4.1	4.5	40.3
Jan Mayen	71 01N	08 28W	131	5	31	21	31	22	39	29	46	38	60	-18	29	2.1	1.7	1.6	1.4	0.9	0.9	1.4	1.8	2.5	3.0	2.2	2.2	21.2
Lerwick, Shetland Island	60 08N	01 11W	269	30	42	35	46	37	50	42	58	49	71	17	30	4.5	3.4	2.9	2.7	2.2	2.2	2.7	3.7	2.9	4.3	4.5	4.5	40.5
Matochikin Shar, Novaya Zemlya	73 16N	56 24E	61	9	8	-6	13	-1	47	36	64	47	68	-41	9	0.6	0.6	0.6	0.4	0.4	0.4	1.5	1.5	0.6	0.6	0.6	0.4	8.9
Ponta Delgada, Azores	37 45N	25 40W	118	8	62	54	64	55	76	64	82	70	85	37	30	4.0	3.5	3.5	2.3	2.5	1.4	1.2	2.9	3.6	3.7	3.0	3.0	32.6
Stornoway, Hebrides	58 11N	06 21W	34	30	44	37	49	39	61	51	76	64	78	11	15	6.4	3.5	3.2	3.2	3.1	2.4	4.3	6.2	4.7	5.5	6.6	6.3	49.1
Thorshavn, Faeroes	62 02N	06 45W	82	50	42	33	45	36	56	47	58	40	70	8	50	6.6	5.2	4.8	3.6	3.4	3.1	4.7	5.9	4.7	6.3	6.6	5.6	56.2
AFRICA																												
Algeria:																												
Adrar	27 52N	00 17W	948	15	69	39	92	60	115	82			124	25	15	*	*	0.1	*	*	*	*	0.1	0.2	0.2	0.2	*	0.6
Algiers	36 46N	03 03E	194	25	59	49	68	55	83	70			107	32	25	4.4	3.3	2.9	1.6	1.8	0.6	0.1	0.2	1.6	3.1	5.1	5.4	30.0
Bone	36 54N	07 46E	66	26	59	46	67	52	85	69			115	32	25	5.6	4.1	2.9	2.2	1.5	0.6	0.1	0.3	1.2	3.0	4.3	5.2	31.0
El Golea	30 35N	02 53E	1,247	15	63	37	84	56	107	79			120	23	15	0.1	0.3	0.5	0.1	0.3	*	0.1	*	0.3	0.4	0.3	0.3	1.9
Fort Flatters	28 06N	06 42E	1,224	15	67	38	90	59	110	78			124	19	15	0.3	0.1	*	0.2	*	0.0	0.0	*	*	*	0.4	0.2	1.1
Tamanrasset	22 42N	05 31E	4,593	15	67	39	86	56	95	71			102	20	15	0.2	*	0.2	0.2	0.1	0.1	0.4	0.1	0.2	*	0.2	0.2	1.5
Touggourt	33 07N	06 04E	226	26	62	38	83	55	107	77			122	26	26	0.2	0.4	0.5	0.2	0.2	0.2	*	0.1	0.3	0.3	0.5	0.3	2.9
Angola:																												
Cangamba	13 41S	19 52E	4,331	7	84	62	89	58	82	46			109	20	7	8.9	7.4	6.8	1.8	0.1	0.0	0.0	*	0.2	1.8	5.1	8.5	40.6
Luanda	08 49S	13 13E	194	59	83	74	85	75	74	65			98	58	59	1.0	1.4	3.0	4.6	0.5	*	0.1	0.1	0.2	1.1	1.1	0.8	12.7
Mocamedes	15 12S	12 09E	10	21	79	65	82	66	74	61			102	44	21	0.3	0.4	0.7	0.5	*	*	*	*	*	*	0.1	0.1	2.1
Nova Lisboa	12 48S	15 45E	5,577	14	78	58	78	57	77	47			90	36	14	8.7	7.8	9.8	5.7	0.4	0.0	0.0	0.0	0.6	5.5	9.6	8.9	57.0
Botswana:																												
Francistown	21 13S	27 30E	3,294	28	90	61	83	56	75	41			107	24	28	4.2	3.1	2.8	0.7	0.2	0.1	*	*	0.2	0.9	3.4	3.4	17.7
Maun	19 59S	23 25E	3,091	20	90	66	87	58	77	42			110	24	20	4.3	3.8	3.5	1.1	0.2	*	*	0.0	*	0.5	1.9	2.8	18.2
Tsabong	26 03S	22 27E	3,156	10	94	65	83	51	71	34			107	15	14	2.0	1.9	1.9	1.3	0.4	0.2	0.1	0.2	0.7	1.1	1.5		11.5

TABLE 7-2. TEMPERATURE AND PRECIPITATION DATA FOR REPRESENTATIVE WORLD - WIDE STATIONS (continued)

COUNTRY AND STATION	LATITUDE ° '	LONGITUDE ° '	ELEVATION FEET	TEMP Length of record YEAR	JAN Max °F	JAN Min °F	APR Max °F	APR Min °F	JULY Max °F	JULY Min °F	OCT Max °F	OCT Min °F	Extreme Max °F	Extreme Min °F	PRECIP Length of record YEAR	Jan IN.	Feb IN.	Mar IN.	Apr IN.	May IN.	June IN.	July IN.	Aug IN.	Sept IN.	Oct IN.	Nov IN.	Dec IN.	YEAR IN.
Cameroon:																												
Ngaoundere	07 17N	13 19E	3,601	9	87	55	87	64	82	63	82	61	102	46	10	*	*	1.1	5.5	7.0	8.4	10.6	9.6	9.2	5.3	0.5	*	57.2
Yaounde	03 53N	11 32E	2,526	11	85	67	85	66	80	66	81	65	96	57	11	0.9	2.6	5.8	6.7	7.7	6.0	2.9	3.1	8.4	11.6	4.6	0.9	61.2
Central African Republic:																												
Bangui	04 22N	18 34E	1,270	5	90	68	91	71	85	69	87	69	101	57	5	1.0	1.7	5.0	5.3	7.4	4.5	8.9	8.1	5.9	7.9	4.9	0.2	60.8
Ndele	08 24N	20 39E	1,939	3	99	67	98	73	86	69	90	68	109	58	3	0.2	1.3	0.6	1.7	8.4	6.1	8.3	10.1	10.7	7.8	0.6	0.0	55.8
Chad:																												
Am Timan	11 02N	20 17E	1,430	3	98	56	105	68	89	70	96	67	113	43	3	0.0	0.0	0.1	1.2	4.3	5.0	7.3	12.3	5.8	1.2	0.0	0.0	37.2
Fort Lamy	12 07N	15 02E	968	5	93	57	107	74	92	72	97	70	114	47	5	0.0	0.0	0.0	0.1	1.2	2.6	6.7	12.6	4.7	1.4	0.0	0.0	29.3
Largeau (Faya)	18 00N	19 10E	837	5	84	54	104	69	109	76	103	72	121	37	5	0.0	0.0	0.0	0.0	*	0.0	*	0.7	*	0.0	0.0	0.0	0.7
Congo, Democratic Republic of the:																												
Albertville	05 54S	29 12E	2,493	5	85	66	83	67	82	58	87	67	92	50	20	4.2	4.7	6.3	8.4	3.3	0.3	0.1	0.3	0.8	2.8	7.9	6.3	45.4
Leopoldville	04 20S	15 18E	1,066	8	87	70	89	71	81	64	88	70	97	58	12	5.3	5.7	7.7	7.7	6.2	0.3	0.1	0.1	1.2	4.7	8.7	5.6	53.3
Luluabourg	05 54S	22 25E	2,198	3	85	68	86	68	85	63	85	68	94	57	14	5.4	5.6	7.7	7.6	3.3	0.8	0.5	2.3	4.6	6.5	9.1	8.9	62.3
Stanleyville	00 26N	25 14E	1,370	8	86	69	88	70	84	67	86	68	97	61	14	2.1	3.3	7.0	6.2	5.4	4.5	5.2	6.5	7.2	8.6	7.8	3.3	67.1
Congo, Republic of:																												
Brazzaville	04 15S	15 15E	1,043	15	88	69	91	71	82	63	89	70	98	54	18	6.3	4.9	7.4	7.0	4.3	0.6	*	*	2.2	5.4	11.5	8.4	58.0
Oursso	01 37N	16 04E	1,132	4	88	69	91	71	85	69	87	69	106	60	4	2.4	3.6	6.4	3.2	5.8	4.6	2.9	3.7	7.9	10.0	5.7	2.4	58.6
Pointe Noire (Loango)	04 39S	11 48E	164	7	85	73	87	74	78	66	83	72	93	59	7	5.4	6.7	6.4	8.0	3.9	0.0	0.0	0.0	0.4	4.1	6.6	6.6	48.1
Dahomey:																												
Cotonou	06 21N	02 26E	23	5	80	74	83	78	78	74	80	75	95	65	10	1.3	1.3	4.6	4.9	10.0	14.4	3.5	1.5	2.6	5.3	2.3	0.5	52.4
Ethiopia:																												
Addis Ababa	09 20N	38 45E	8,038	15	75	43	77	50	69	50	75	45	94	32	37	0.5	1.5	2.6	3.4	3.4	5.4	11.0	11.8	7.5	0.8	0.6	0.2	48.7
Asmara	15 17N	38 55E	7,628	9	74	44	78	51	71	53	72	53	88	31	17	*	*	0.4	1.5	1.5	1.3	6.7	5.0	1.3	0.3	0.4	*	18.4
Diredawa	09 02N	41 45E	3,937	8	81	58	91	69	90	68	89	67	100	49	8	0.8	0.8	3.3	3.0	2.8	1.5	4.3	3.8	2.2	0.5	0.3	0.8	24.1
Gambela	08 15N	34 35E	1,345	26	98	64	98	71	87	69	92	67	111	48	30	0.2	0.4	1.4	3.2	5.9	6.7	8.5	9.5	7.3	3.5	1.8	0.4	48.8
French Territory of Afars and Issas (F.T.A.I.):																												
Djibouti	11 36N	43 09E	23	16	84	73	90	79	106	87	92	80	117	63	46	0.4	0.5	1.0	0.5	0.2	*	0.1	0.3	0.3	0.4	0.9	0.5	46.0
Gabon:																												
Libreville	00 23N	09 26E	115	11	87	73	89	73	83	68	86	71	99	62	21	9.8	9.3	13.2	13.4	9.6	0.5	0.1	0.7	4.1	13.6	14.7	9.8	98.8
Mayoumba	03 25S	10 38E	200	8	84	73	86	73	78	68	82	72	91	60	8	6.5	9.3	6.2	10.2	2.3	0.1	0.0	0.2	2.6	9.3	10.7	4.6	62.0
Gambia:																												
Bathurst	13 21N	16 40W	90	9	88	59	91	65	86	74	89	72	106	45	9	0.1	0.1	*	*	0.4	2.3	11.1	19.7	12.2	4.3	0.7	0.1	51.0
Ghana:																												
Accra	05 33N	00 12W	88	17	87	73	88	76	81	73	85	74	100	59	65	0.6	1.3	2.2	3.2	5.6	7.0	1.8	0.6	1.4	2.5	1.4	0.9	28.5
Kumasi	06 40N	01 37W	942	10	88	66	89	71	82	70	86	70	100	51	10	0.8	2.3	5.7	5.1	7.5	7.9	4.3	3.1	6.8	7.1	3.7	0.8	55.2

Station	Lat.	Long.	Elev.												Jan	Feb	Mar	Apr	May	Jun	Jul	Aug	Sep	Oct	Nov	Dec	Ann.
Guinea:																											
Conakry	09 31N	13 43W	23	7	88	72	90	73	83	72	87	73	96	63	10	0.1	0.1	0.4	0.9	6.2	22.0	51.4	41.5	26.9	14.6	4.8	0.4	169.0
Kouroussa	10 39N	09 53W	1,217	9	93	60	99	73	87	69	90	69	109	39	10	0.3	0.3	0.9	2.8	5.3	9.7	11.7	13.6	13.4	6.6	1.3	0.4	66.4
Ifni (now in Morocco):																											
Sidi Ifni	29 27N	10 11W	148	14	66	52	71	59	75	64	62		124	40	14	1.0	0.6	0.5	0.6	0.1	*	*	*	0.1	0.4	0.9	1.8	6.1
Ivory Coast:																											
Abidjan	05 19N	04 01W	65	13	88	73	90	75	83	74	85	74	96	59	10	1.6	2.1	3.9	4.9	14.2	19.5	8.4	2.1	2.8	6.6	7.9	3.1	77.1
Bouake	07 42N	05 00W	1,194	12	91	68	92	70	85	68	89	76	104	57	10	0.4	1.5	4.1	5.8	5.3	6.0	3.1	4.6	8.2	5.2	1.5	1.0	46.7
Kenya:																											
Mombasa	04 03S	39 39E	52	45	87	75	86	76	81	74	84	74	96	61	54	1.0	0.7	2.5	7.7	12.6	4.7	3.5	2.5	3.4	3.8	3.8	2.4	47.3
Nairobi	01 16S	36 48E	5,971	15	77	54	75	58	69	55	76	55	87	41	17	1.5	2.5	4.9	8.3	6.2	1.8	0.6	0.9	1.2	2.1	4.3	3.4	37.7
Liberia:																											
Monrovia	06 18N	10 48W	75	6	89	71	90	72	80	72	86	72	97	62	4	0.2	0.1	0.4	4.4	13.4	36.1	24.2	18.6	29.9	25.2	8.2	2.9	174.9
Libya:																											
Benghazi	32 06N	20 04E	82	46	63	50	74	58	84	71	80	66	109	37	46	2.6	1.6	0.8	0.2	*	*	0.0	0.0	0.1	0.7	1.8	2.6	10.5
Cufra	24 12N	23 21E	1,276	7	69	43	90	62	101	75	91	64	122	26	7	1.6	0.0	0.0	0.0	0.0	0.0	0.0	*	*	*	0.0	*	*
Sabhah	27 01N	14 26E	1,457	3	64	41	89	60	102	74	101	64	120	24	10	*	*	0.0	*	0.1	0.1	0.0	*	0.1	0.0	*	*	0.3
Tripoli	32 54N	13 11E	72	47	61	47	72	57	85	71	80	65	114	33	56	3.2	1.8	1.1	0.4	0.2	0.1	*	*	0.4	1.6	2.6	3.7	15.1
Malagasy Republic:																											
Diego Suarez	12 17S	49 17E	100	11	88	75	88	75	84	72	86	65	98	63	31	10.6	9.5	7.6	2.2	0.3	0.2	0.3	0.2	0.3	0.7	1.1	5.8	38.7
Tananarive	18 55S	47 33E	4,500	44	79	61	76	58	68	54	80	54	92	34	62	11.8	11.0	7.0	2.1	0.7	0.3	0.4	0.3	0.7	2.4	5.3	11.3	53.4
Tulear	23 20S	43 41E	20	27	92	72	89	64	81	58	86	65	108	43	15	3.1	3.2	1.4	0.3	0.7	0.4	0.2	0.4	0.3	0.3	1.4	1.7	13.5
Malawi:																											
Karonga	09 57S	33 56E	1,596	8	86	71	85	70	81	66	91	66	99	51	8	7.1	7.0	10.8	6.2	1.7	*	*	*	0.0	0.3	1.1	4.7	38.3
Zomba	15 23S	35 19E	3,141	27	80	65	78	62	72	53	85	64	95	41	29	12.1	9.9	10.1	2.7	0.7	0.3	0.3	0.4	0.2	1.0	4.3	10.9	52.9
Mali:																											
Araouane	18 54N	03 33W	935	8	81	48	110	67	111	79	103	70	130	37	10	*	*	0.0	0.0	*	0.2	0.5	0.1	0.6	0.1	0.1	*	1.7
Bamako	12 39N	07 58W	1,116	11	91	61	103	76	89	71	93	71	117	47	10	*	*	0.1	0.6	2.9	5.4	11.0	13.7	8.1	1.7	0.6	*	44.1
Gao	16 16N	00 03W	902	15	83	58	105	77	97	78	100	78	116	44	19	*	0.0	*	0.1	0.4	1.0	2.9	5.4	1.5	0.2	*	0.0	11.5
Mauritania:																											
Atar	20 31N	13 04W	761	7	84	54	97	67	106	81	98	67	117	39	10	*	0.0	0.0	*	*	0.1	0.1	1.2	1.1	0.1	0.1	*	2.8
Nema	16 36N	07 16W	883	9	86	62	105	79	99	79	101	79	120	47	10	0.1	*	*	*	0.7	1.1	2.3	4.7	2.1	0.7	*	0.1	11.6
Nouakchott	18 07N	15 36W	69	5	85	57	90	64	89	71	91	71	115	44	10	*	0.1	*	*	0.1	0.1	1.1	4.1	0.9	0.4	*	*	6.2
Morocco:																											
Casablanca	33 35N	07 39W	164	48	63	45	69	52	79	58	76	58	110	31	40	2.1	1.9	2.2	1.4	0.9	0.2	*	*	0.3	1.5	2.6	2.8	15.9
Marrakech	31 36N	08 01W	1,509	35	65	40	79	52	101	57	83	57	120	27	31	1.0	1.1	1.3	1.2	0.6	0.3	*	0.1	0.3	0.9	1.2	1.2	9.4
Rabat	34 00N	06 50W	213	35	63	52	71	58	75	58	77	58	118	32	29	2.6	2.5	2.6	1.7	1.1	0.3	*	*	0.4	1.9	3.3	3.4	19.8
Tangier	35 48N	05 49W	239	35	60	47	65	51	80	59	72	59	106	28	35	4.5	4.2	4.8	3.5	1.7	0.6	0.3	*	0.9	3.9	5.8	5.4	35.3
Mozambique:																											
Beira	19 50S	34 51E	28	37	89	75	86	71	77	58	87	64	109	48	39	10.9	8.4	10.1	4.2	2.2	1.3	0.8	1.2	1.5	0.8	5.2	9.2	59.9
Chicoa	15 36S	32 21E	899	8	96	65	93	63	86	57	101	55	117	32	8	7.8	5.7	4.4	0.6	*	*	*	*	*	1.1	2.6	5.2	27.4
Lourenco Marques	25 58S	32 36E	194	42	86	71	83	66	76	55	82	64	114	45	42	5.1	4.9	4.9	2.1	1.1	0.8	0.5	1.1	0.8	1.9	3.2	3.8	29.9
Niger:																											
Agades	16 59N	07 59E	1,706	8	86	50	105	70	104	68	101	68	115	40	10	0.0	0.0	*	*	0.2	0.3	1.9	3.7	0.2	0.0	0.0	0.0	6.8
Bilma	18 41N	12 55E	1,171	9	81	45	101	63	108	75	101	62	116	29	10	0.0	0.0	*	*	*	0.0	0.1	0.5	0.1	0.0	0.0	0.0	0.9
Niamey	13 31N	02 06E	709	10	93	58	108	77	94	74	101	74	114	47	10	*	*	0.3	0.3	1.3	3.2	5.2	7.4	3.7	0.5	0.0	0.0	21.6

See footnotes at end of table.

TABLE 7-2. TEMPERATURE AND PRECIPITATION DATA FOR REPRESENTATIVE WORLD - WIDE STATIONS (continued)

COUNTRY AND STATION	LATITUDE ° '	LONGITUDE ° '	ELEVATION FEET	Temp. Length of record YEAR	JAN. Max °F	JAN. Min °F	APR. Max °F	APR. Min °F	JULY Max °F	JULY Min °F	OCT. Max °F	OCT. Min °F	Extreme Max °F	Extreme Min °F	Precip. Length of record YEAR	Jan IN.	Feb IN.	Mar IN.	Apr IN.	May IN.	June IN.	July IN.	Aug IN.	Sept IN.	Oct IN.	Nov IN.	Dec IN.	YEAR IN.
Nigeria:																												
Enugu	06 27N	07 29E	763	11	90	72	91	74	83	71	87	71	99	55	33	0.7	1.1	2.6	5.9	10.4	11.4	7.6	6.7	12.8	9.8	2.1	0.5	71.5
Kaduna	10 35N	06 26E	2,113	18	89	59	95	72	83	68	89	66	105	46	34	*	0.1	0.5	2.5	5.9	7.1	8.5	11.9	10.6	2.9	0.1	*	50.1
Lagos	06 27N	03 24E	10	32	88	74	89	77	83	74	85	74	104	60	47	1.1	1.8	4.0	5.9	10.6	18.1	11.0	2.5	5.5	8.1	2.7	1.0	72.3
Maiduguri	11 51N	13 05E	1,162	15	90	54	104	72	90	73	96	68	112	43	40	*	*	*	0.3	1.6	2.7	7.1	8.7	4.2	0.7	*	0.0	25.3
Portuguese Guinea:																												
Bolama	11 34N	15 26W	62	31	88	67	91	73	84	74	87	74	106	59	37	*	*	*	*	0.8	7.8	23.1	27.6	16.9	8.0	1.6	0.1	85.9
Rhodesia:																												
Bulawayo	20 09S	28 37E	4,405	15	81	61	79	56	70	45	85	59	99	28	50	5.6	4.3	3.3	0.7	0.4	0.1	*	*	0.2	0.8	3.2	4.8	23.4
Salisbury	17 50S	31 08E	4,831	15	78	60	78	55	70	44	83	58	95	32	50	7.7	7.0	4.6	1.1	0.5	0.1	*	0.1	0.2	1.1	3.8	6.4	32.6
Senegal:																												
Dakar	14 42N	17 29W	131	25	79	64	81	65	88	76	89	76	109	53	26	*	*	*	*	*	0.7	3.5	10.0	5.2	1.5	0.1	0.3	21.3
Kaolack	14 08N	16 04W	20	9	93	60	103	68	91	75	93	74	114	48	10	*	*	*	*	0.3	2.6	6.9	10.7	7.0	2.7	0.1	*	30.3
Sierra Leone:																												
Freetown/Lungi	08 37N	13 12W	92	8	87	73	88	76	82	73	85	72	98	62	8	0.4	0.2	1.2	3.1	9.5	14.3	29.2	36.5	22.3	14.2	5.5	1.2	137.6
Somalia:																												
Berbera	10 26N	45 02E	45	30	84	68	89	77	107	88	92	76	117	58	30	0.3	0.1	0.2	0.5	0.3	*	*	0.1	*	0.1	0.2	0.2	2.0
Mogadiscio	02 02N	45 21E	39	13	86	73	90	78	83	73	86	76	97	59	21	*	*	*	2.3	2.3	3.8	2.5	1.9	1.0	0.9	1.6	0.5	16.9
South Africa, Republic of:																												
Capetown	33 54S	18 32E	56	19	78	60	72	53	63	45	70	52	103	28	18	0.6	0.3	0.7	1.9	3.1	3.3	3.5	2.6	1.7	1.2	0.7	0.4	20.0
Durban	29 50S	31 02E	16	15	81	69	78	64	72	52	75	62	107	39	78	4.3	4.8	5.1	3.0	2.0	1.3	1.1	1.5	2.8	4.3	4.8	4.7	39.7
Kimberley	28 48S	24 46E	3,927	19	91	64	77	52	65	36	83	54	103	20	57	2.4	2.5	3.1	1.5	0.7	0.2	0.2	0.3	0.6	1.0	1.6	2.0	16.1
Port Elizabeth	33 59S	25 36E	190	14	78	61	73	55	67	45	70	54	104	31	84	1.2	1.3	1.9	1.8	0.7	1.8	1.9	2.3	2.3	2.2	2.2	1.7	22.7
Port Nolloth	29 14S	16 52E	23	20	67	53	66	50	62	45	64	49	107	31	64	0.1	0.1	0.2	0.2	0.3	0.3	0.3	0.3	0.2	0.1	0.1	0.1	2.3
Pretoria	25 45S	28 14E	4,491	13	81	60	75	50	66	37	80	55	96	24	12	5.0	4.3	4.5	1.7	0.9	0.6	0.3	0.2	0.8	2.2	5.2	5.2	30.9
Walvis Bay	22 56S	14 30E	24	20	73	59	75	55	70	47	67	51	104	25	20	*	0.2	0.3	0.1	0.1	*	*	0.1	*	*	*	*	0.9
Southwest Africa:																												
Keetmanshoop	26 35S	18 08E	3,295	17	95	65	85	57	70	42	87	55	108	26	45	0.8	1.1	1.4	0.6	0.2	*	*	*	0.1	0.2	0.3	0.4	5.2
Windhoek	22 34S	17 06E	5,669	30	85	63	77	55	68	43	84	59	97	25	60	3.0	2.9	3.1	1.6	0.3	*	*	*	0.1	0.4	0.9	1.9	14.3
Spanish Sahara:																												
Semara	26 46N	11 31W	1,509	6	73	47	88	58	99	66	88	61	121	37	6	0.1	*	0.0	*	*	0.0	0.0	*	1.0	*	0.4	0.0	1.5
Villa Cisneros	23 42N	15 52W	35	12	71	56	74	60	78	65	80	65	107	48	14	*	*	0.0	*	0.1	0.0	*	0.2	1.4	0.1	0.2	1.0	3.0
Sudan:																												
El Fasher	13 38N	25 21E	2,395	17	88	50	102	64	96	70	99	64	113	33	17	*	*	*	0.3	0.3	0.7	4.5	5.3	1.2	0.2	0.0	0.0	12.2
Khartoum	15 37N	32 33E	1,279	46	90	59	105	72	101	77	104	75	118	41	46	*	*	*	0.1	0.1	0.3	2.1	2.8	0.7	0.2	*	0.0	6.2
Port Sudan	19 37N	37 13E	18	30	81	68	89	77	106	83	93	76	117	50	40	0.2	0.1	*	*	*	*	*	0.1	*	0.4	1.7	0.9	3.7
Wadi Halfa	21 55N	31 20E	410	39	75	46	98	62	106	74	98	67	127	28	39	*	*	*	*	*	0.0	*	*	*	*	*	*	*
Wau	07 42N	28 03E	1,443	38	96	64	99	72	89	69	93	69	115	50	38	0.2	0.2	0.9	2.6	5.3	6.5	7.5	8.2	6.6	4.9	0.6	*	43.3

Station	Lat.	Long.	Elev.													Jan	Feb	Mar	Apr	May	Jun	Jul	Aug	Sep	Oct	Nov	Dec	Ann.
Tanzania:																												
Dares Salaam	06 50S	39 18E	47	44	83	77	86	73	83	66	85	69	96	59	49	2.6	2.6	5.1	11.4	7.4	1.3	1.2	1.0	1.2	1.6	2.9	3.6	41.9
Iringa	07 47S	35 42E	5,330	14	76	59	75	59	72	52	80	57	90	42	24	6.8	5.1	7.1	3.5	0.5	*	*	*	0.1	0.2	1.5	4.5	29.3
Kigoma	04 53S	29 38E	2,903	26	80	67	81	67	83	63	84	69	100	53	18	4.8	5.0	5.9	5.1	1.7	0.2	0.7	0.2	0.7	1.9	5.6	5.3	36.5
Togo:																												
Lome	06 10N	01 15E	72	5	85	72	86	74	80	71	83	72	94	58	15	0.6	0.9	1.9	4.6	5.7	8.8	2.8	0.4	1.4	2.4	1.1	0.4	31.0
Tunisia:																												
Gabes	33 53N	10 07E	7	50	61	43	74	54	89	71	81	62	122	27	50	0.9	0.7	0.8	0.4	0.3	*	*	0.1	0.5	1.2	1.2	0.6	6.7
Tunis	36 47N	10 12E	217	50	58	43	70	51	90	68	77	59	118	30	50	2.5	2.0	1.6	1.4	0.7	0.3	0.1	0.3	1.3	2.0	1.9	2.4	16.5
Uganda																												
Kampala	00 20N	32 36E	4,304	15	83	65	79	64	77	62	81	63	97	53	15	1.8	2.4	5.1	6.9	5.8	2.9	1.8	3.4	3.6	3.8	4.8	3.9	46.2
Lira	02 15N	32 54E	3,560	14	91	61	86	64	81	61	86	61	100	50	14	0.7	1.0	3.5	6.9	7.9	4.9	6.4	10.0	8.3	6.1	3.2	1.8	60.7
United Arab Republic::																												
Alexandria	31 12N	29 53E	105	45	65	51	74	59	85	73	83	68	111	37	61	1.9	0.9	0.4	0.1	*	*	*	*	*	0.2	1.3	2.2	7.0
Aswan	24 02N	32 53E	366	46	74	50	96	66	106	79	98	71	124	35	11	*	*	*	*	*	0.0	0.0	0.0	0.0	*	*	*	*
Cairo	29 52N	31 20E	381	42	65	47	83	57	96	70	86	65	117	34	42	0.2	0.2	0.2	0.1	0.1	0.0	0.0	0.0	*	*	0.1	0.2	1.1
Upper Volta:																												
Bobo Dioulasso	11 10N	04 15W	1,411	11	92	58	99	71	87	69	90	70	115	46	10	0.1	0.2	1.1	2.1	4.6	4.8	9.8	12.0	8.5	2.5	0.7	0.0	46.4
Ouagadougou	12 22N	01 31W	991	10	92	60	103	79	91	74	95	74	118	48	15	*	0.1	0.5	0.6	3.3	4.8	8.0	10.9	5.7	1.3	*	0.0	35.2
Zambia:																												
Balovale	13 34S	23 06E	3,577	8	82	65	84	61	81	47	91	64	108	38	9	8.5	6.9	5.8	1.2	*	0.0	0.0	*	0.3	2.3	4.4	8.9	38.3
Kasama	10 12S	31 11E	4,544	10	79	61	79	60	76	50	87	62	95	39	10	10.7	9.9	10.9	2.8	0.5	*	*	*	*	0.8	6.4	9.5	51.5
Lusaka	15 25S	28 19E	4,191	10	78	63	79	59	73	49	88	64	100	39	10	9.1	7.5	5.6	0.7	0.1	*	*	0.0	*	0.4	3.6	5.9	32.9
ATLANTIC ISLANDS:																												
Funchal, Madeira Island	32 38N	16 55W	82	30	66	56	67	58	75	66	74	65	103	40	30	2.5	2.9	3.1	1.3	0.7	0.2	*	*	1.0	3.0	3.5	3.3	21.5
Georgetown, Ascension Island	07 56S	14 25W	55	29	85	73	88	75	84	72	83	71	95	65	45	0.2	0.4	0.7	1.1	0.5	0.5	0.5	0.4	0.3	0.3	0.2	0.1	5.2
Hutts Gate, St. Helena	15 57S	05 40W	2,062	30	68	60	69	61	62	55	61	54	82	50	30	2.1	3.1	4.2	3.1	2.8	3.2	4.3	2.6	2.2	1.7	1.2	1.6	32.1
Las Palmas, Canary Islands	28 11N	15 28W	20	45	70	58	71	61	77	67	79	67	99	46	48	1.4	0.9	0.5	*	0.2	*	*	0.4	0.2	1.1	2.1	1.6	8.6
Porto da Praia, Cape Verde Is.	14 54N	23 31W	112	25	77	68	79	69	83	75	85	76	94	56	25	0.1	*	*	*	0.0	0.2	3.8	4.5	1.2	0.3	0.3	0.1	10.2
Santa Isabel, Fernando Po	03 46N	08 46E	—	2	87	67	89	70	84	69	86	70	102	61	16	*	2.5	4.2	7.2	9.4	11.1	7.4	6.6	9.6	10.4	3.5	1.7	74.9
Sao Tome, Sao Tome	00 20N	06 43E	16	10	86	73	86	73	84	71	84	71	91	56	10	3.2	4.2	5.9	5.0	5.3	1.1	1.1	*	0.9	4.3	4.6	3.5	38.0
Tristan da Cunha	37 03S	12 19W	75	5	66	59	64	57	57	50	59	51	75	38	5	3.5	3.5	6.4	4.7	7.1	5.9	6.1	6.9	7.9	5.8	4.3	4.0	66.1
INDIAN OCEAN ISLANDS:																												
Agalega Island	10 26S	56 40E	10	3	86	77	87	77	83	75	84	75	91	69	2	5.9	10.1	4.9	6.9	13.2	8.9	8.7	3.2	1.8	4.2	7.0	10.0	84.7
Cocos (Keeling) Island	12 05S	96 53E	15	36	86	77	85	78	82	76	84	76	94	68	38	5.4	7.7	8.5	10.4	7.9	9.0	8.7	4.8	3.7	3.3	4.2	4.6	78.2
Heard Island	53 01S	73 23E	16	5	41	35	39	33	34	27	35	28	58	13	5	5.8	5.8	5.7	6.1	5.8	3.9	3.6	2.2	2.5	3.7	4.0	5.1	54.3
Hellburg, Reunion Island	21 04S	55 22E	3,070	11	74	59	73	56	65	48	69	51	84	40	11	22.4	8.0	7.2	7.2	16.4	4.1	3.1	3.0	2.0	2.3	3.5	12.9	90.5
Port Victoria, Seychelles	04 37S	55 27E	15	60	83	76	86	77	81	75	83	75	92	67	64	15.2	10.5	9.2	9.2	9.2	9.2	3.3	2.7	5.1	6.1	9.1	13.4	92.5
Royal Alfred Observatory, Mauritius	20 06S	57 32E	181	40	86	73	82	70	75	62	80	64	95	50	43	8.5	7.8	8.7	5.0	3.8	2.6	2.3	2.5	1.4	1.6	1.8	4.6	50.6

TABLE 7-2. TEMPERATURE AND PRECIPITATION DATA FOR REPRESENTATIVE WORLD - WIDE STATIONS (continued)

ASIA – FAR EAST

COUNTRY AND STATION	LATITUDE ° '	LONGITUDE ° '	ELEVATION FEET	TEMP Length of record YEAR	JAN. Max °F	JAN. Min °F	APR. Max °F	APR. Min °F	JULY Max °F	JULY Min °F	OCT. Max °F	OCT. Min °F	Extreme Max °F	Extreme Min °F	PRECIP Length of record YEAR	Jan IN	Feb IN	Mar IN	Apr IN	May IN	Jun IN	Jul IN	Aug IN	Sep IN	Oct IN	Nov IN	Dec IN	YEAR IN
China																												
Canton	23 10N	113 20E	59	26	65	49	77	65	91	77	85	67	101	31	36	0.9	1.9	4.2	6.8	10.6	10.6	8.1	8.5	6.5	3.4	1.2	0.9	63.6
Chanasha	28 15N	112 58E	161	14	45	35	70	56	94	78	75	59	109	16	26	1.9	3.7	5.3	5.7	8.2	8.7	4.4	4.3	2.7	3.0	2.7	1.5	52.1
Chungking	29 30N	106 33E	855	27	51	42	73	59	93	76	71	61	111	28	60	0.7	0.8	1.5	3.8	5.7	7.1	5.6	4.7	5.8	4.3	1.9	0.8	42.9
Hankow	30 35N	114 17E	75	29	46	34	69	55	93	78	74	60	108	9	55	1.8	1.9	3.6	5.8	7.0	9.0	7.0	4.1	3.0	3.1	1.9	1.2	49.4
Harbin	45 45N	126 38E	476	35	7	-14	54	31	84	65	54	31	102	-43	38	0.2	0.2	0.4	0.9	1.7	3.7	6.6	4.7	2.3	1.2	0.5	0.2	22.6
Kashgar	39 24N	76 07E	4,296	27	33	12	71	48	92	68	71	43	106	-15	18	0.6	0.1	0.5	0.2	0.3	0.2	0.4	0.3	0.1	0.1	0.2	0.3	3.2
Kunming	25 02N	102 43E	6,211	32	61	37	76	51	77	62	70	53	91	22	31	0.4	0.5	0.7	0.8	4.3	6.3	8.8	8.6	5.0	3.0	1.7	0.4	40.5
Lanchow	36 06N	103 55E	5,105	8	33	7	65	40	84	61	62	39	100	-3	4	0.2	0.2	0.2	0.5	0.7	0.8	3.3	5.1	2.2	3.0	0.0	0.3	14.1
Mukden	41 47N	123 24E	138	40	20	-2	60	36	87	69	62	39	103	-28	42	0.2	0.2	0.7	1.2	2.6	3.8	7.0	6.3	2.9	1.7	0.9	0.4	28.2
Shanghai	31 12N	121 26E	16	56	47	32	67	49	91	75	75	56	104	10	81	1.9	2.4	3.3	3.6	3.8	7.0	5.8	5.5	5.2	2.9	2.1	1.5	45.0
Tientsin	39 10N	117 10E	13	24	33	16	68	45	90	73	68	48	109	-3	25	0.2	0.1	0.4	0.5	1.1	2.4	7.6	6.0	1.7	0.6	0.4	0.2	21.0
Urumchi	43 45N	87 40E	2,972	6	13	-7	60	36	82	58	50	31	112	-30	6	0.6	0.3	0.5	1.5	1.1	1.5	0.7	1.0	0.6	1.7	1.6	0.4	11.5
Hong Kong:	22 18N	114 10E	109	50	64	56	75	67	87	78	81	73	97	32	50	1.3	1.8	2.9	5.4	11.5	15.5	15.0	14.2	10.1	4.5	1.7	1.2	85.1
Japan:																												
Kushiro	43 02N	144 12E	315	41	30	8	44	31	66	55	58	40	87	-19	41	1.8	1.4	2.8	3.6	3.8	4.1	4.4	4.9	6.6	4.0	3.1	2.0	42.9
Miyako	39 38N	141 59E	98	30	43	23	58	37	77	62	66	46	99	1	30	2.9	3.0	3.2	3.5	4.5	5.0	5.0	7.2	9.5	6.8	3.0	2.6	56.2
Nagasaki	32 44N	129 53E	436	59	49	36	66	50	85	73	72	58	98	22	59	2.8	3.3	4.9	7.3	6.7	12.3	10.1	6.9	9.8	4.5	3.7	3.2	75.5
Osaka	34 47N	135 26E	49	60	47	32	65	47	87	73	65	49	102	19	60	1.7	2.3	3.8	5.2	4.9	7.4	5.9	4.4	7.0	5.1	3.0	1.9	52.6
Tokyo	35 41N	139 46E	19	60	47	29	63	46	83	70	69	55	101	17	60	1.9	2.9	4.2	5.3	5.8	6.5	5.6	6.0	9.2	8.2	3.8	2.2	61.6
Korea:																												
Pusan	35 10N	129 07E	6	36	43	29	62	47	81	71	70	54	97	7	36	1.7	1.4	2.7	5.5	5.2	7.9	11.6	5.1	6.8	2.9	1.6	1.2	53.6
Pyongyang	39 01N	125 49E	94	43	27	8	61	38	84	69	65	43	100	-19	43	0.6	0.4	1.0	1.8	2.6	5.0	9.3	9.0	4.4	1.8	1.6	0.8	36.4
Seoul	37 31N	126 55E	34	22	32	15	62	41	84	70	67	45	99	-12	22	1.2	0.8	1.5	3.0	3.2	5.1	14.8	10.5	4.7	1.6	1.8	1.0	49.2
Mongolia:																												
Ulan Bator	47 54N	106 56E	4,287	13	-2	-27	45	18	71	50	44	17	97	-48	15	*	*	0.1	0.2	0.3	1.0	2.9	1.9	0.8	0.2	0.2	0.1	7.7
Taiwan:																												
Tainan	22 57N	120 12E	53	13	72	55	82	67	89	77	86	70	95	39	13	0.7	0.7	1.1	3.2	6.3	15.6	16.0	15.8	8.4	1.2	0.9	0.6	70.5
Taipei	25 04N	121 32E	21	12	66	53	77	64	92	76	80	68	101	32	12	3.8	5.3	4.3	5.3	6.9	8.8	8.8	8.7	8.2	5.5	4.2	2.9	72.7
Union of Soviet Socialist Republics:																												
Alma-Ata	43 16N	76 53E	2,543	19	23	7	56	38	81	60	55	35	100	-30	27	1.3	0.9	2.2	4.0	3.7	2.6	1.4	1.2	1.0	2.0	1.9	1.3	23.5
Chita	52 02N	113 30E	2,218	10	-10	-27	42	19	75	51	38	18	99	-52	24	0.1	0.1	0.1	0.4	1.1	1.8	3.3	3.3	1.2	0.5	0.2	0.2	12.3
Dubinka	69 07N	87 00E	141	5	-23	-31	6	-10	59	47	19	11	84	-62	5	0.3	0.4	0.3	0.3	0.6	1.9	1.5	2.1	1.8	0.9	0.4	0.3	10.7
Irkutsk	52 16N	104 19E	1,532	10	3	-15	42	20	70	50	41	21	98	-58	38	0.5	0.4	0.3	0.6	1.3	2.2	3.1	2.8	1.7	0.7	0.6	0.6	14.9
Kazalinsk	45 46N	62 06E	207	10	16	5	58	27	90	65	57	35	108	-27	19	0.4	0.4	0.5	0.5	0.6	0.2	0.2	0.3	0.3	0.4	0.5	0.6	4.9

Khabarovsk	48 28N	135 03E	165	7	-2	-13	41	28	75	63	48	34	91	-46	8	0.3	0.2	0.3	0.7	2.0	3.5	4.1	3.3	3.0	0.7	0.6	0.5	19.2
Kirensk	57 47N	108 07E	938	18	-14	-28	38	15	74	51	10	-4	95	-71	19	0.8	0.5	0.5	0.5	1.0	1.8	2.1	2.1	1.7	1.0	1.0	1.0	14.0
Krasnoyarsk	56 01N	92 52E	498	10	3	-10	34	23	67	55	34	26	103	-47	8	0.1	0.2	0.1	0.2	1.0	1.4	1.2	2.1	1.7	0.9	0.5	0.4	9.8
Markovo	64 45N	170 50E	85	15	-19	-29	5	-8	59	47	16	9	84	-72	16	0.2	0.2	0.3	0.1	0.3	0.8	1.0	1.9	1.1	0.4	0.4	0.3	7.0
Narym	58 50N	81 39E	197	13	-7	-18	35	19	71	56	35	25	94	-61	14	0.8	0.5	0.8	0.5	1.3	2.6	2.4	2.7	1.7	1.4	1.1	0.9	16.8
Okhotsk	59 21N	143 17E	18	19	-6	-17	29	10	57	48	33	21	78	-50	25	0.1	0.1	0.2	0.4	0.9	1.6	2.2	2.6	2.4	1.0	0.2	0.1	11.8
Omsk	54 58N	73 20E	279	19	-1	-14	39	21	74	56	40	27	102	-56	22	0.6	0.3	0.3	0.5	1.2	2.0	2.0	2.0	1.1	1.0	0.7	0.8	12.5
Petropavlovsk	52 53N	158 42E	286	7	23	11	35	25	56	47	46	34	84	-29	35	3.0	2.2	3.4	2.5	2.2	2.0	3.1	3.2	3.8	3.9	3.6	3.0	35.9
Salehkard	66 31N	66 35E	60	18	-13	-21	18	4	61	49	26	20	85	-65	27	0.3	0.3	0.3	0.3	0.7	1.3	1.9	2.0	1.5	0.7	0.5	0.4	10.2
Semipalatinsk	50 24N	80 13E	709	10	8	-7	45	26	81	57	46	30	101	-47	10	0.9	0.5	0.5	0.6	1.2	1.5	1.1	1.3	0.7	1.2	1.1	1.0	11.6
Sverdlovsk	56 49N	60 38E	894	21	6	-5	42	26	70	54	37	28	94	-45	29	0.5	0.4	0.5	0.7	1.9	2.7	2.6	2.7	1.6	1.2	1.1	0.8	16.7
Tashkent	41 20N	69 18E	1,569	19	37	21	65	47	92	64	65	41	106	-19	19	2.1	1.1	2.6	2.3	1.4	0.5	0.2	0.1	0.1	1.2	1.5	1.6	14.7
Verkhoyansk	67 34N	133 51E	328	24	-54	-63	19	-10	66	47	12	-3	98	-90	44	0.2	0.2	0.1	0.2	0.3	0.9	1.1	1.0	0.5	0.3	0.3	0.2	5.3
Vladivostok	43 07N	131 55E	94	14	13	0	46	34	71	60	55	41	92	-22	53	0.3	0.4	0.7	1.2	2.1	2.9	3.3	4.7	4.3	1.9	1.2	0.6	23.6
Yakutsk	62 01N	129 43E	535	19	-45	-53	27	6	73	54	23	11	97	-84	22	0.3	0.2	0.1	0.3	0.4	1.1	1.6	1.3	1.1	0.5	0.4	0.3	7.4

ASIA – SOUTHEAST

Brunei:																												
Brunei	04 55N	114 55E	10	5	85	76	87	77	87	76	86	77	99	70	12	14.6	7.6	7.8	9.8	10.9	9.5	9.0	7.3	11.8	14.5	15.2	13.0	131.0
Burma:																												
Mandalay	21 59N	96 06E	252	20	82	55	101	77	93	78	89	73	111	44	20	0.1	0.1	0.2	1.2	5.8	6.3	2.7	4.1	5.4	4.3	2.0	0.4	32.6
Moulmein	16 26N	97 39E	150	43	89	65	95	77	83	74	88	75	103	52	60	0.2	0.2	0.5	3.0	19.9	37.1	47.5	44.2	27.1	8.5	1.7	0.3	190.2
Cambodia:																												
Phnom Penh	11 33N	104 51E	39	37	88	71	95	76	90	76	87	76	105	55	49	0.3	0.4	1.4	3.1	5.7	5.8	6.0	6.1	8.9	9.9	5.5	1.7	54.8
Indonesia:																												
Jakarta	06 11S	106 50E	26	80	84	74	87	75	87	73	87	74	98	66	78	11.8	11.8	8.3	5.8	4.5	3.8	2.5	1.7	2.6	4.4	5.6	8.0	70.8
Manokwari	00 53S	134 03E	10	5	86	73	86	74	86	74	87	74	93	68	40	12.0	9.4	13.2	11.1	7.8	7.2	5.4	5.6	4.9	4.7	6.5	10.3	98.1
Mapanget	01 32N	124 55E	264	21	85	73	86	73	87	73	89	72	97	65	63	18.6	13.8	12.2	8.0	6.4	6.5	4.8	4.0	3.3	4.9	8.9	14.7	106.1
Penfui	10 10S	123 39E	335	21	87	75	89	72	88	70	92	72	101	58	63	15.2	13.7	9.2	2.6	1.2	0.4	0.2	0.0	0.0	0.7	3.3	9.1	55.7
Pontiansk	00 00N	109 20E	13	20	87	74	89	75	89	74	89	75	96	68	63	10.8	8.2	9.5	10.9	11.1	8.7	6.5	8.0	9.0	14.4	15.3	12.7	125.1
Tabing	00 52S	100 21E	19	21	87	74	87	75	87	74	86	74	94	68	63	13.9	10.1	12.2	14.5	12.8	11.7	10.5	13.7	16.2	20.1	20.5	19.2	175.4
Tarakan	03 19N	117 33E	20	19	85	73	86	75	87	74	87	74	94	67	31	10.9	10.2	14.0	13.9	13.5	12.6	10.3	12.4	11.6	14.3	15.2	13.4	152.3
Laos:																												
Vientiane	17 58N	102 34E	559	13	83	58	95	73	89	75	88	71	108	32	27	0.2	0.6	1.5	3.9	10.5	11.9	10.5	11.5	11.9	4.3	0.6	0.1	67.5
Malaya, Fed.:																												
Kuala Lumpur	03 06N	101 42E	111	19	90	72	91	74	90	72	89	73	99	64	19	6.2	7.9	10.2	11.5	8.8	5.1	3.9	6.4	8.6	9.8	10.2	7.5	96.1
Singapore	01 18N	103 50E	33	39	86	73	88	75	88	75	87	74	97	66	64	9.9	6.8	7.6	7.4	6.8	6.8	6.7	7.7	7.0	8.2	10.0	10.1	95.0
North Borneo:																												
Sanda Kan	05 54N	118 03E	38	45	85	74	89	76	89	75	88	75	99	70	46	19.0	10.9	8.6	4.5	6.2	7.4	6.7	7.9	9.3	10.2	14.5	18.5	123.7
Philippine Islands:																												
Davao	07 07N	125 38E	88	15	87	72	91	73	88	73	89	73	97	65	34	4.8	4.5	5.2	5.8	9.2	9.1	6.5	6.5	6.7	7.9	5.3	6.1	77.6
Manila	14 31N	121 00E	49	61	86	69	93	73	88	75	88	74	101	58	75	0.9	0.5	0.7	1.3	5.1	10.0	17.0	16.6	14.0	7.6	5.7	2.6	82.0
Sarawak:																												
Kuching	01 29N	110 20E	85	5	85	72	90	73	90	72	89	73	98	64	19	24.0	20.1	12.9	11.0	10.3	7.1	7.7	9.2	8.6	10.5	14.1	18.2	153.7

TABLE 7-2. TEMPERATURE AND PRECIPITATION DATA FOR REPRESENTATIVE WORLD - WIDE STATIONS (continued)

COUNTRY AND STATION	LATI-TUDE	LONGI-TUDE	ELE-VATION FEET	TEMP Length of record YEAR	JAN. Max	JAN. Min	APR. Max	APR. Min	JULY Max	JULY Min	OCT. Max	OCT. Min	Extreme Max	Extreme Min	PRECIP Length of record YEAR	January	February	March	April	May	June	July	August	September	October	November	December	YEAR
Thailand:																												
Bangkok	13 44N	100 30E	53	10	89	67	95	78	90	76	88	76	104	50	10	0.2	1.1	1.1	2.3	5.2	6.0	6.9	9.2	14.0	9.9	1.8	0.1	57.8
Viet Nam																												
Hanoi	21 03N	105 52E	20	12	68	58	80	70	92	79	84	72	108	41	12	0.8	1.2	2.5	3.6	4.1	11.2	11.9	15.2	10.0	3.5	2.6	2.8	69.4
Saigon	10 49N	106 39E	33	31	89	70	95	76	88	75	88	74	108	57	33	0.6	0.1	0.5	1.7	8.7	13.0	12.4	10.6	13.2	10.6	4.5	2.2	78.1

ASIA – MIDDLE EAST

COUNTRY AND STATION	LATI-TUDE	LONGI-TUDE	ELE-VATION FEET	TEMP Length of record YEAR	JAN. Max	JAN. Min	APR. Max	APR. Min	JULY Max	JULY Min	OCT. Max	OCT. Min	Extreme Max	Extreme Min	PRECIP Length of record YEAR	January	February	March	April	May	June	July	August	September	October	November	December	YEAR
Aden:																												
Riyan	14 39N	49 19E	83	13	82	67	88	74	92	77	88	72	111	57	13	0.3	0.1	0.6	0.2	*	0.1	0.1	0.1	*	*	0.7	0.3	2.5
Afghanistan:																												
Kabul	34 30N	69 13E	5,955	9	36	18	66	43	92	61	73	42	104	- 6	45	1.3	1.5	3.6	3.3	0.9	0.2	0.1	0.1	*	0.4	0.6	0.6	12.6
Kandhar	31 36N	65 40E	3,462	7	56	31	83	50	102	66	85	44	111	14	7	3.1	1.7	0.8	0.3	0.2	*	0.1	*	0.0	*	*	0.8	7.0
Ceylon:																												
Colombo	06 54N	79 52E	22	25	86	72	88	76	85	77	85	75	99	59	40	3.5	2.7	5.8	9.1	14.6	8.8	5.3	4.3	6.3	13.7	12.4	5.8	92.3
East Pakistan:																												
Dacca	23 46N	90 23E	24	60	77	56	92	74	88	79	88	75	108	43	61	0.3	1.2	2.4	5.4	9.6	12.4	13.0	13.3	9.8	5.3	1.0	0.2	73.9
India:																												
Ahmadabad	23 03N	72 37E	180	45	85	58	104	75	93	79	97	68	118	36	45	*	0.1	0.1	*	0.4	3.7	12.2	8.1	4.2	0.4	0.1	*	29.3
Bangalore	12 57N	77 40E	2,937	60	80	57	93	69	81	66	88	69	102	46	60	0.2	0.3	0.4	1.6	4.2	2.9	3.9	5.0	6.7	5.9	2.7	0.4	34.2
Bombay	19 06N	72 51E	27	60	88	62	93	74	88	75	91	67	110	46	60	0.1	0.1	0.1	*	0.7	19.1	24.3	13.4	10.4	2.5	0.5	0.1	71.2
Calcutta	22 32N	88 20E	21	60	80	55	97	76	90	79	89	74	111	44	60	0.4	1.2	1.4	1.7	5.5	11.7	12.8	12.9	9.9	4.5	0.8	0.2	63.0
Cherrapunji	25 15N	91 44E	4,309	35	60	46	71	59	72	65	72	61	87	33	35	0.7	2.1	7.3	26.2	50.4	106.1	96.3	70.1	43.3	19.4	2.7	0.5	425.1
Hyderabad	17 27N	78 28E	1,741	50	85	59	101	75	87	73	88	68	112	43	45	0.3	*	0.5	1.1	1.1	4.4	6.0	5.3	6.5	2.5	1.1	0.3	29.6
Jalpaiguri	26 32N	88 43E	272	50	74	50	90	68	89	77	87	70	104	36	55	0.3	0.7	1.3	3.7	11.8	25.9	32.2	25.3	21.2	5.6	0.5	0.2	128.7
Lucknow	26 45N	80 52E	400	50	74	47	101	71	92	80	91	67	119	34	60	0.8	0.7	0.3	0.3	0.8	4.5	12.0	11.5	7.4	1.3	0.2	0.3	40.1
Madras	13 04N	80 15E	51	60	85	67	95	78	96	79	90	75	113	57	60	1.4	0.4	0.6	0.6	1.0	1.9	3.6	4.6	4.7	12.0	14.0	5.5	50.0
Mormugao	15 22N	73 49E	157	10	86	70	88	79	83	75	88	75	98	59	60	*	*	*	0.7	2.6	29.6	31.2	15.9	9.5	3.8	1.3	0.2	94.8
New Delhi	28 35N	77 12E	695	10	71	43	97	68	95	80	93	64	115	31	75	0.9	0.7	0.5	0.3	0.5	2.9	7.1	6.8	4.6	0.4	0.1	0.4	25.2
Silchar	24 49N	92 48E	95	60	78	52	88	69	90	77	88	72	103	41	53	0.8	2.1	7.9	14.3	15.6	21.7	19.7	19.7	14.4	6.5	1.4	0.4	124.5
Indian Ocean Islands:																												
Port Blair, Andaman Is.	11 40N	92 43E	261	60	84	72	89	75	84	74	84	74	97	62	60	1.8	1.1	1.1	2.4	15.1	21.7	15.4	16.3	17.4	12.5	10.5	7.9	123.2
Amini Divi, Laccadive Is.	11 07N	72 44E	13	29	86	74	92	80	86	77	86	77	99	65	30	0.7	*	*	1.5	3.7	14.3	12.0	7.7	6.3	5.8	2.6	1.3	56.0
Minicoy, Maldive Is.	08 18N	73 00E	9	20	85	73	87	80	85	76	85	76	98	63	50	1.8	0.7	0.9	2.3	7.0	11.6	8.9	7.8	6.3	7.3	5.5	3.4	63.5
Car Nicobar, Nicobar Is.	09 09N	92 49E	47	13	86	77	90	77	85	77	86	75	95	66	30	3.9	1.2	2.1	3.5	12.5	12.4	9.3	10.2	12.9	11.6	11.4	7.8	98.8

This table lists geographic coordinates, elevation, and climatic data for stations. Column headers are not printed on this page (see note below). Data columns 1–12 are whole-number values; columns 13–25 are decimal values (13–24 appear to be monthly precipitation, 25 the annual total). An asterisk (*) denotes a trace/negligible value.

Station	Lat	Long	Elev	1	2	3	4	5	6	7	8	9	10	11	12	13	14	15	16	17	18	19	20	21	22	23	24	25
Iran:																												
Abadan	30 21N	48 13E	10	12	64	44	90	62	112	81	98	63	127	24	10	1.5	1.7	0.6	0.8	0.1	0.0	0.0	0.0	0.0	0.1	1.0	1.8	7.6
Isfahan	32 37N	51 41E	5,238	45	47	25	72	46	98	67	78	47	108	-4	45	0.7	0.6	0.8	0.6	0.3	*	0.1	*	*	0.1	0.4	0.7	4.4
Kermanshah	34 19N	47 07E	4,331	15	45	23	68	38	99	56	79	38	108	-13	15	2.6	2.3	2.8	2.2	1.6	*	*	*	*	0.4	2.0	2.4	16.4
Rezaiyeh	37 32N	45 05E	4,364	3	32	17	67	45	91	64	67	47	99	-11	3	1.9	2.3	2.0	1.7	1.2	0.5	*	0.1	0.1	1.5	0.8	1.6	13.8
Tehran	35 41N	51 19E	3,937	24	45	27	71	49	99	72	76	53	109	-5	33	1.8	1.5	1.8	1.4	0.5	0.1	0.1	0.1	0.1	0.3	0.8	1.2	9.7
Iraq:																												
Baghdad	33 20N	44 24E	111	15	60	39	85	57	110	76	92	61	121	18	15	0.9	1.0	1.1	0.5	0.1	*	*	*	*	0.1	0.8	1.0	5.5
Basra	30 34N	47 47E	8	10	64	45	85	63	104	81	94	64	123	24	10	1.4	1.1	1.2	1.2	0.2	0.2	*	*	*	*	1.4	0.8	7.3
Mosul	36 19N	43 09E	730	26	54	35	77	49	109	72	88	51	124	12	29	2.8	3.1	2.1	1.9	0.7	0.0	*	*	*	0.2	1.9	2.4	15.2
Israel:																												
Haifa	32 48N	35 02E	23	16	65	49	77	58	88	75	85	68	112	27	30	6.9	4.3	1.6	1.0	0.2	*	*	*	*	1.0	3.7	7.3	26.2
Jerusalem	31 47N	35 13E	2,654	19	55	41	73	50	87	63	81	59	107	26	50	5.1	4.7	2.9	0.9	0.1	0.0	0.0	0.0	*	0.3	2.2	3.5	19.7
Tel Aviv	32 06N	34 46E	33	10	64	50	70	57	82	72	79	65	102	34	10	4.9	2.7	2.0	0.7	0.1	0.0	0.0	0.0	0.1	0.4	4.1	6.1	21.1
Jammu/Kashmir:																												
Srinagar	33 58N	74 46E	5,458	50	41	24	67	45	88	64	74	41	106	-4	50	2.9	2.8	3.6	3.7	2.4	1.4	2.3	2.4	1.5	1.2	0.4	1.3	25.9
Jordan:																												
Amman	31 58N	35 59E	2,547	25	54	39	73	49	89	65	81	57	109	21	25	2.7	2.9	1.2	0.6	0.2	0.0	0.0	0.0	*	0.2	1.3	1.8	10.9
Kuwait:																												
Kuwait	29 21N	48 00E	16	14	61	49	83	68	103	86	91	73	119	33	10	0.9	0.9	1.1	0.2	*	0.0	0.0	0.0	0.0	0.1	0.6	1.1	5.1
Lebanon:																												
Beirut	33 54N	35 28E	111	62	62	51	72	58	87	73	81	69	107	30	71	7.5	6.2	3.7	2.2	0.7	0.1	*	*	0.2	2.0	5.2	7.3	35.1
Nepal:																												
Katmandu	27 42N	85 22E	4,423	27	65	36	84	53	84	69	80	56	99	27	9	0.6	0.9	2.3	4.8	9.7	14.7	13.6	6.1	1.5	0.3	0.1	—	56.2
Oman and Muscat:																												
Muscat	23 37N	58 35E	15	23	77	66	90	78	97	87	93	80	116	51	38	1.1	0.7	0.4	0.4	*	0.0	*	*	0.0	0.1	0.4	0.7	3.9
Pakistan (West):																												
Karachi	24 48N	66 59E	13	43	77	55	90	73	91	81	91	72	118	39	59	0.5	0.4	0.3	0.1	0.1	0.7	3.2	1.6	0.5	0.1	0.1	0.2	7.8
Multan	30 11N	71 25E	400	60	68	42	95	68	102	86	94	64	122	29	60	0.4	0.4	0.4	0.3	0.3	0.6	2.0	1.8	0.5	0.1	0.1	0.2	7.1
Rawalpindi	33 35N	73 03E	1,676	60	62	38	86	59	98	77	89	57	118	25	60	2.5	2.5	2.7	1.9	1.3	2.3	8.1	9.2	3.9	0.6	0.3	1.2	36.5
Saudi Arabia:																												
Dhahran	26 16N	50 10E	78	10	69	54	90	69	107	86	95	73	120	40	10	1.1	0.6	0.4	0.2	0.1	0.0	0.0	0.0	0.0	0.0	0.2	0.9	3.5
Jidda	21 28N	39 10E	20	5	84	66	91	70	99	79	95	73	117	49	5	0.2	*	*	*	*	0.0	*	*	*	*	1.0	1.2	2.5
Riyadh	24 39N	46 42E	1,938	3	70	46	89	64	107	78	94	61	113	19	3	0.1	0.8	0.9	1.0	0.4	*	0.0	0.0	0.0	0.0	*	*	3.2
Syria:																												
Deir Ez Zor	35 21N	40 09E	699	5	53	35	80	52	105	78	86	56	114	16	8	1.6	0.8	0.3	0.1	0.1	0.0	0.0	0.0	0.0	0.2	1.5	0.9	6.2
Damascus	33 30N	36 20E	2,362	13	53	36	75	48	96	64	81	54	113	21	7	1.7	1.7	0.3	0.5	0.1	0.1	0.0	0.0	0.7	0.4	1.6	1.6	8.6
Aleppo	36 14N	37 08E	1,280	8	50	34	75	48	97	69	81	54	117	9	10	3.5	2.5	1.5	1.1	0.3	0.3	0.0	0.0	*	1.0	2.2	3.3	15.5
Trucial Kingdoms:																												
Sharjah	25 20N	55 24E	18	11	74	54	86	65	100	82	92	71	118	37	12	0.9	0.9	0.4	0.2	0.0	0.0	0.0	0.0	0.0	0.0	0.4	1.4	4.2

See footnotes at end of table.

TABLE 7-2. TEMPERATURE AND PRECIPITATION DATA FOR REPRESENTATIVE WORLD - WIDE STATIONS (continued)

COUNTRY AND STATION	LATI-TUDE (° ')	LONGI-TUDE (° ')	ELE-VATION (FEET)	Length of record (YEAR)	JAN. Max °F	JAN. Min °F	APR. Max °F	APR. Min °F	JULY Max °F	JULY Min °F	OCT. Max °F	OCT. Min °F	Extreme Max °F	Extreme Min °F	Length of record (YEAR)	January IN.	February IN.	March IN.	April IN.	May IN.	June IN.	July IN.	August IN.	September IN.	October IN.	November IN.	December IN.	YEAR IN.
Turkey:																												
Adana	36 59N	35 18E	82	21	57	39	74	51	93	71	84	58	109	19	31	4.3	4.0	2.5	1.6	2.0	0.7	0.2	0.2	0.7	1.9	2.4	3.8	24.3
Ankara	39 57N	32 53E	2,825	26	39	24	63	40	86	59	69	44	104	-13	24	1.3	1.2	1.3	1.3	1.9	1.0	0.5	0.4	0.7	0.9	1.2	1.9	13.6
Erzurum	39 54N	41 16E	6,402	16	24	8	50	32	78	53	59	37	93	-22	16	1.4	1.6	2.0	2.5	3.1	2.1	1.3	0.9	1.1	2.3	1.8	1.1	21.2
Izmir	38 27N	27 15E	92	39	55	39	70	49	92	69	76	55	108	12	58	4.4	3.3	3.0	1.7	1.3	0.6	0.2	0.2	0.8	2.1	3.3	4.8	25.5
Samsun	41 17N	36 19E	131	24	50	38	59	45	79	65	69	56	103	20	27	2.9	2.6	2.7	2.3	1.8	1.5	1.5	1.3	2.4	3.2	3.5	2.4	29.1
Yemen:																												
Kamaran I.	15 20N	42 37E	20	26	82	74	89	79	98	85	93	82	105	66	21	0.2	0.2	0.1	0.1	0.1	*	0.5	0.7	0.1	0.1	0.4	0.9	3.4

AUSTRALIA & PACIFIC ISLANDS

COUNTRY AND STATION	LATI-TUDE (° ')	LONGI-TUDE (° ')	ELE-VATION (FEET)	Length of record (YEAR)	JAN. Max °F	JAN. Min °F	APR. Max °F	APR. Min °F	JULY Max °F	JULY Min °F	OCT. Max °F	OCT. Min °F	Extreme Max °F	Extreme Min °F	Length of record (YEAR)	January IN.	February IN.	March IN.	April IN.	May IN.	June IN.	July IN.	August IN.	September IN.	October IN.	November IN.	December IN.	YEAR IN.
Australia:																												
Adelaide	34 57S	138 32E	20	86	86	61	73	55	59	45	73	51	118	32	104	0.8	0.7	1.0	1.8	2.7	3.0	2.6	2.6	2.1	1.7	1.1	1.0	21.1
Alice Springs	23 48S	133 53E	1,791	62	97	70	81	54	67	39	88	58	111	19	30	1.7	1.3	1.1	0.4	0.6	0.5	0.3	0.3	0.3	0.7	1.2	1.5	9.9
Bourke	30 05S	145 58E	361	63	99	70	82	55	65	40	85	56	125	25	72	1.4	1.5	1.1	1.1	1.0	1.1	0.9	0.8	0.8	0.9	1.2	1.4	13.2
Brisbane	27 25S	153 05E	17	53	85	69	79	61	68	49	80	60	110	35	91	6.4	6.3	5.7	3.7	2.8	2.6	2.2	1.9	1.9	2.5	3.7	5.0	44.7
Broome	17 57S	122 13E	56	41	92	79	93	72	82	58	91	72	113	40	50	6.3	5.8	3.9	1.2	0.6	0.9	0.2	0.1	*	*	0.6	3.3	22.9
Burketown	17 45S	139 33E	30	31	93	77	91	69	82	55	93	70	110	40	53	8.2	6.3	5.2	1.2	0.2	0.3	*	*	*	0.4	1.5	4.4	27.5
Canberra	35 18S	149 11E	1,886	23	82	55	67	44	52	33	68	43	109	14	25	1.9	1.7	2.2	1.6	1.8	2.1	1.8	2.2	1.6	2.2	1.9	2.0	23.0
Carnarvon	24 53S	113 40E	13	43	88	72	84	66	71	51	78	61	118	37	57	0.4	0.7	0.7	0.6	1.5	2.4	1.6	0.7	0.2	0.1	*	0.2	9.1
Cloncurry	20 40S	140 30E	622	32	99	77	90	67	77	51	95	68	127	35	59	4.4	4.2	2.4	0.7	0.5	0.6	0.3	0.1	0.3	0.5	1.3	2.7	18.0
Esperance	33 50S	121 55E	14	44	77	60	72	54	62	45	68	50	117	31	60	0.7	0.7	1.2	1.8	3.3	4.1	4.0	3.8	2.7	2.2	1.0	0.9	26.4
Laverton	28 40S	122 23E	1,510	30	96	69	81	57	64	41	82	55	115	25	30	0.8	0.8	1.6	0.8	0.9	0.7	0.6	0.5	0.2	0.3	0.8	0.8	8.8
Melbourne	37 49S	144 58E	115	88	78	57	68	51	56	42	67	48	114	27	88	1.9	1.8	2.2	2.3	2.1	2.1	1.9	1.9	2.3	2.6	2.3	2.3	25.7
Mundiwindi	23 52S	120 10E	1,840	15	101	64	87	61	70	41	89	58	112	22	15	1.0	1.9	2.0	0.8	0.6	0.9	0.1	0.3	0.3	0.5	0.5	1.2	10.1
Perth	31 56S	115 58E	64	44	85	63	76	57	63	48	70	53	112	31	63	0.3	0.4	0.8	1.7	5.1	7.1	6.7	5.7	3.4	2.2	0.8	0.5	34.7
Port Darwin	12 25S	130 52E	104	58	90	77	92	76	87	67	93	77	105	55	70	15.2	12.3	10.0	3.8	0.6	0.1	*	0.1	0.5	2.0	4.7	9.4	58.7
Sydney	33 52S	151 02E	62	87	78	65	71	58	60	46	71	56	114	35	87	3.5	4.0	5.0	5.3	5.0	4.6	4.6	3.0	2.9	2.8	2.9	2.9	46.5
Thursday Island	10 35S	142 13E	200	31	87	78	87	76	82	73	86	76	98	64	49	18.2	15.8	13.9	8.0	1.6	0.5	0.4	0.2	0.1	0.3	1.5	7.0	67.5
Townsville	19 15S	146 46E	18	31	87	76	84	70	75	59	83	71	110	39	67	10.9	11.2	7.2	1.3	1.3	1.4	0.9	0.5	0.7	1.3	1.9	5.4	45.7
William Creek	28 55S	136 21E	247	39	96	69	80	55	65	41	84	56	119	25	30	0.5	0.6	0.3	0.3	0.3	0.5	0.2	0.3	0.3	0.5	0.5	0.7	5.0
Windorah	25 26S	142 36E	390	29	101	74	86	59	70	43	91	61	116	26	50	1.4	1.6	1.6	0.9	0.8	0.8	0.5	0.4	0.5	0.6	0.9	1.4	11.4
Tasmania:																												
Hobart	42 53S	147 20E	177	70	71	53	63	48	52	40	63	46	105	28	100	1.9	1.5	1.8	1.9	1.8	2.2	2.1	1.9	2.1	2.3	2.4	2.1	24.0

Station	Lat	Long	Elev (ft)	Yrs (T)	Jan max	Jan min	Apr max	Apr min	Jul max	Jul min	Oct max	Oct min	Hi	Lo	Yrs (P)	Jan	Feb	Mar	Apr	May	Jun	Jul	Aug	Sep	Oct	Nov	Dec	Ann
New Zealand																												
Auckland	37 00S	174 47E	23	36	73	60	67	56	56	46	63	52	90	33	92	3.1	3.7	3.2	3.8	5.0	5.4	5.7	4.6	4.0	4.0	3.5	3.1	49.1
Christchurch	43 29S	172 32E	118	52	70	53	62	45	50	35	62	44	96	21	64	2.2	1.7	1.9	1.9	2.6	2.6	2.7	1.9	1.8	1.7	1.9	2.2	25.1
Dunedin	45 55S	170 12E	4	77	66	50	59	45	48	37	59	42	94	23	77	3.4	2.8	3.0	2.8	3.2	3.2	3.1	3.0	2.7	3.0	3.2	3.5	36.9
Wellington	41 17S	174 46E	415	66	69	56	63	51	53	42	60	48	88	29	79	3.2	3.2	3.2	3.8	4.6	4.6	5.4	4.6	4.0	4.0	3.5	3.5	47.4
PACIFIC ISLANDS:																												
Canton, Phoenix Is.	02 46S	171 43W	9	12	88	78	89	78	90	78	90	78	98	70	30	2.6	2.2	3.6	4.3	2.6	2.5	1.2	1.1	1.0	1.6	1.9	2.6	29.4
Guam, Marianas Is.	13 33N	144 50E	361	30	84	72	87	73	86	76	86	75	95	54	30	5.9	3.5	2.6	3.0	3.5	5.9	9.0	12.8	13.4	13.1	10.3	6.1	88.5
Honolulu, Hawaii	21 20N	157 55W	7	30	79	66	80	68	85	73	87	72	93	56	30	3.8	3.3	3.3	2.9	1.0	0.3	0.4	0.4	0.9	1.0	2.2	3.0	21.9
Iwo Jima, Bonin Is.	24 47N	141 19E	353	15	71	64	77	69	86	78	84	76	95	46	17	3.2	2.1	3.7	4.9	6.4	7.6	4.8	4.6	6.5	4.8	4.3	4.3	52.8
Madang, New Guinea	05 12S	145 47E	19	12	87	75	88	74	84	74	88	75	98	62	20	12.1	11.9	14.9	16.9	15.1	10.8	7.6	4.8	5.3	10.0	13.3	14.5	137.2
Midway Is.	28 13N	177 23W	29	21	69	62	73	64	79	72	79	69	92	46	20	4.6	3.1	3.7	1.9	1.3	2.9	3.9	3.7	3.7	3.6	6.6	4.2	40.7
Naha, Okinawa	26 12N	127 39E	96	30	67	56	76	64	80	73	81	72	96	41	30	5.3	5.4	6.1	8.9	10.0	7.1	6.6	5.9	5.9	4.3	4.3	4.3	82.8
Noumea, New Caledonia	22 16S	166 27E	246	24	86	72	83	70	76	62	80	65	99	52	52	3.7	5.1	5.7	5.2	4.4	3.7	3.6	2.5	2.0	2.4	2.4	2.6	43.5
Pago Pago, Samoa	14 19S	170 43W	29	2	87	75	87	76	83	74	85	75	98	67	41	24.5	20.5	19.2	16.5	15.4	12.3	10.0	8.2	13.1	14.9	19.2	19.8	193.6
Ponape, Caroline Is.	06 58N	158 13E	123	30	86	75	87	76	87	75	87	75	96	67	30	11.1	9.7	14.6	20.0	20.3	16.7	16.2	16.3	15.8	16.0	16.9	18.3	191.9
Port Moresby, New Guinea	09 29S	147 09E	126	20	89	76	89	74	83	73	87	76	98	64	38	7.0	7.6	6.7	4.2	2.5	1.1	1.0	0.7	1.4	1.9	4.4	4.4	39.8
Rabaul, New Guinea	04 13S	152 11E	28	19	90	73	89	72	84	70	87	70	100	65	24	14.8	10.4	10.2	10.0	5.2	3.3	3.5	5.4	7.1	10.0	10.1	10.1	88.8
Suva, Fiji Is.	18 08S	178 26E	20	43	86	74	84	73	79	68	81	70	98	55	43	11.4	10.7	14.5	12.2	6.7	4.9	8.3	7.7	8.3	9.8	12.5	12.5	117.1
Tahiti, Society Is.	17 33S	149 36W	7	23	89	72	89	72	86	68	87	73	93	61	27	13.2	11.5	6.5	6.8	3.2	2.6	1.9	2.3	3.4	6.5	11.9	11.9	74.7
Tulagi, Solomon Is.	09 05S	160 10E	8	20	88	76	88	76	86	76	87	76	96	68	37	14.3	15.8	15.0	10.0	8.1	6.8	7.6	8.0	8.7	10.0	10.4	10.4	123.4
Wake Is.	19 17N	166 39E	11	30	82	73	83	74	83	77	86	77	92	64	30	1.1	1.4	1.5	1.9	2.0	4.6	7.1	5.2	5.3	3.1	1.8	1.8	36.9
Yap, Caroline Is.	09 31N	138 08E	62	30	85	77	87	77	86	76	87	76	97	69	30	7.9	4.6	5.4	6.4	9.5	10.7	13.8	14.7	14.0	13.2	11.2	10.2	121.6
ANTARCTICA																												
Byrd Station	80 01S	119 32W	5,095	6	10	-2	-11	-30	-25	-45	-15	-33	31	-82	6	0.4	0.4	0.2	0.3	0.4	0.5	0.7	0.7	0.3	0.3	0.0	0.3	4.9
Ellsworth	77 44S	41 07W	139	6	22	12	-10	-25	-21	-35	-2	-15	36	-70	6	0.3	0.2	0.3	0.6	0.6	0.2	0.2	0.5	0.4	0.3	0.2	0.2	3.6
McMurdo Station	77 53S	166 48W	8	10	30	21	-1	-13	-1	-24	2	-12	42	-59	10	0.5	0.7	0.4	0.4	0.4	0.3	0.2	0.2	0.4	0.2	0.3	0.3	4.3
South Pole Station	89 59S	000 00W	9,186	5	-16	-23	-66	-79	-67	-81	-55	-64	6	-107	5	*	0.1	0.0	0.0	0.0	0.0	0.0	0.0	0.0	*	0.0	*	0.1
Wilkes	66 16S	110 31E	31	7	34	28	17	9	8	-3	16	6	46	-35	7	0.5	0.4	1.7	1.1	1.4	1.2	1.3	0.8	1.5	1.2	0.8	0.3	12.2

NOTES

1. "Length of Record" refers to average daily maximum and minimum temperatures and precipitation. A standard period of the 30 years from 1931-1960 had been used for locations in the United States and some other countries. The length of record of extreme maximum and minimum temperatures includes all available years of data for a given location and is usually for a longer period.

2. * — Less than 0.05".

3. Except for Antarctica, amounts of solid precipitation such as snow or hail have been converted to their water equivalent. Because of the frequent occurrence of blowing snow, it has not been possible to determine the precise amount of precipitation actually falling in Antarctica. The values shown are the average amounts of solid snow accumulating in a given period as determined by snow markers. The liquid content of the accumulation is undetermined.

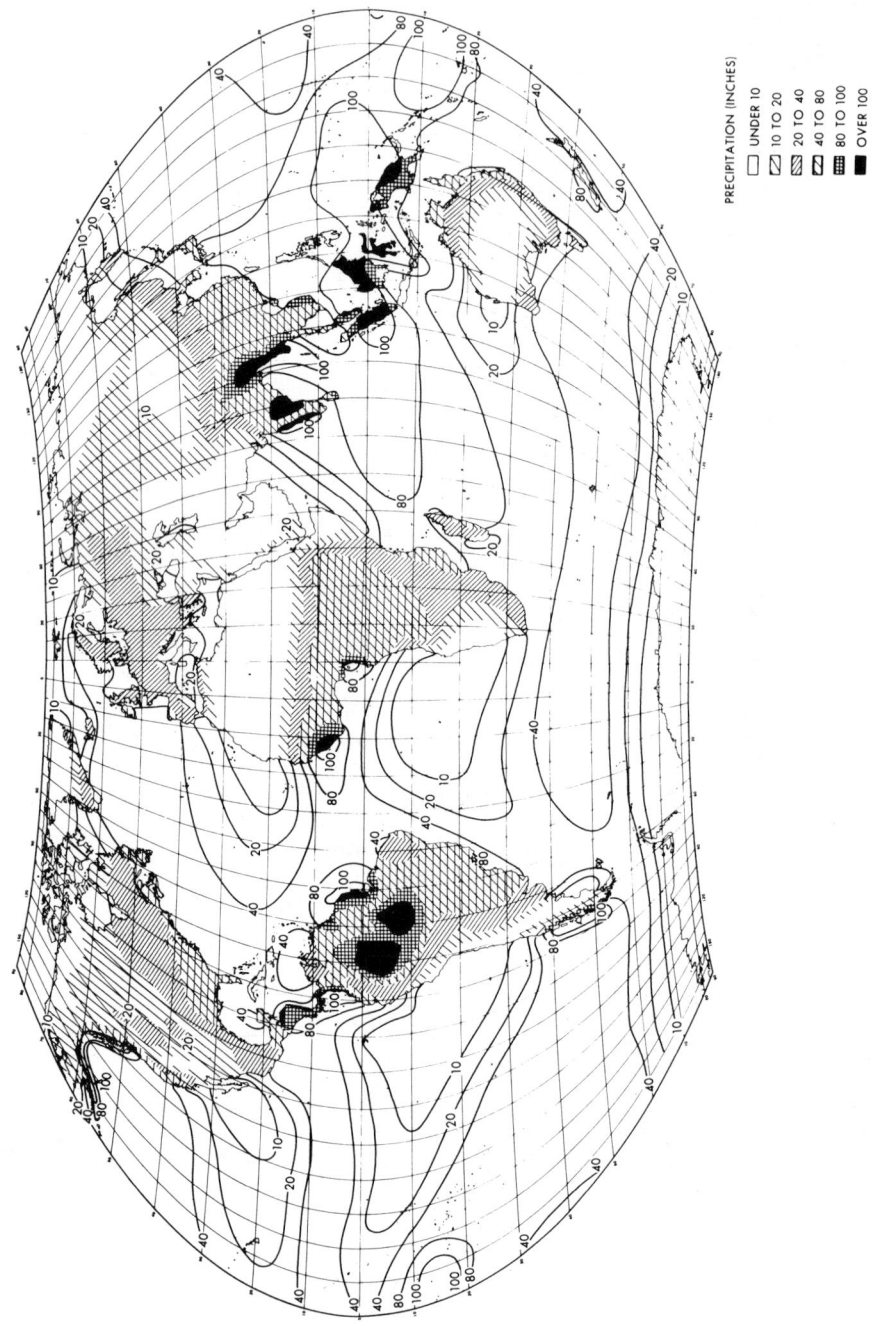

FIGURE 7-1. ANNUAL WORLD-WIDE PRECIPITATION

(Source: U.S. Department of Commerce, Environmental Science Services Administration, 1969)

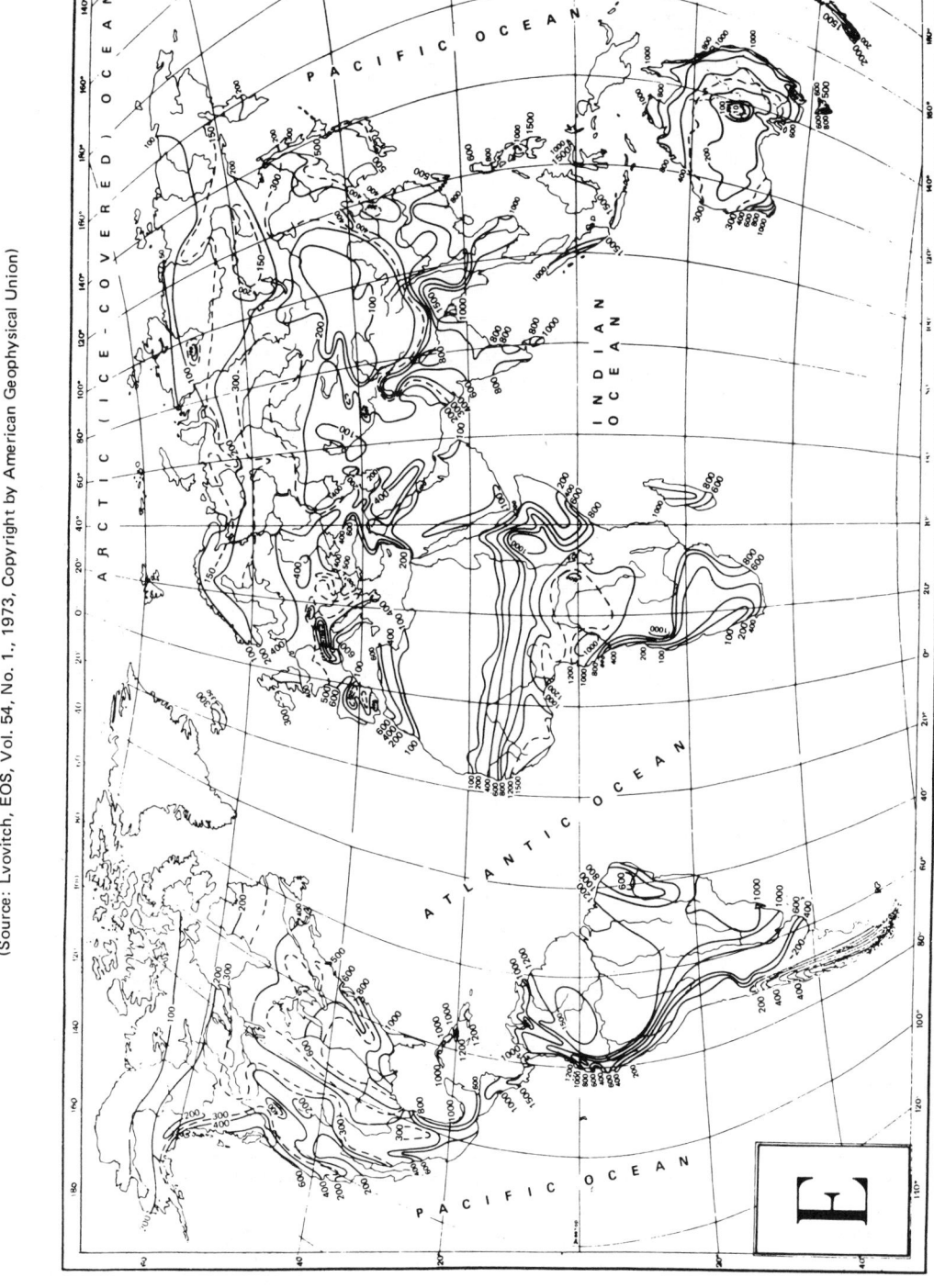

FIGURE 7-2. ANNUAL WORLD-WIDE EVAPORATION

(Source: Lvovitch, EOS, Vol. 54, No. 1., 1973, Copyright by American Geophysical Union)

TABLE 7-3. WORLD-WIDE EXTREMES OF ANNUAL PRECIPITATION AND TEMPERATURE
(Source: U.S. Department of Commerce, 1969)

EXTREMES OF AVERAGE ANNUAL PRECIPITATION

Area	Greatest amount inches	Place	Elevation feet	Years of record
Oceania	460.0	Mt. Waialeale, Kauai, Hawaii	5,075	32
Asia	450.0	Cherrapunji, India	4,309	74
Africa	404.6	Debundscha, Cameroon	30	32
South America	353.9	Quibdo, Colombia	240	10-16
North America	262.1	Henderson Lake, B.C., Canada	12	14
Europe	182.8	Crkvica, Yugoslavia	3,337	22
Australia	179.3	Tully, Queensland	—	31

Area	Least amount inches	Place	Elevation feet	Years of record
South America	0.03	Arica, Chile	95	59
Africa	less than 0.1	Wadi Halfa, Sudan	410	39
Antarctica	0.8 [1]	South Pole Station	9,186	10
North America	1.2	Batagues, Mexico	16	14
Asia	1.8	Aden, Arabia	22	50
Australia	4.05	Mulka, South Australia	—	34
Europe	6.4	Astrakhan, USSR	45	25
Oceania	8.93	Puako, Hawaii	5	13

TEMPERATURE EXTREMES

Area	Highest °F	Place	Elevation feet	Date
Africa	136	Azizia, Libya	380	Sep. 13, 1922
North America	134	Death Valley, California	−178	July 10, 1913
Asia	129	Tirat Tsvi, Israel	−722	June 21, 1942
Australia	128	Cloncurry, Queensland	622	Jan. 16, 1889
Europe	122	Seville, Spain	26	Aug. 4, 1881
South America	120	Rivadavia, Argentina	676	Dec. 11, 1905
Oceania	108	Tuguegarao, Philippines	72	Apr. 29, 1912
Antarctica	58	Esperanza, Palmer Pen.	26	Oct. 20, 1956

Area	Lowest °F	Place	Elevation feet	Date
Antarctica	−127	Vostok	11,220	Aug. 24, 1960
Asia	− 90	Oymykon, USSR	2,625	Feb. 6, 1933
Greenland	− 87	Northice	7,690	Jan. 9, 1954
North America	− 81	Snag, Yukon, Canada	1,925	Feb. 3, 1947
Europe	− 67	Ust' Shchugor, USSR	279	January [2]
South America	− 27	Sarmiento, Argentina	879	June 1, 1907
Africa	− 11	Ifrane, Morocco	5,364	Feb. 11, 1935
Australia	− 8	Charlotte Pass, N.S.W.	— [4]	July 22, 1947 [3]
Oceania	14	Haleakala Summit, Maui	9,750	Jan. 2, 1961

[1] The value given is the average amount of solid snow accumulating in one year as indicated by snow markers. The liquid content of the snow is undetermined.
[2] Exact date unknown; lowest in 15-year period.
[3] and earlier date.
[4] Elevation unknown.

FIGURE 7-3. DIAGRAM SHOWING WATER AVAILABILITY ON EARTH

(Source: Doxiadis, Water for Peace, 1967)

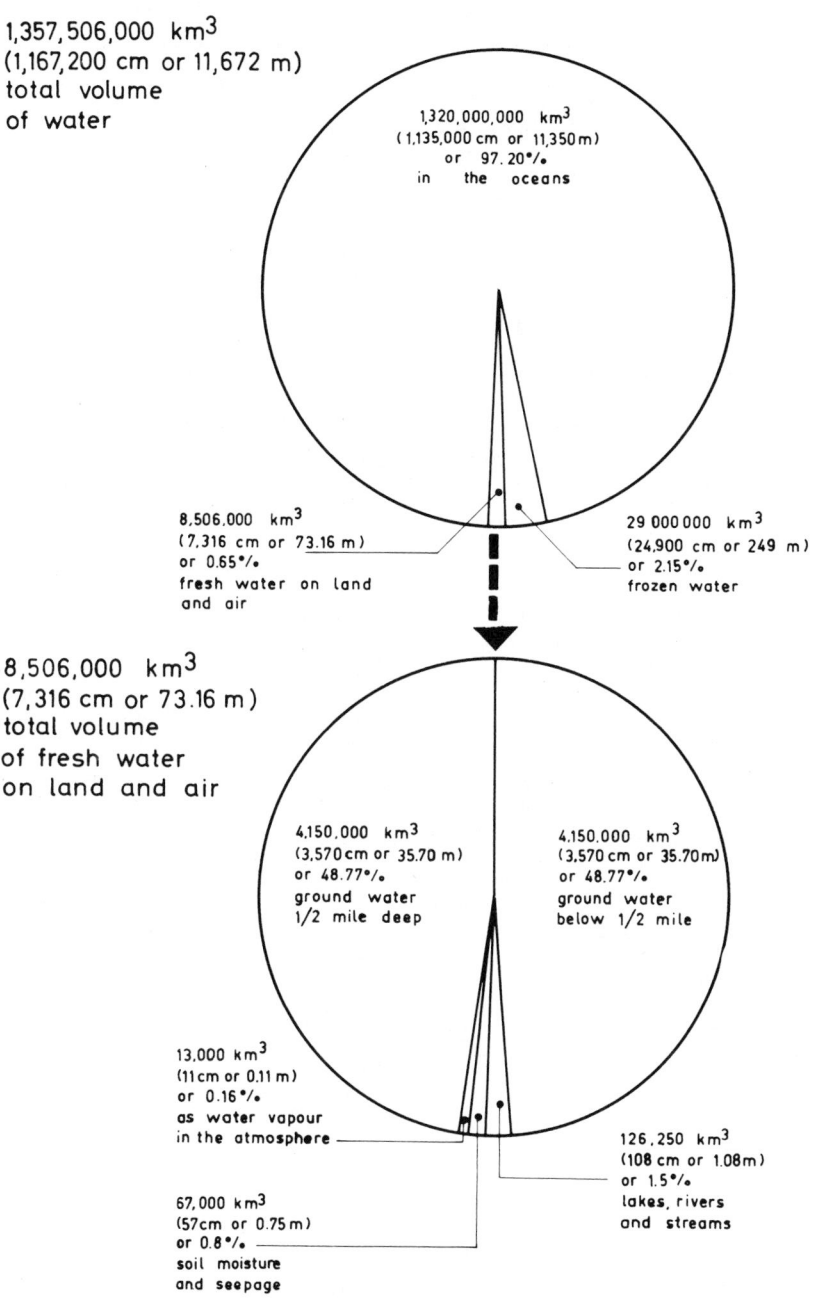

1,357,506,000 km³
(1,167,200 cm or 11,672 m)
total volume
of water

1,320,000,000 km³
(1,135,000 cm or 11,350 m)
or 97.20°/o
in the oceans

8,506.000 km³
(7,316 cm or 73.16 m)
or 0.65°/o
fresh water on land
and air

29 000 000 km³
(24,900 cm or 249 m)
or 2.15°/o
frozen water

8,506,000 km³
(7,316 cm or 73.16 m)
total volume
of fresh water
on land and air

4,150,000 km³
(3,570 cm or 35.70 m)
or 48.77°/o
ground water
1/2 mile deep

4,150,000 km³
(3,570 cm or 35.70 m)
or 48.77°/o
ground water
below 1/2 mile

13,000 km³
(11 cm or 0.11 m)
or 0.16°/o
as water vapour
in the atmosphere

126,250 km³
(108 cm or 1.08 m)
or 1.5°/o
lakes, rivers
and streams

67,000 km³
(57 cm or 0.75 m)
or 0.8°/o
soil moisture
and seepage

note: figures in brackets indicate the height that the relevant quantities of water would reach if they were placed on the whole non-frozen land area of the earth which is 116,400,000 km²

TABLE 7-4. ESTIMATED WORLD WATER SUPPLY AND BUDGET

(Source: Nace, U.S. Geological Survey, 1967)

Water item	Volume (thousands)		Percent of total water
	Cubic miles	Cubic kilometers	
Water in land areas:			
Fresh-water lakes	30	125	0.009
Saline lakes and inland seas	25	104	.008
Rivers (average instantaneous volume)	.3	1.25	.0001
Soil moisture and vadose water	16	67	.005
Ground water to depth of 4,000 m (about 13,100 ft)	2,000	8,350	.61
Icecaps and glaciers	7,000	29,200	2.14
Total in land area (rounded)	9,100	37,800	2.8
Atmosphere	3.1	13	.001
World ocean	317,000	1,320,000	97.3
Total, all items (rounded)	326,000	1,360,000	100
Annual evaporation: [1]			
From world ocean	85	350	0.025
From land areas	17	70	.005
Total	102	420	0.031
Annual precipitation:			
On world ocean	78	320	0.024
On land areas	24	100	.007
Total	102	420	0.031
Annual runoff to oceans from rivers and icecaps ...	9	38	0.003
Ground-water outflow to oceans [2]	.4	1.6	.0001
Total	9.4	39.6	0.0031

[1] Evaporation (420,000 km3) is a measure of total water participating annually in the hydrologic cycle.

[2] Arbitrarily set equal to about 5 percent of surface runoff.

TABLE 7-5. WORLD WATER BALANCE, BY CONTINENT

(Source: Lvovitch, M.I., EOS, Vol. 54, No. 1, Jan. 1973, Copyright by American Geophysical Union)

Water balance elements	Europe *	Asia	Africa	North America **	South America	Australia †	Total land area ‡
Area, millions of km2	9.8	45.0	30.3	20.7	17.8	8.7	132.3
in mm							
Precipitation	734	726	686	670	1,648	736	834
Total river runoff	319	293	139	287	583	226	294
Groundwater runoff	109	76	48	84	210	54	90
Surface water runoff	210	217	91	203	373	172	204
Total soil moisture	524	509	595	467	1,275	564	630
Evaporation	415	433	547	383	1,065	510	540
in km3							
Precipitation	7,165	32,690	20,780	13,910	29,355	6,405	110,303
Total river runoff	3,110	13,190	4,225	5,960	10,380	1,965	38,830
Groundwater runoff	1,065	3,410	1,465	1,740	3,740	465	11,885
Surface water runoff	2,045	9,780	2,760	4,220	6,640	1,500	26,945
Total soil moisture	5,120	22,910	18,020	9,690	22,715	4,905	83,360
Evaporation	4,055	19,500	16,555	7,950	18,975	4,440	71,475
relative values							
Groundwater runoff as percent of total runoff	34	26	35	32	36	24	31
Coefficient of groundwater discharge into rivers	0.21	0.15	0.08	0.18	0.16	0.10	0.14
Coefficient of runoff	0.43	0.40	0.23	0.31	0.35	0.31	0.36

* Including Iceland.

** Excluding the Canadian archipelago and including Central America.

† Including Tasmania, New Guinea and New Zealand, only within the limits of the continent: P - 440 mm, R - 47 mm, U - 7 mm, S - 40 mm, W - 400 mm, E - 393 mm.

‡ Excluding Greenland, Canadian archipelago and Antarctica.

TABLE 7-6. WORLD -WIDE STABLE RUNOFF, BY CONTINENT

(Source: Lvovitch, M.I., EOS, Vol. 54, No. 1, Jan. 1973, Copyright by American Geophysical Union)

	Stable runoff, km3 [1]				Total river runoff [2]	Total stable runoff as percent of total river runoff
	Of under-ground origin	Regu-lated by lakes	Regulated by water reservoirs	Total		
Europe	1,065	60	200	1,325	3,110	43
Asia	3,410	35	560	4,005	13,190	30
Africa	1,465	40	400	1,905	4,225	45
North America	1,740	150	490	2,380	5,960	40
South America	3,740	—	160	3,900	10,380	38
Australia [3]	465	—	30	495	1,965	25
Total land area except polar zones	11,885	285	1,840	14,010	38,830	36

[1] Excluding flood flows.

[2] Including flood flow.

[3] Including Tasmania, New Guinea and New Zealand.

TABLE 7-7. WORLD-WIDE PER CAPITA WATER RESOURCES, BY CONTINENT
(Source: Lvovitch, M.I., E OS, Vol. 54, No. 1, Jan. 1973, Copyright by American Geophysical Union)

	Popula-tion in millions (1969)	Annual volume of river runoff, km^3		Runoff volume per capita, m^3/year	
		Total	Stable runoff	Total	Stable runoff
Europe	642	3,110	1,325	4,850	2,100
Asia	2,047	13,190	4,005	6,440	1,960
Africa	345	4,225	1,905	12,250	5,500
North America	312	5,960	2,380	19,100	7,640
South America	185	10,380	3,900	56,100	21,100
Australia [*]	18	1,965	495	10,900	2,750
Total land area	3,549	38,830	14,010	10,940	3,950

[*] Including New Guinea and New Zealand.

TABLE 7-8. MEAN ANNUAL RUNOFF IN SELECTED COUNTRIES OF THE WORLD
(Source: Framji and Mahajan, 1969, Irrigation and Drainage in the World, New Delhi)

Country	Average annual runoff	
	10^{9}m^3	Million acre-feet [1]
Brazil	5,190	4,209
USSR	4,340	3,518
China (Mainland)	2,620	2,124
Canada	2,267	1,839
India	1,678	1,356
USA	1,630	1,321
Pakistan	1,439	1,167
Burma	1,069	867
Argentina	750	608
Venezuela	700	567
Mexico	390	316
Australia	345	280
Malagasy Republic	337	273
New Zealand	300	243
Thailand	220	178
France	200	162
Guyana	197	160
Chile	177	143
Italy	159	129
UAR	84	68

[1] Rounded.

FIGURE 7-4. ANNUAL WORLD-WIDE TOTAL RIVER RUNOFF

(Source: Lvovitch, EOS, Vol. 54, No. 1, 1973. Copyright by American Geophysical Union)

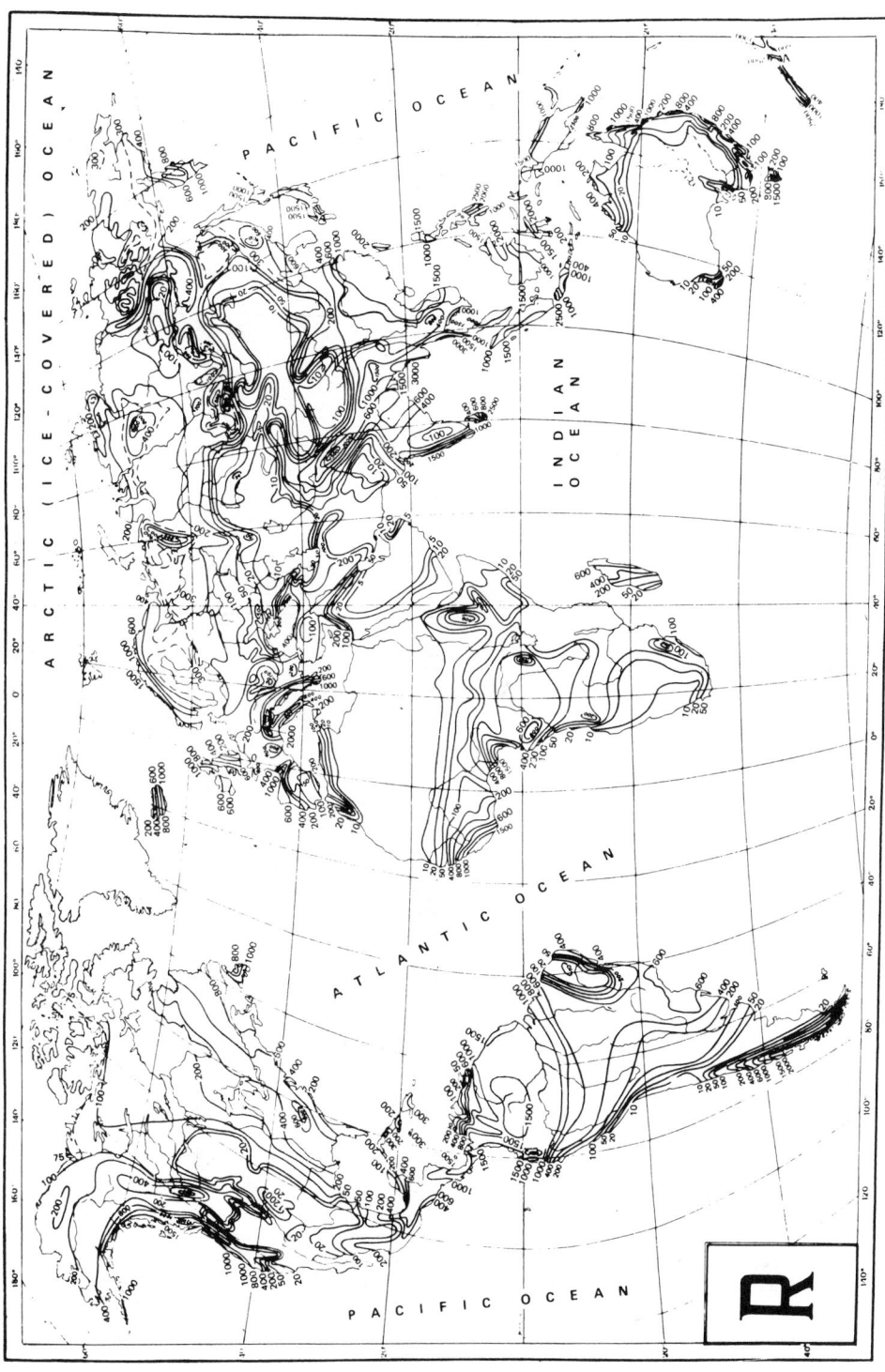

TABLE 7-9. LARGE RIVERS OF THE WORLD

(Source: Young, L.L., U.S. Geological Survey, 1964)

River	Country	Drainage area (thousands of sq mi)	Average discharge at mouth (thousands of cfs)	Rank
North America				
Mississippi [1]	U.S.A. and Canada	1,244	611	7
St. Lawrence	U.S.A. and Canada	498	500	11
Mackenzie	Canada	697	280	17
Columbia	U.S.A. and Canada	258	256	19
Yukon	Canada	360	180	24
Frazer	Canada	92	113	32
Nelson	Canada	414	80	37
Mobile	U.S.A.	42	58	43
Susquehanna	U.S.A.	28	38	48
South America				
Amazon	Brazil	2,231	7,500[2]	1
Orinoco	Venezuela	340	600	8
Parana	Argentina	890	526	10
Tocantins	Brazil	350	360	16
Magdalena	Colombia	93	265	18
Uruguay	(3)	90	136	26
Sao Francisco	Brazil	260	100	34
Africa				
Congo	Congo	1,550	1,400	2
Zambezi	Mozambique	500	250	20
Niger	Nigeria	430	215	22
Nile	Egypt	1,150	100	33
Asia				
Yangtze	China	750	770	3
Brahmaputra	Bangladesh	361	700	4
Ganges	India	409	660	5
Yenisei	U.S.S.R.	1,000	614	6
Lena	U.S.S.R.	936	547	9
Irrawaddy	Burma	166	479	12
Ob	U.S.S.R.	959	441	13
Mekong	Thailand	310	390	14
Amur	U.S.S.R.	712	388	15
Indus	West Pakistan	358	196	23
Kolyma	U.S.S.R.	249	134	27
Sankai (Si)	China	46	127	28
Godavari	India	115	127	29
Hwang Ho (Yellow)	China	260	116	31
Pyasina	U.S.S.R.	74	90	36
Krishna	India	119	69	39
Indigirka	U.S.S.R.	139	64	40
Salween	Burma	108	53	44
Shatt-al Arab [4]	Iraq	209	51	45
Yana	U.S.S.R.	95	35	49
Europe				
Danube	Romania	315	218	21
Pechora	U.S.S.R.	126	144	25
Dvina (Northern)	U.S.S.R.	139	124	30
Neva	U.S.S.R.	109	92	35
Rhine	Netherlands and Germany	56	78	38
Dnepr	U.S.S.R.	194	59	41
Rhone	France	37	59	42
Po	Italy	27	51	46
Vistula	Poland	76	38	47

[1] Includes Atchafalaya River.
[2] Department of Interior News Release, Feb. 24, 1964.
[3] Argentina and Uruguay.
[4] Tigris, Euphrates and Karun.

TABLE 7-10. LONGEST RIVERS OF THE WORLD
(Source: Copyright Encyclopaedia Brittanica, 15th edition, 1974; Reprinted by permission)

name	outflow	length		rank in world (first 100)
		miles	kilo-meters	
World*				
Nile†	Mediterranean Sea	4,132	6,650	1
Amazon	South Atlantic Ocean	4,000	6,437	2
Mississippi	Gulf of Mexico	3,741	6,020	3
Yangtze	East China Sea	3,716	5,980	4
Yenisey	Kara Sea	3,442	5,540	5
Ob	Gulf of Ob	3,362	5,410	6
Huang Ho (Yellow)	Gulf of Chihli	3,011	4,845	7
Congo	South Atlantic Ocean	2,900	4,700	8
Amur	Sea of Okhotsk	2,761	4,444	9
Lena	Laptev Sea	2,734	4,400	10
Africa				
Nile†	Mediterranean Sea	4,132	6,650	1
Congo	South Atlantic Ocean	2,900	4,700	8
Niger	Bight of Biafra	2,600	4,200	14
Zambezi	Mozambique Channel	2,200	3,500	23
Kasai	Congo River	1,338	2,153	59
Orange	South Atlantic Ocean	1,300	2,100	63
White Nile (al-Bahr al-Abyad)	Nile River	1,180	1,900	81
Lualaba	Congo River	1,100	1,770	92
Limpopo	Mozambique Channel	1,100	1,770	92
Juba	Indian Ocean	1,030	1,658	98
Sénégal	South Atlantic Ocean	1,015	1,633	100
Okavango (Cubango)	Okavango Swamp	1,000	1,600	
America, North				
Mississippi-Missouri-Red Rock	Gulf of Mexico	3,741	6,020	3
Mackenzie-Slave-Peace	Beaufort Sea	2,635	4,241	13
St. Lawrence-Great Lakes	Gulf of Saint Lawrence	2,500	4,000	16
Mississippi	Gulf of Mexico	2,348	3,779	19
Missouri	Mississippi River	2,315	3,726	20
Yukon-Nisutlin	Bering Sea	1,979	3,185	26
Rio Grande	Gulf of Mexico	1,885	3,035	28
Nelson-Saskatchewan	Hudson Bay	1,600	2,570	43
Yukon	Bering Sea	1,587	2,554	44
Arkansas	Mississippi River	1,450	2,333	53
Colorado	Gulf of California	1,440	2,320	54
Ohio-Allegheny	Mississippi River	1,306	2,102	62
Red	Mississippi River	1,222	1,966	73
Columbia	North Pacific Ocean	1,214	1,954	76
Brazos	Gulf of Mexico	1,210	1,947	77
Saskatchewan	Lake Winnipeg	1,205	1,939	78
Peace	Slave River	1,195	1,923	79
Churchill	Hudson Bay	1,000	1,600	
Snake	Columbia River	1,000	1,600	
America, South				
Amazon-Ucayali-Apurimac	South Atlantic Ocean	4,000	6,437	2
Paraná	Rio de la Plata	2,485	3,998	18
Madeira-Mamoré-Guaporé	Amazon River	2,082	3,350	24
Juruá	Amazon River	2,040	3,283	25
Purus	Amazon River	1,956	3,148	27
São Francisco	South Atlantic Ocean	1,800	2,900	31
Ucayali-Apurimac	Amazon River	1,701	2,738	37
Tocantins	Pará River	1,700	2,700	38
Araguaia	Tocantins River	1,632	2,627	41
Paraguay	Paraná River	1,584	2,550	45
Pilcomayo	Paraguay River	1,550	2,500	49
Orinoco	South Atlantic Ocean	1,337	2,151	60
Negro	Amazon River	1,250	2,000	68
Tapajós-Teles Pires	Amazon River	1,238	1,992	69
Xingu	Amazon River	1,230	1,980	70
Marañón-Huallaga	Amazon River	1,184	1,905	80
Putumayo (Iça)	Amazon River	1,151	1,852	85
Japurá (Caquetá)	Amazon River	1,148	1,848	86
Guaporé (Iténez)	Mamoré River	1,087	1,749	94
Parnaiba	South Atlantic Ocean	1,056	1,700	97
Uruguay	Rio de la Plata	1,001	1,612	
Ucayali	Amazon River	1,000	1,600	
Solimões	Amazon River	1,000	1,600	

TABLE 7-10. LONGEST RIVERS OF THE WORLD (continued)

name	outflow	length		rank in world (first 100)
		miles	kilo-meters	
Asia				
Yangtze	East China Sea	3,716	5,980	4
Yenisey-Baikal-Selenga	Kara Sea	3,442	5,540	5
Ob-Irtysh	Gulf of Ob	3,362	5,410	6
Huang Ho (Yellow)	Gulf of Chihli	3,011	4,845	7
Amur-Argun	Sea of Okhotsk	2,761	4,444	9
Lena	Laptev Sea	2,734	4,400	10
Ob-Katun	Gulf of Ob	2,696	4,338	11
Irtysh-Chorny Irtysh	Ob River	2,640	4,248	12
Yenisey	Kara Sea	2,543	4,092	15
Mekong	South China Sea	2,500	4,000	16
Ob	Gulf of Ob	2,287	3,680	22
Syrdarya-Arabelsu	Aral Sea	1,876	3,019	29
Nizhnaya Tunguska	Yenisey River	1,857	2,989	30
Brahmaputra	Jamuna River	1,800	2,900	31
Indus	Arabian Sea	1,800	2,900	31
Amur	Sea of Okhotsk	1,755	2,824	35
Euphrates	Shatt al-Arab	1,700	2,700	38
Vilyuy	Lena River	1,647	2,650	40
Amu Darya-Pyandzh	Aral Sea	1,578	2,540	46
Kolyma-Kulu	East Siberian Sea	1,562	2,513	47
Ganges	Padma River	1,557	2,506	48
Ishim	Irtysh River	1,522	2,450	50
Salween	Gulf of Martaban	1,500	2,400	52
Aldan	Lena River	1,412	2,273	55
Olenyok	Laptev Sea	1,411	2,270	56
Syrdarya	Aral Sea	1,374	2,212	57
Kolyma (Kolima)	East Siberian Sea	1,323	2,129	61
Chu Chiang (Pearl)-Hsi	South China Sea	1,300	2,100	63
Irrawaddy	Andaman Sea	1,300	2,100	63
Tarim	Lop Nor Basin	1,261	2,030	66
Chulym-Bely Iyus	Ob River	1,257	2,023	67
Vitim-Vitimkan	Lena River	1,229	1,978	71
Indigirka-Khastakh	East Siberian Sea	1,228	1,977	72
Hsi	South China Sea	1,216	1,957	74
Sungari	Amur River	1,215	1,955	75
Tigris	Shatt al-Arab	1,180	1,900	81
Podkamennya Tunguska	Yenisey River	1,159	1,865	84
Vitim	Lena River	1,141	1,837	87
Chulym	Ob River	1,118	1,799	90
Angara	Yenisey River	1,105	1,779	91
Indigirka	East Siberian Sea	1,072	1,726	95
Han	Yangtze River	1,059	1,705	96
Khatanga	Laptev Sea	1,017	1,636	99
Tobol-Kokpektysay	Irtysh River	1,012	1,628	
Ket	Ob River	1,007	1,621	
Argun	Amur River	1,007	1,620	
Europe				
Volga	Caspian Sea	2,293	3,690	21
Danube	Black Sea	1,770	2,850	34
Ural	Caspian Sea	1,509	2,428	51
Dnepr	Black Sea	1,367	2,200	58
Don	Sea of Azov	1,162	1,870	83
Pechora	Barents Sea	1,124	1,809	88
Kama	Volga River	1,122	1,805	89
Oceania				
Darling	Murray River	1,702	2,739	36
Murray	Great Australian Bight	1,609	2,589	42

*Longest system draining through river listed; for specific rivers comprising systems, see continent entries, except Nile, see footnote †.

†System measured comprises the Nile, White Nile, Mountain Nile (Bahr el-Jebel), Albert Nile, Lake Albert, Victoria Nile, Lake Victoria, and Kagera River.

TABLE 7-11. RIVERS OF THE WORLD RANKED BY SEDIMENT YIELD

(Source: Holeman, Water Resources Research, 1968. Copyright by Am. Geophysical Union)

| River | Drainage basin, 10^3km^2 | Average annual suspended load | | Average discharge at mouth, 10^3cfs |
		Metric tons x 10^6	Metric tons/km^2	
Yellow	673	1,887	2,804	53
Ganges	956	1,451	1,518	415
Brahmaputra	666	726	1,090	430
Yangtze	1,942	499	257	770
Indus	969	435	449	196
Ching	57	408	7,158	2
Amazon	5,776	363	63	6,400
Mississippi	3,222	312	97	630
Irrawaddy	430	299	695	479
Missouri	1,370	218	159	69
Lo	26	190	7,308	—
Kosi	62	172	2,774	64
Mekong...................	795	170	214	390
Colorado	637	135	212	5.5
Red	119	130	1,092	138
Nile	2,978	111	37	100

TABLE 7-12. LARGEST LAKES OF THE WORLD RANKED BY AREA AND VOLUME

(Source: Beeton, Center for Great Lakes Studies, Univ. of Wisconsin, 1973)

Lake	Surface mi^2	Rank in area	Volume mi^3	Rank in volume	Maximum depth ft	Mean depth ft
Caspian	168,450	1	19,035	1	3,103	597
Superior	31,700	2	2,916	4	1,330	489
Victoria	26,557	3	637	8	259	131
Aral	23,932	4	215	14	223	51
Huron	23,000	5	827	7	731	195
Michigan	22,300	6	1,161	6	923	279
Tanganyika	13,124	7	4,659	3	4,821	1,876
Baikal	12,159	8	5,581	2	5,313	2,427
Malawi	11,888	9	2,009	5	2,316	895
Great Bear	11,736	10	529	9	1,464	238
Erie	9,910	11	116	15	210	62
Great Slave	9,804	12	373	12	2,015	204
Winnipeg	9,468	13	76	—	62	43
Ontario	7,340	14	393	11	802	283

TABLE 7-13. MAJOR LAKES OF THE WORLD
(Source: National Geographic Society)

Name	Continent	Area sq mi	Length mi	Depth feet	Elev. feet
Caspian Sea.	Asia-Europe	143,550	760	3,264	−92
Superior.	North America	31,800	350	1,333	600
Victoria	Africa	26,828	250	265	3,720
Aral Sea.	Asia	25,300	280	223	174
Huron	North America	23,000	206	750	579
Michigan	North America	22,400	307	923	579
Tanganyika	Africa	12,700	420	4,710	2,534
Great Bear	North America	12,275	192	1,356	512
Baykal.	Asia	11,780	395	5,315	1,493
Nyasa	Africa	11,430	360	2,226	1,550
Great Slave	North America	10,980	298	2,015	512
Erie	North America	9,910	241	210	570
Winnipeg	North America	9,464	266	60	713
Ontario	North America	7,600	193	802	245
Ladoga	Europe	6,835	120	738	13
Balkhash	Asia	6,720	373	85	1,115
Chad.	Africa	6,300	175	24	787
Maracaibo.	South America	5,127	96	115	S.L.
Onega	Europe	3,710	145	361	108
Volta	Africa	3,276	250	. . . . .	. . . . . .
Titicaca	South America	3,200	122	922	12,506
Athabasca	North America	3,120	208	407	699
Nicaragua	North America	3,100	102	230	105
Eyre.	Australia	2,970	90	4	−52
Rudolf	Africa	2,473	154	200	1,230
Reindeer	North America	2,467	143	. . . . .	1,150
Issyk Kul	Asia	2,355	115	2,303	5,279
Torrens	Australia	2,230	130	. . . . .	106
Vänern	Europe	2,156	91	328	144
Winnipegosis	North America	2,103	141	38	833
Albert.	Africa	2,075	100	54	2,030
Kariba.	Africa	2,050	175	390	1,590
Nettilling	North America	1,956	67	. . . . .	100
Nipigon	North America	1,870	72	540	855
Gairdner	Australia	1,840	90	. . . . .	112
Manitoba	North America	1,817	140	12	814
Urmia	Asia	1,815	90	49	4,180
Mweru	Africa	1,770	76	84	3,010
Kyoga.	Africa	1,710	50	25	3,400
Khanka	Asia	1,700	55	33	226
Lake of the Woods. . . .	North America	1,695	72	69	1,060
Koko (Tsing)	Asia	1,625	68	125	10,515
Dubawnt	North America	1,600	69	. . . . .	764
Great Salt.	North America	1,500	75	48	4,200
Tungt'ing	Asia	1,430	75	. . . . .	36
Van Gölü	Asia	1,419	80	82	5,643

TABLE 7-14. HYDROLOGIC DATA FOR CLOSED LAKES

(Source: Langbein, W.B., U.S. Geological Survey, 1961)

[These lakes occupy topographic sinks with no discharges by surface streams or seepage and with a groundwater gradient toward the lake.]

Lake	Drainage area sq mi	Evaporation ft per year Gross	Net*	Coefficient of variation of lake area**	Response time† years	Overflow expressed as depth over tributary area feet	Salinity Date	Ppm	Mean depth ft	Lake area sq mi
Devils Lake, N. Dak.	3,000	2.5	1.2	0.40	14	2.5	1899	8,470	13	45
							1923	15,210	10	26
							1948	25,000	4.5	14
							1952	8,680	10	20
Basin Lake, Saskatchewan	105	2.25	1.0	.07	25	—	1938-41	11,900	20	16
Quill Lakes, Saskatchewan	2,700	2.0	.75	.15	20	—	1938-41	25,000	10	230
Redberry Lake, Saskatchewan	120	2.25	1.0	.038	50	—	1938-41	14,000	43	27
Great Salt Lake, Utah	21,000	3.3	2.7	.125	9	1.0	1877	138,000	18	2,200
							1932	276,000	13	1,300
Sevier Lake, Utah	16,000	3.7	3.2	.35	3	4	1872	86,400	8	188
Pyramid Lake, Nev.	2,650	4.2	3.7	.04	65	9	1882	3,486	167	200
Walker Lake, Nev.	3,500	4.2	3.8	.075	45	13	1882	2,500	120	110
Mono Lake, Calif.	600	4.1	3.3	.043	35	200	1882	51,170	61	85
Elsinore Lake, Calif.	717	4.5	3.2	.68	3.0	.3	1949	8,880	5	5
Owens Lake, Calif.	2,900	5.5	5.0	.10	10	12	1876	60,000	24	105
							1905	213,700	11	76
Omak Lake, Wash.	100	3.2	2.2	.067	30	19	(1)	5,704	50	5.5
Lake Abert, Oregon	900	3.5	2.5	.5	6	11	1902	76,000	5	50
							1912	30,000	–	–
							1956-59	20,000		60
Summer Lake, Oregon	330	3.5	2.5	1.0	2	61	1901	36,000	10	30
Harney Lake, Oregon	5,300	3.3	2.5	.8	2	1.0	1912	18,000	3	–
Lake Eyre, Australia	550,000	7.5	7.0	2.5	1.5	10		22,380	4.8	47
							1950	40,000 2)	8.5	3,100
							1951	240,000 2)	2.8	740
Lake Corangamite, Australia	1,300	4.0	2.0	.30	10	3.8	1933	105,000	3.5	74
							1950	50,000	6.0	88
							1956	12,000	12	140
Aral Sea, USSR	625,000	3.0	2.6	.10	35	10	—	10,700	52	25,000
Caspian Sea, Asia	1,400,000	3.3	2.8	.015	300	22	—	11,000	600	170,000
Dead Sea, Palestine	12,000	5.1	4.8	.03	40	175	—	220,000	460	390
Lake of Urmia, Iran	20,000	3.0	2.5	.11	9	—	—	148,000	16	1,800
Lake Van, Turkey	6,000	3.3	2.0	.02	150	53	1944	22,400 2)	175	1,450
Tuz Golu, Turkey	4,400	3.4	2.4	.5	1	86	1959	250,000	2	650
Elton Lake, USSR	—	—	3.0	1.0	1.0	—	—	300,000	2.3	110
Baskuntschak Lake, USSR	—	—	3.0	2.0	.5	—	—	260,000	1.15	50

* Net evaporation is gross evaporation minus precipitation.

** Coefficient of variation of lake area is equal to the standard deviation of lake volume divided by the area of the lake.

† Response time is the ratio of a change in lake volume to the corresponding change in rate of discharge.

1) Before 1924

2) Milligrams per liter.

TABLE 7-15. MAJOR DAMS AND RESERVOIRS IN THE WORLD

(Source: U.S. Bureau of Reclamation, 1969)

[This table lists dams having a height exceeding 492 feet, or a total volume of embankment exceeding 20 million cubic yards.]

Name of dam	River and basin	Nearest city	State or province	Country	Type 2	Structural height meters	Length of crest meters	Volume content of dam m3x103	Gross capacity of reservoir m3x106	Year of completion 3
Akosombo-Main	Volta	Accra	Eastern Region	Ghana	R	141	640	7,890	148,000	1965
Almendra	Tormes-Douro	Almendra	Salamanca	Spain	A	203	567	2,600	2,648	1970
Alpe Gera	Comor-Adda-Po	Sondrio	Lombardia	Italy	G	178	520	1,735	65	1965
Amir Kabir 4	Karadj-Caspian Sea	Karadj	Markazi	Iran	A	180	390	750	205	1961
Auburn	N. Fork American-Sacramento	Auburn	California	U.S.A.	MA	209	1,067	4,588	2,837	UC
Balimela	Sileru	Jeypore	Orissa	India	E	70	4,633	22,650	3,823	UC
Beas	Beas-Indus	Talwara Township	Punjab	India	E	134	1,524	33,792	19,576	UC
W.A.C. Bennett 5	Peace-MacKenzie	Hudson Hope	British Columbia	Canada	E	183	2,040	43,729	70,100	1968
Bhakra	Sutlent-Indus	Chandigarh	Punjab	India	G	226	518	4,130	9,868	1963
Bhumiphol (Yanhee)	Ping-Chao Phraya	Tak	North	Thailand	GA	154	486	998	12,200	1964
Bratsk	Angara	Bratsk	Irkutsk	U.S.S.R.	GE	120	5,215	17,000	169,400	1967
Bukharma	Irtish	Ust-Kamenogorsk	Altay	U.S.S.R.	G 6	90	380	1,170	53,000	1960
Cabora Basa	Zambezi	Tete	Mozambique	Portugal	A	168	303	450	159,600	UC
Canelles	Noguera Ribagorzana-Ebro	Estopinan	Huesca	Spain	A	150	210	333	678	1960
Castaic	Castaic CR.-Santa Clara	Castaic	California	U.S.A.	E	104	1,585	33,642	226	UC
Charvak	Chirchik-Sir Darya	Tashkent	Uzbekistan	U.S.S.R.	E	168	762	18,500	2,000	UC
Chirkey	Sulak-Caspian Sea	Makhachkala	North Caucasus	U.S.S.R.	A	233	338	1,226	2,780	UC
Cochiti	Rio Grande	Sante Fe	New Mexico	U.S.A.	E	77	8,595	31,425	633	UC
Contra	Verzasca-Ticino-Po	Locarno	Ticino	Switzerland	A	230	380	660	86	1965
Curnera	Rein de Curnera-Rhine	Sedrun	Grischun	Switzerland	A	152	340	562	40	1967
Dneprodzerzhinsk	Dnieper	Dneprodzershinsk	Ukraine	U.S.S.R.	GE	35	36,292	27,430	2,450	1964
Don Pedro (new)	Tuolume-San Joaquin	La Grange	California	U.S.A.	ER	178	579	12,815	2,504	1970
Dworshak	N. Fork Clearwater-Columbia	Orofino	Idaho	U.S.A.	G	219	1,002	4,970	4,259	UC
Emosson	Barberine	Martigny	Valais	Switzerland	A	180	555	1,100	225	UC
Flaming Gorge	Green-Colorado	Vernal	Utah	U.S.A.	GA	153	392	754	4,674	1964
Fort Peck	Missouri	Fort Peck	Montana	U.S.A.	E	76	6,409	96,029	23,930	1940
Fort Randall	Missouri	Pickstown	South Dakota	U.S.A.	E	50	3,261	38,381	7,243	1956
Gardiner 8	South Saskatchewan	Outlook	Saskatchewan	Canada	E	68	5,090	65,548	9,876	1968
Garrison	Missouri	Riverdale	North Dakota	U.S.A.	E	64	3,444	50,843	30,221	1956
Gatum	Chagres	Colon		Panama	E	35	2,347	17,553	5,433	1912
Gepatsch	Faggenbach-Inn	Landeck	Tyrol	Austria	E	153	630	7,100	140	1964
Glen Canyon	Colorado	Page	Utah-Arizona	U.S.A.	A	216	475	3,747	33,304	1964
Gorky	Volga-Caspian Sea	Gorky	Volga Area	U.S.S.R.	E,G	32	12,905	44,320	8,700	1955
Gokcekaya	Sakarya	Eskisshir	Eskisshir	Turkey	A	158	466	650	910	UC
Goschenalp	Goschenerreuss-Rhine	Goscheren	Uri	Switzerland	E	155	540	9,350	75	1960
Grand Coulee	Columbia	Coulee City	Washington	U.S.A.	G	168	1,272	8,093	11,597	1942
Grand Dixence	Dixence-Rhone	Heremence	Valais	Switzerland	G	284	700	5,957	400	1962
Gran Suarna	Navia	Navia Suarga	Lugo	Spain	MA	152	350	675	700	UC

Dam	Year	River	Reservoir	Location	Country	Purpose				
Guri	1968	Caroni-Orinoco	Sto Tomede Guayana	Bolivar	Venezuela	G,E,R	106[9]	690	3,762	17,700
High Aswan (Saad-El-Aalii)	UC	Nile	Aswan	Aswan	United Arab Republic	E,R	111	3,820	42,620	164,000
Hirakud	1956	Mahanadi	Sambalpur	Orissa	India	G,E	61	4,800	19,180	8,141
Hoover	1936	Colorado	Boulder City	Nevada-Arizona	U.S.A.	A	221	379	3,364	38,296
Hungry Horse	1953	S.F. Flathead-Columbia	Columbia Falls	Montana	U.S.A.	AG	172	645	2,361	4,278
Idikki	UC	Periyar	Thodupuzha	Kerala	India	MA	171	366	466	1,460
Ihla Solteria	UC	Parana-Rio de la Plata	Pereira Barreto	Sao Paulo	Brazil	EG	80	6,185	25,125	21,200
Inguri	UC	Inguri	Zugdidi	Georgia	U.S.S.R.	A	272	690	3,800	1,100
Irkutsk	1956	Angara	Irkutsk	Central Siberia	U.S.S.R.	G,E	44	2,740	12,410	46,000
Iroquois	1958	St. Lawrence	Cornwall	Ontario	Canada	G	23	812		29,960
Ivankovo	1937	Volga-Caspian Sea	Kalinin	Kalinin	U.S.S.R.	E,G	30	9,570	15,450	1,120
Jayakwadi	UC	Godavari		Maharas	India	E	37	9,904	11,780	2,605
Jari	1967	Jari	Jhelum	Azad Kashmir	Pakistan	E	71	1,738	32,440	494
Daniel Johnson [7]	1968	Manicougan-St. Lawrence	Baie Comeau	Quebec	Canada	MA	214	1,306	2,255	141,975
Kakhovka	1955	Dnieper	Kakhovka	Ukraine	U.S.S.R.	E,G	37	1,640	35,640	18,200
Kanev	UC	Dnieper	Kanev	Ukraine	U.S.S.R.	E	25	16,140	37,860	2,620
Kapchagay	UC	Ili	Alma-Ata	Kazakhstan	U.S.S.R.	E	50	2,360	7,910	28,100
Kariba	1959	Zambesi	Lusaka	Boundary	Rhodesia-Zambia	G	128	617	1,032	160,368
Keban	UC	Firat (Euphrates)	Keban	Elazig	Turkey	R,G	207	1,097	15,000	31,000
Kiev	1964	Dnieper	Kiev	Ukraine	U.S.S.R.	E	19	54,100	44,370	3,730
King Paul (Kremasta)	1965	Acheloos	Agrinion	Central Greece	Greece	E,R	160	460	7,800	4,750
Krasnoyarsk	UC	Yenisei	Krasnoyarsk	Krasnoyarsk	U.S.S.R.	G	124	1,065	4,350	73,300
Krememchug	1961	Dnieper	Kremenchug	Ukraine	U.S.S.R.	E,G	30	10,891	27,743	13,500
Kurobegawa No. 4	1964	Kurobe	Omachi	Toyama	Japan	A	186	489	1,360	199
Las Portas	UC	Camba	Villarino Conso	Orensa	Spain	A	152	484	748	752
Luzzone	1963	Brenno di Luzzone-Ticino	Olivone	Ticino	Switzerland	A	208	530	1,350	87
Mangla	1967	Jhelum	Jhelum	Rawalpindi	Pakistan	E	115	3,353	65,651	6,358
Marimbondo	UC	Grande	Fronteira	Minas	Brazil	E	90	3,650	18,614	6,400
Mauvoisin	1958	Drance de Bagnes-Rhone	Fionnay	Valais	Switzerland	A	237	520	2,030	180
Mica	UC	Columbia	Revelstoke	British Columbia	Canada	R	244	792	32,109	24,691
Mingechaur	1953	Kura	Mingechaur	Azerbaijan	U.S.S.R.	E	80	1,550	15,600	16,000
Missi Falls Control South	UC	Churchill	Thompson	Manitoba	Canada	RE	27	1,981	1,070	37,037[11]
Mohamed Reza Shah Pahlavi [10]	1963	Dez-Karun	Andimeshk	Khuezestan	Iran	A	203	212	465	3,350
Montejaque	1924	Gaduares	Montejaque	Malaga	Spain	A	74	87	27,000	40,000
Monteynard	1962	Drac-Isere-Rhone	Monestier	Isere	France	A	155	215	455	240
Mossyrock	1968	Cowlitz-Columbia	Mossyrock	Washington	U.S.A.	MA	184	533	945	1,600
Mratinje	UC	Piva-Drina-Danube	Foca	Crna Cora	Yugoslavia	A	220	261	780	850
Nagawado	1971	Nan & Azusa	Matsumoto	Nagano Pref.	Japan	A	155	356	672	123
New Bullards Bar	1968	North Yuba-Sacramento	Marysville	California	U.S.A.	A	194	671	2,064	1,147
New Melones	UC	Stanislaus-San Joaquin	Modesto	California	U.S.A.	R	191	488	12,211	2,960
Nurek	UC	Vakhsh	Nurek	Tadjikistan	U.S.S.R.	E	310	730	58,000	10,400
Oahe	1963	Missouri	Pierre	South Dakota	U.S.A.	E	75	2,835	70,340	29,110

TABLE 7-15. MAJOR DAMS AND RESERVOIRS IN THE WORLD [1] (continued)

Name of dam	Year of completion [3]	River and basin	Nearest city	State or province	Country	Type [2]	Structural height meters	Length of crest meters	Volume content of dam m3x103	Gross capacity of reservoir m3x106
Okutadami	1961	Tadami	Koide	Fukushima Niigate	Japan	G	157	480	1,640	601
Oroville	1968	Feather-Sacramento	Oroville	California	U.S.A.	E	236	2,316	59,639	4,299
Owen Falls	1954	Lake Victoria-Nile	Jinja		Uganda	G	31	831		204,800
Place Moulin	1965	Buthier-Dora Baltea	Aosta	Valle d'Aosto	Italy	AG	155	663	1,500	100
Reza Shah Kabir	UC	Karoun		Khuzestan	Iran	A	200	380		
Roselend	1961	Doronde Beaufort-Rhone	Albertville	Savoie	France	A,B	150	806	945	187
Ross	1949	Skagit	Rockport	Washington	U.S.A.	A,B	165	396	695	1,733
Rybinsk	1941	Volga-Caspian Sea	Rybinsk	Yaroslavl	U.S.S.R.	GE	30	628	2,545	25,400
Sakuma	1956	Tenryu	Toyohashi	Shizouka Aichi	Japan	G	156	294	1,120	327
Sanmen Hsia	1962	Hwang Ho-Yellow	Loyang	Honan	China	G	107	839		65,000
Santa Giustina	1950	Noce-Adige	Bolzano	Trento	Italy	A	153	124	112	183
San Luis	1967	San Luis-San Joaquin	Los Banos	California	U.S.A.	E	116	5,669	59,386	2,603
Saratov	UC	Volga-Caspian Sea	Saratov	Volga	U.S.S.R.	E	40	1,260	14,563	13,400
Sayansk	UC	Yenisei	Kizil	Krasnoyarsk	U.S.S.R.	A	236	1,068	9,117	31,300
Shasta	1945	Sacramento	Redding	California	U.S.A.	G	183	1,055	6,660	5,551
Speccheri	1958	Leno Di Vallarsa-Adige	Trento	Veneto	Italy	A	157	192	150	10
Swift	1958	Lewis-Columbia	Vancouver	Washington	U.S.A.	E	156	640	11,789	932
Tachien	UC	Tachia	Taipei		Taiwan	A	180	260	430	232
Talbingo	1971	Tumut	Tumut	New South Wales	Australia	R	162	701		921
Tankiangkow	1962	Tan & Han			China	G	130			51,600
Tarbela	UC	Indus	R. Pindi		Pakistan	ER	143	2,750	149,961	13,287
Tignes	1952	Isere-Rhone	Albertville	Savoie	France	A	181	430	635	230
Toktogul	UC	Naryn-Syr Darya	Naryn	Kirghizia	U.S.S.R.	A	215	412	2,660	19,500
Trinity	1962	Trinity Klemath	Lewiston	California	U.S.A.	E	164	792	22,162	3,084
Tsimlyansk	1952	Don	Rostov	Rostov	U.S.S.R.	E,G	39	13,232	33,891	21,850
Tuttle Creek	1962	Big Blue-Missouri	Manhattan	Kansas	U.S.A.	ER	48	2,286	16,056	2,920
Twin Buttes	1963	Concho-Colorado Texas	San Angelo	Texas	U.S.A.	E	41	12,941	16,394	790
Ust-Ilim	UC	Angara	Bratsk	Irkutsk	U.S.S.R.	GE	105	3,565	13,062	59,300
Vaiont	1961	Vaiont-Piave	Belluno	Veneto	Italy	MA	262	190	352	169
Verkhne-Svirskaya	1952	Svir	Leningrad	Leningrad	U.S.S.R.	E,G	32	541	1,520	17,500
Vidraru	1965	Arges-Danube	Curtea de Arges	Arges	Rumania	A	166	305	500	465
Volga-22nd Congress U.S.S.R.	1958	Volga-Caspian Sea	Volgograd	Volga Area	U.S.S.R.	E,R,G	44	3,974	25,247	33,500
Volga-V.I. Lenin	1955	Volga-Caspian Sea	Kuibyshev	Volga Area	U.S.S.R.	E,G	45	3,781	33,869	58,000
Yellowtail	1966	Bighorn-Missouri	Hardin	Montana	U.S.A.	A	160	451	1,182	1,696
Zervreila	1957	Valserrhein-Rhine	Vals	Graubunden	Switzerland	A	151	504	656	100
Zeuzier	1957	Lienne-Rhone	Sion	Valais	Switzerland	A	156	280	300	50
Zeya	UC	Zeya	Zeya	East Siberia	U.S.S.R.	G	113	705	8,000	68,000

1 This table lists dams having a height exceeding 150 meters (492 feet); or a total volume of embankment exceeding 15 million cubic meters (20 million cubic yards); a reservoir capacity of 24,670x106 cubic meters (20 million acre-feet).

2 A—Arch; B—Buttress; E—Earth; G—Gravity; MA—Multi-arch; R—Rockfill.

3 Usually beginning of storage or substantial completion. UC—Under Construction.

4 Formerly Karadj.

5 Formerly Portage Mountain.

6 Hollow Gravity.

7 Formerly Manicougan No. 5.

8 Formerly South Saskatchewan.

9 Ultimate 156 m, Capacity 55,000x106m3 6,000 mw ult.

10 Formerly Dez.

11 Storage provided by construction of 3 dams.

TABLE 7-16. WORLD'S LARGEST HYDROELECTRIC GENERATING PLANTS
(Source: Mermel, T.W., U.S. Bureau of Reclamation, 1971)

[Plants at one location and source of water for which provision is made in intakes and substructure for an installed rated capacity of one million kilowatts or more. UC is under construction.]

	Name of dam	Rated capacity in Megawatts Present	Ultimate	Year of initial operation
1.	Grand Coulee, U.S.A.	2,025	9,771	1941
2.	Guri, Venezuela	524	6,500	1967
3.	Sayansk, U.S.S.R.	—	6,400	UC
4.	Krasnoyarsk, U.S.S.R.	5,080	6,096	1968
5.	Churchill Falls, Canada	—	5,225	UC
6.	Sukhovo, U.S.S.R.	—	5,225	UC
7.	Bratsk, U.S.S.R.	4,500	4,600	1961
8.	Ust-Illimsk, U.S.S.R.	720	4,300	UC
9.	Cabora Basa, Mozambique	—	4,000	UC
10.	Ilha Solteira, Brazil	—	3,200	UC
11.	John Day, U.S.A.	2,160	2,700	UC
12.	Nurek, U.S.S.R.	—	2,700	UC
13.	Volga - 22nd Congress, U.S.S.R.	2,543	2,560	1958
14.	Mica, Canada	—	2,500	UC
15.	Volga - V.I. Lenin, U.S.S.R.	2,100	2,300	1955
16.	W.A.C. Bennett, Canada 1)	1,150	2,270	UC
17.	Iron Gate, Rumania-Yugoslavia	—	2,160	UC
18.	Saad-El-Aali (High Aswan Dam), U.A.R.	1,750	2,100	1967
19.	Tarbela, Pakistan	—	2,100	UC
20.	Chief Joseph, U.S.A.	1,024	2,073	1956
21.	Robert Moses-Niagara, U.S.A.	1,950	1,950	1961
22.	St. Lawrence Power Dam, Canada-U.S.A.	1,880	1,880	1958
23.	The Dalles, U.S.A.	1,119	1,813	1957
24.	Kemano, Canada	835	1,670	1954
25.	Beauharnois, Canada	1,586	1,641	1951
26.	Cheboksary, U.S.S.R.	—	1,632	UC
27.	Inguri, U.S.S.R.	—	1,600	UC
28.	Kariba, Rhodesia-Zambia	600	1,500	1959
29.	Liukiahsia, China	—	1,500	1963
30.	Tumut - 3, Australia	—	1,500	UC
31.	Talbingo, Australia	—	1,500	UC
32.	McNary, U.S.A.	986	1,406	1953
33.	Jupia, Brazil	—	1,400	1961
34.	Marimbondo, Brazil	—	1,400	UC
35.	Sir Adam Beck No. 2, Canada	900	1,370	1954
36.	Hoover, U.S.A.	1,345	1,345	1936
37.	Daniel Johnson, Canada 2)	165	1,353	1970
38.	Wanapum, U.S.A.	831	1,330	1963
39.	Saratov, U.S.S.R.	—	1,290	1967
40.	Priest Rapids, U.S.A.	789	1,262	1959
41.	Castaic, U.S.A. 3)	—	1,250	UC
42.	Keban, Turkey	620	1,240	UC
43.	Kettle Rapids, Canada	714	1,224	UC
44.	Rocky Reach, U.S.A.	712	1,215	1961
45.	Furnas, Brazil	900	1,200	1963
46.	Toktogul, U.S.S.R.	—	1,200	UC
47.	El Chocan, Argentina	—	1,200	UC
48.	Manicouagan No. 3, Canada	—	1,176	UC
49.	Sanmen Hsia, China	—	1,100	
50.	Nizhne-Kamskaya, U.S.S.R.	—	1,090	UC
51.	Dworshak, U.S.A.	—	1,060	UC
52.	Bersimis No. 1, Canada	1,050	1,050	1956
53.	Bhakra, India	450	1,050	
54.	Zeya, U.S.S.R.	—	1,020	UC
55.	Manicouagan No. 2, Canada	—	1,016	1965
56.	Votkinsk, U.S.S.R.	—	1,000	1961
57.	Mangla, Pakistan	300	1,000	UC
58.	Chirkey, U.S.S.R.	—	1,000	UC
59.	Kaniji, Nigeria	—	1,000	UC
60.	Northfield Mountain, U.S.A. 3)	—	1,000	UC

1) Formerly Portage Mountain.   2) Formerly Manicouagan No. 5.   3) Pumped storage.

TABLE 7-17. WORLD HYDROELECTRIC AND THERMAL POWER GENERATING CAPACITY AND PRODUCTION, BY CONTINENT

(Source: Federal Power Commission, 1968)

[Data as of year end 1968]

Geographical division	Installed capacity (MW) [1]			Energy production (Gwh.) [2]			Population (1,000)	Kwh. per capita
	Hydro	Thermal	Total	Hydro	Thermal	Total		
North America	79,772	272,230	352,502	373,468	1,261,731	1,635,199	269,261	6,073
Central America	483	577	1,060	2,104	1,875	3,979	15,563	256
West Indies	187	3,643	3,830	487	14,507	14,994	23,825	629
South America	11,396	12,623	24,019	47,914	41,226	89,140	180,017	495
Europe [3]	120,920	330,682	451,602	436,084	1,337,665	1,773,749	692,902	2,560
Africa	4,942	13,697	18,639	18,702	51,100	69,802	329,216	212
Asia	35,424	70,645	106,069	136,562	299,738	436,300	1,939,831	225
Oceania	5,683	10,653	16,336	18,133	39,468	57,601	18,041	3,193
World	258,807	715,250	974,057	1,033,454	3,047,310	4,080,764	3,468,656	1,176

[1] Megawatts = thousand kilowatts.
[2] Gigawatt-hours = million kilowatt-hours.
[3] Includes all of USSR.

TABLE 7-18. WORLD POTENTIAL AND DEVELOPED HYDROELECTRIC POWER CAPACITY

(Source: M.K. Hubbert, 1974)

Region	Potential hydroelectric power (10^3 megawatts)	Percent of total	Developed hydroelectric power capacity 1967 (10^3 megawatts)	Percent developed
North America	313	11	76	23.0
South America	577	20	10	1.7
Western Europe	158	6	90	57.0
Africa	780	27	5	0.6
Middle East	21	1	1	4.8
Southeast Asia	455	16	6	1.3
Far East	42	1	20	48.0
Australia	45	2	5	11.0
USSR, China and satellites	466	16	30	6.4
World	2,857	100	243	8.5

TABLE 7-19. MAJOR FLOODS AND TIDAL WAVES OF THE WORLD
(Copyright The World Almanac, 1975; amended)

Date	Location	No. of deaths
1887	Hwang-ho River, China	900,000
1889 (May 31)	Johnstown, Pennsylvania	2,200
1900 (Sep. 8)	Galveston, Texas	5,000
1903 (Jun. 15)	Heppner, Oregon	325
1911	Yangtze River, China	100,000
1913 (Mar. 25-27)	Ohio, Indiana	732
1913 (Dec. 1-5)	Brazos River, Texas	177
1915 (Aug. 17)	Galveston, Texas	275
1927	Mississippi River Valley	214
1928 (Mar. 13)	Collapse of St. Francis Dam, Santa Paula, California	450
1928 (Sep. 13)	Lake Okeechobee, Florida	2,000
1937 (Jan. 22)	Ohio, Mississippi Valleys	250
1939	Northern China	200,000
1947	Honshu Island, Japan	1,900
1951 (Aug.)	Manchuria	1,800
1953 (Jan. 31)	Western Europe	2,000
1954 (Aug. 17)	Farahzad, Iran	2,000
1955 (Oct. 7-12)	India, Pakistan	1,700
1959 (Nov. 1)	Western Mexico	2,000
1959 (Dec. 2)	Frejus, France	412
1960 (May 23-24)	Hawaii, Japan, Okinawa	237
1960 (Oct. 10)	East Pakistan	6,000
1960 (Oct. 31)	East Pakistan	4,000
1962 (Feb. 17)	German North Sea coast	343
1962 (Sep. 27)	Barcelona, Spain	445
1963 (Oct. 9)	Dam collapse, Vaiont, Italy	1,800
1963 (Nov. 14-15)	Haiti	500
1964 (Nov. 12)	South Vietnam	7,000
1965 (Jun. 11)	Sanderson, Texas	10
1966 (Nov. 4-6)	Florence, Venice, Italy	113
1967 (Jan. 18-24)	Eastern Brazil	894
1967 (Mar. 19)	Rio de Janeiro, Brazil	436
1967 (Nov. 26)	Lisbon, Portugal	457
1968 (Aug. 7-14)	Gujarat state, India	1,000
1968 (Oct. 7)	Northeastern India	780
1969 (Jan. 18-26)	Southern California	91
1969 (Mar. 17)	Mundau Valley, Alagoas, Brazil	218
1969 (Jul. 4)	Northern Ohio	41
1969 (Oct. 1-8)	Tunisia	500
1969 (Aug. 25)	Western Virginia	189
1969 (Sep. 15)	South Korea	250
1970 (May 20)	Central Romania	160
1970 (Jul. 22)	Himalayans, India	500
1971 (Feb. 26)	Rio de Janeiro, Brazil	130
1972 (Feb. 26)	Collapse of mine waste dam, Buffalo Creek, West Virginia	118
1972 (Jun. 9)	Rapid City, South Dakota	236
1972 (Aug. 7)	5 - week Philippines	454

TABLE 7-20. MAJOR WATERFALLS OF THE WORLD

(Source: National Geographic Society)

[Height--total drop in one or more leaps; †--falls of more than one leap; *--falls that diminish greatly seasonally; **--falls that reduce to a trickle or are dry for part of each year. If river names not shown, they are same as the falls. R.--river; L.--lake; (C)--Cascade-type.]

Name and Location	Ft.	Name and Location	Ft.
AFRICA		France—†Gavarnie (C).	1,385
Angola		**Great Britain—Wales**	
Duque de Braganca, Lucala R.	344	Pistyll Cain, Afon Gain R.	150
Ruacana, Cunene R.	406	Pistyll Rhaiadr.	240
Ethiopia		Scotland	
Baratieri, Ganale Dorya R.	459	Glomach	370
Dal Verme, Ganale Dorya R.	98	Iceland—Detti, Jokul R.	144
Fincha	508	Gull, Hvita R.	101
*Tesissat, Blue Nile R.	140	Italy—Toce (C).	470
Lesotho		**Norway—**	
Maletsunyane	630	†Eastern Mardalsfoss	1,696
Rhodesia-Zambia		Highest fall.	974
*Victoria, Zambezi R.	355	Western Mardalsfoss	1,535
South Africa		(Both on L. Eikesdal).	
*Aughrabies, Orange R.	400	Skjeggedal	525
Howick, Umgeni R.	311	Skykkje, Skykkjua R.	820
†Tugela (5 falls)	3,110	Vettis, Morkedöla R.	1,214
Highest fall ·	1,350	Highest fall.	889
Tanzania-Zambia		Vöring, Bjoreia R.	597
*Kalambo	726	**Sweden**	
Uganda		†Handöl, Handöl Cr.	345
Murchison, Victoria Nile R.	130	†*Stora Sjöfallet, Lule R.	130
Zambia		Tannforsen, Are R.	120
Chirombo, Ieisa R.	880	**Switzerland**	
		†Giétroz (Glacier) (C).	1,640
ASIA		†Diesbach	394
India—**Cauvery	330	†Giessbach	1,312
†**Gersoppa (Jog), Sharavati R.	830	Handegg, Aare R.	151
Japan		Iffigen.	394
**Kegon, L. Chuzenji.	330	Pissevache, La Salanfe R.	213
Yudaki, L. Yuno.	335	†Reichenbach	656
		Rhine	65
AUSTRALASIA		†Simmen, Simme R.	459
Australia		Stäuber	590
New South Wales		Staubbach	984
†Wentworth	518	†Trümmelbach.	1,312
Highest fall	360		
Wollomombi.	1,100	**NORTH AMERICA**	
Queensland.		**Canada**	
Coomera	210	British Columbia	
Tully	450	†Takakkaw (Daly Glacier)	1,650
New Zealand		Highest fall	1,200
*Bowen (from Glaciers).	540	Panther, Nigel Cr.	600
Helena	890	Labrador	
Stirling	505	Churchill Falls, Churchill R.	245
†Sutherland, Arthur R.	1,904	Mackenzie District	
		Virginia, S. Nahanni R.	315
EUROPE		Quebec	
Austria—Upper Gastein	207	Montmorency	251
Lower Gastein	280	**Canada-United States**	
(Both on Ache R.)		Ontario-New York	
†Golling, Schwarzbach R.	200	Niagara: American	193
Krimml (Krimmler)	1,250	Horseshoe.	186

TABLE 7-20. MAJOR WATERFALLS OF THE WORLD (continued)

Name and Location	Ft.	Name and Location	Ft.
United States		Washington	
Arizona		Fairy Falls	700
Mooney, Havasu Cr.	220	Mt. Rainer Nat. Pk	
California		Narada, Paradise R.	168
Feather, Fall R.	640	Sluiskin, Paradise R.	300
Illilouette.	370	Palouse	198
Nevada, Merced R.	594	Snoqualmie	270
**Ribbon.	1,612	Wisconsin	
Silver Strand	1,170	Manitou, Black R.	165
Vernal, Merced R.	317	Wyoming	
†Yosemite	2,425	Yellowstone National Pk.	
Bridalveil	620	Tower	132
*Yosemite (upper)	1,430	Yellowstone (upper)	109
*Yosemite (lower)	320	Yellowstone (lower)	308
Colorado		**Mexico—El Salto**	
Seven	266	**Juanacatlán, Rio Grande de Santiago	66
Georgia			
†Tallulah	251	**SOUTH AMERICA**	
Idaho		**Argentina-Brazil**	
Henry's Fork (upper)	96	†Iguazú	230
Henry's Fork (lower)	70	**Brazil**—Glass	1,325
**Shoshone, Snake R.	195	Herval	400
**Twin, Snake R.	125	Paulo Afonso, São Francisco R.	275
Kentucky		Patos-Maribondo, Rio Grande	115
Cumberland	68	Urubupunga, Alto Paraná R.	40
Maryland		**Brazil-Paraguay**	
Great Potomac R. (C)	90	Sete Quedas, or Guaira Alto Paraná R.	130
Minnesota		**Colombia**—Tequendama,	
**Minnehaha	54	Bogotá R.	427
Montana		Catarata de Candelas, Cusiana R.	984
Missouri	75	**Ecuador**	
New Jersey		Agoyan, Pastaza R.	200
**Passaic	70	**Guyana**	
New York		Kaieteur, Potaro R.	741
Taughannock.	215	King Edward VIII, Semang R.	840
Oregon		King George VI, Utshi R.	1,600
†Multnomah	620	†Marina, Ipobe R.	500
Highest fall	542	Highest fall	300
Tennessee		**Venezuela**—†Angel	3,212
Fall Creek	256	Highest fall	2,648
Rock House Creek	125	Cuquenán	2,000

TABLE 7-21. GLACIAL ICE COVERAGE OF THE WORLD

(Source: Huberty and Flock, Natural Resources, McGraw-Hill, Copyright 1959)

Land Area	Square Miles
Continental Europe	3,880
Continental Asia	43,270
Continental North America	30,900
Continental South America	9,600
South polar regions	5,020,450
North polar regions	721,150
Africa	8
New Zealand	386
New Guinea	6
Total	5,829,650

TABLE 7-22. RUNOFF FROM POLAR GLACIERS INTO THE OCEAN

(Source: Lvovitch, M.I., 1970)

[Runoff in km^3]

Authors	Antarctica	Greenland	Total
G. Wust (1920)	3,500	1,000	4,500
W. Meinardus (1934)	640	—	—
M.I. Lvovitch (1945,1964)	1,100	600	1,700
A.N. Krenke, V.M. Kotlyakov (1969)	2,200	600	2,800

TABLE 7-23. DIMENSIONS OF THE OCEANS

(Source: U.S. Naval Oceanographic Office, 1966)

Ocean	Area (10^9m^2)	Mean depth (meters)	Volume (10^{15}m^3)	Maximum depth (meters)
Arctic	14,090	1,205	17.0	a) 4,880 (4,280 at North Pole)
North Pacific	83,462	3,858	322.0	b) 11,500
South Pacific	65,521	3,891	254.9	c) 10,850
North Atlantic	46,772	3,285	153.6	d) 9,200
South Atlantic	37,364	4,091	152.8	e) 8,260
Indian	81,602	4,284	349.6	f) 7,450
Antarctic	32,249	3,730	120.3	g) —

a) Estimated by U.S. Navy, 1958 (Hydrographic Office Publication No. 9)
b) Marianas Trench (U.S. Navy's TRIESTE, January 1960)
c) Tonga, South Pacific (McGraw-Hill Encyclopedia, 1962 Year Book)
d) Puerto Rican Trench, Western Atlantic (McGraw-Hill Encyclopedia, 1962 Year Book)
e) South Sandwich Islands Trench (McGraw-Hill Encyclopedia, 1962 Year Book)
f) Java Trench, South of Java (McGraw-Hill Encyclopedia, 1962 Year Book)
g) Not yet determined

TABLE 7-24. DIMENSIONS OF INDIVIDUAL SEAS

(Source: U.S. Naval Oceanographic Office, 1966)

Sea	Area (10^9 m2)	Mean depth (meters)	Volume (10^{12} m3)
Tributary to Arctic Ocean			
Norwegian Sea	1,383	1,742	2,408
Greenland Sea	1,205	1,444	1,740
Barents Sea	1,405	229	322
White Sea	90	89	8
Kara Sea	883	118	104
Laptev Sea	650	519	338
East Siberian Sea	901	58	53
Chukchi Sea	582	88	51
Beaufort Sea	476	1,004	478
Baffin Bay	689	861	593
Tributary to North Atlantic			
North Sea	600	91	55
Baltic Sea	386	86	33
Mediterranean Sea	2,516	1,494	3,758
Black Sea	461	1,166	537
Caribbean Sea	2,754	2,491	6,860
Gulf of Mexico	1,543	1,512	2,332
Gulf of St. Lawrence	238	127	30
Hudson Bay	1,232	128	158
Tributary to South Atlantic			
Gulf of Guinea	1,533	2,996	4,592
Tributary to Indian Ocean			
Red Sea	450	558	251
Persian Gulf	241	40	10
Arabian Sea	3,863	2,734	10,561
Bay of Bengal	2,172	2,586	5,616
Andaman Sea	602	1,096	660
Great Australian Bight	484	950	459
Tributary to North Pacific			
Gulf of California	177	818	145
Gulf of Alaska	1,327	2,431	3,226
Bering Sea	2,304	1,598	3,683
Okhotsk Sea	1,590	859	1,365
Japan Sea	978	1,752	1,713
Yellow Sea	417	40	17
East China Sea	752	349	263
Sulu Sea	420	1,139	478
Celebes Sea	472	3,291	1,553
In both North and South Pacific			
South China Sea	3,685	1,060	3,907
Makassar Strait	194	967	188
Molukka Sea	307	1,880	578
Ceram Sea	187	1,209	227
Tributary to South Pacific			
Java Sea	433	46	20
Bali Sea	119	411	49
Flores Sea	121	1,829	222
Savu Sea	105	1,701	178
Banda Sea	695	3,064	2,129
Ceram Sea	187	1,209	227
Timor Sea	615	406	250
Arafura Sea	1,037	197	204
Coral Sea	4,791	2,394	11,470

TABLE 7-25. TEMPERATURES AND SALINITIES OF THE OCEANS

(Source: U.S. Naval Oceanographic Office, 1966)

[Temperatures in degrees Centigrade; salinities in parts per thousand]

NORTH ATLANTIC		Temp.	Salinity
1.	North Polar water	−1 to +2	34.9
2.	Subarctic water	+3 to +5	34.7 to 34.9
3.	North Atlantic central water	+4 to +17	35.1 to 36.2
4.	North Atlantic deep water	+3 to +4	34.9 to 35.0
5.	North Atlantic bottom water	+1 to +3	34.8 to 34.9
6.	Mediterranean water	+6 to +10	35.3 to 36.4
SOUTH ATLANTIC			
1.	South Atlantic central water	+5 to +16	34.3 to 35.6
2.	Antarctic intermediate water	+3 to +5	34.1 to 34.6
3.	Subantarctic water	+3 to +9	33.8 to 34.5
4.	Antarctic circumpolar water	+0.5 to + 2.5	34.7 to 34.8
5.	South Atlantic deep and bottom water	0 to +2	34.5 to 34.9
6.	Antarctic bottom water	−0.4	34 to 36
INDIAN OCEAN			
1.	Equatorial water	4 to 16	34.8 to 35.2
2.	Indian central water	6 to 15	34.5 to 35.4
3.	Antarctic intermediate water	2 to 6	34.4 to 34.7
4.	Subantarctic water	2 to 8	34.1 to 34.6
5.	Indian Ocean deep and antarctic circumpolar water	0.5 to 2	34.7 to 34.75
6.	Red Sea water	9	35.5
SOUTH PACIFIC			
1.	Eastern South Pacific water	9 to 16	34.3 to 35.1
2.	Western South Pacific water	7 to 16	34.5 to 35.5
3.	Antarctic intermediate water	4 to 7	34.3 to 34.5
4.	Subantarctic water	3 to 7	34.1 to 34.6
5.	Pacific deep water and Antarctic circumpolar water	(−1) to 3	34.6 to 34.7
NORTH PACIFIC			
1.	Subarctic water	2 to 10	33.5 to 34.4
2.	Pacific equatorial water	6 to 16	34.5 to 35.2
3.	Eastern North Pacific water	10 to 16	34.0 to 34.6
4.	Western North Pacific water	7 to 16	34.1 to 34.6
5.	Arctic intermediate water	6 to 10	34.0 to 34.1
6.	Pacific deep water and Arctic circumpolar water	(−1) to 3	34.6 to 34.7

TABLE 7-26. COMPOSITION OF SEA WATER

(Source: Hem, J.D., U.S. Geological Survey, 1970; after Goldberg, 1963)

Constituent	Concentration (mg/l)	Principal forms in which constituent occurs	Constituent	Concentration (mg/l)	Principal forms in which constituent occurs
Cl	19,000	Cl^-	U	.003	$UO_2(CO_3)_3^{-4}$
Na	10,500	Na^+	Mn	.002	Mn^{+2}, $MnSO_4$ aq
SO4	2,700	SO_4^{-2}	Ni	.002	Ni^{+2}, $NiSO_4$ aq
Mg	1,350	Mg^{+2}, $MgSO_4$ aq	V	.002	$VO_2(OH)_3^{-2}$
Ca	400	Ca^{+2}, $CaSO_4$ aq	Ti	.001	
K	380	K^+	Co	.0005	Co^{+2}, $CoSO_4$ aq
HCO3	142	HCO_3^-, H_2CO_3 aq, CO_3^{-2} [a]	Cs	.0005	Cs^+
Br	65	Br^-	Sb	.0005	
Sr	8	Sr^{+2}, $SrSO_4$ aq	Ce	.0004	
SiO2	6.4	H_4SiO_4 aq, $H_3SiO_4^-$	Ag	.0003	$AgCl_2^-$, $AgCl_3^{-2}$
B	4.6	H_3BO_3 aq, $H_2BO_3^-$	La	.0003	
F	1.3	F^-	Y	.0003	
N	.5	NO_3^-, NO_2^-, NH_4^+ [b]	Cd	.00011	Cd^{+2}, $CdSO_4$ aq
Li	.17	Li^+	W	.0001	WO_4^{-2}
Rb	.12	Rb^+	Ge	.00007	$Ge(OH)_4$ aq, $H_3GeO_4^-$
P	.07	HPO_4^{-2}, $H_2PO_4^-$, PO_4^{-3}, H_3PO_4 aq	Cr	.00005	
			Th	.00005	
I	.06	IO_3^-, I^-	Sc	.00004	
Ba	.03	Ba^{+2}, $BaSO_4$ aq	Ga	.00003	
Al	.01	c)	Hg	.00003	$HgCl_3^-$, $HgCl_4^{-2}$
Fe	.01	$Fe(OH)_3(c)$	Pb	.00003	Pb^{+2}, $PbSO_4$ aq
Mo	.01	MoO_4^{-2}	Bi	.00002	
Zn	.01	Zn^{+2}, $ZnSO_4$ aq	Nb	.00001	
Se	.004	SeO_4^{-2}	Au	.000004	$AuCl_4^-$
As	.003	$HAsO_4^{-2}$, $H_2AsO_4^-$, H_3AsO_4 aq, H_3AsO_3 aq	Be	.0000006	
			Pa	2×10^{-9}	
Cu	.003	Cu^{+2}, $CuSO_4$ aq	Ra	1×10^{-10}	Ra^{+2}, $RaSO_4$ aq
Sn	.003				

a) Reported HCO3 also includes some carbon present in organic compounds.
b) Total N also includes some dissolved nitrogen gas.
c) Probably present as $Al(OH)_4^-$, AlF^{+2}, and AlF_2+ (Hem, 1968).

TABLE 7-27. APPROXIMATE MINERAL CONTENT OF ONE CUBIC MILE OF SEA WATER

(Source: Smith, The Sun, the Sea, and Tomorrow; Potential Sources of Food, Energy and Minerals from the Sea, Charles Scribners, 1954)

Mineral	Weight, in tons	Mineral	Weight, in tons
Sodium Chloride	120,000,000	Fluorine	6,400
Magnesium Chloride	18,000,000	Barium	900
Magnesium Sulfate	8,000,000	Iodine	100 to 12,000
Calcium Sulfate	6,000,000	Arsenic	50 to 350
Potassium Sulfate	4,000,000	Rubidium	200
Calcium Carbonate	550,000	Silver	up to 45
Magnesium Bromide	350,000	Copper, Manganese, Zinc, Lead	10 to 30
Bromine	300,000	Gold	up to 25
Strontium	60,000	Radium	about 1/6 (ounce)
Boron	21,000	Uranium	7

TABLE 7-28. WATER BALANCE OF THE OCEANS
(Source: Budyko, M.I., 1970)

	Precipitation (cm/year)	Evaporation (cm/year)	Runoff from continents to ocean (cm/year)	Water exchange with other oceans
Atlantic Ocean	89	124	23	−12
Pacific Ocean	133	132	7	8
Indian Ocean	117	132	8	− 7
World Ocean	114	126	12	

TABLE 7-29. DESALTING PLANTS OF ONE MILLION GALLONS PER DAY CAPACITY OR GREATER
(Source: Office of Saline Water, 1970)

Country	Location	Process [1]	Status [2]	Plant capacity 1,000 gpd
United States (USA)				
California	San Diego	MSF	O	1,000
New Mexico	Roswell	VC	O	1,000
Texas	Freeport	VTE	O	1,000
Florida	Siesta Key	ED	O	1,200
Pennsylvania	Clairton	MSF	O	1,440
Texas	Texas City	MSF	O	2,160
California	San Diego	MSF	O	2,600
Florida	Key West	MSF	O	2,620
United States Territories				
Virgin Islands	Virgin Island Water & Power Auth.	VTE	O	1,000
Virgin Islands	St. Thomas	MSF	O	1,000
Virgin Islands	St. Croix	MSF	O	1,500
Virgin Islands	Virgin Islands Water & Power Auth.	MSF	O	2,500
Puerto Rico	Penualas	MSF	O	2,500
Virgin Islands	V.I. Water & Power Auth.-St. Croix	VTE	O	1,000
North America except USA and its territories				
Mexico	Rosarita	MSF	O	7,500
Caribbean				
Curaçao	Mundo Nobo	ST	O	1,074
Bahamas/Br.	Freeport	MSF	O	1,267
Bahamas/Br.	Nassau, New Providence Island	MSF	O	1,440
Bahamas/Br.		MSF	C	2,400
Curaçao	Shell Petroleum	MSF	O	1,584
Cuba	Guantanamo	MSF	O	2,250
Aruba/Neth. Ant.	Balashi Govt.	ST	O	2,688
Aruba/Neth. Ant.	Balashi Govt.	MSF	C	1,584
Curaçao	Mundo Nobo	MSF	O	3,440
Curaçao	Mundo Nobo	MSF	C	1,060
Curaçao	Emmastad	MSF	C	1,590
Antigua	St. Johns	MSF	C	1,200

TABLE 7-29. DESALTING PLANTS OF ONE MILLION GALLONS PER DAY CAPACITY OR GREATER

(continued)

Country	Location	Process [1]	Status [2]	Plant capacity 1,000 gpd
South America				
Venezuela	(Port) Cardon	MSF	O	1,440
Europe (Continental)				
Netherlands	Europoort	VTE	C	1,140
Italy	Italsider	MSF	O	1,200
Italy	Brindisi	MSF	C	2,550
Malta	Valetta	MSF	O	1,200
Italy	Brindisi	MSF	O	1,320
Netherlands	Terneuzen	MSF	O	7,650
Malta	Valetta	MSF	O	4,400
England and Ireland				
England	Kent Oil Refinery	ST	O	1,295
Channel Islands	Jersey	MSF	O	1,800
Australia				
Asia				
Middle East				
Israel	Eilat A	MSF	O	1,000
Kuwait	Shuwaikh C	MSF	O	1,260
Kuwait	Shuwaikh D	MSF	O	1,260
Qatar	Doha	MSF	O	1,800
Qatar	Doha Ced	MSF	O	2,300
Kuwait	Shuwaikh E	MSF	O	2,400
Kuwait	Shuwaikh F	MSF	O	2,400
Kuwait	Shuaiba B	MSF	O	2,400
Kuwait	Shuaiba A	MSF	O	3,600
Kuwait	Shuaiba	MSF	O	4,800
Kuwait	Shuwaikh	MSF	O	4,800
Kuwait	Shuwaikh	MSF	O	4,800
Kuwait	Shuwaikh G	MSF	C	4,800
Saudi Arabia	Jidda	MSF	C	5,000
Oman/TR.	Abu Dhabi	MSF	C	7,200
Africa				
Libya	Benghazi	ED	C	5,100
Morocco/Sp.	Ceuta	MSF	O	1,055
Canary Islands	Las Palmas	MSF	C	5,284
Union of Soviet Socialist Republic				
USSR	Shevchenko	VTE	O	1,200
USSR	Shevchenko	VTE	O	3,500
USSR	Shevchenko	VTE	O	3,500
USSR	Shevchenko	VTE	O	3,500
Total	(59 Plants)			148,551

[1] Type of Process
Distillation
MSF — Multi - stage Flash
VTE — Long Tube Vertical
ST — Submerged Tube
VC — Vapor Compression
Membrane
ED — Electrodialysis

[2] Status
C — Plant under construction
O — Plant in operation

TABLE 7-30. DESALTING PLANTS OF 25,000 GALLONS PER DAY CAPACITY OR GREATER

(Source: Office of Saline Water, 1970)

Area	Number of Plants	Plant Capacity Million gallons per day
United States (USA)	301	43.1
United States Territories	14	10.3
North America except USA and its Territories	12	8.5
Caribbean	31	24.1
South America	23	4.1
England and Ireland	63	15.9
Europe (Continental)	97	31.7
Union of Soviet Socialist Republics (USSR)	9	12.6
Middle East	78	69.1
Africa	50	19.7
Asia	27	4.2
Australia	7	1.3
Totals	712	244.6

TABLE 7-31. DISTRIBUTION OF DESALTING PLANTS BY TYPE OF PROCESS

(Source: Office of Saline Water, 1970)

Process	Number of Plants	Plant Capacity Million gallons per day	Percent of Total Capacity
Total all processes.	712	244.6	100.0
Distillation			
Single Stage Flash (F)	49	6.6	2.7
Multi-Stage Flash (MSF)	196	158.4	64.9
Vertical Tube Evaporator (VTE).	102	30.5	12.3
Submerged Tube (ST)	293	33.5	13.7
Vapor Compression (VC)	22	2.9	1.2
Membrane			
Electrodialysis (ED).	44	12.2	5.0
Reverse Osmosis (RO)	3	0.2	0.1
Freezing			
Vacuum Freezing—Vapor Compression (VFVC) . . .	3	0.3	0.1

TABLE 7-32. SIZE RANGES OF DESALTING PLANTS

(Source: Office of Saline Water, 1970)

Size Range Thousands of gallons per day	Number of Plants	Plant Capacity Millions of gallons per day
25-99	359	18.4
100-499	263	50.7
500-999	32	21.6
1,000-4,999	51	108.7
5,000 and greater	7	45.2
Total	712	244.6

TABLE 7-33. WORLD –WIDE GROWTH OF DESALTING CAPACITY 1961 TO 1968
(Source: Office of Saline Water,)

Year Ending	Municipal Water Use Total Capacity (MGD)	Industrial and Other Water Use [1] Total Capacity (MGD)	Total
1961	17.6	42.2	59.8
1962	20.9	45.5	66.4
1963	28.4	50.4	78.8
1964	32.5	53.5	86.0
1965	39.3	58.9	98.2
1966	52.6	101.6	154.2
1967	102.2	115.3	217.5
1968	121.4	125.8	247.2
Historical Annual Growth in percent	32	17	23
Projection to 1975	835	415	1,250
Projected Annual Growth in percent	32	19	26

[1] This sector is primarily industrial; other water uses include uses for military, demonstration, irrigation, tourism and combined municipal and industrial uses including the large plant No. 9762 in the USSR (31.7 mgd).

FIGURE 7-5. WORLD POPULATION PROJECTION
(Source: Doxiadis, Water for Peace, 1967)

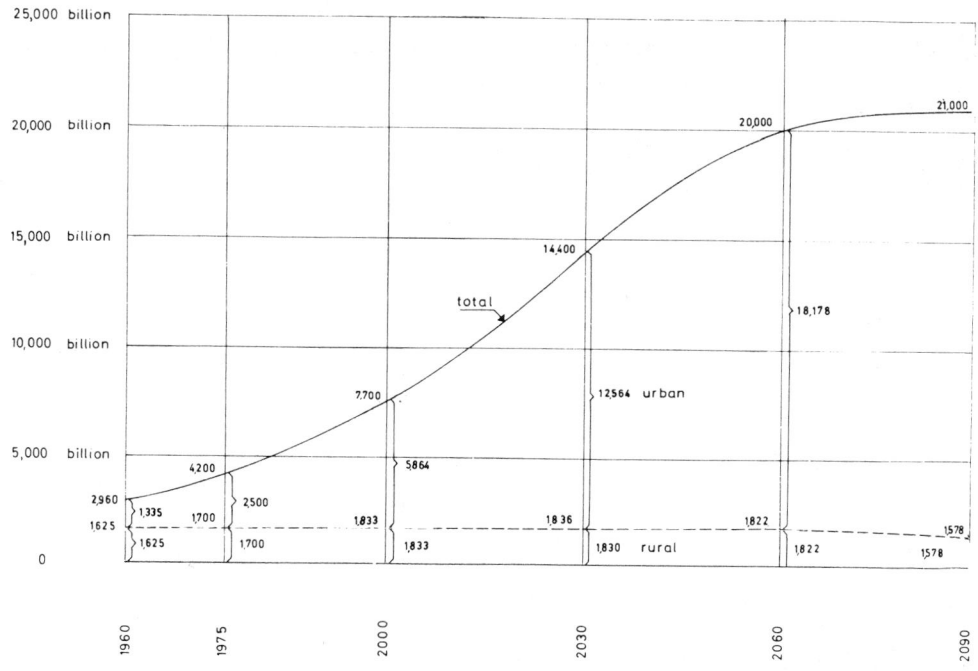

FIGURE 7-6. PROJECTION OF WORLD-WIDE PER CAPITA WATER DEMAND

(Source: Doxiadis, Water for Peace, 1967)

[in m³ per capita per year]

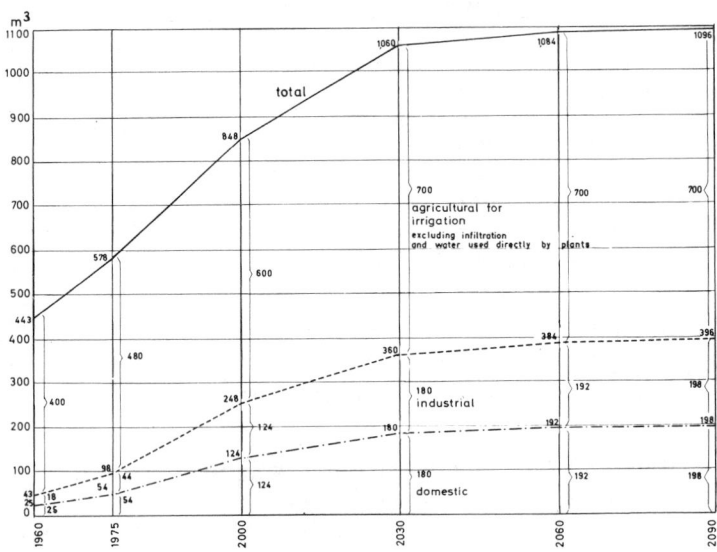

FIGURE 7-7. PROJECTION OF WORLD-WIDE TOTAL WATER DEMAND

(Source: Doxiadis, Water for Peace, 1967)

[in km³ = 10⁹m³ per year]

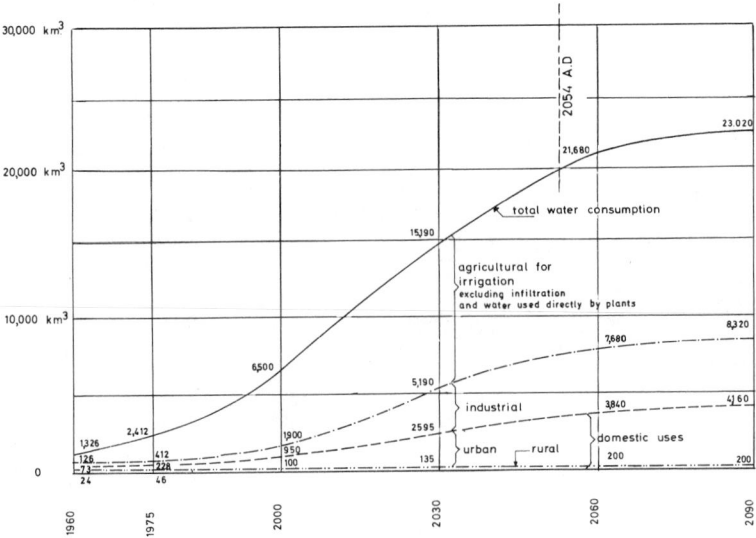

FIGURE 7-8. URBAN AND RURAL POPULATION IN INDUSTRIALIZED AND DEVELOPING COUNTRIES, 1970

(Source: van Damme. WHO International Reference Center for Community Supply, 1973)
[By region]

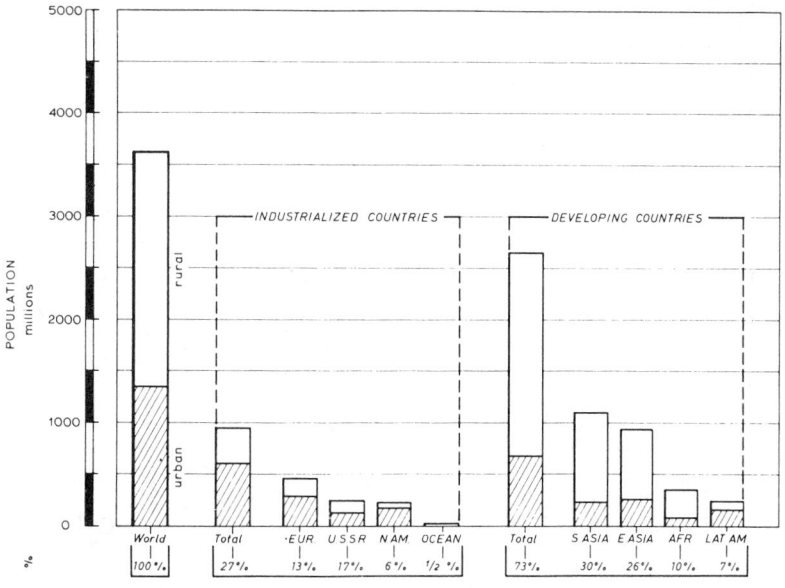

FIGURE 7-9. POPULATION SERVED WITH WATER IN DEVELOPING COUNTRIES (WHO MEMBER STATES), 1970

(Source: van Damme, WHO International Reference Center for Community Supply, 1973)
[By region: AFR - Africa, AMR - Americas, EMR - Eastern Mediterranean, EUR - Europe (Algeria, Morocco, Tunisia), SEAR - Southeast Asia, WPR - Western Pacific]

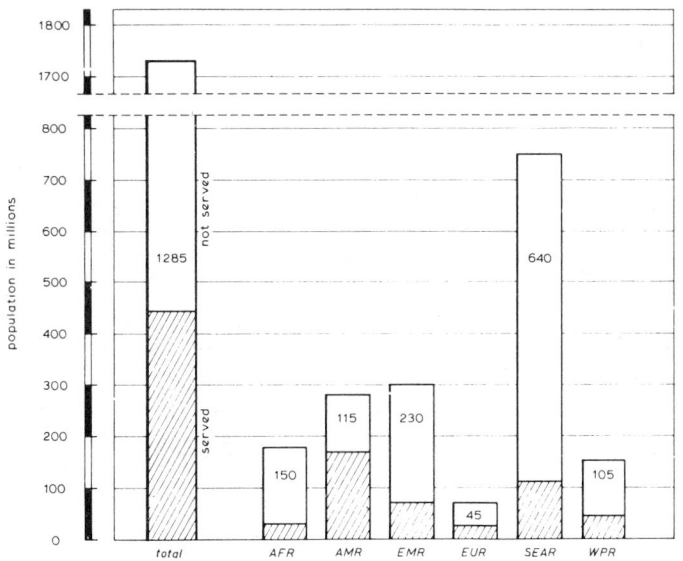

FIGURE 7-10. URBAN AND RURAL POPULATION SERVED WITH WATER IN DEVELOPING COUNTRIES (WHO MEMBER STATES), 1970

(Source: van Damme, WHO International Reference Center for Community Supply, 1973)

[By region: AFR - Africa, AMR - Americas, EMR - Eastern Mediterranean, EUR - Europe (Algeria, Morocco, Tunisia), SEAR - Southeast Asia, WPR - Western Pacific]

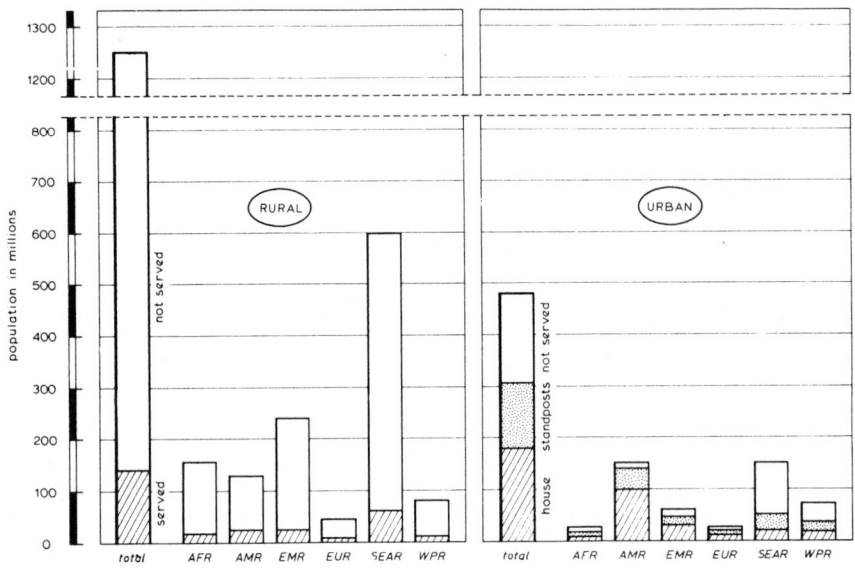

FIGURE 7-11. TRENDS IN COMMUNITY WATER-SUPPLY SITUATION IN DEVELOPING COUNTRIES (WHO MEMBER STATES), 1962–80

(Source: van Damme, WHO International Reference Center for Community Supply, 1973)

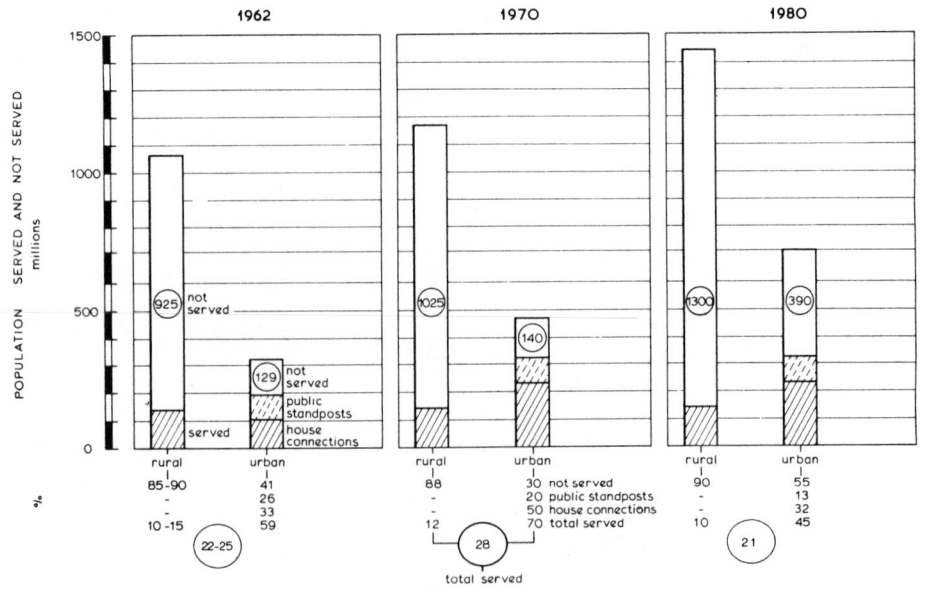

TABLE 7-34. COMMUNITY WATER SUPPLY SERVICES IN DEVELOPING COUNTRIES

(Source: World Health Organization, 1973)

[Data as of December 31, 1970]

Region and Country	Urban population supplied						Rural population with reasonable access		Total population supplied	
	By house connections		By public standposts		Total urban					
	N'000	%	N'000	%	N'000	%	N'000	%	N'000	%
Summary for all developing countries:										
Africa	8,876	29	11,921	39	20,797	68	16,717	11	37,514	21
Americas	95,410	60	26,724	17	122,134	76	29,549	24	151,683	54
Eastern Mediterranean	38,093	59	16,726	26	54,819	84	31,255	18	86,074	33
European Region	12,406	50	5,426	22	17,832	73	18,400	44	36,232	55
South-East Asia	56,391	36	26,798	17	83,189	53	61,095	9	144,284	17
Western Pacific	25,107	65	3,668	10	28,775	75	16,067	21	44,842	40
Total	236,283	49	91,263	19	327,546	68	173,083	14	500,629	29
Africa										
Botswana	16	46	19	54	35	100	149	25	184	29
Burundi	15	15	60	62	75	77	–	–	75	2
Cameroon	150	13	750	64	900	77	1,000	21	1,900	32
Central African Republic	16	4	34	9	50	13	–	–	50	3
Chad	30	11	170	65	200	76	780	22	980	26
Congo	80	28	198	69	278	98	46	7	324	34
Dahomey	33	9	313	86	346	94	455	19	801	29
Gabon	5	5	1	1	6	6	1	n	7	1
Gambia	10	27	26	70	36	97	9	3	45	12
Ghana	652	22	1,483	51	2,135	73	870	14	3,005	33
Guinea	337	75	100	22	437	97	–	–	437	11
Ivory Coast	260	28	656	70	916	97	1,000	29	1,916	44
Kenya	1,000	90	72	77	1,072	97	240	2	1,312	12
Lesotho	5	19	22	81	27	100	–	–	27	3
Liberia	60	43	50	57	140	100	67	6	207	17
Madagascar	236	25	594	63	830	87	45	1	875	12
Mali	160	26	20	3	180	29	–	–	180	3
Mauritania	80	91	6	7	86	98	114	10	200	17
Niger	40	12	180	55	220	68	570	16	790	20
Nigeria	2,810	22	4,650	36	7,460	58	3,586	8	11,046	20
Senegal	300	29	722	69	1,022	98	2,178	74	3,200	81
Sierra Leone	102	27	180	53	282	75	26	1	308	12
Togo	34	13	214	84	248	97	86	5	334	18
Uganda	400	58	216	31	616	89	1,600	20	2,216	25
United Republic of Tanzania	100	11	400	44	500	54	1,200	10	1,700	13
Upper Volta	40	20	100	49	140	68	1,300	25	1,440	25
Zaire	1,205	41	400	14	1,605	55	750	5	2,355	13
Zambia	700	71	255	25	955	97	645	19	1,600	37

TABLE 7-34. COMMUNITY WATER SUPPLY SERVICES IN DEVELOPING COUNTRIES (continued)

Region and Country	Urban population supplied						Rural population with reasonable access		Total population supplied	
	By house connections		By public standposts		Total urban					
	N'000	%	N'000	%	N'000	%	N'000	%	N'000	%
Americas										
Argentina	11,800	60	900	5	12,700	64	654	14	13,354	55
Barbados	85	75	15	13	100	88	138	96	238	93
Bolivia	542	33	1,009	62	1,551	95	53	2	1,604	34
Brazil	28,700	53	12,600	23	41,300	77	12,000	30	53,300	57
Chile	4,200	58	800	11	5,000	69	250	9	5,250	53
Colombia	9,493	73	2,000	15	11,493	88	2,680	31	14,173	65
Costa Rica	611	91	46	7	657	98	639	55	1,296	71
Dominican Republic	934	55	291	17	1,225	72	271	10	1,496	34
Ecuador	1,498	61	312	13	1,810	74	244	6	2,054	33
El Salvador	540	37	446	31	986	68	436	21	1,422	40
Guatemala	725	45	858	35	1,583	98	425	12	2,008	39
Guyana	200	75	60	23	260	98	309	63	569	75
Haiti	159	17	457	48	616	64	—	—	616	12
Honduras	475	65	233	32	708	97	193	10	901	33
Jamaica	475	62	24	3	499	65	647	52	1,146	57
Mexico	18,840	64	3,290	11	22,130	75	5,770	26	27,900	54
Nicaragua	296	34	227	26	523	60	170	14	693	34
Panama	611	87	66	9	677	96	308	39	985	66
Paraguay	162	17	25	3	187	20	81	5	268	11
Peru	3,580	51	620	9	4,200	60	500	7	4,700	34
Trinidad and Tobago	297	54	59	11	356	65	666	100	1,022	95
Uruguay	1,947	85	216	9	2,163	95	60	10	2,223	77
Venezuela	5,400	72	1,520	20	6,920	92	1,285	38	8,205	75
Eastern Mediterranean										
Afghanistan	125	10	200	15	325	25	110	1	435	3
Bahrain	125	89	13	9	138	98	77	99	215	98
Cyprus	275	99	n	n	275	99	300	87	575	92
Democratic Yemen	234	62	100	26	334	88	400	43	734	57
Egypt	11,170	75	2,830	19	14,000	94	18,000	93	32,000	93
Ethiopia	600	31	900	47	1,500	79	n	n	1,800	6
Iran	6,491	55	3,238	27	9,729	82	2,000	12	11,729	40
Iraq	3,600	77	966	21	4,566	98	300	6	4,866	49
Jordan	972	88	108	10	1,080	98	742	59	1,822	77
Kuwait	120	29	280	67	400	96	n	n	400	54
Lebanon	2,100	95	n	n	2,100	95	850	85	2,950	92
Libyan Arab Republic	650	54	200	17	850	71	300	42	1,150	60
Pakistan	5,194	34	6,270	41	11,464	76	1,720	3	13,184	20
Qatar	50	91	4	7	54	98	10	40	64	80
Saudi Arabia	1,500	79	330	17	1,830	97	2,000	34	3,830	49

Somalia	12	2	108	19	120	21	300	13	420	15
Sudan	1,140	71	10	1	1,150	72	1,800	12	2,950	18
Syrian Arab Republic	2,455	89	244	9	2,699	98	1,746	50	4,445	71
Tunisia	1,200	53	850	38	2,050	91	500	17	2,550	49
Yemen	80	23	75	22	155	45	100	2	255	4
European Region										
Algeria	4,500	73	1,000	16	5,500	89	—	—	5,500	39
Morocco	2,200	39	3,000	53	5,200	91	2,900	28	8,100	51
Turkey	5,706	45	1,426	11	7,132	65	15,500	66	22,632	63
South-East Asia										
Bangladesh	750	16	1,100	24	1,850	41	30,000	43	31,850	43
Burma	369	7	1,581	30	1,950	37	2,995	13	4,945	18
India	46,400	39	19,900	17	66,300	56	25,000	6	91,300	16
Indonesia	5,000	23	2,500	12	7,500	35	—	—	7,500	6
Mongolia	98	20	182	37	280	58	—	—	280	21
Nepal	13	2	300	56	313	59	10	n	323	3
Sri Lanka	920	36	800	31	1,720	67	90	1	1,810	14
Thailand	2,841	52	435	8	3,275	60	3,000	10	6,216	17
Western Pacific										
Fiji	144	100	—	—	144	100	50	12	194	35
Khmer Republic	572	64	296	33	868	98	2,400	38	3,268	45
Korea, Republic of	10,430	84	549	4	10,979	88	6,857	34	17,836	55
Laos	300	65	148	32	448	97	1,000	39	1,448	48
Malaysia	3,257	72	881	19	4,138	91	55	1	4,193	38
Philippines	7,350	55	1,312	10	8,662	65	5,060	20	13,722	35
Singapore	1,586	74	—	—	1,586	74	n	n	1,586	74
Viet-Nam, Republic of	1,450	33	475	11	1,925	44	645	5	2,570	14
Western Samoa	18	49	7	19	25	68	—	—	25	16

— Data not available.
n Nil or magnitude negligible.

TABLE 7-35. PER CAPITA CONSUMPTION OF DRINKING WATER IN DEVELOPING COUNTRIES

(Source: World Health Organization, 1973)

[Data as of December 31, 1970; in liters per day]

Region and country	Present consumption						Future consumption					
	Urban				Rural		Urban				Rural	
	With house connections		With public standposts				With house connections		With public standposts			
	Min.	Max.	Min.	Max.	Min.	Max.	Min.	Max.	Min.	Max.	Min.	Max.
Africa .... ⎫	65	290	20	45	15	35	90	275	30	60	20	50
Americas .... ⎪	160	380	25	50	70	190	195	375	30	55	120	195
Eastern Mediterranean ⎪	95	245	30	60	40	85	160	310	55	95	70	115
European Region.... ⎬ Average 1)	65	210	25	40	20	65	90	250	35	45	35	70
South-East Asia .... ⎪	75	165	25	50	30	70	125	225	45	85	40	85
Western Pacific..... ⎭	85	365	30	95	30	95	230	375	50	85	70	145
Average............	90	280	25	55	35	90	150	300	40	70	60	110
Africa												
Botswana	90	1,820	–	–	10	45	90	320	–	45	20	45
Burundi	100	350	10	40	–	–	150	350	40	70	20	40
Cameroon..........	100	180	18	34	10	20	120	200	30	50	20	30
Central African Republic	50	300	–	20	–	–	75	220	15	20	–	–
Chad	60	400	8	25	5	15	150	400	25	45	20	40
Congo	75	100	50	75	10	30	75	100	50	75	–	–
Dahomey	10	125	10	30	10	20	80	150	25	50	20	40
Gambia	60	220	50	150	22	50	90	310	–	55	–	–
Ghana	36	120	22	36	22	100	115	180	20	55	20	45
Guinea	100	150	40	60	–	–	100	150	40	60	–	–
Ivory Coast	20	130	20	40	10	20	40	150	20	40	20	40
Kenya	20	200	5	15	10	20	50	300	20	30	15	75
Lesotho	55	270	–	–	27	54	55	270	35	70	35	70
Liberia	95	190	20	40	20	40	115	285	40	80	40	95
Madagascar	40	250	10	24	4	10	80	250	–	25	10	40
Mali	10	25	–	–	–	–	40	160	30	50	–	–
Mauritania	20	200	20	50	10	50	100	300	50	100	30	100
Niger	100	300	1	2	3	10	–	–	–	–	–	–
Nigeria	45	230	45	70	45	45	90	230	45	90	45	70
Senegal	76	96	18	22	–	–	100	125	20	24	–	–
Togo	60	100	–	–	–	–	–	–	–	–	–	–
Uganda	50	500	5	15	5	10	70	700	20	30	10	15
U.R. of Tanzania	80	110	40	80	25	50	100	150	55	100	20	40
Upper Volta	50	250	5	50	5	20	75	300	10	75	10	50
Zaire	30	250	10	30	20	40	100	300	20	50	5	10
Zambia	200	700	50	90	10	50	130	700	50	90	40	50

	1	2	3	4	5	6	7	8	9	10	11	12
Americas												
Argentina	300	600	—	—	100	200	200	350	—	—	100	200
Barbados	230	1,730	23	68	23	910	135	570	23	68	135	570
Bolivia	60	150	10	25	60	100	150	250	30	—	80	150
Brazil	80	500	10	50	20	75	100	500	30	50	20	75
Chile	180	400	10	20	100	100	250	500	—	—	100	100
Colombia	113	275	—	—	40	200	115	300	—	—	80	150
Costa Rica	175	275	—	—	120	150	200	300	—	—	150	250
Dominican Republic	320	375	55	95	95	130	130	340	55	95	95	130
Ecuador	140	200	—	—	70	140	—	—	—	40	—	—
El Salvador	17	295	40	40	60	100	130	400	40	—	60	100
Guatemala	150	150	—	—	25	25	200	200	—	—	60	60
Guyana	270	360	—	—	135	270	450	550	—	—	360	550
Haiti	150	200	20	40	—	—	175	225	20	40	20	40
Honduras	20	270	—	—	45	140	160	270	—	—	90	135
Jamaica	320	390	45	70	20	320	340	570	45	90	45	450
Mexico	100	350	20	50	50	250	100	350	20	50	50	250
Nicaragua	130	220	40	60	75	150	240	300	40	60	95	150
Panama	190	300	—	—	40	80	210	340	—	—	60	90
Paraguay	160	350	10	30	100	200	160	350	10	30	100	200
Peru	90	400	25	30	80	100	150	300	30	50	600	100
Uruguay	120	250	—	—	100	180	126	262	—	—	105	190
Venezuela	200	300	—	—	150	300	400	600	—	—	150	200
Eastern Mediterranean												
Afghanistan	60	70	20	30	15	20	60	100	30	50	30	50
Bahrain	220	420	23	140	110	340	230	360	—	—	140	280
Cyprus	145	275	—	—	90	145	185	320	—	—	145	185
Democratic Yemen	50	180	10	23	10	18	140	230	18	36	50	70
Egypt	100	260	30	40	30	40	250	350	150	250	40	60
Ethiopia	20	100	5	10	5	10	40	100	10	20	5	15
Iran	75	150	—	25	40	75	150	190	—	—	110	150
Iraq	90	200	—	—	65	130	160	360	—	—	90	145
Jordan	60	120	—	—	30	60	80	150	—	—	40	80
Kuwait	150	220	70	220	—	—	180	410	150	220	—	—
Lebanon	150	200	—	—	80	125	200	250	—	—	100	150
Pakistan	70	180	20	60	20	100	150	220	20	70	50	100
Qatar	150	300	80	110	40	80	230	300	80	150	80	150
Saudi Arabia	50	400	25	50	25	50	150	250	25	50	100	200
Somalia	n	250	20	50	—	—	250	—	60	—	—	—
Sudan	45	900	23	32	14	42	110	1,140 [2]	—	—	18	45
Syrian Arab Republic	150	200	—	—	50	—	—	250	—	—	—	75
Tunisia	100	150	5	10	—	—	150	—	5	10	—	—
Yemen	50	80	30	50	20	40	—	—	—	—	30	60

TABLE 7-35. PER CAPITA CONSUMPTION OF DRINKING WATER IN DEVELOPING COUNTRIES (continued)

Region and country	Present consumption						Future consumption					
	Urban				Rural		Urban				Rural	
	With house connections		With public standposts				With house connections		With public standposts			
	Min.	Max.	Min.	Max.	Min.	Max.	Min.	Max.	Min.	Max.	Min.	Max.
European Region												
Algeria	20	200	10	30	10	60	80	200	50	60	50	60
Morocco	60	260	10	20	–	70	100	300	20	30	20	80
Turkey	120	170	60	70	50	60	–	–	–	–	–	–
South-East Asia												
Bangladesh	45	70	15	25	10	20	70	135	25	45	25	45
Burma	100	180	45	100	22	60	150	220	70	120	50	100
India	50	270	–	–	25	100	90	270	–	–	45	130
Indonesia	50	150	5	20	–	–	86	150	–	100	30	60
Mongolia	24	150	5	60	–	–	187	420	–	–	–	100
Nepal	60	100	40	60	40	60	100	200	60	100	60	100
Sri Lanka	170	220	30	50	20	70	170	220	30	50	20	70
Thailand	120	180	–	–	50	100	150	200	–	–	50	80
Western Pacific												
Fiji	140	260	–	–	–	–	–	270	–	–	9	90
Khmer Rep.	40	400	15	140	–	15	–	–	–	–	–	–
Korea, Rep. of	150	250	–	–	40	80	200	350	–	–	80	120
Laos	50	300	40	80	20	40	100	200	50	150	50	100
Malaysia	18	410	–	–	14	230	250	250	–	–	23	110
Philippines	110	540	–	–	40	110	360	1,100	–	–	180	360
Singapore	–	220	–	–	–	–	–	315	–	–	–	–
Viet-Nam, Rep. of	–	150	–	60	–	–	–	300	–	60	–	–
Western Samoa	–	770	–	–	–	–	–	220	–	50	–	100

1) Averages rounded to nearest 5 liters.
2) Estimation includes garden watering.
– Data not available.
n Nil or magnitude negligible.

TABLE 7-36. SURVEILLANCE OF DRINKING WATER QUALITY IN DEVELOPING COUNTRIES

(Source: World Health Organization, 1973)
[Data as of December 31, 1970]

Region and country	Agency responsible for water quality control — Public health authority — Only	Public health authority — With another agency	Other agency only	Water quality standards — Adopted, based on — WHO standards — Adapted to suit country needs	WHO standards — Adopted in toto	National standards prepared before	Other standards	Not adopted — Not contemplated	Contemplated in near future	In preparation	Extent and frequency of bacteriological examinations — Every supply regularly U[2]	Every supply regularly R[2]	Some supplies regularly U	Some supplies regularly R	Every supply occasionally U	Every supply occasionally R	Some supplies occasionally U	Some supplies occasionally R	No examination U	No examination R
NUMBER OF DEVELOPING COUNTRIES:																				
Africa	12	8	3	5	5	1	2	7	3	5	8	2	11	3	8	2	8	13	—	9
Americas	13	7	1	7	4	3	2	2	3	1	8	2	13	6	2	—	10	16	—	6
Eastern Mediterranean	9	7	1	7	3	—	1	1	6	2	6	3	8	4	5	2	9	13	2	3
European Region	3	—	—	1	1	—	—	—	—	2	—	—	2	1	1	1	2	2	—	—
South-East Asia	5	1	1	1	2	—	2	—	1	2	2	—	4	2	1	1	5	8	—	1
Western Pacific	6	1	1	2	—	1	1	2	1	2	6	1	1	4	—	2	3	5	—	1
Total	48	24	7	23	14	5	8	12	14	14	30	8	39	20	17	7	37	57	2	20
TYPES OF SURVEILLANCE:																				
Africa																				
Botswana	x							x							x	x				x
Burundi	x							x					x							x
Cameroon	x								x				x		x					x
Central African Rep.	x							x									x			x
Chad	x					x							x							x
Congo	x						x						x							x
Dahomey	x							x					x							x
Gabon		x					x				x	x								
Gambia		x		x								x	x							
Ghana			x		x									x	x			x		
Guinea	x				x						x		x				x	x		
Ivory Coast	x								x		x		x				x	x		
Kenya	x									x	x		x				x	x		
Lesotho		x			x						x	x					x	x		
Liberia		x		x							x	x					x	x		
Madagascar	x									x	x		x							x

TABLE 7-36. SURVEILLANCE OF DRINKING WATER QUALITY IN DEVELOPING COUNTRIES (continued)

Region and country	Public health authority: Only	Public health authority: With another agency	Other agency only	WHO standards: Adapted to suit country needs	WHO standards: Adopted in toto	WHO standards: National standards prepared before	Other standards	Not adopted: Not contemplated	Not adopted: Contemplated in near future	Not adopted: In preparation	Every supply regularly U2)	Every supply regularly R2)	Some supplies regularly U	Some supplies regularly R	Every supply occasionally U	Every supply occasionally R	Some supplies occasionally U	Some supplies occasionally R	No examination U	No examination R
Africa (continued)																				
Mali		x						x									x			x
Mauritania	x							x					x					x		x
Niger			x		x						x							x		
Nigeria		x		x											x			x		
Senegal		x							x		x			x						
Sierra Leone 1)		x						x1)	x1)		x					x	x			x
Togo										x										
Uganda	x			x											x			x		
U.R. of Tanzania			x							x	x							x		
Zaire	x				x								x		x		x	x		
Zambia				x											x			x		
Americas																				
Argentina		x				x					x		x					x		
Barbados	x							x			x	x								
Bolivia									x				x					x		x
Brazil		x											x				x	x		
Chile		x		x									x		x		x	x		
Colombia	x			x									x				x	x		x
Costa Rica	x				x									x			x	x		
Dominican Republic		x					x						x					x		
Ecuador	x			x									x				x	x		
El Salvador	x					x					x				x			x		
Guatemala	x			x										x	x		x	x		
Guyana	x								x				x	x			x	x		
Haiti	x							x			x		x				x			x
Honduras	x				x												x	x		x
Jamaica	x				x						x						x	x		x

Mexico . . . .

Nicaragua . . . .

Panama . . . .

Paraguay . . . .

Peru . . . .

Uruguay . . . .

Venezuela . . . .

Eastern Mediterranean

Afghanistan . . . .

Bahrain . . . .

Cyprus . . . .

Democratic Yemen . . . . .

Egypt . . . .

Ethiopia . . . .

Iran . . . .

Iraq . . . .

Jordan . . . .

Kuwait . . . .

Lebanon . . . .

Libyan Arab Rep. . . . .

Pakistan . . . .

Qatar . . . .

Saudi Arabia . . . .

Somalia . . . .

Sudan . . . .

Syrian Arab Rep. . . . .

Tunisia . . . .

Yemen . . . .

European Region

Algeria . . . .

Morocco . . . .

Turkey . . . .

TABLE 7-36. SURVEILLANCE OF DRINKING WATER QUALITY IN DEVELOPING COUNTRIES (continued)

Region and country	Agency responsible for water quality control			Water quality standards							Extent and frequency of bacteriological examinations									
	Public health authority		Other agency only	Adopted, based on				Not adopted			Every supply regularly		Some supplies regularly		Every supply occasionally		Some supplies occasionally		No examination	
	Only	With another agency		WHO standards			Other standards	Not contemplated	Contemplated in near future	In preparation										
				Adapted to suit country needs	Adopted in toto	National standards prepared before					U[2]	R[2]	U	R	U	R	U	R	U	R
South-East Asia																				
Bangladesh	x				x								x					x		
Burma	x				x								x				x	x		
India	x						x				x			x				x		
Indonesia	x									x							x	x		
Mongolia	x						x				x			x				x		x
Nepal			x						x						x		x	x		
Sri Lanka										x			x				x	x		
Thailand		x		x									x				x	x		
Western Pacific																				
Fiji	x										x					x				
Khmer Republic	x							x								x	x	x		
Korea, Republic of	x					x					x			x			x			
Laos			x							x	x			x						
Malaysia		x							x		x			x			x	x		x
Philippines	x						x						x	x				x		
Singapore		x		x							x	x								
Viet-Nam, Republic of	x			x							x							x		
Western Samoa	x							x									x	x		x

1) Sierra Leone contemplated water quality standards for Freetown only.
2) U – Urban
   R – Rural

TABLE 7-37. CONSTRAINTS IN CONSTRUCTION OF COMMUNITY WATER SUPPLY AND SEWAGE DISPOSAL SYSTEMS IN DEVELOPING COUNTRIES[1]

(Source: World Health Organization, 1973)
[Data as of December 31, 1970]

Region and country	Community water supply								Sewage disposal										
	Internal financing insufficient	External assistance insufficient	Insufficient local production of material	Inappropriate administrative framework	Inappropriate financial framework	Inadequate legal framework	Lack of trained personnel	Other 2	Internal financing insufficient	External assistance insufficient	Insufficient local production of material	Inappropriate administrative framework	Inappropriate financial framework	Inadequate legal framework	Lack of trained personnel	Lack of design capability	Inadequate supervision of construction	Lack of national organization responsible for program	Other 3
Africa																			
Botswana	1	2	6	4	5	7	3	—	—	—	—	—	—	—	—	—	—	—	—
Burundi	1	2	6	5	4	8	3	7	2	—	—	—	—	—	5	—	—	1	1
Cameroon	4	8	2	7	6	5	1	3	1	4	8	9	10	11	5	6	7	3	1
Central African Republic	2	5	6	3	4	7	3	—	—	—	—	—	—	—	—	4	—	—	—
Chad	1	2	4	—	—	—	3	—	—	2	3	—	—	—	—	—	—	—	—
Congo	2	3	7	1	5	6	4	—	1	—	4	2	5	—	—	—	—	3	—
Dahomey	1	4	3	2	5	6	—	—	—	—	—	—	—	—	—	—	—	—	—
Gabon	2	1	—	—	—	—	3	—	—	—	—	—	—	—	—	—	—	—	—
Gambia	2	1	—	—	—	5	1	—	—	—	—	—	—	—	—	—	—	—	—
Ghana	1	4	3	—	—	—	—	—	—	—	—	—	—	—	—	—	—	—	—
Guinea	1	1	3	—	—	5	3	—	2	1	3	—	—	—	6	4	5	—	—
Ivory Coast	2	2	—	4	—	—	1	—	5	10	9	2	4	3	8	6	7	1	—
Kenya	2	4	7	3	5	6	—	—	5	3	10	6	8	7	2	11	12	1	4,9
Lesotho	2	4	3	5	—	—	1	—	—	—	2	—	—	—	3	5	—	—	—
Liberia	4	1	2	—	5	—	3	—	1	4	—	—	—	—	—	—	—	—	—
Madagascar	1	4	2	—	—	3	3	—	—	—	—	—	—	—	—	—	—	—	—
Mali	2	2	3	—	—	—	4	—	1	1	6	7	8	9	3	1	1	—	—
Mauritania	2	1	6	4	5	7	3	—	1	2	9	4	6	—	3	5	4	—	—
Mauritius	—	—	—	—	—	—	—	—	—	8	9	7	8	—	4	2	7	—	—
Niger	2	4	7	3	6	5	1	—	2	3	—	—	—	—	—	5	6	5	—
Nigeria	1	4	3	6	5	7	2	—	—	—	—	—	—	—	—	—	—	1	—
Senegal	1	—	1	—	—	—	1	—	—	—	—	—	—	—	—	—	—	—	—
Sierra Leone	2	1	5	6	4	7	3	—	—	—	—	—	—	—	—	—	—	—	—
Togo	1	2	3	7	4	6	5	—	3	4	—	—	—	—	—	—	—	—	—
Uganda	1	2	4	5	6	7	3	—	—	—	1	—	—	—	2	6	5	—	—
United Republic of Tanzania	1	2	4	—	—	—	3	—	—	—	—	—	—	—	—	—	—	—	—

TABLE 7-37. CONSTRAINTS IN CONSTRUCTION OF COMMUNITY WATER SUPPLY AND SEWAGE DISPOSAL SYSTEMS IN DEVELOPING COUNTRIES [1]
(continued)

Region and country	Community water supply								Sewage disposal										
	Internal financing insufficient	External assistance insufficient	Insufficient local production of material	Inappropriate administrative framework	Inappropriate financial framework	Inadequate legal framework	Lack of trained personnel	Other 2	Internal financing insufficient	External assistance insufficient	Insufficient local production of material	Inappropriate administrative framework	Inappropriate financial framework	Inadequate legal framework	Lack of trained personnel	Lack of design capability	Inadequate supervision of construction	Lack of national organization responsible for program	Other 3
Africa (continued)																			
Upper Volta	1	4	2	5	7	6	3	—	1	1	1	—	1	—	—	1	—	—	—
Zaire	1	2	6	3	4	7	5	8	1	6	—	3	5	7	4	7	7	7	—
Zambia	4	8	7	2	6	5	1	3	1	3	2	3	5	7	4	7	7	7	—
Americas																			
Argentina	1	6	—	2	5	3	4	—	1	6	—	2	5	3	4	—	—	—	—
Barbados	4	7	6	2	5	3	1	—	—	—	—	—	—	—	—	—	—	—	—
Bolivia	1	5	2	3	4	—	5	—	1	3	—	2	4	5	5	—	—	—	2
Brazil	1	2	6	4	3	7	6	—	1	2	6	4	3	7	—	—	—	—	—
Chile	4	5	7	1	3	2	5	—	—	—	—	—	—	—	5	—	—	—	—
Colombia	1	7	8	3	4	6	4	2	1	7	8	3	4	6	5	8	—	—	2
Costa Rica	1	2	6	7	5	3	6	—	1	5	—	3	2	4	6	9	9	10	—
Dominican Republic	2	5	3	4	1	7	6	—	7	6	7	2	3	4	8	8	5	—	1
Ecuador	1	7	2	4	5	3	7	—	1	5	—	4	2	3	—	9	9	6	1
El Salvador	1	2	3	4	5	6	7	—	1	2	7	4	3	5	6	—	5	3	—
Guatemala	1	7	4	3	6	2	1	—	2	4	—	4	5	10	5	8	6	—	11
Guyana	1	—	—	—	2	—	3	—	4	8	7	1	3	—	8	7	8	1	—
Haiti	1	3	6	4	2	5	4	—	1	—	—	2	5	4	4	7	—	6	—
Honduras	2	7	—	1	—	5	2	—	6	10	9	5	3	1	—	7	—	3	—
Jamaica	5	6	6	1	3	4	5	—	—	—	9	5	5	—	6	—	—	—	—
Mexico	1	3	7	2	3	4	—	—	1	3	—	2	3	2	4	4	9	—	2
Nicaragua	1	2	—	—	—	2	7	4	1	4	5	3	3	6	5	8	—	—	—
Panama	1	2	2	4	4	4	5	8	1	6	7	—	2	2	5	—	—	—	—
Paraguay	3	6	4	3	3	2	4	—	4	9	8	7	3	5	2	1	—	—	—
Peru	3	6	7	1	2	5	1	8	9	9	—	1	1	1	1	1	1	—	—
Uruguay	6	6	5	1	1	4	—	—	1	4	—	1	1	3	4	1	—	1	—
Venezuela	1	7	6	2	5	4	3	—	—	4	—	—	2	—	4	—	—	4	—

East Mediterranean

Country																					
Afghanistan	3	7	4	1	5	6	2	5	7	—	1	3	3	8	5	7	2	4	5	10	9	—
Bahrain	1	2	4	7	6	3	5	7	2	—	7	—	—	—	—	—	—	—	—	—	—	
Cyprus	1	2	6	4	5	3	7	4	8	—	4	—	—	—	—	—	—	—	—	—	—	
Democratic Yemen	2	1	4	8	6	7	3	6	5	5	—	—	—	—	—	—	—	—	—	—	—	
Egypt	1	2	3	5	4	7	6	4	8	8	—	—	—	—	—	—	—	—	—	—	—	
Ethiopia	1	2	3	5	6	5	4	5	3	—	1	2	2	4	6	5	4	1	6	4	—	1,2,3
Iran	2	4	7	1	5	6	3	5	2	—	1	—	4	2	5	3	1	6	4	—	5	—
Iraq	6	3	1	3	5	4	7	7	2	—	—	1	2	—	—	—	3	1	—	1	—	2
Jordan	1	6	1	4	3	2	6	6	7	—	—	—	—	10	7	5	4	1	2	3	9	—
Lebanon	—	—	4	—	—	1	2	3	6	2	—	—	—	—	8	—	1	5	5	1	—	—
Libyan Arab Republic	7	4	5	2	6	3	1	6	4	—	10	—	6	8	7	5	4	1	2	3	9	—
Pakistan	1	7	6	3	4	8	5	4	3	2	—	—	—	—	—	—	1	1	7	4	—	—
Qatar	—	6	5	2	2	4	2	3	2	—	—	—	—	—	1	5	5	1	7	4	—	—
Saudi Arabia	4	—	2	2	—	2	—	—	1	—	1	2	9	10	11	5	6	7	8	4	3	3
Somalia	3	4	2	3	5	1	—	6	4	—	—	2	9	—	—	—	—	—	—	—	—	
Sudan	4	1	2	6	6	7	5	5	2	—	1	3	—	—	—	7	6	7	8	4	3	3
Syrian Arab Republic	1	2	3	3	—	2	6	7	4	—	—	—	—	—	—	—	—	—	—	—	—	
Tunisia	2	3	7	5	3	4	2	6	2	—	1	10	11	—	—	5	6	7	8	4	3	3
Yemen	5	4	1	1	—	2	6	2	—	—	—	—	—	—	—	—	—	—	—	—	—	

European Region

Country																					
Morocco	3	4	7	2	6	1	2	—	5	—	1	3	1	—	7	5	4	6	2	—	—	—
Turkey	1	2	3	—	—	—	5	—	1	—	—	—	—	—	—	—	4	5	6	2	—	—

South-East Asia

Country																					
Bangladesh	1	2	2	6	4	5	7	—	5	—	1	1	3	3	7	4	5	8	9	10	—	—
Burma	—	3	4	3	2	5	—	6	—	6	1	—	4	4	4	4	4	4	9	3	5	—
India	1	4	4	6	5	2	7	2	6	—	1	1	2	—	3	4	—	—	—	5	—	—
Indonesia	1	4	1	6	6	1	7	—	4	—	1	3	8	10	2	6	4	7	9	5	—	—
Mongolia	—	1	3	5	6	2	7	5	5	—	1	1	7	6	5	—	8	9	10	3	3	—
Nepal	1	4	5	1	2	3	2	2	3	—	1	4	6	7	3	4	9	8	10	2	2	2
Sri Lanka	6	5	4	5	6	7	3	7	3	—	4	2	—	3	2	—	—	—	—	1	—	—
Thailand	1	3	7	4	2	2	5	6	—	—	—	—	—	—	—	—	—	5	—	—	—

Western Pacific

Country																					
Fiji	2	3	—	—	6	—	2	1	7	—	7	3	2	10	9	9	5	3	4	8	3	—
Khmer Republic	—	8	—	6	6	7	4	—	1	1	2	4	5	3	11	5	9	4	6	8	7	1
Korea, Rep. of	1	—	—	—	—	2	7	1	—	—	1	6	7	2	3	7	5	9	10	10	10	1
Laos	2	1	3	5	6	4	4	—	—	—	10	—	—	—	—	5	3	—	—	—	—	
Malaysia	—	—	3	2	4	1	7	—	5	5	10	9	—	9	4	3	6	8	1	—	—	2
Philippines	1	4	3	6	5	7	2	6	6	—	—	7	4	5	7	5	2	3	1	—	—	
Singapore	—	—	—	—	—	—	—	—	—	—	1	2	5	4	6	—	1	—	2	3	8	—
Viet-Nam, Rep. of	2	3	—	—	—	—	—	—	4	—	1	2	—	—	4	4	1	—	3	—	3	—
Western Samoa	1	—	—	—	4	—	—	—	—	—	1	2	—	5	6	7	—	—	—	—	—	

Footnotes on next page.

Footnotes:

— Nil or magnitude negligible.

1) Numbers indicated in the table give the rank of importance (number 1 showing greatest importance).

2) Other Constraints: Community Water Supply

Africa
> Burundi: Lack of coordination between various organizations dealing with water supply (REGIDESO, AIDR, FED, PNUD, bilateral assistance).
> Cameroon: Constraint not specified.
> Zaire: Health education and economic conditions of the population.
> Zambia: Living accommodation and supporting services.

Americas
> Colombia: Lack of well prepared projects.
> Nicaragua: Lack of health education outside urban areas.
> Panama: Dispersion of rural population.
> Peru: Minimum standard of water quality.

East Mediterranean
> Democratic Yemen: Inability of citizens to meet water rates.
> Ethiopia: High cost of water to consumer.
> Pakistan: Constraint not specified.

South-East Asia
> India: Low priority in development program.

Western Pacific
> Khmer Republic: War situation.
> Malaysia: Lack of effective competition in the consulting engineering field.

3) Other Constraints: Sewage Disposal

Africa
> Central African Republic: Lack of facilities for water distribution.
> Kenya: 4th constraint: Lack of records on existing facilities.
> 9th constraint: Education of the public.

Americas
> Colombia: Lack of good projects.
> Dominican Republic: There are no programs.
> Guyana: High cost of facilities due to topography and soil conditions.
> Mexico: Failure of cooperation.

East Mediterranean
> Ethiopia: 1st constraint: No need felt.
> 2nd constraint: Most dwellings are without running water.
> 3rd constraint: The necessary investment was given low priority in national development plans.
> Iraq: Necessity not realized.
> Tunisia: No national authority responsible for the operation of treatment plants.

South-East Asia
> Nepal: No need expressed by communities.

Western Pacific
> Khmer Republic: War situation.
> Malaysia: Priority at the national level too low.

TABLE 7-38. POPULATION SERVED BY SEWAGE DISPOSAL FACILITIES IN DEVELOPING COUNTRIES

(Source: World Health Organization, 1973)
[Data as of December 21, 1970; by type of service]

Region and country	Urban											Rural with adequate disposal		Total	
	Connected to public sewerage system					Household systems				Total urban					
	Conventional treatment	Oxidation ponds	Without treatment	Total		Pit privy, septic tank	Buckets	Total							
	N'000	N'000	N'000	N'000	%	N'000	N'000	N'000	%	N'000	%	N'000	%	N'000	%
Africa	696	159	347	1,202	11	3,431	953	4,384	40	5,586	51	13,534	18	19,120	22
Americas	2,933	1,614	45,699	50,246	34	46,041	20	46,061	31	96,307	65	25,595	22	121,902	46
Eastern Mediterranean ..	1,023	164	751	1,938	8	21,274	300	21,574	86	23,512	94	14,704	21	38,216	40
European Region.......	267	20	2,976	3,263	27	1,148	355	1,503	13	4,766	40	848	5	5,614	19
South-East Asia	4,468	500	36,659	41,627	26	31,950	43,220	75,170	48	116,797	74	23,055	3	139,852	16
Western Pacific........	1,341	19	8,633	9,993	26	14,182	6,099	20,281	53	30,274	80	3,870	5	34,144	31
Total	10,728	2,476	95,065	108,269	27	118,026	50,947	168,973	42	277,242	69	81,606	8	358,848	25
Africa															
Burundi	—	—	14	14	14	80	—	80	82	94	97	N.A.	N.A.	94	3
Central African Republic ..	—	—	1	1	0	391	—	397	100	398	100	15	1	413	27
Chad	—	1	—	1	0	28	—	28	11	29	11	13	0	42	1
Dahomey	—	—	45	45	12	240	60	300	81	345	93	17	1	362	13
Guinea	14	—	46	60	13	391	—	391	87	451	100	60	2	511	13
Ivory Coast	—	—	110	110	12	65	45	110	12	220	23	—	—	220	5
Kenya	440	75	—	515	47	341	250	591	53	1,106	100	4,453	45	5,559	50
Liberia	30	1	—	32	26	79	10	89	74	121	100	100	9	221	19
Madagascar	10	—	20	30	3	350	570	920	97	950	100	—	—	950	4
Mali	—	—	—	—	—	390	—	390	63	390	100	—	—	390	8
Mauritania	60	—	—	60	68	28	—	28	32	88	100	—	—	88	7
Mauritius	—	35	61	96	23	90	8	98	23	194	46	449	100	643	74
Niger	—	—	—	—	—	30	—	30	9	30	9	2	0	32	1
Uganda	100	26	24	150	22	435	6	441	64	591	85	7,000	87	7,591	87
Upper Volta ,,,,,,,,,	—	—	—	—	—	195	—	195	95	195	95	—	—	195	4
Zaire	—	—	25	25	1	200	—	200	7	225	8	875	6	1,100	6
Zambia	42	21	—	63	6	92	4	96	10	159	16	550	16	709	16

TABLE 7-38. POPULATION SERVED BY SEWAGE DISPOSAL FACILITIES IN DEVELOPING COUNTRIES (continued)

Region and country	Urban — Connected to public sewerage system — Conventional treatment N'000	Oxidation ponds N'000	Without treatment N'000	Total N'000	%	Household systems — Pit privy, septic tank N'000	Buckets N'000	Total N'000	%	Total urban N'000	%	Rural with adequate disposal N'000	%	Total N'000	%
Americas															
Argentina	1,000	200	5,000	6,200	31	10,000	—	10,000	51	16,200	82	4,200	89	20,400	83
Bolivia	—	100	243	343	21	90	—	90	6	433	27	127	4	560	12
Brazil	1,200	120	14,280	15,600	29	29,880	—	29,880	56	45,480	85	10,384	26	55,864	59
Colombia	100	100	7,717	7,817	60	2,000	—	2,000	15	9,817	75	3,060	35	12,877	59
Costa Rica	57	2	151	210	31	232	—	232	34	442	66	467	40	909	50
Dominican Republic	117	24	136	277	16	796	N.A.	796	47	1,073	63	1,444	54	2,517	58
Ecuador	N.A.	50	1,333	1,383	57	N.A.	—	N.A.	—	1,383	67	N.A.	—	1,383	22
El Salvador	—	—	524	524	36	393	N.A.	393	27	917	64	272	13	1,189	34
Guatemala	N.A.	9	716	725	45	83	N.A.	83	5	808	50	515	14	1,323	26
Guyana	—	—	66	66	25	154	—	154	58	220	83	450	92	670	89
Haiti	—	—	—	—	—	669	—	669	70	669	70	43	1	712	13
Honduras	15	2	403	420	58	30	—	30	4	450	62	173	9	623	23
Jamaica	94	—	—	94	12	396	—	396	52	490	64	1,249	100	1,739	86
Mexico	N.A.	700	7,072	7,772	26	N.A.	N.A.	N.A.	—	7,772	26	N.A.	—	7,772	15
Nicaragua	—	2	285	287	33	N.A.	N.A.	N.A.	—	287	33	90	8	377	18
Panama	N.A.	—	482	482	68	118	N.A.	118	17	600	85	520	66	1,120	75
Paraguay	—	N.A.	131	131	14	N.A.	—	N.A.	—	131	14	N.A.	—	131	5
Peru	250	300	2,950	3,500	50	300	—	300	4	3,800	54	1,000	15	4,800	35
Uruguay	150	5	1,060	1,215	53	900	20	920	40	2,135	93	66	11	2,201	76
Venezuela	50	N.A.	3,150	3,200	43	N.A.	N.A.	N.A.	—	3,200	43	1,535	45	4,735	43
Eastern Mediterranean															
Afghanistan	15	—	—	15	1	1,278	—	1,278	99	1,293	100	2,400	15	3,693	21
Ethiopia	—	—	155	155	8	1,500	—	1,500	79	1,655	87	1,800	8	3,455	14
Iran	200	—	300	500	4	11,056	300	11,356	96	11,856	100	8,500	50	20,356	71
Iraq	178	14	20	212	5	4,270	—	4,270	91	4,482	96	20	—	4,502	46
Libyan Arab Republic	210	—	10	220	18	800	—	800	67	1,020	85	300	42	1,320	69
Saudi Arabia	—	150	150	150	8	800	—	800	42	950	50	700	12	1,650	21
Tunisia	420	—	266	686	30	1,570	—	1,570	70	2,256	100	984	34	3,240	63
European Region															
Algeria	N.A.	N.A.	100	100	2	380	N.A.	380	6	480	8	480	6	960	7
Morocco	207	20	2,876	3,163	55	768	355	1,123	20	4,286	75	368	4	4,654	29

South-East Asia

Bangladesh	—	—	500	500	11	2,000	1,200	3,200	70	3,700	81	100	0	3,800	5
Burma	58	—	205	263	5	1,600	600	2,200	42	2,463	47	7,200	32	9,663	34
India	4,000	500	35,500	40,000	34	14,000	40,000	54,000	46	94,000	80	5,000	1	99,000	18
Indonesia	410	—	50	460	2	10,000	—	10,000	47	10,460	49	4,250	4	14,710	2
Nepal	—	—	40	40	8	150	20	170	32	210	39	5	—	215	2
Sri Lanka	—	—	364	364	14	600	1,400	2,000	78	2,364	92	4,000	39	6,364	50
Thailand	—	—	—	—	—	3,600	—	3,600	66	3,600	66	2,500	8	6,100	17

Western Pacific

Fiji	14	5	17	36	28	75	—	75	59	111	87	400	95	511	93
Khmer Republic	—	—	739	739	83	150	—	150	17	889	100	120	2	1,009	14
Korea, Republic of	—	—	3,840	3,840	31	340	3,490	3,830	31	7,670	61	N.A.	—	7,670	24
Malaysia	192	4	202	398	9	1,617	1,509	3,126	69	3,524	77	3,259	51	6,783	62
Philippines	130	—	350	480	4	11,185	—	11,185	84	11,665	88	N.A.	—	11,665	30
Singapore	1,000	—	—	1,200	47	170	800	970	46	1,970	93	—	—	1,970	93
Viet-Nam, Republic of	5	10	3,485	3,500	79	608	300	908	21	4,408	100	N.A.	—	4,408	24
Western Samoa	—	—	—	—	—	37	—	37	100	37	100	91	79	128	84

— Nil or magnitude negligible.
N.A. — Data not available.

TABLE 7-39. ANNUAL INVESTMENT IN 1970 FOR CONSTRUCTION OF COMMUNITY WATER SUPPLY AND SEWAGE DISPOSAL FACILITIES IN DEVELOPING COUNTRIES [1]

(Source: World Health Organization, 1973)

[In thousands of US $; including External, National and Local Capital, Material and Labor]

Region and country	Community water supply construction			Sewage disposal system construction		
	Urban	Rural	Total	Urban	Rural	Total
Africa	72,200	19,890	92,090	7,373	2,150	9,523
Americas	262,753	46,172	308,925	65,440	5,299	70,739
Eastern Mediterranean ..	197,656	36,534	234,190	28,970	100	29,070
European Region	27,481	66,816	94,297	21,601	460	22,061
South-East Asia	141,926	43,618	185,544	29,550	4,280	33,830
Western Pacific	62,540	3,992	66,532	13,260	100	13,360
Total	764,556	217,022	931,578	166,194	12,389	178,583
Africa						
Botswana	380	140	520			
Burundi.............	74	80	154	25	N.A.	25
Cameroon	150	—	150			
Central African Republic	250	100	350	—	—	—
Chad	587	1,100	1,687	248	18	266
Dahomey	850	1,000	1,850	60	4	64
Gabon	1,400	590	1,990			
Gambia.............	294	20	314			
Ghana	3,000	5,700	8,700			
Guinea	N.A.	N.A.	N.A.	2,400	—	2,400
Ivory Coast...........	1,248	48	1,296	—	—	—
Kenya	1,100	1,800	2,900	272	—	272
Lesotho	200	170	370			
Liberia	8,500	2,000	10,500	206	—	206
Madagascar	864	100	964			
Mali	1,000	200	1,200	—	—	—
Mauritania	35	20	55	200	N.A.	200
Mauritius	N.A.	N.A.	N.A.	475	—	475
Niger	468	1,764	2,232	160	2,000	2,160
Nigeria	22,860	2,580	25,440			
Senegal	1,600	—	4,600			
Sierra Leone	1,480	160	1,640			
Togo	292	—	292			
Uganda	9,000	—	9,000	2,500	100	2,600
United Rep. of Tanzania	1,500	300	1,800			
Upper Volta	2,840	238	3,078	739	—	739
Zaire	328	20	348			
Zambia	8,900	1,760	10,660	88	28	116
Americas						
Argentina	18,500	4,300	22,800	11,500	550	12,050
Barbados	445	—	445			
Bolivia	6,000	400	6,400	2,283	N.A.	2,283
Brazil	98,400	—	98,400	N.A.	N.A.	N.A.
Chile	15,000	— 2)	15,000			
Colombia	34,000	2,200	36,200	8,550	325	8,875
Costa Rica	2,136	401	2,537	400	36	436
Dominican Republic....	1,954	1,435	3,389	6,000	170	6,170
Ecuador	2,200	29	2,229	357	—	357
El Salvador	1,098	400	1,498	76	—	76
Guatemala	2,341	982	3,323	674	—	674
Guyana	25	850	875	—	—	—
Haiti	976	—	976	—	—	—
Honduras	395	256	651	546	7	553
Jamaica	4,679	3,754	8,433	2,186	1,700	3,886
Mexico	39,000	16,500	55,500	10,000	—	10,000
Nicaragua	492	1,567	2,059	252	8	260
Panama	493	558	1,051	5,401	165	5,566
Paraguay	390	90	480	65	115	180
Peru	10,334	2,900	13,234	8,300	250	8,550
Trinidad and Tobago....	1,300	1,550	2,850			
Uruguay	6,013	—	6,013	850	—	850
Venezuela...........	16,582	8,000	24,582	8,000	1,973	9,973

TABLE 7-39. ANNUAL INVESTMENT IN 1970 FOR CONSTRUCTION OF COMMUNITY WATER SUPPLY AND SEWAGE DISPOSAL FACILITIES IN DEVELOPING COUNTRIES [1] (continued)

Region and country	Community water supply construction			Sewage disposal system construction		
	Urban	Rural	Total	Urban	Rural	Total
Eastern Mediterranean						
Afghanistan	100	50	150	—	—	—
Bahrain	801	260	1,061			
Cyprus	440	549	989			
Democratic Yemen	100	70	170			
Egypt	4,375	5,625	10,000			
Ethiopia	6,000	100	6,100	60	—	60
Iran	13,159	3,200	16,359			
Iraq	12,000	1,960	13,960	3,210	N.A.	3,210
Jordan	600	500	1,100			
Kuwait	60	—	60			
Lebanon	815	900	1,715			
Libyan Arab Republic ..	1,438		1,438			
Pakistan	48,228	500	48,728			
Qatar	450	20	490			
Saudi Arabia	92,500		92,500	25,000	100	25,100
Somalia	4,600	1,100	5,700			
Sudan	3,900	20,000	23,900			
Syrian Arab Republic ...	2,000	1,200	3,200			
Tunisia	6,090	500	6,590	700	—	700
Yemen	—	—	—			
European Region						
Algeria	13,820	3,450	17,270	5,700	220	5,920
Morocco	1,520	1,200	2,720	15,901	1,240	16,141
Turkey	12,141	62,166	74,307			
South-East Asia						
Bangladesh	10,000	4,000	14,000	1,500	—	1,500
Burma	—	—	—			
India	114,300	38,000	152,300	20,000	3,000	23,000
Indonesia	6,000	100	6,100	50		50
Nepal	1,000	100	1,100			
Sri Lanka	2,520	168	2,688	—	80	80
Thailand	8,106	1,250	9,356	8,000	1,200	9,200
Western Pacific						
Fiji	545	186	731	250	100	350
Khmer Republic	22	2	24	55	N.A.	55
Korea, Republic of	13,850	860	14,710	7,700	N.A.	7,700
Laos	300	—	300			
Malaysia		5,500	5,500			
Philippines	7,186	2,775	9,961	400	—	400
Singapore	3,243	—	3,243	4,400	N.A.	4,400
Viet-Nam, Republic of	31,882	—	31,882	455	N.A.	455
Western Samoa	12	169	181			

[1] Amounts shown as investments on construction of sewage disposal systems are likely lower than the actual amounts invested, owing to under-reporting. This under-reporting is estimated to be of lesser magnitude in the case of water supplies.

[2] Investments over the past 5 years US $ 6 million, momentarily frozen.

— Nil or magnitude negligible.

N.A. — Not available.

TABLE 7-40. COMMUNITY WATER SUPPLY IN DEVELOPING COUNTRIES

(Source: World Health

Region and country	Urban							
	House connections				Public standposts			
	Population to be served [1]		Cost		Population to be served [1]		Cost	
	1980 N'000	Increase over 1970 N'000	US $ per consumer [2]	Total millions US $	1980 N'000	Increase over 1970 N'000	US $ per consumer [2]	Total millions US $
Africa	31,561	22,685	53	1,196.2	21,041	9,299	28	260.9
Americas	186,933	95,963	40	3,861.3	26,074	–	–	–
Eastern Mediterranean . .	61,927	24,248	30	735.4	40,121	24,558	11	274.4
European Region.	25,140	12,734	12	1,528.1	16,759	11,333	25	283.3
South-East Asia	143,948	87,557	16	1,387.0	95,966	69,168	9	618.5
Western Pacific	36,577	11,489	22	253.5	24,384	20,316	20	415.4
Total	486,086	254,676	35	8,961.5	224,345	135,074	14	1,853.5
Africa								
Botswana	33	17	200	3.4	22	3	180	0.5
Burundi	104	89	52	4.6	69	9	17	0.2
Cameroon	1,241	1,091	100	109.1	827	77	25	1.9
Central African Republic	416	400	12	4.8	278	244	25	6.1
Chad	269	239	52	12.4	180	10	17	0.2
Congo	284	204	56	11.4	189	–	–	–
Dahomey	421	388	12	4.7	281	–	9	–
Gabon	88	83	56	4.6	59	58	25	1.5
Gambia	32	22	22	0.5	21	–	–	–
Ghana	3,146	2,494	40	99.3	2,097	614	25	15.4
Guinea	503	166	80	13.8	336	236	25	5.9
Ivory Coast	1,013	753	70	45.2	675	19	60	1.1
Kenya	1,234	234	50	11.7	822	750	25	18.8
Lesotho.	9	4	56	0.2	6	–	–	–
Liberia	119	59	50	3.0	79	–	5	–
Madagascar	934	698	170	118.7	623	29	4	0.1
Mali	590	430	38	16.3	393	373	25	9.3
Mauritania.	92	12	300	3.6	62	56	280	15.7
Niger	376	336	56	18.8	251	71	25	1.8
Nigeria	12,785	9,975	30	299.3	8,523	3,873	30	116.2
Senegal	939	639	56	35.8	626	–	–	–
Sierra Leone	340	238	120	28.6	226	46	70	32
Togo	291	257	56	14.4	194	–	–	–
Uganda	808	408	56	22.8	539	323	3	1.0
United Rep. of Tanzania	980	880	40	35.2	654	254	25	6.4
Upper Volta	211	171	200	34.2	141	41	5	0.2
Zaire	3,184	1,979	100	197.9	2,122	1,722	25	43.1
Zambia	1,119	419	100	41.9	746	491	25	12.3
Americas								
Argentina	19,290	7,490	100	749.0	900	–	–	–
Barbados	109	24	80	1.9	15	–	–	–
Bolivia	1,585	1,043	30	31.3	1,009	–	–	–
Brazil	64,222	35,522	24	852.5	12,600	–	–	–
Chile	7,843	3,643	75	273.2	800	–	–	–
Colombia	18,642	9,149	50	457.5	2,000	–	–	–
Costa Rica	1,034	423	91	38.5	46	–	–	–
Dominican Republic	2,342	1,408	50	70.4	291	–	–	–
Ecuador	3,185	1,687	25	42.2	312	–	–	–
El Salvador	1,531	991	40	39.6	446	–	–	–
Guatemala.	1,744	1,019	30	30.6	858	–	–	–
Guyana	372	172	55	9.5	60	–	–	–

—TARGETS FOR 1980 AND ESTIMATED COSTS [1]

Organization, 1973)

Urban			Rural				Total		
Total urban			Easy access to safe water				Total		
Population to be served		Cost	Population to be served [1]		Cost		Population to be served		Cost
1980 N'000	Increase over 1970 N'000	Total millions US $	1980 N'000	Increase over 1970 N'000	US $ per consumer [2]	Total millions US $	1980 N'000	Increase over 1970 N'000	Total millions US $
52,602	31,984	1,457.1	46,841	30,913	20	631.7	99,443	62,897	2,088.8
213,007	95,963	3,861.3	59,971	31,378	24	749.3	272,978	127,341	4,610.6
102,048	48,806	1,010.8	54,077	37,165	13	465.6	156,125	85,971	1,476.4
41,899	24,067	1,811.4	12,029	2,671	20	53.4	53,928	26,738	1,864.8
239,914	156,725	2,605.5	218,458	163,269	8	1,244.2	458,372	319,994	2,249.7
60,961	32,205	668.9	22,354	8,405	6	52.1	83,315	40,610	721.0
710,431	389,750	10,815.0	413,730	273,801	12	3,196.3	1,124,161	663,551	14,011.3
55	20	3.9	187	38	15	0.6	242	58	4.5
173	98	4.8	1,130	1,130	22	24.9	1,303	1,228	29.7
2,068	1,168	111.0	1,342	342	20	6.8	3,410	1,510	117.8
694	644	10.9	320	320	20	6.4	1,014	964	17.3
449	249	12.6	1,100	320	22	7.0	1,549	569	19.6
473	204	11.4	186	140	20	2.8	659	344	14.2
702	388	4.7	725	270	6	1.6	1,427	658	6.3
147	141	6.1	98	97	20	1.9	245	238	8.0
53	22	0.5	102	93	46	4.3	155	115	4.8
5,243	3,108	115.2	1,889	1,019	50	51.0	7,132	4,127	166.2
839	402	19.2	1,061	1,061	20	21.2	1,900	1,463	40.4
1,688	772	46.3	992	—	10	—	2,680	772	46.3
2,056	984	30.5	3,327	3,087	15	46.3	5,383	4,071	76.8
15	4	0.2	320	320	20	20	335	324	6.6
198	59	30	316	249	10	2.5	514	308	5.5
1,557	727	118.8	1,965	1,420	1	1.4	3,522	2,147	120.2
983	803	25.6	1,421	1,421	20	28.4	2,404	2,224	54.0
154	68	19.3	343	229	50	11.5	497	297	30.8
627	407	20.6	1,181	611			1,808	1,018	20.6
21,308	13,848	415.5	13,137	9,551	13	124.2	34,445	23,399	539.7
1,565	639	35.8	897	—	—	—	2,462	639	35.8
566	284	31.8	716	690	60	41.4	1,282	974	73.2
485	257	14.4	502	416	20	8.3	987	673	22.7
1,347	731	23.8	2,539	939	14	13.2	3,886	1,670	37.0
1,634	1,134	55.7	4,024	2,824	20	56.5	5,658	3,958	112.2
352	212	34.4	1,625	325	20	6.5	1,977	537	40.9
5,306	3,701	241.0	4,359	3,609	40	144.4	9,665	7,310	385.4
1,865	910	54.2	1,037	392	31	12.2	2,902	1,302	664
20,190	7,490	749.0	1,555	901	40	36.0	21,745	8,391	785.0
124	24	1.9	134	—	150	—	258	24	1.9
2,594	1,043	31.3	1,163	1,110	20	22.2	3,757	2,153	53.5
76,822	35,522	852.5	21,223	9,223	16	147.6	98,045	44,745	1,000.1
8,643	3,643	273.2	808	558	70	39.1	9,451	4,201	312.3
20,642	9,149	457.5	4,771	2,091	27	56.5	25,413	11,240	514.0
1,080	423	38.5	1,109	470	60	28.2	2,189	893	66.7
2,633	1,408	70.4	1,188	917	22	20.2	3,821	2,325	90.6
3,497	1,687	42.2	1,613	1,369	12	16.4	5,110	3,056	58.6
1,977	991	39.6	1,241	805	27	21.7	3,218	1,796	61.3
2,602	1,019	30.6	1,734	1,309	15	19.6	4,336	2,328	50.2
432	172	9.5	432	123	40	4.9	864	295	14.4

TABLE 7-40. COMMUNITY WATER SUPPLY IN DEVELOPING COUNTRIES

Region and country	Urban							
	House connections				Public standposts			
	Population to be served [1]		Cost		Population to be served [1]		Cost	
	1980 N'000	Increase over 1970 N'000	US $ per consumer[2]	Total millions US $	1980 N'000	Increase over 1970 N'000	US $ per consumer[2]	Total millions US $
Americas								
Haiti	912	753	30	22.6	457	—	—	—
Honduras	1,014	539	40	21.6	233	—	—	—
Jamaica..............	915	440	72	31.7	24	—	—	—
Mexico	38,025	19,785	28	554.0	3,290	—	—	—
Nicaragua	908	612	30	18.4	227	—	—	—
Panama	1,044	433	40	17.3	66	—	—	—
Paraguay	868	706	45	31.8	25	—	—	—
Peru	8,068	4,488	25	112.2	620	—	—	—
Trinidad and Tobago ...	600	303	40	12.1	59	—	—	—
Uruguay	2,537	590	28	16.5	216	—	—	—
Venezuela...........	10,143	4,743	90	426.9	1,520	—	—	—
Eastern Mediterranean								
Afghanistan	1,150	1,025	10	10.3	767	567	6	3.4
Bahrain..............	125	—	6	—	83	70	11	0.8
Cyprus	221	—	50	—	148	148	11	1.6
Democratic Yemen	393	159	20	3.2	262	162	12	1.9
Egypt	13,524	2,354	20	47.1	9,016	6,186	11	68.0
Ethiopia	1,762	1,162	100	116.2	1,174	274	40	11.0
Iran	11,424	4,933	32	157.9	7,616	4,378	7	30.6
Iraq	4,651	1,051	35	36.8	3,101	2,135	11	23.5
Jordan	1,034	62	30	1.9	690	582	11	6.4
Kuwait	612	492	32	15.7	408	128	11	1.4
Lebanon	1,740	—	19	—	1,160	1,160	11	12.8
Libyan Arab Republic...	1,032	382	32	12.2	688	488	11	5.4
Pakistan	14,760	9,566	25	239.2	9,840	3,570	15	53.6
Qatar	52	2	55	0.1	34	30	60	1.8
Saudi Arabia..........	2,090	590	32	18.9	262	1,064	11	11.7
Somalia	507	495	15	7.4	1,394	230	1	0.2
Sudan	1,549	409	45	18.4	338	1,022	15	15.3
Syrian Arab Republic ...	2,748	293	32	9.4	1,032	1,588	11	17.5
Tunisia	2,138	938	32	30.0	1,832	575	11	6.3
Yemen	415	335	32	10.7	276	201	11	2.2
European Region								
Algeria	6,589	2,089	120	250.7	4,392	2,392	25	84.8
Morocco	5,743	3,543	120	425.2	3,829	829	25	20.7
Turkey	12,808	7,102	120	852.2	8,538	7,112	25	177.8
South-East Asia								
Bangladesh	4,694	3,944	9	35.5	3,129	2,029	7	14.2
Burma	4,639	4,270	15	64.1	3,093	1,512	9	13.6
India	106,016	59,616	15	894.2	70,677	50,777	9	457.0
Indonesia	20,002	15,002	20	300.0	13,334	10,834	9	97.5

—TARGETS FOR 1980 AND ESTIMATED COSTS [1] (continued)

			Rural						
	Total urban		Easy access to safe water				Total		
Population to be served		Cost	Population to be served [1]		Cost		Population to be served		Cost
1980 N'000	Increase over 1970 N'000	Total millions US $	1980 N'000	Increase over 1970 N'000	US $ per consumer [2]	Total millions US $	1980 N'000	Increase over 1970 N'000	Total millions US $
1,369	753	22.6	1,613	1,613	15	24.2	2,982	2,366	46.8
1,247	539	21.6	988	795	25	19.9	2,235	1,334	41.5
939	440	31.7	844	197	84	16.5	1,783	647	48.2
41,315	19,785	554.0	12,650	6,880	23	158.2	53,965	26,665	712.2
1,135	612	18.4	602	432	35	15.1	1,737	1,044	33.5
1,110	433	17.3	566	258	33	8.5	1,676	691	25.8
893	706	31.8	675	594	40	23.8	1,568	1,300	55.6
8,688	4,488	112.2	2,839	2,339	20	46.8	11,527	6,827	159.0
659	303	12.1	484	—	—	—	1,143	303	12.1
2,753	590	16.5	194	134	23	3.1	2,947	724	19.6
11,663	4,743	426.9	1,545	260	80	20.8	13,208	5,003	447.7
1,917	1,592	13.7	5,095	4,985	1	5.0	7,012	6,577	18.7
208	70	0.8	24	—	—	—	232	70	0.8
369	148	1.6	77	—	30	—	446	148	1.6
655	321	5.1	276	—	25	—	931	321	5.1
22,540	8,540	115.1	5,894	—	12	—	28,434	8,540	115.1
2,936	1,436	127.2	7,239	7,239	20	144.8	10,175	8,675	272.0
19,040	9,311	188.5	5,088	3,088	10	30.9	24,128	12,399	219.4
7,752	3,186	60.3	1,605	1,305	42	54.8	9,357	4,491	115.1
1,724	644	8.3	397	—	15	—	2,121	644	8.3
1,020	620	17.1	172	172	13	2.2	1,192	792	19.3
2,900	1,160	12.8	300	—	25	—	3,200	1,160	12.8
1,720	870	17.6	232	—	—	—	1,952	870	17.6
24,600	13,136	292.8	16,450	14,730	9	132.6	41,050	27,866	425.4
86	32	1.9	6	—	13	—	92	32	1.9
2,352	1,654	30.6	1,786	—	—	—	4,138	1,654	30.6
1,901	725	7.6	716	416	13	5.4	2,617	1,141	13.0
1,887	1,431	33.7	4,935	3,135	20	62.7	6,822	4,566	96.4
3,780	1,881	26.9	1,090	—	13	—	4,870	1,881	26.9
3,970	1,513	36.3	900	400	13	5.2	4,870	1,913	41.5
691	536	12.9	1,795	1,695	13	22.0	2,486	2,231	24.9
10,981	5,481	335.5	2,315	2,315	20	46.3	13,296	7,796	381.8
9,572	4,372	445.9	3,256	356	20	7.1	12,828	4,728	453.0
21,346	14,214	1,030.0	6,458	—	—	—	27,804	14,214	1,030.0
7,823	5,973	49.7	24,093	—	1	—	31,916	5,973	49.7
7,732	5,782	77.7	6,937	3,942	6	23.7	14,669	9,724	101.4
176,693	110,393	1,351.2	137,461	112,460	8	899.7	314,154	222,853	2,250.9
33,336	25,836	397.5	32,863	32,863	4	131.5	66,199	58,699	529.0

TABLE 7-40. COMMUNITY WATER SUPPLY IN DEVELOPING COUNTRIES

Region and country	Urban							
	House connections				Public standposts			
	Population to be served [1]		Cost		Population to be served [1]		Cost	
	1980 N'000	Increase over 1970 N'000	US $ per consumer [2]	Total millions US $	1980 N'000	Increase over 1970 N'000	US $ per consumer [2]	Total millions US $
South-East Asia								
Mongolia.............	470	372	15	5.6	314	132	9	1.2
Nepal	527	514	10	5.1	352	52	10	0.5
Sri Lanka	2,262	1,342	15	20.1	1,508	708	9	6.4
Thailand	5,338	2,497	25	62.4	3,559	3,124	9	28.1
Western Pacific								
Fiji	125	—	70	—	83	83	20	1.7
Khmer Republic	863	291	10	2.9	575	279	10	2.8
Korea, Rep. of	12,201	1,771	11	19.5	8,134	7,585	20	151.7
Laos	517	217	16	3.5	344	196	40	7.8
Malaysia	4,455	1,198	83	99.4	2,970	2,089	20	41.8
Philippines	12,808	5,458	16	87.3	8,539	7,227	20	144.5
Singapore	1,605	19	16	0.3	1,070	1,070	20	21.4
Viet-Nam, Rep. of......	3,967	2,517	16	40.3	2,645	2,170	20	43.4
Western Samoa	36	18	16	0.3	24	17	20	0.3

[1] For all regions except the Americas the targets set were: the supply of 60% of the urban population with house connections, the supply of 40% of the urban population with public standposts and the supply of 25% of the rural population with easy access to safe water.

For the Americas the minimum targets adopted by PAHO were followed: the reduction of the percentage of population not supplied in 1970 by 50% for city dwellers and by 30% for the rural population for each country.

[2] For non-reporting countries the regional median was employed, but summaries were arrived at by dividing the estimated total cost by the additional population to be served in 1980.

—TARGETS FOR 1980 AND ESTIMATED COSTS [1] (continued)

Urban			Rural				Total		
Total urban			Easy access to safe water				Total		
Population to be served		Cost	Population to be served [1]		Cost		Population to be served		Cost
1980 N'000	Increase over 1970 N'000	Total millions US $	1980 N'000	Increase over 1970 N'000	US $ per consumer[2]	Total millions US $	1980 N'000	Increase over 1970 N'000	Total millions US $
784	504	6.8	245	245	8	8.0	1,029	749	8.8
879	566	5.6	3,355	3,345	15	50.2	4,234	3,911	55.8
3,770	2,050	26.5	3,087	2,997	21	62.9	6,857	5,047	89.4
8,897	5,621	90.5	10,417	7,417	10	74.2	19,314	13,038	164.7
208	83	1.7	129	79	7	0.6	337	162	2.3
1,438	570	5.7	2,111	—	10	—	3,549	570	5.7
20,335	9,356	171.2	5,254	—	8	—	25,589	9,356	171.2
861	413	11.3	774	—	—	—	1,635	413	11.3
7,425	3,287	141.2	1,782	1,727	7	12.1	9,207	5,014	153.3
21,347	12,685	231.8	8,429	3,369	5	16.8	29,776	16,054	248.6
2,675	1,089	21.7	—	—	—	—	2,675	1,089	21.7
6,612	4,687	83.7	3,840	3,195	7	22.4	10,452	7,882	106.1
60	35	0.6	35	35	7	0.2	95	70	0.8

— Category not applicable.

TABLE 7-41. SEWAGE DISPOSAL IN DEVELOPING COUNTRIES

(Source: World Health

	Urban							
	Connected to public sewerage system				Household systems			
	Population to be served [1]		Cost		Population to be served [1]		Cost	
Region and country	1980 N'000	Increase over 1970 N'000	US $ per user [2]	Total millions US $	1980 N'000	Increase over 1970 N'000	US $ per user [2]	Total millions US $
Africa	7,831	6,629	35	231.2	11,744	7,360	13	96.3
Americas	121,510	71,264	26	1,840.5	46,061	–	–	–
Eastern Mediterranean ..	15,440	13,502	72	975.9	24,274	2,801	23	64.4
European Region.	8,221	4,958	29	143.8	12,332	10,829	5	54.1
South-East Asia.	95,652	54,025	16	864.4	143,478	68,308	9	586.2
Western Pacific	24,040	15,066	46	692.9	36,060	15,780	12	191.5
Total	272,694	165,444	29	4,748.7	273,922	105,078	9	992.5
Africa								
Burundi	69	55	35	1.9	104	24	6	0.1
Central African Republic	278	277	35	9.7	416	19	10	0.2
Chad	180	179	35	6.3	269	241	9	2.2
Dahomey	281	236	60	14.2	421	121	17	2.1
Guinea	336	276	20	5.5	503	112	13	1.5
Ivory Coast	675	565	35	19.8	1,013	903	13	11.7
Kenya	832	307	19	5.8	1,234	643	8	5.1
Liberia	79	47	65	3.1	119	30	7	0.2
Madagascar	623	593	43	25.5	934	14	13	0.2
Mali	393	393	35	13.8	590	200	20	4.0
Mauritania	62	2	35	0.1	92	64	28	1.8
Mauritius	234	138	35	4.8	351	253	13	3.3
Niger	251	251	35	8.8	376	346	28	9.7
Uganda	539	389	25	9.7	808	367	5	1.8
Upper Volta	141	141	35	4.9	211	16	20	0.3
Zaire	2,122	2,097	35	73.4	3,184	2,984	13	38.8
Zambia	746	683	35	23.9	1,119	1,023	13	13.3
Americas								
Argentina	12,538	6,338	25	158.5	10,000	–	–	–
Bolivia	1,072	729	35	25.5	90	–	–	–
Brazil	42,815	27,215	20	544.3	29,880	–	–	–
Colombia	15,517	7,700	12	92.4	2,000	–	–	–
Costa Rica	563	353	37	13.1	232	–	–	–
Dominican Republic	1,239	962	49	47.1	796	–	–	–
Ecuador	2,770	1,387	40	55.5	–	–	–	–
El Salvador	1,229	705	10	7.1	393	–	–	–
Guatemala	1,491	766	16	12.3	83	–	–	–
Guyana	204	138	25	3.5	154	–	–	–
Haiti	468	468	25	11.7	669	–	–	–
Honduras	873	453	15	6.8	30	–	–	–
Jamaica	429	335	38	12.7	396	–	–	–
Mexico	22,259	14,487	24	347.7	–	–	–	–
Nicaragua	718	431	36	15.5	–	–	–	–
Panama	871	389	100	38.9	118	–	–	–
Paraguay	594	463	25	11.6	–	–	–	–
Peru	6,946	3,446	33	113.7	300	–	–	–
Uruguay	1,838	623	20	12.5	920	–	–	–
Venezuela	7,076	3,876	80	310.1	–	–	–	–

—TARGETS FOR 1980 AND ESTIMATED COSTS [1]

Organization, 1973)

Urban			Rural				Total		
Total			with adequate disposal				Total		
Population to be served		Cost	Population to be served [1]		Cost		Population to be served		Cost
1980 N'000	Increase over 1970 N'000	Total millions US$	1980 N'000	Increase over 1970 N'000	US$ per user [2]	Total millions US$	1980 N'000	Increase over 1970 N'000	Total millions US$
19,575	13,989	327.5	23,575	15,943	5	84.6	43,150	29,932	412.1
167,571	71,264	1,840.5	57,181	32,192	6	185.2	224,752	103,456	2,025.7
39,687	16,303	1,040.3	21,945	9,220	11	106.0	61,632	25,523	1,146.3
20,553	15,787	197.9	5,571	4,723	3	14.2	26,124	20,510	212.1
239,130	122,333	1,450.6	218,212	196,333	3	561.1	457,342	318,666	2,011.7
60,100	30,846	884.4	21,580	19,514	4	81.4	81,680	50,360	965.8
546,616	270,522	5,741.2	348,064	277,925	4	1,032.5	894,680	548,447	6,773.7
173	79	2.0	1,130	1,130	3	3.4	1,303	1,209	5.4
694	296	9.9	320	305	10	3.1	1,014	601	13.0
449	420	85	1,100	1,087	3	3.3	1,549	1,507	11.8
702	357	16.3	725	708	12	8.5	1,427	1,065	24.8
839	388	7.0	1,061	1,001	5	5.0	1,900	1,389	12.0
1,688	1,468	31.5	992	992	5	5.0	2,680	2,460	36.5
2,056	950	10.9	3,327	—	6	—	5,383	950	10.9
198	77	3.3	316	216	3	0.6	514	293	3.9
1,557	607	25.7	1,965	1,965	5	9.8	3,522	2,572	35.5
983	593	17.8	1,421	1,421	2	2.8	2,404	2,054	20.6
154	66	1.9	343	343	10	3.4	497	409	5.3
585	391	8.1	134	—	—	—	719	391	8.1
627	597	18.5	1,181	1,179	10	11.8	1,808	1,776	30.3
1,347	756	11.5	2,539	—	2	—	3,886	756	11.5
352	157	5.2	1,625	1,625	5	8.1	1,977	1,782	13.3
5,306	5,081	112.2	4,359	3,484	5	17.4	9,665	8,565	129.6
1,865	1,706	37.2	1,037	487	5	2.4	2,902	2,193	39.6
22,538	6,338	158.5	3,594	—	5	—	26,132	6,338	158.5
1,162	729	25.5	1,222	1,095	3	3.3	2,384	1,824	28.8
72,695	27,215	544.3	19,975	9,591	3	28.8	92,670	36,806	573.1
17,517	7,700	92.4	5,076	2,016	3	6.0	22,593	9,716	98.4
795	353	13.1	939	472	2	0.9	1,734	825	14.0
2,035	962	47.1	2,183	739	4	3.0	4,218	1,701	50.1
2,770	1,387	55.5	1,415	1,415	7	9.9	4,185	2,802	65.4
1,622	705	7.1	1,083	811	3	2.4	2,705	1,516	9.5
1,574	766	12.3	1,807	1,292	14	18.1	3,381	2,058	30.4
358	138	3.5	548	98	7	0.7	906	236	4.2
1,137	468	11.7	1,666	1,623	7	11.4	2,803	2,091	23.1
903	453	6.8	962	789	2	1.6	1,865	1,242	8.4
825	335	12.7	1,273	24	6	0.1	2,098	359	12.8
22,259	14,487	347.7	7,873	7,873	7	55.1	30,132	22,360	402.8
718	431	15.5	544	454	1	0.5	1,262	885	16.0
989	389	38.9	750	230	2	0.5	1,739	619	39.4
594	463	11.6	605	605	7	4.2	1,199	1,068	15.8
7,246	3,446	113.7	3,336	2,336	12	28.0	10,582	5,782	141.7
2,758	623	12.5	199	133	35	4.7	2,957	756	17.2
7,076	3,876	310.1	2,131	596	10	6.0	9,207	4,472	316.1

TABLE 7-41. SEWAGE DISPOSAL IN DEVELOPING COUNTRIES

Region and country	Urban							
	Connected to public sewerage system				Household systems			
	Population to be served [1]		Cost		Population to be served [1]		Cost	
	1980 N'000	Increase over 1970 N'000	US $ per user [2]	Total millions US $	1980 N'000	Increase over 1970 N'000	US $ per user [2]	Total millions US $
Eastern Mediterranean								
Afghanistan	767	752	75	56.4	1,150	—	17	—
Ethiopia	1,174	1,019	45	45.9	1,762	262	43	11.3
Iran	7,616	7,116	75	533.7	11,424	68	12	0.8
Iraq	3,101	2,889	81	234.0	4,651	381	18	6.9
Libyan Arab Republic...	688	468	75	35.1	1,032	232	23	5.3
Saudi Arabia	262	112	100	11.2	2,090	1,290	28	36.1
Tunisia	1,832	1,146	52	59.6	2,138	568	7	4.0
European Region								
Algeria	4,392	4,292	29	124.5	6,589	6,209	5	31.0
Morocco	3,829	666	29	19.3	5,743	4,620	5	23.1
South-East Asia								
Bangladesh	3,129	2,629	16	42.1	4,694	1,494	2	3.0
Burma	3,093	2,830	16	45.3	4,639	2,439	14	34.1
India	70,677	30,677	16	490.8	106,016	52,016	7	364.1
Indonesia	13,334	12,874	16	206.0	20,002	10,002	14	140.0
Nepal	352	312	16	5.0	527	357	9	3.2
Sri Lanka	1,508	1,144	16	18.3	2,262	262	7	1.8
Thailand	3,559	3,559	16	56.9	5,338	1,738	23	40.0
Western Pacific								
Fiji	83	47	40	1.9	125	50	32	1.6
Khmer Republic	575	—	46	—	863	713	3	2.1
Korea, Rep. of	8,134	4,294	46	197.5	12,201	8,371	6	50.2
Malaysia	2,970	2,572	45	115.7	4,455	1,329	5	6.6
Philippines	8,539	8,059	46	370.7	12,808	1,623	12	19.5
Singapore	1,070	70	85	6.0	1,605	635	7	4.4
Viet-Nam, Rep. of......	2,645	—	46	—	3,967	3,059	35	107.1
Western Samoa	34	24	46	1.1	36	—	—	—

[1] For all regions except the Americas the targets assumed were: to have 40% of the urban population connected to the public sewerage system, 60% of the urban population provided with household systems and 25% of the urban population provided with adequate sewerage facilities.

For the Americas the minimum targets adopted by PAHO were followed: the reduction of the percentage of population not served in 1970 by 30% for each country.

[2] The cost per user was estimated for each country concerned, but summaries were arrived at by dividing the estimated total cost by the additional population to be served in 1980.

— Category not applicable.

—TARGETS FOR 1980 AND ESTIMATED COSTS [1] (continued)

Urban			Rural				Total		
Total			with adequate disposal				Total		
Population to be served		Cost	Population to be served [1]		Cost		Population to be served		Cost
1980 N'000	Increase over 1970 N'000	Total millions US $	1980 N'000	Increase over 1970 N'000	US $ per user [2]	Total millions US $	1980 N'000	Increase over 1970 N'000	Total millions US $
1,917	752	56.4	5,095	2,695	5	13.5	7,012	3,447	69.9
2,936	1,281	57.2	7,239	5,439	15	81.6	10,175	6,720	138.8
19,040	7,184	534.5	5,088	—	7	—	24,128	7,184	534.5
7,752	3,270	240.9	1,605	—	1	—	9,357	3,270	240.9
1,720	700	40.4	232	—	—	—	1,952	700	40.4
2,352	1,402	47.3	1,786	1,086	10	10.9	4,138	2,488	58.2
3,970	1,714	63.6	900	—	2	—	4,870	1,714	63.6
10,981	10,501	155.5	2,315	1,835	3	5.5	13,296	12,336	161.0
9,572	5,286	42.4	3,256	2,888	3	8.7	12,828	8,174	51.1
7,823	4,123	45.1	24,093	23,993	2	48.0	31,916	28,116	93.1
7,732	5,269	79.4	6,937	—	5	—	14,669	5,269	79.4
176,693	82,693	854.9	137,460	132,460	2	264.9	314,153	215,153	1,119.8
33,336	22,876	346.0	32,863	28,613	5	143.1	66,199	51,489	489.1
879	669	8.2	3,355	3,350	3	10.1	4,234	4,019	18.3
3,770	1,406	20.1	3,087	—	3	—	6,857	1,406	20.1
8,897	5,297	96.9	10,417	7,917	12	95.0	19,314	13,214	191.9
208	97	3.5	129	—	—	—	337	97	3.5
1,438	713	2.1	2,111	1,991	2	4.0	3,549	2,704	6.1
20,335	12,665	247.7	5,254	5,254	1	5.3	25,589	17,919	253.0
7,425	3,901	122.3	1,782	—	3	—	9,207	3,901	122.3
21,347	9,682	390.2	8,429	8,429	4	33.7	29,776	18,111	423.9
2,675	705	10.4	—	—	—	—	2,675	705	10.4
6,612	3,059	107.1	3,840	3,840	10	38.4	10,452	6,899	145.5
60	24	1.1	35	—	—	—	95	24	1.1

TABLE 7-42. AREAS UNDER IRRIGATION, BY COUNTRY

(Source: FAO Production Year Book, 1968 and Framji and Mahajan, 1969)

Name of country	Total area million ha	Area cultivated million ha	Area irrigated million ha	Percentage of cultivated area to total area	Percentage of irrigated area to total area	Percentage of irrigated area to cultivated area
Afghanistan	64.750	7.770	0.813	12.00	1.26	10.5
Albania	2.875	0.501	0.156	17.43	5.43	31.1
Algeria	238.174	4.200	0.245	1.76	0.10	5.8
Argentina	277.666	27.200	1.147	9.79	0.41	4.2
Australia	769.500	13.961	1.271	1.81	0.165	9.1
Austria	8.385	3.935	0.004	46.93	0.05	0.1
Belgium	3.051	0.895	0.009	29.33	0.30	1.0
Bolivia	109.858	0.728	0.040	0.66	0.036	5.5
Botswana	57.000	0.073	0.002	0.13	0.004	2.7
Brazil	851.196	70.000	0.141	8.23	0.017	0.2
Bulgaria	11.091	4.822	0.960	43.48	8.65	19.9
Burma	67.058	8.715	0.753	12.88	1.12	8.6
Cambodia	18.100	2.500	0.074	13.81	0.40	3.0
Canada	997.618	25.267	0.627	2.58	0.06	2.5
Ceylon	6.561	1.665	0.333	25.38	5.07	20.0
Chile	4.177	2.816	1.300	3.80	1.75	46.2
China, Mainland	956.100	109.354	74.000	11.44	7.73	67.7
China, Republic of	3.596	0.896	0.537	24.91	1.945	59.9
Colombia	113.617	4.400	0.180	3.87	0.16	4.1
Costa Rica	5.070	0.622	0.026	12.26	0.51	4.2
Cuba	11.452	1.970	0.347	17.20	3.03	17.6
Cyprus	0.925	0.466	0.091	50.38	9.83	19.5
Czechoslovakia	12.780	7.21	0.140	56.42	1.1	1.9
Denmark	4.304	2.682	0.110	62.3	2.56	4.1
Dominican Republic	4.877	0.958	0.154	19.64	3.16	16.0
Ecuador	28.356	3.670	0.146	12.94	0.52	4.0
El Salvador	2.100	0.600	0.020	28.57	0.95	3.3
Ethiopia	122.190	12.525	0.030	10.25	0.025	0.24
Fiji	1.827	0.280	—	15.33	—	—
Finland	33.701	2.746	0.009	8.15	0.027	0.33
France	55.000	20.000	2.500	36.36	4.55	12.50
Germany (Fed. Repub)	24.853	14.300	0.270	57.58	0.09	1.89
Ghana	23.854	2.850	0.007	11.95	0.03	0.25
Greece	13.194	3.800	0.560	28.8	4.24	14.7
Guatemala	10.889	1.47	0.032	13.5	0.29	2.2
Guyana	21.497	0.304	0.188	1.41	0.875	61.8
Haiti	2.775	0.370	0.047	13.33	1.69	12.7
Honduras	11.500	3.008	0.035	26.0	0.304	1.17
Hungary	9.303	6.928	0.205	74.55	2.20	3.0
India	327.634	137.91	37.640	42.0	11.5	27.3
Indonesia	149.156	14.000	3.797	9.38	2.55	27.1
Iran	164.800	6.843	3.107	4.15	1.89	45.4
Iraq	44.444	7.496	4.000	16.87	9.0	53.4
Israel	2.070	0.407	0.153	19.66	7.4	37.6
Italy	30.122	27.540	3.150	91.5	10.5	11.4
Ivory Coast	32.246	2.056	0.009	6.27	0.028	0.44
Jamaica	1.142	0.221	0.071	19.35	6.22	32.1
Japan	36.966	5.996	3.390	16.2	9.17	56.6
Jordan	9.661	1.140	0.050	11.81	0.52	4.38
Kenya	58.265	1.696	0.004	2.91	0.007	0.236
Korea (Rep. of)	9.850	2.312	0.763	23.47	7.75	33.0
Laos	23.680	2.000	0.023	8.45	0.097	1.15
Lebanon	1.004	0.306	0.054	30.4	5.38	17.65
Libya	175.954	2.375	0.167	1.35	0.096	7.03

TABLE 7-42. AREAS UNDER IRRIGATION, BY COUNTRY (continued)

Name of country	Total area million ha	Area cultivated million ha	Area irrigated million ha	Percentage of cultivated area to total area	Percentage of irrigated area to total area	Percentage of irrigated area to cultivated area
Malagasy Republic	59.579	1.000	0.900	1.68	1.51	90.0
Malaysia	33.263	3.458	0.239	10.4	0.72	6.91
Malawi	11.759	1.308	0.002	11.1	0.017	0.15
Mali	120.162	1.221	—	1.016	—	—
Malta	0.032	0.016	0.0015	50.0	4.69	9.38
Mexico	196.400	15.000	3.300	7.64	1.68	22.0
Morocco	44.505	7.858	0.265	17.66	0.595	3.37
Nepal	14.080	2.023	0.059	14.37	4.2	2.91
Netherlands	4.100	2.570	0.051	62.68	1.24	1.983
New Zealand	26.868	0.809	0.081	3.02	0.3	10.0
Nicaragua	13.970	1.812	0.029	12.99	0.21	1.60
Nigeria(North-Eastern State)	27.195		0.0024		0.009	
(North-Western State)	16.835	0.0024	0.0024	0.014	0.014	100.00
Norway	32.422	0.845		2.59		
Pakistan	94.671	21.770	11.971	30.39	12.65	41.6
Panama	7.565	0.564	0.014	7.46	0.185	2.48
Peru	128.522	2.171	0.120	1.69	0.093	5.53
Philippines	29.968	8.296	0.960	27.65	3.20	11.57
Poland	31.252	15.435	0.209	48.39	0.67	1.35
Portugal	9.196	6.753	0.639	73.43	6.95	9.46
Puerto Rico	0.886	0.295	0.037	33.3	4.2	12.5
Rhodesia	38.936	1.631	0.034	4.19	0.087	2.1
Romania, Soc. Rep.	23.75	10.500	0.600	44.2	2.53	5.71
Saudi Arabia	225.33	0.373	0.162	0.166	0.072	43.4
Senegal	19.667	2.300	0.120	11.69	0.61	5.22
Sierra Leone	7.174	3.664	0.0008	51.0	0.011	0.0218
Somalia	63.766	0.957	0.165	1.50	0.259	17.24
South Africa (Union of)	122.100	12.058	0.607	9.88	0.497	5.03
Spain	50.475	20.482	2.300	40.99	4.56	11.2
Sudan	250.581	7.100	0.790	2.83	0.313	11.1
Suriname	18.145	0.055	0.025	0.303	0.137	45.5
Swaziland	1.736	0.138	0.028	7.95	1.61	20.3
Sweden	44.999	3.158	0.025	7.02	0.056	0.792
Switzerland	4.129	1.059	0.023	25.65	0.558	2.17
Syria	18.518	3.127	0.507	16.89	2.74	16.2
Tanzania	93.970	11.702	0.040	12.45	0.043	0.342
Thailand	51.400	7.300	1.900	14.20	3.70	26.0
Tunisia	16.415	4.334	0.076	26.40	0.463	1.75
Turkey	78.058	23.539	1.724	30.16	2.21	7.32
Uganda	23.589	4.640	0.003	19.69	0.013	0.065
Union of Soviet Socialist Republic (USSR)	2,240.000	2.255	9.900	10.06	0.442	4.39
United Arab Republic (UAR)	100.000	2.940	2.94	2.94	2.940	100.00
United Kingdom(UK)	24.281	7.330	0.113	30.15	0.465	1.542
United States of America (USA)	936.333	176.000	16.932	18.79	1.81	9.62
Uruguay	18.693	2.000	0.027	10.70	0.144	1.35
Venezuela	91.205	5.219	0.362	5.72	0.397	6.94
Vietnam(Rep. of)	17.326	2.750	0.269	15.87	1.55	9.78
Yemen	19.500	—	—	—	—	—
Yugoslavia	25.654	7.507	0.150	29.50	0.585	1.98
Zambia	75.272	2.064	0.006	2.74	0.0080	0.291

TABLE 7-43. WATER REQUIREMENTS FOR SELECTED INDUSTRIES IN THE WORLD

(Source: Dept. of Economic and Social Affairs, United Nations, 1969)

[Water requirements for unit of product produced]

Industry, product and country	Unit of product (ton, except as specified)	Water required per unit (litres)
Food Products		
Bread or pastry, Belgium		1,100
Bread, United States		2,100 to 4,200
*Bread, Cyprus		600
Canned food:		
Belgium:		
Fish, canned		400
Fish, preserved		1,500
Fruit		15,000
Vegetables		8,000 to 80,000
Canada:		
*Fruits and vegetables		10,000 to 50,000
Cyprus:		
*Citrus/tomato juice		2,800
*Grapefruit sections		16,000
*Peaches/pears		10,000 to 15,000
*Grapes		30,000
*Tomatoes, whole		2,000
*Tomato paste		21,000
*Peas		10,000
*Carrots		16,000
*Spinach		30,000
Israel:		
*Citrus fruits	ton of raw citrus	4,000
*Vegetables		10,000 to 15,000
United States		
Apricots		21,200
Asparagus		20,500
Beans, green		9,300
Beans, lima		69,800
Beets, corn and peas		7,000
Grapefruit juice		2,800
Grapefruit sections		15,600
Peaches and pears		18,100
Pork and beans		9,300
Pumpkin and squash		7,000
Sauerkraut		950
Spinach		49,400
Succotash		34,800
Tomato products		20,500
Tomatoes, whole		2,200
*Industry average, fruits, vegetables and juices (1965)		24,000
Meat:		
*Meat freezing, Cyprus	ton of carcass	500
Meat freezing, New Zealand		3,000 to 8,600
*Meat packing, United States	ton of prepared meat	23,000
*Meat packing, Canada	ton of carcass	8,800 to 34,000
Meat products, Belgium	ton of prepared meat	200
Sausage factory, Finland		20,000 to 35,000
*Sausage factory, Cyprus		25,000
Slaughtering, Finland	ton, live weight	4,000 to 9,000
*Slaughtering, Cyprus	ton of carcass	10,000
*Meat preserving, Israel	ton of prepared meat	10,000

TABLE 7-43. WATER REQUIREMENTS FOR SELECTED INDUSTRIES IN THE WORLD (continued)

Industry, product and country	Unit of product (ton, except as specified)	Water required per unit (litres)
Fish:		
*Fresh and frozen fish, Canada		30,000 to 300,000
*Canned fish, Canada		58,000
*Canning and preserving fish, Israel	ton of raw fish	16,000 to 20,000
Poultry:		
*Poultry, Canada	ton	6,000 to 43,000
*Chickens, Israel	ton of dressed chicken	33,000
*Chickens, United States	per bird	25
*Turkeys, United States	per bird	75
Milk and milk products:		
Butter:		
*New Zealand		20,000
Cheese:		
*Cyprus		10,000
*New Zealand		2,000
*United States		27,500
Milk:		
Belgium	1,000 litres	7,000
Finland		2,000 to 5,000
*Israel		2,700
Sweden		2,000 to 4,000
*United States		3,000
Milk powder:		
*New Zealand		45,000
South Africa		200,000
*Whey, United States		10,000
*Dairy products, general, Canada		12,200
*Ice cream, United States		10,000
*Yogourt, Cyprus		20,000
Sugar:		
*Denmark	ton of sugar beets	4,800 to 15,800
Finland	ton of sugar beets	10,000 to 20,000
*France	ton of sugar beets	10,900
*Germany, Federal	ton of sugar beets	10,400 to 14,000
*Great Britain	ton of sugar beets	14,900
*Israel	ton of sugar beets	1,800
*Italy	ton of sugar beets	10,500 to 12,500
*Republic of China	ton of sugar cane	15,000
*United States	ton of sugar beets (range)	3,200 to 8,300
*United States	ton of sugar beets (average)	6,000
Beverages:		
Beer:		
Belgium	kilolitre	7,000 to 20,000
*Canada	kilolitre	10,000 to 20,000
*Cyprus	kilolitre (incl. cleaning bottles)	22,000 to 30,000
Finland	kilolitre	10,000 to 20,000
*France	kilolitre	14,500
*Israel	kilolitre	13,500
*United Kingdom	kilolitre	6,000 to 10,000
United States	kilolitre	15,200
*Whiskey, United States	kilolitre of proof spirit	2,600 to 76,000
*Distilled spirits, Israel	kilolitre	30,000
*Wine, France	kilolitre	2,900
*Wine, Israel	kilolitre	500

TABLE 7-43. WATER REQUIREMENTS FOR SELECTED INDUSTRIES IN THE WORLD (continued)

Industry, product and country	Unit of product (ton, except as specified)	Water required per unit (litres)		
Miscellaneous Food Products:				
Chocolate, confectionery, Belgium		15,000	to	17,000
Gelatin (edible), United States		55,100	to	83,500
Maize (wet milling), United States	litre of maize	15.0	to	25.5
Maize syrup, United States	litre of maize	3.8	to	4.3
*Wheat milling, Cyprus		2,000		
*Wheat milling, Israel		700	to	1,300
Potato flour, Finland	ton of potatoes	10,000	to	20,000
*Potato starch, Canada	ton of starch	80,000	to	150,000
*Macaroni, Cyprus		1,200		
Molasses, Belgium	hectolitre of raw material	1,000	to	12,000
Molasses, United States	hectolitre of 100 proof	840		
Pulp and Paper				
Groundwood pulp:				
Finland	ton of wood-pulp	30,000	to	40,000
Sulphate pulp:				
*China, Republic of	ton of bleached pulp	340,000		
*China, Republic of	ton of unbleached pulp	230,000		
Finland	ton of pulp	250,000	to	350,000
*Sweden	ton of unbleached pulp	75,000	to	300,000
*Sweden	ton of bleached pulp	170,000	to	500,000
Sulphite pulp:				
Finland	ton of bleached pulp	450,000	to	500,000
Finland	ton of unbleached pulp	250,000	to	300,000
*Sweden	ton of bleached pulp	300,000	to	700,000
*Sweden	ton of unbleached pulp	140,000	to	500,000
Wood pulp:				
*Sweden	ton of dry pulp	50,000	to	100,000
South Africa		150,000		
Blotting paper, Sweden		350,000	to	400,000
Craft, printing and fine paper, Finland		375,000		
*Printing paper, Republic of China		340,000		
*Newsprint, Republic of China		190,000		
*Newsprint, Canada		165,000	to	200,000
*Fine paper, Republic of China		800,000		
Fine paper, Sweden		900,000	to	1,000,000
Newsprint paper, Sweden		200,000		
Packing and cartridge paper, Sweden		125,000		
Press paper, Finland		200,000		
Printing paper, Sweden		500,000		
Cardboard, Finland		125,000		
Paperboard, United States		62,000	to	376,000
Paper and cardboard, Belgium		180,000		
Strawboard, United States		109,000		
Wallboard, Finland		125,000		
*Wallboard, Sweden		50,000		
*Industry average, United States	ton of pulp and paper	236,000		
*Industry average, United Kingdom	ton of paper and board	90,000[1]		
*Industry average, France	ton of pulp and paper	150,000		
Petroleum and Synthetic Fuels				
Aviation gasoline, United States	kilolitre	25,000		
*Aviation gasoline, Republic of China	kilolitre	25,000		
Gasoline, United States	kilolitre	7,000	to	10,000
*Gasoline, Republic of China	kilolitre	8,000		

[1] Does not include cooling water for power generating plants.

TABLE 7-43. WATER REQUIREMENTS FOR SELECTED INDUSTRIES IN THE WORLD (continued)

Industry, product and country	Unit of product (ton, except as specified)	Water required per unit (litres)
Petroleum and Synthetic Fuels—Continued		
Gasoline, polymerization, United States	kilolitre	34,000
Kerosene, Belgium	ton	40,000
Synthetic gasoline, United States	kilolitre	377,000
Oilfields, United States	kilolitre of crude petroleum	4,000
Oil refineries:		
*Belgium		
*China, Republic of	ton of crude petroleum	30,500
Sweden	ton of crude petroleum	10,000
*United States		
Synthetic fuel:		
From coal:		
South Africa		50,100
United States	kilolitre	265,500
From natural gas, United States	kilolitre	88,900
From shale, United States	kilolitre	20,800
Chemicals		
Acetic acid, United States		417,000 to 1,000,000
Alcohol, 100 proof, United States	litre	138
Alcohol, 190 proof, United States	litre	52 to 100
Alumina (Bayer process), United States		26,300
Ammonia, synthetic, United States	ton of liquid NH_3	129,000
*Ammonia (Naphtha, reforming), Japan		255,000
Ammonium nitrate, Belgium		52,000
Ammonium sulphate, United States		835,000
Calcium carbide, United States		125,000
Calcium metaphosphate, United States		16,700
Carbon dioxide, United States		83,500
*Caustic soda and chlorine, Canada		125,000
*Caustic soda (Solvay process), United States		60,500
*Caustic soda (Dual process), Federal Republic of Germany		160,000
*Caustic soda (Dual process), Republic of China		200,000
*Caustic soda (Solvay process), Republic of China		150,000
Cellulose nitrate, United States		41,700
Charcoal and wood chemicals, United States	ton of crude $CaAc_2$	271,000
*Chlorine, Federal Republic of Germany		12,600
*Ethylene, Israel		16,000
*Gases, compressed and liquified, Canada	cubic metre	60 to 70
Glycerine, United States		4,600
Gunpowder, United States		401,000 to 835,000
Hydrochloric acid (salt process), United States	ton of 20 Be HCl	12,100
Hydrochloric acid (synthetic process), United States	ton of 20 Be HCl	2,000 to 4,200
Hydrogen, United States		2,750,000
Lactose, United States		835,000 to 918,000
Magnesium carbonate, basic,	ton of basic $MgCO_3$	18,000
United States	ton of $MgCO_3$	163,000
Oxygen, United States	cubic metre of O_2	243
*Polyethylene, Federal Republic of Germany		231,000 (incl. 225,000 cooling water)
*Polyethylene, Israel		8,400
Potassium chloride (sylvinite), United States		167,000 to 209,000
Smokeless powder, United States		209,000
Soap, Belgium		37,000
*Soap, Cyprus		4,500
Soap (laundry), United States		960 to 2,100
Soda ash (ammonia soda process), 58 percent, United States		62,600 to 75,100
Sodium chlorate, United States		250,000

TABLE 7-43. WATER REQUIREMENTS FOR SELECTED INDUSTRIES IN THE WORLD (continued)

Industry, product and country	Unit of product (ton, except as specified)	Water required per unit (litres)
Chemicals—Continued		
Sodium silicate, United States	ton of 40 Be water-glass	670
Stearine, soap and washing agents, Sweden	ton of fat	70,000 to 200,000
Sulphuric acid, Belgium		20,000 to 25,000
Sulphuric acid (chamber process), United States	ton of 100 percent H_2SO_4	10,400
Sulphuric acid (contact process), United States	ton of 100 percent H_2SO_4	2,700 to 20,300
*Sulfuric acid, Federal Republic of Germany	ton of SO_3	83,500
Textiles		
Steeping, dressing, scouring and bleaching:		
Steeping flax, Belgium		30,000 to 40,000
Dressing flax, Sweden		30,000 to 40,000
Scouring wool, Belgium		240,000 to 250,000
Washing wool, Sweden		10,000
Bleaching textiles, Belgium		180,000
Dyeing:		
Textiles, Belgium		200,000
*Textiles, France (range)		52,000 to 560,000
*Textiles, France (average)		180,000
Finishing:		
Wet finishing of textiles, Belgium		100,000 to 150,000
Dyeing and finishing:		
*Cotton yarn, Israel		60,000 to 180,000
*Synthetic yarn, Israel		90,000 to 180,000
*Wool yarn, Israel		70,000 to 140,000
*Fabrics, Israel		60,000 to 100,000
Mills:		
Cotton:		
Finland		50,000 to 150,000
Sweden		10,000 to 250,000
*Canada	square yard	1.0
Wool:		
Finland	ton of cloth or yarn	150,000 to 350,000
Sweden	ton of wool	400,000
Synthetic fibres:		
Artificial silk, Sweden		2,000,000
Rayon:		
Belgium		2,000,000
Finland		1,000,000 to 2,000,000
Rayon staple, Belgium		550,000
*Industrial duck products, Canada		22,000
*Carpets, Canada	square yard	20
Mining and Quarrying		
Gold, South Africa	ton of ore	1,000
Iron ore (brown), United States		4,200
*Bauxite, United States	ton of ore	300
Sulfur, United States		12,500
Copper, Finland		3,750
*Copper, Israel		3,100
*Gravel, Israel		400
Limestone and by-products, Belgium		200 to 6,500

TABLE 7-43. WATER REQUIREMENTS FOR SELECTED INDUSTRIES IN THE WORLD (continued)

Industry, product and country	Unit of product (ton, except as specified)	Water required per unit (litres)		

Iron and Steel Products

Belgium:				
Blast furnace, no recycling		58,000	to	73,000
Blast furnace, with recycling		50,000		
*Finished and semi-finished steel,				
no recycling		61,000		
Finished and semi-finished steel,				
with recycling		27,000		
Canada:				
*Pig iron		130,000		
*Open hearth steel		22,000		
France:				
*Smelting		46,000		
*Martin process (Open hearth)		15,000		
*Thomas process (Bessemer converter)		10,000		
*Electric furnace steel		40,000		
*Rolling mills		30,000		
Germany, Federal Republic:				
*Steel works		8,000	to	12,000
South Africa:				
Steel		12,500		
Sweden:				
Iron and steel works		10,000	to	30,000
United States (average):				
*Fully integrated mills		86,000		
*Rolling and drawing mills		14,700		
*Blast furnace smelting		103,000		
*Electrometallurgical ferroalloys		72,000		
*Industry, consumptive use (est.)		3,800		

Miscellaneous Products

*Automobiles, United States	vehicle	38,000		
Boilers, steam, United States	horsepower-hour	15		
*Casein, New Zealand		55,000		
Cement, Portland:				
*Belgium		1,900		
*Cyprus (dry process)		550		
Finland		2,500		
*United States (wet process)		900		
Ceramics and tiles, Belgium		1,800	to	2,000
**Coal:				
*Ruhr (Fed. Rep. of Germany)		1,000 (min.) to		1,750 (avg.)
*Great Britain		less than		3,000
*Netherlands		2,650		

**Includes generation of electricity. If this is not included, the quantities above are reduced by about one-half

Coal, Belgium		5,000	to	6,000
Coal, coke and by-product coke,				
United States		6,300	to	15,000
Coal washing, United States		840		
Condensers, surface, United States	pound of condensed steam	9.1 to		27.3
Distilling, grain:				
Belgium	hectolitre of grain treated	6,000	to	7,000
United States	hectolitre of grain treated	6,450		
Distilling, Sweden	kilolitre of 100 percent alcohol	15,000	to	100,000

TABLE 7-43. WATER REQUIREMENTS FOR SELECTED INDUSTRIES IN THE WORLD (continued)

Industry, product and country	Unit of product (ton, except as specified)	Water required per unit (litres)		
Miscellaneous Products—Continued				
Electric power (conventional thermal):				
Sweden	ton of coal	200,000	to	400,000
South Africa	kilowatt-hour (consumptive use)	5		
*United States	kilowatt-hour	200		
*Republic of China	kilowatt-hour	230		
Explosives:				
Sweden		800,000		
United States		835,000		
Fertilizer plant, Finland	ton of saltpetre (25 percent nitrogen)	270,000		
Glass, Belgium		68,000		
Laundry:				
*Cyprus	ton of washed goods	45,000		
Finland	ton of washed goods	20,000		
Sweden	ton of washed goods	30,000	to	50,000
Leather, South Africa		50,100		
Leather factory, Finland	ton of hides	50,000	to	125,000
*Leather tanning, United States	sq. meter of hide	20	to	2,550 (range)
*Leather tanning, United States	sq. meter of hide	440	(average)	
*Leather tanning, Cyprus	sq. meter of small animal skins	110		
Non-ferrous metals, raw and semi-finished, Belgium		80,000		
Rock wool, United States		16,700	to	20,900
Rubber, synthetic, United States:				
Butadiene		83,500	to	2,750,000
Buna S		125,000	to	2,630,000
Grade GR-S		117,000	to	2,800,000
Starch:				
Belgium	ton of maize	13,000	to	18,000
Sweden	ton of potatoes	10,000		

*Figures based on newer data (post-1960).
Other figures based on older data (pre-1950)

TABLE 7-44. INTERNATIONAL HYDROLOGICAL DECADE STATIONS OF THE WORLD

(Source: IHD, EOS, Copyright American Geophysical Union, 1974)

[Status as of 1974; data on ground-water stations not available]

CONTINENT	RIVER STATIONS			LAKE STATIONS			PAN EVAPORATION STATIONS		LYSIMETER STATIONS	
	Number of countries	Number of rivers	Number of stations	Number of countries	Number of lakes	Number of stations	Number of countries	Number of stations	Number of countries	Number of stations
Africa	88	284	362	8	14	18	15	150	9	26
America North	9	205	225	4	113	161	6	144	2	16
South	7	87	103	1	1	1	4	55		
Asia	16	344	371	5	28	30	14	193	5	36
Europe	26	437	539	12	54	55	19	191	12	60
Oceania	4	59	59	1	2	2	2	12		
Totals	90	1,416	1,659	31	212	267	60	745	28	168

TABLE 7-45. CURRENT UNITED NATIONS DEVELOPMENT PROGRAMME
WATER RESOURCES PROJECTS

(Source: Alagappan, United Nations, 1974)

Country or region	Project	UNDP contribution US $
	GROUND-WATER EXPLORATION AND DEVELOPMENT	
1. Argentina	Development of Water Resources in North West Argentina (NOA HIDRICO) (ARG/72/006)	241,000
2. Bolivia	Ground-Water Development in the Altiplano (BOL/68/514)	1,124,050
3. Bolivia	Hydrologic Study of Valle Alto, Cochabamba (BOL/73/008)	411,000
4. Botswana	Head of the Hydrogeological Division (OPAS) (BOT/72/021)	86,000
5. Cameroon	Ground-Water Investigation and Pilot Development (CMR/71/516)	966,200
6. Chile	Water Resources Development in the Norte Grande (CHI/69/535)	996,800
7. Chad	Rural Water Supply (CHD/71/510)	
8. Ethiopia	Strengthening of the Geological Survey (ETH/71/537)	
9. Guatemala	Ground-Water Investigation in Guatemala (GUA/72/011)	724,900
10. India	Ground-Water Surveys in Rajasthan and Gujarat (IND/71/614)	644,900
11. Lebanon	(LEB/70/014) Small scale Ground-Water Survey	
12. Mali	Ground-Water Exploration and Development (MLI/74/001)	
13. Mali	Ground-Water Exploration and Development (MLI/73/004)	
14. Mauritania	Establishment of a Ground-Water Service (MAU/67/502)	1,531,000
15. Nicaragua	Ground-Water Investigations in Central Pacific Coastal Regions (NIC/71/508) (Phase II)	754,400
16. Nicaragua	Rehabilitation of Water Supplies to Affected Communities (NIC/73/007)	128,550
17. Pakistan	Ground-Water Investigations in Selected Areas of Baluchistan (PAK/73/032)	2,082,000
18. Paraguay	Ground-Water Development in the Chaco (Phase II) (PAR/72/004)	548,300
19. Somalia	Mineral and Ground-Water Survey (SOM/71/528)	
20. Somalia	Mineral and Water-Well Drilling	
21. Sudan	Water Resources Planning (SUD/72/003)	620,000
22. Sudan	Water-Well Drilling (SUD/68/001)	
23. Togo	Ground-Water Exploration in the Coastal Region (TOG/71/511)	424,200
24. Tunisia	Intensification of Ground-Water Exploration in Northern and Central Tunisia (TUN/69/528)	319,300
24. Tunisia	(TUN/73/010)	30,000
25. - 30. Sahel	Six projects in : Mauritania, Senegal, Mali, Upper Volta, Chad, Niger	2,825,000 (UNICEF $850,000)
31. Africa Regional	Ground-Water Resources and Well Drilling & Testing (Trans Saharan Road-REG 251)	2,223,000
32. Yemen	Assistance in Rural Development in Hadramout Valley and Northern Areas (multi-sectoral)	217,500

TABLE 7-45. CURRENT UNITED NATIONS DEVELOPMENT PROGRAMME
WATER RESOURCES PROJECTS (continued)

Country or region	Project	UNDP contribution US $
	SURFACE WATER PROJECTS AND FEASIBILITY STUDIES	
1. Burma	Development of the Sittang River Valley (BUR/68/513)	1,096,000
2. Egypt	Assistance to the Hydraulic Research and Experiment Station (EGY/73/023)	66,200 (prep. asst.)
3. Jamaica	Water Resources Planning (JAM/73/012)	60,700
4. Malawi	Study to Define Irrigable Areas of Lake Malawi (MLW/72/006)	
5. Paraguay	Navigability of the Paraguay River (Phase II) (PAR/73/001; also 73/058)	795,850
6. Sierra Leone	Pilot Project for the Determination of Surface Water Resources (SIL/72/004)	86,500
7. Sri Lanka	Water Resources Development Board (68/001)	43,000
8. Tanzania	Coordination of Water Master Planning (URT/73/004)	30,000 (prep. budget)
9. Upper Volta	Water and Water Resources Project (UPV/72/039)	300,000
10. Upper Volta	Study of the Volta Noire Development	700,000
11. Yugoslavia	Computer Control of the Water Resources of the Morava River Basin	320,000
12. Africa Regional	Hydrological and Topographical Studies of the Gambia River Basin (RAF/70/060)	605,230
13. Africa Regional	Planning the Development of the Kagera River Basin (RAF/71/147)	1,885,913
14. Europe Regional	Integrated Development of the Vardar/Axios River Basin (REM/71/203)	1,143,100
15. Latin America	Multi-Purpose Water Development of the Yaguaron River Basin (BRA/71/561) (URU/71/518)	520,000
16. Iran	Coordination of Water Resources Development (IRA/73/015)	202,400
	WATER ADMINISTRATION AND NATIONAL WATER INSTITUTES	
1. Afghanistan	Establishment of a Water Management Department (AFG/68/518)	1,416,200
2. Argentina	Water Economy, Law and Administration Research & Training Institute (ARG/71/544)	707,200
3. Ethiopia	National Water Resources Commission (ETH/72/001)	780,750
4. India	Establishment of National Institute of Hydrology	
5. India	Hydromechanics Division	
6. India	Applied Earth Sciences Division, DWPRS, Poona (IND/73/041)	500,000
7. Indonesia	Institute of Hydraulic Engineering (INS/70/527)	1,779,600
8. Lesotho	Water Law, Legislation and Administration: Research & Drafting (LES/72/057)	10,000
9. India	Coastal Engineering Research Centre and Development of Hydraulic Instrumentation (IND/71/601)	936,300
10. Israel	Coastal and Ocean Engineering Research Centre (ISR/73/031)	20,000
	NON-CONVENTIONAL SOURCES OF WATER	
1. Kuwait	Water Resources Development Centre (Phase II)	Funds-in-Trust

TABLE 7-46. WATER SUPPLY AND SEWERAGE PROJECTS FINANCED BY THE INTERNATIONAL BANK FOR RECONSTRUCTION AND DEVELOPMENT (WORLD BANK) AND THE INTERNATIONAL DEVELOPMENT ASSOCIATION, 1962-75

(Source: World Bank, 1975)

[IBRD Loans and IDA Credits for Water Supply and Sewerage Projects in million US dollars]

Fiscal year	Country	IBRD loan	IDA credit	City	Type of project [1]	Total project cost [2]	IBRD	IDA	Joint loans	Total project cost %	Cancellations or refunds	Percentage disbursed (at June 30, 1974)
1962	China		9	Taipei	W	9.7		4.4		45	0.4	100
	Iceland	311		Reykjavik	W	6.2	2.0			32		100
	Jordan		18	Amman	W	2.9	2.0	2.0		69	0.5	100
	Total					18.8	2.0	6.4			0.9	
1963	Nicaragua		26	Managua	W	4.8		3.0		63		100
1964	Pakistan		41	Dacca	WS	50.1		26.0		52	26.0	—
			43	Chittagong	WS	43.0		24.0		56	24.0	—
	Jordan		43		W	5.0		3.5		70	1.0	100
	Total					98.1		53.5			51.0	
1965	Philippines	386		Manila	W	48.2	20.2			42	0.6	100
	Singapore	405		Singapore I	W	13.7	6.8			50		100
	Total					61.9	27.0					
1966	Burundi		85	Bunjumbura	W	1.6		1.1		69		100
	Venezuela	444		Caracas	W	54.1	21.3		4.2	39		100
	Total					55.7	21.3	1.1				
1967	Pakistan		106	Lahore	WS	5.6		1.8	1.7 (Sweden)	63		100
1968	Singapore	503		Singapore II	W	16.0	8.0			50		100
	Colombia	536		Bogota I	W	48.2	14.0		4.6(US & Germany)	29		97
	Jamaica	598		Kingston	W	9.1	5.0			55		80
	Total					73.3	27.0					
1969	Singapore	547		Singapore III	S	22.4	6.0			27		100
	Malaysia	561		Kuala Lumpur	W	7.7	3.6			47		95
	Tunisia	581		Tunis and Others I	W	32.8	15.0		5.0 (Sweden)	61		66
	Cameroon	604		Yaounde, Douala	W	6.7	5.0		1.4 (France)	96		96
	Total					69.6	29.6		6.4			
1970	Ghana	682	160	Accra-Tema	WS	5.9		3.5		59		95
	Colombia			Cali	WS	37.5	18.5			49		35
	Tunisia		209	Tunis & Others II	W	19.2		10.5	3.5 (Sweden)	73		39
	Total					62.6	18.5	14.0				

TABLE 7-46. WATER SUPPLY AND SEWERAGE PROJECTS FINANCED BY THE INTERNATIONAL BANK FOR RECONSTRUCTION AND DEVELOPMENT (WORLD BANK) AND THE INTERNATIONAL DEVELOPMENT ASSOCIATION, 1962-75 (continued)

Fiscal year	Country	IBRD loan	IDA credit	City	Type of project 1)	Total project cost 2)	IBRD	IDA	Joint loans	Total project cost %	Cancellations or refunds	Percentage disbursed (at June 30, 1974)
1971	Botswana	776		Shashe Multi-purpose	W & other mining infrastructure	58.9	32.0		37.8 (Canada, USA, UK & Others)	68		92
	Botswana		233	Gaborone-Lobatse	W	3.3		3.0		91		72
	Brazil	757		Sao Paolo	W	59.8	22.0			37		18
	Brazil	758		Sao Paolo	S	81.5	15.0			18		9
	Colombia	738		Palmira	WS	3.8	2.0			53	3.0	32
	Colombia	741		Bogota II	W	118.0	88.0			75		15
	Cyprus	741		Nicosia	S	6.8	3.5			52		57
	Cyprus	730		Famagusta	S	3.2	1.9			56	0.8	60
	Kenya	714		Nairobi	W	13.4	8.3			62		60
	Yugoslavia	777		Ibar	Multipurpose	93.3	45.0			48		7
	Total					442.0	271.7	3.0			3.8	
1972	Nicaragua	808		Managua	W	10.0	6.9			69		34
	Ethiopia	818		Addis Ababa	WS	13.0	10.8			83		66
	Turkey	844		Istanbul	W	85.1	37.0			44		3
	Tunisia	858	329	Various Cities	Tourism, infrastructure WS	55.0	14.0	10.0	10.9 (Germany)	62		1
	Total					163.1	68.7	10.0				
1973	Morocco	850		Casa-Rabat	W	104.9	48.0			46		33
	Colombia	860		Medium Cities	W	15.9	9.1			59		4
	Israel	869		78 Cities, 2 localities	S	75.0	30.0			40		0
	Bangladesh		367	Chittagong 3)	W	23.8		7.0		29		64
	Bangladesh		368	Dacca 3)	WS	41.6		13.2		32		65
	Gabon	895		Libreville	W	19.9	9.5		4.3 (France)	48		3
	India		390	Bombay	WS	158.2	55.0			35		1
	Jordan		385	Amman	WS	11.6		8.7		75		17
	Malaysia	908		Kuala Lumpur	W	26.2	13.5			51		0
	Syria		401	Damascus	WS	32.6		15.0		46		0
	Mexico	909		Mexico City	W	194.0	90.0			46		0
	Total					703.7	255.1	43.9				
1974	Thailand	1,021		Bangkok	W	214.0	55.0			26		0
	Tunisia	989		Sfax & Others	W	42.6	23.0			54		0
	Ecuador	1,030		Guayaquil	W	38.3	23.2			61		0
	Ghana		499	Accra-Tema	W	44.7		10.4	12.3 (Af. DB & Canada)	51		0

TABLE 7-46. WATER SUPPLY AND SEWERAGE PROJECTS FINANCED BY THE INTERNATIONAL BANK FOR RECONSTRUCTION AND DEVELOPMENT (WORLD BANK) AND THE INTERNATIONAL DEVELOPMENT ASSOCIATION, 1962-75 (continued)

Fiscal year	Country	IBRD loan	IDA credit	City	Type of project 1)	Total project cost 2)	IBRD	IDA	Joint loans	Total project cost %	Cancellations or refunds	Percentage disbursed (at June 30, 1974)
	Brazil	1,009		State of Minas Gerais	WS	92.0	36.0			40		0
	Nepal		470	Kathmandu	WS	10.4		7.8		75		0
	Singapore	918		Singapore	S	29.5	12.0			41		3
	Yemen Arab Republic		464	Sana'a	W	6.8		6.3		91		0
	Total					478.3	149.2	24.5				
1975	Indonesia	1,049		Five Cities	W	25.4	14.5					0
	Yugoslavia	1,066		Dubrovnic	WS	9.6	6.0					0
	Ivory Coast	1,076		Abidjan	SD	16.7	9.0	7.7				0
	Colombia	1,072		Multi-City	WS	52.9	27.0					0
	Total					104.6	56.5	7.7				
	Grand Total 4)					2,342.1	872.6	168.9			60.5	

1) W=Water Supply; S=Sewerage; D=Drainage.
2) Appraisal Report Estimates.
3) Reappraised from previous Credit Nos. 41 and 42.
4) In the years 1962-1974, 55 operations in 31 countries with total loans and credits (net of cancellations or refunds) was US $916.8 million.

TABLE 7-47. INTERNATIONAL STANDARDS FOR DRINKING WATER
(Source: World Health Organization, 1971)

TENTATIVE LIMITS FOR TOXIC SUBSTANCES IN DRINKING WATER

Substance	Upper limit of concentration mg/l
Arsenic (as As)	0.05
Cadmium (as Cd)	0.01
Cyanide (as CN)	0.05
Lead (as Pb)	0.1
Mercury (total as Hg)	0.001
Selenium (as Se)	0.01

RECOMMENDED CONTROL LIMITS FOR FLUORIDES IN DRINKING WATER

Annual average of maximum daily air temperature in °C	Recommended control limits for fluorides (as F) in mg/l	
	Lower	Upper
10 −12	0.9	1.7
12.1−14.6	0.8	1.5
14.7−17.6	0.8	1.3
17.7−21.4	0.7	1.2
21.5−26.2	0.7	1.0
26.3−32.6	0.6	0.8

SUBSTANCES AND CHARACTERISTICS AFFECTING THE ACCEPTIBILITY OF WATER FOR DOMESTIC USE

Substance or characteristic	Undesirable effect that may be produced	Highest desirable level	Maximum permissible level
Substances causing discoloration	Discoloration	5 units [a]	50 units [a]
Substances causing odors	Odors	Unobjectionable	Unobjectionable
Substances causing tastes	Tastes	Unobjectionable	Unobjectionable
Suspended matter	Turbidity Possibly gastrointestinal irritation	5 units [b]	25 units [b]
Total solids	Taste Gastrointestinal irritation	500 mg/l	1,500 mg/l
pH range	Taste Corrosion	7.0 to 8.5	6.5 to 9.2
Anionic detergents [c]	Taste and foaming	0.2 mg/l	1.0 mg/l
Mineral oil	Taste and odor after chlorination	0.01 mg/l	0.30 mg/l
Phenolic compounds (as phenol)	Taste, particularly in chlorinated water	0.001 mg/l	0.002 mg/l
Total hardness	Excessive scale formation	2 mEq/ [d,e,] (100 mg/l CaCO3)	10 mEq/l (500 mg/l CaCO3)
Calcium (as Ca)	Excessive scale formation	75 mg/l	200 mg/l

TABLE 7-47. INTERNATIONAL STANDARDS FOR DRINKING WATER (continued)

SUBSTANCES AND CHARACTERISTICS AFFECTING THE ACCEPTIBILITY OF WATER FOR DOMESTIC USE

Substance or characteristic	Undesirable effect that may be produced	Highest desirable level	Maximum permissible level
Chloride (as Cl)	Taste; corrosion in hot-water systems	200 mg/l	600 mg/l
Copper (as Cu)	Astringent taste; discoloration and corrosion of pipes, fittings and utensils	0.05 mg/ l	1.5 mg/l
Iron (total as Fe)	Taste; discoloration; deposits and growth of iron bacteria; turbidity	0.1 mg/l	1.0 mg/l
Magnesium (as Mg)	Hardness; taste; gastrointestinal irritation in the presence of sulfate	Not more than 30 mg/l if there are 250 mg/l of sulfate; if there is less sulfate, magnesium up to 150 mg/l may be allowed	150 mg/l
Manganese (as Mn)	Taste; discoloration; deposits in pipes; turbidity	0.05 mg/l	0.5 mg/l
Sulfate (as SO4)	Gastrointestinal irritation when magnesium or sodium are present	200 mg/l	400 mg/l
Zinc (as Zn)	Astringent taste; opalescence and sand-like deposits	5.0 mg/l	15 mg/l

a) On the platinum-cobalt scale.

b) Turbidity units.

c) Different reference substances are used in different countries.

d) If the hardness is much less than this, other undesirable effects may be caused, for example, heavy metals may be dissolved out of pipes.

e) I mEq/l of hardness-producing ion=50 mg CaCO3/l=5.0 French degrees of hardness=2.8 (approx) German degrees of hardness=3.5 (approx) English degrees of hardness.

TABLE 7-48. EUROPEAN STANDARDS FOR DRINKING WATER

(Source: World Health Organization, 1970)

LIMITS OF TOLERANCE FOR TOXIC SUBSTANCES IN PIPED SUPPLIES

Substance	Upper limit of concentration
Lead (as Pb)	0.1 mg/l [a]
Arsenic (as As)	0.05 mg/l
Selenium (as Se)	0.01 mg/l
Chromium (as Cr hexavalent)	0.05 mg/l
Cadmium [b] (as Cd)	0.01 mg/l
Cyanide (as CN)	0.05 mg/l

[a] 0.1 mg/l of lead (as Pb) should be the upper limit in the supply, but certain undertakings still use lead piping, and in these instances the concentration of lead in the water after prolonged contact with the pipes may be higher. In no instance should the concentration of lead (as Pb) exceed 0.3 mg/l after 16 hours' contact with the pipes. If the limit of 0.3 mg/l is regularly exceeded it will be necessary to take steps either to change the piping or to treat the water. Lead is used as a stabilizer in some plastic pipes, and the possibility of its being leached out must be borne in mind.

[b] Cadium has been included in this list because of the possibility of its being dissolved out of plastic pipes. Mercury and tin may also be dissolved out of plastic pipes and the possibility of such pipes giving rise to unpleasant colors and tastes in water should also be noted.

CONSTITUENTS IN WATER WHICH, IF PRESENT IN EXCESSIVE AMOUNTS, MAY GIVE RISE TO TROUBLE

Substance	Nature of trouble which may arise	Approximate level above which trouble may arise
Phenolic compounds (as phenol)	Taste, particularly in chlorinated water	Less than 0.001 mg/l [a]
Fluoride (as F)	Fluorosis	See table below
Nitrate (as NO3)	Danger of infantile methaemoglobinaemia if the water is consumed by infants	Recommended: less than 50 mg/l Acceptable: 50 to 100 mg/l Not recommended: more than 100 mg/l [b,c]
Copper (as Cu)	Astringent taste; discoloration and corrosion of pipes, fittings and utensils	0.05 mg/l at pumping station; 3.0 mg/l after 16 hours' contact with new pipes
Iron (total as Fe)	Taste; discoloration; deposits and growth of iron bacteria; turbidity	0.1 mg/l as the water enters the distribution system [c,d]
Manganese (as Mn)	Taste; discoloration; deposits in pipes; turbidity	0.05 mg/l
Zinc (as Zn)	Astringent taste; opalescence and sand-like deposits	5.0 mg/l
Magnesium (as Mg)	Hardness; taste	Not more than 30 mg/l if there are 250 mg/l of sulfate; if there is less sulfate, magnesium up to 125 mg/l may be allowed
Sulfate (as SO4)	Gastrointestinal irritation when combined with magnesium or sodium	250 mg/l

TABLE 7-48. EUROPEAN STANDARDS FOR DRINKING WATER (continued)

CONSTITUENTS IN WATER WHICH, IF PRESENT IN EXCESSIVE AMOUNTS, MAY GIVE RISE TO TROUBLE

Substance	Nature of trouble which may arise	Approximate level above which trouble may arise
Hydrogen sulfide (as H$_2$S) e)	Taste and odor	0.05 mg/l
Chloride (as Cl)	Taste; corrosion in hot-water systems	200 mg/l. This limit may be exceeded in certain existing conditions but in no circumstances should the level exceed 600 mg/l

a) This limit is justified at present in that it is low enough not to give rise to unpleasant tastes in chlorinated water. Attention should be given to the control of phenolic compounds in water; some phenolic compounds are capable of being toxic when ingested over a long period of time.

b) If the nitrate content is within the acceptable range and the water is otherwise chemically and bacteriologically satisfactory, it may not give rise to trouble, but physicians in the area should be warned of the possibility of infantile methaemoglobinaemia occurring. More information is required on the exact circumstances under which infantile methaemoglobinaemia occurs and the mechanism by which it is produced.

c) It is advisable that a special sample be collected for examination for nitrate and iron. The sample should be "fixed" at the time of collection by adding 1 ml of concentrated sulfuric acid for each liter of water.

d) In small installations in which removal of iron would be uneconomic, or where the iron is present in a stable form, a level of up to 0.3 mg/l can be permitted.

e) This examination should be carried out as soon after the collection of the sample as is practicable.

RECOMMENDED CONTROL LIMITS OF FLUORIDE IN DRINKING WATER

Annual average of maximum daily air temperature in °C	Recommended control limits of fluoride (as F) in mg/l	
	Lower	Upper
10.0–12.0 a)	0.9	1.7
12.1–14.6 b)	0.8	1.5
14.7–17.6 b)	0.8	1.3
17.7–21.4 c)	0.7	1.2
21.5–26.2 c)	0.7	1.0

a) Generally suitable for Northern Europe.
b) Generally suitable for Central Europe.
c) Generally suitable for Southern Europe.

TABLE 7-48. EUROPEAN STANDARDS FOR DRINKING WATER (continued)

SUBSTANCES OF WHICH THE LEVEL SHOULD PREFERABLY BE CONTROLLED

Substance	Nature of trouble which may arise	Approximate level above which trouble may arise
Anionic detergents [a]	Taste and foaming	0.2 mg/l
Ammonia (as NH_4)	Growth of organisms, danger of corrosion in pipes; difficulties in chlorination	0.05 mg/l [b]
Free carbon dioxide (as CO_2)	Damage to pipes; danger of bringing toxic metals into solution	For aggressive carbon dioxide - zero [c]
Dissolved oxygen [d]	Taste and odor; corrosion; growth of organisms - if the concentration of dissolved oxygen is less than 5 mg/l the formation of a protective layer will be hampered, thus causing all the free carbonic acid of a non-aggressive water to be corrosive to iron piping	Preferably at least 5 mg/l
Total hardness	Excessive scale formation; danger of dissolving heavy metals if the level of hardness is below the recommended limit	Limits of hardness 2 to 10 mEq/l (100 to 500 mg/l $CaCO_3$) [e]

[a] Different reference substances are used in different countries.

[b] In deep groundwater sources where iron is present, this limit may be exceeded in exceptional situations.

[c] The examination for free carbon dioxide should preferably be carried out at the time of collection of the sample. If it is not possible to do this a special sample should be taken. The sampling bottle should be completely filled and the sample kept cool with ice until it is examined.

[d] The levels for all substances in this table, with the exception of dissolved oxygen, are those which it is preferable not to exceed. For dissolved oxygen the concentration should preferably be kept above the level given.

[e] It is recommended that hardness be expressed in units, one unit of hardness being 1 mEq/l of hardness — producing ion. (1 mEq/l of hardness- producing iron = 50 mg $CaCO_3$/l = 5.0 French degrees of hardness = 2.8 (approx) German degrees of hardness = 3.5 (approx) English degrees of hardness).

TABLE 7-49. U.S. PUBLIC HEALTH SERVICE DRINKING WATER STANDARDS
(Source: U.S. Public Health Service, 1962)

Category A — The following chemical substances should not be present in a water supply in excess of the listed concentrations where other more suitable supplies are or can be made available:

Substance	Concentration in mg/l
Alkyl Benzene Sulfonate (ABS)	0.5
Arsenic (As)	0.01
Chloride (Cl)	250.
Copper (Cu)	1.
Carbon Chloroform Extract (CCE)	0.2
Cyanide (CN)	0.01
Iron (Fe)	0.3
Manganese (Mn)	0.05
Nitrate (NO$_3$) [a]	45.
Phenols	0.001
Sulfate (SO$_4$)	250.
Total Dissolved Solids (TDS)	500.
Zinc (Zn)	5.

Category B — The presence of the following substances in excess of the concentrations listed shall constitute grounds for rejection of the supply:

Substance	Concentration in mg/l
Arsenic (As)	0.05
Barium (Ba)	1.0
Cadmium (Cd)	0.01
Chromium (Hexavalent) (Cr^{+6})	0.05
Cyanide (CN)	0.2
Fluoride (F)	0.6 to 1.7 [b]
Lead (Pb)	0.05
Selenium (Se)	0.01
Silver (Ag)	0.05

Category C — Radioactivity

Source	Recommended limits micromicrocuries per liter
Radium — 226	3
Strontium — 90	10
Gross beta activity	1,000

See footnotes on next page.

Footnotes for Table 7-49:

a) In areas in which the nitrate content of water is known to be in excess of the listed concentration, the public should be warned of the potential dangers of using the water for infant feeding.

b) Varies with air temperature; see table below.

Annual average of maximum daily air temperatures [1]	Recommended control limits — Fluoride concentrations in mg/l		
(degrees F)	Lower	Optimum	Upper
50.0-53.7	0.9	1.2	1.7
53.8-58.3	0.8	1.1	1.5
58.4-63.8	0.8	1.0	1.3
63.9-70.6	0.7	0.9	1.2
70.7-79.2	0.7	0.8	1.0
79.3-90.5	0.6	0.7	0.8

[1] Based on temperature data obtained for a minimum of five years.

TABLE 7-50. U.S. ENVIRONMENTAL PROTECTION AGENCY PROPOSED NATIONAL INTERIM PRIMARY DRINKING WATER STANDARDS [1]

(Source: National Water Well Assoc., 1975)

I. Maximum Contaminant Levels for Inorganic Chemicals

Contaminant	Level (mg/l)	Contaminant	Level (mg/l)
Arsenic	0.05	Lead	0.05
Barium	1.	Mercury	0.002
Cadmium	0.010	Nitrate	10.
Chromium	0.05	Selenium	0.01
Cyanide	0.2	Silver	0.05

Fluorides

When the annual average of the maximum daily air temperatures for the location in which the public water system is situated is the following, the corresponding concentration of fluoride shall not be exceeded:

Temperature (in degrees F)	(degrees C)	Level (mg/l)
50.0-53.7	10.0-12.0	2.4
53.8-58.3	12.1-14.6	2.2
58.4-63.8	14.7-17.6	2.0
63.9-70.6	17.7-21.4	1.8
70.7-79.2	21.5-26.2	1.6
79.3-90.5	26.3-32.5	1.4

II. Maximum Contaminant Levels for Organic Chemicals

The maximum contaminant level for the total concentration of organic chemicals, is 0.7 mg/l.

III. Maximum Contaminant Levels for Pesticides

Chlorinated Hydrocarbons	Level (mg/l)
Chlordane	0.003
Endrin	0.0002
Heptachlor	0.0001
Heptachlor Epoxide	0.0001
Lindane	0.004
Methoxychlor	0.1
Toxaphene	0.005

Chlorophenoxys	
2, 4-D	0.1
2, 4, 5-TP Silvex	0.01

[1] Due to become effective in December 1976.

WORLD
REFERENCES CITED

U.S. Department of State, 1972, World Data Handbook, General Foreign Policy Series 264, Washington, D.C.

U.S. Department of Commerce, 1969, Climates of the World, Environmental Science Services Administration, Washington, D.C.

Lvovitch, M.I., 1971, The Global Water Balance, EOS, Transactions American Geophysical Union, Vol. 54, No. 1, Jan. 1973.

Doxiadis, C.A., 1967, Water and Environment, International Conference on Water for Peace, Washington, D.C.

Nace, Raymond, 1967, Are We Running Out of Water?, U.S. Geol. Survey Circ. 536.

Framji, K.K., and Mahajan, I.K., 1969, Irrigation and Drainage in the World, International Commission on Irrigation and Drainage, New Delhi.

Young, L.L., 1964, Summary of Developed and Potential Water Power of the United States and Other Countries of the World, 1955-62, U.S. Geol. Survey Circ. 483, Washington, D.C.

Encyclopaedia Britannica, 15th Edition, 1974, Chicago, Ill. 60611.

Holeman, J., 1968, Sediment Yield of Major Rivers of the World, Water Resources Research, Vol. 4, No. 4, August.

Beeton, A.M., 1973, Eutrophication Problems in the St. Lawrence - Great Lakes, First World Congress in Water Resources, International Water Resources Association, Chicago, Ill.

Langbein, W.B., 1961, Salinity and Hydrology of Closed Lakes, U.S. Geol. Survey Prof. Paper 412, Washington, D.C.

Federal Power Commission, 1968, World Power Data, FPC - P. 40, Washington, D.C.

Hubbert, M.K., World Potential and Developed Water Power Capacity, Water Resources Bulletin, Vol. 10, No. 2, April 1974.

Huberty, M.R., and Flock, W.L., 1959, Natural Resources, 2nd Ed., McGraw-Hill, New York.

Lvovitch, M.I., World Water Balance, Proc. Reading Symposium, July 1970, IASH - UNESCO - WMO.

U.S. Naval Oceanographic Office, 1966, Handbook of Oceanographic Tables, Washington, D.C.

Hem, J.D., 1970, Study and Interpretation of the Chemical Characteristics of Natural Water, U.S. Geol. Survey Water - Supply Paper 1473, Washington, D.C.

Mermel, T.W., 1971, The World's Highest Dams, Greatest Manmade Lakes, Largest Hydroelectric Plants, U.S. Bureau of Reclamation.

Budyko, M.I., The Water Balance of the Oceans, World Water Balance, Proc. Reading Symposium, July 1970, IASH-UNESCO-WMO.

Office of Saline Water, 1969, Desalting Plants Inventory Report No. 2; 1970, Desalting Plants Inventory Report No. 3, Washington, D.C.

van Damme, J.M.G., 1973, Needs and Problems in Water Supply in Developing Countries, WHO, International Reference Centre for Community Water Supply, The Hague, The Netherlands.

World Health Organization, 1973, World Health Statistics Report, Vol. 26, No. 11.

IHD Bulletin, EOS, Transactions American Geophysical Union, Vol. 56, No. 1, January 1975.

Alagappan, A., 1974, Data on U.N. Water Resources Projects, United Nations, New York (Private Communication).

World Bank, 1975, Washington, D.C. (Private Communication).

World Health Organization, 1971, International Standards for Drinking Water, 3rd Edition, Geneva.

World Health Organization, 1970, European Standard for Drinking Water, 2nd Edition, Geneva.

U.S. Department of Health, Education and Welfare, Public Health Service, 1962, Drinking Water Standards, Washington, D.C.

The World Almanac, 1975, Newspaper Enterprise Association, Inc., New York.

The Well Log, National Water Well Association, Vol. 5, No. 4, April 1975.

EXPLANATORY NOTES

The following symbols have been used in the tables throughout the book:

A full stop (.) is used to indicate decimals.

A comma (,) is used to distinguish thousands and millions.

The term "billion" signifies a thousand million.

Details and percentages in tables do not necessarily add to totals, because of rounding.

The following abbreviations appear in the tables:

ac. ft	acre-feet
bgd	billion gallons per day
cl	centiliter
cfs	cubic feet per second
cusec	cubic feet per second
gpd	gallons per day
GWh	Giga watt-hour (million kilowatt-hours)
ha	hectare
hm	hectometer
in	inch
kg	kilogram
km	kilometer
kl	kiloliter
kWh	kilowatt-hour
l/d	liters per day
l/s	liters per second
l.s.	land surface
m	meter
m3/s	cubic meters per second
mg/l	milligrams per liter
mm	millimeter
mg	million gallons
mgd	million gallons per day
mg. ft	morgen feet
MAF	million acre-feet
MW	megawatt
ppm	parts per million
TDS	total dissolved solids

The following acronyms have been used:

AID	Agency for International Development
CEPAL	Comision Economica para America Latina (see ECLA)
CIDA	Canadian International Development Agency
COPLANARH	Comision del Plan Nacional de Aprovechamiento de los Recursos Hidraulicos (Venezuela)
ECA	Economic Commission for Africa
ECAFE	Economic Commission for Asia and the Far East
ECE	Economic Commission for Europe
ECLA	Economic Commission for Latin America
EXIMBANK	Export-Import Bank
FAO	Food and Agriculture Organization
ICID	International Commission on Irrigation and Drainage
IBRD	International Bank for Reconstruction and Development (World Bank)
IDB	Inter-American Development Bank
IHD	International Hydrological Decade
IHP	International Hydrological Program
UNDP	United Nations Development Programme
UNESCO	United Nations Educational, Scientific and Cultural Organization
WHO	World Health Organization
WMO	World Meteorological Organization

CONVERSION FACTORS

VOLUME

1 U.S. gallon (g)	= 3.785 liters
1 Imperial gallon	= 1.201 U.S. gallons 4.546 liters
1 cubic foot (ft3)	= 7.481 U.S. gallons 28.32 liters
1 acre-foot (acre-ft)	= 3.259×10^5 U.S. gallons 1,234 m3
1 liter (l)	= 0.2642 U.S. gallons 1,000 cm3
1 cubic meter (m3)	= 264.2 U.S. gallons 1,000 liters
1 cubic hectometer (hm3)	= 1×10^6 m3
1 million U.S. gallons (mg)	= 3.069 acre-feet 3,785.4 m3
1 morgen foot (South Africa)	= 2,610.7 m3

FLOW RATE

1 U.S. gallon per minute (gpm)	= 0.0631 l/s 5.42 m3/day
1 million U.S. gallons per day (mgd)	= 43.7 l/s 3,785 m3/day
1 million Imperial gallons per day	= 52.60 l/s
1 billion U.S. gallons per day (bgd)	= 3.785 hm3/day
1 cubic foot per second (cfs)	= 449 gpm 28.3 l/s
1 acre-foot per day	= 14.2 l/s
1 liter per second (l/s)	= 15.9 gpm 86.4 m3/day
1 cubic meter per second (m3/s)	= 22.8 mgd 35.3 cfs
1 cubic meter per day (m3/d)	= 0.183 gpm
1 cubic hectometer per day (hm3/d)	= 264.2 mgd
1 quinaria (Ancient Rome)	= 0.47 - 0.48 l/s

TRANSMISSIVITY

1 sq m/day	= 80.5 gpd/ft

LENGTH

1 inch (in)	= 2.54 cm
1 foot (ft)	= 30.48 cm
1 mile (mi)	= 1.609 km
1 centimeter (cm)	= 0.3937 in 10 mm
1 meter (m)	= 39.37 in 3.2808 ft 100 cm
1 kilometer (km)	= 0.621 mi 1,000 m 10 hm

AREA

1 acre	= 0.4047 hectare
1 square mile (mi2)	= 2.590 km2
1 hectare (ha)	= 2.471 acres
1 square kilometer (km2)	= 247.1 acres 0.3861 mi2 100 hectares

WEIGHT

1 pound (lb)	= 0.4536 kg
1 short ton	= 2,000 lb 0.9072 metric ton
1 metric ton	= 1,000 kg
1 kilogram (kg)	= 2.205 lb

INDEX